New Technology of Unconventional Metallurgy

非常规冶金新技术

彭金辉　张利波　夏洪应
巨少华　陈　菓　许　磊　著

北　京
冶　金　工　业　出　版　社
2015

New Technology of Unconventional Metallurgy

Jinhui Peng Libo Zhang Hongying Xia
Shaohua Ju Guo Chen Lei Xu

Beijing
Metallurgical Industry Press
2015

Published and distributed by
Metallurgical Industry Press
39 Songzhuyuan North Alley, Beiheyan St.
Beijing 100009, P. R. China

图书在版编目(CIP)数据

非常规冶金新技术 = New Technology of Unconventional Metallurgy：英文/彭金辉等著．—北京：冶金工业出版社，2015.3
ISBN 978-7-5024-6866-8

Ⅰ.①非… Ⅱ.①彭… Ⅲ.①冶金—技术—英文
Ⅳ.①TF1

中国版本图书馆 CIP 数据核字(2015)第 045939 号

出 版 人 谭学余
地　　址 北京市东城区嵩祝院北巷 39 号 邮编 100009 电话 (010)64027926
网　　址 www.cnmip.com.cn 电子信箱 yjcbs@cnmip.com.cn
责任编辑 张熙莹 美术编辑 彭子赫 版式设计 孙跃红
责任校对 王永欣 责任印制 牛晓波
ISBN 978-7-5024-6866-8
冶金工业出版社出版发行；各地新华书店经销；三河市双峰印刷装订有限公司印刷
2015 年 3 月第 1 版，2015 年 3 月第 1 次印刷
787mm×1092mm 1/16；32.25 印张；779 千字；502 页
128.00 元

冶金工业出版社 投稿电话 (010)64027932 投稿信箱 tougao@cnmip.com.cn
冶金工业出版社营销中心 电话 (010)64044283 传真 (010)64027893
冶金书店 地址 北京市东四西大街 46 号(100010) 电话 (010)65289081(兼传真)
冶金工业出版社天猫旗舰店 yjgy.tmall.com
(本书如有印装质量问题，本社营销中心负责退换)

Preface

Unconventional metallurgy is a new technology developed in recent years, mainly focusing on major pressing issues of comprehensive utilization of mineral resources, energy saving and emission reduction, and deep processing of value – added metallurgical products in metallurgical industry, using modern technology to improve traditional metallurgical process, developing an efficient and environment – friendly new metallurgical reactor, deepening metallurgical theory and process under the outside fields and extreme conditions, obtaining the original innovations and intellectual property in the fields of unconventional metallurgy, enhancing the scientific and technological innovation in metallurgical industry, and promoting the leading role of the metallurgical industry on the national economy and social development. At present main research interests include: (1) novel technology of microwave metallurgy; (2) application of microwave technology in materials science and chemical engineering; (3) novel technology of ultrasonic metallurgy and micro – fluidics metallurgy.

As a green heating mode, microwaves heat materials directly through internal energy dissipation of materials, with advantages of selective heating, fast heat rate and high efficiency, and furthermore can lower the reaction temperature, shorten the reaction time, resulting in the energy saving and consumption reducing, is one of the effective way to realize clean production in metallurgical industry. So, it is of great importance to develop new microwave metallurgical technology. In recent years, the authors and members of the research group have conducted in – depth research on microwave metallurgy technologies, built up the microwave heating network models; determined the dielectric properties and temperature rising characteristics of metallurgical materials in microwave field, investigated changes of material properties by microwave drying, reduction, calcining, roasting and leaching; and carried out the application of microwave new technology in materials science and chemical engineering, promoted the comprehensive utilization of the strategic non – ferrous metals refractory resources in China.

The cavitation effects, mechanical effects and thermal effects of ultrasonic can make the ultrasonic wave produce fast and intensive mechanic motion when propagation in a liquid, forming bubbles or holes, when the bubbles shrink rapidly, they would result in local high temperature and high pressure accompanied by an intense shock wave, promoting phase boundary and interface updating and disturbing, accelerating the heat and mass transfer, which has been widely used in metallurgical, chemical, solution purification and other fields. The authors and members of the research group investigated the typical metallurgical units of ultrasonic metallurgy, studying the leaching kinetics of zinc residue under ultrasonic field, and compared with conventional leaching process, showing that the ultrasound technique can make up for the shortcomings of traditional hydrometallurgy technology, strengthening the leaching process and reducing the processing time, is a new and effective way.

Micro – fluid metallurgical technology refers to techniques of control, manipulation and detection of complex fluids under the microscopic size, realizing the micro – and nano – level mixing, mass transfer and reaction in a micro – channel. The authors and members of the research group investigated the typical metallurgical units of micro – fluidic metallurgy, studying the extraction, separation of In, Fe and Zn by micro – fluids, realizing the separation of In and Fe, Zn, shortening the reaction time, and enhancing the security of the process; for the separation extraction of nickel and cobalt ions, the micro fluidic single – stage extraction efficiency increased more than 10%, total extraction stages can be reduced by 4 levels. Combined with the advantages of fast mixing mass transfer rate, reacting homogenously, continuous stable of micro – fluidic technology, realizing the continuous rapid preparation process of nano - powders through micro - channel mixing. Micro - fluidic metallurgical technology has been developing rapidly, promoting industrial upgrading of the metallurgical industry by using its advantages of high efficiency, low consumption and safety, to transform unit processes of the traditional metallurgical industries with disadvantages of low efficiency and high consumption, such as the extraction, heat transfer and mixing.

The book is divided into four chapters. The Chapter Ⅰ discusses the application of microwave metallurgical technology in the drying, reduction, calcining, roasting, leaching, comprehensive utilization of the metallurgical material and process simulation of interaction between microwave field and materials; the Chapter Ⅱ introduces the application of microwave technique in material science and chemical engineering, such as novel materials preparation by microwave sintering, and the comprehensive utilization of waste/spent catalyst and the regeneration of activated carbon; the Chapter Ⅲ describes the effect of ultrasonic metallurgical technique on the process of zinc residue leaching and model of leaching kinetics; the Chapter Ⅳ introduces the application of micro - fluid metallurgical technique in the extraction separation and the synthesis of nano - powders. The book is informative, illustrated, reader - friendly and practical, not only has practical value, but also make it easier for the reader's understanding of new technologies and innovation.

Authors refer to a number of books and literatures in the process of writing the book, the authors pay the deep respect and gratitude to whom concerned the book and gave suggestions and comments. Special thanks go to the National Science Foundation of China, Fund of Ministry of Science and Technology and Ministry of Education, Yunnan Provincial Natural Science Foundation and Supports of Corporate. With these supports, the authors can carry out the uninterrupted research on unconventional metallurgical technique, and included the results in the book. This book was completed under the guidance of me and members of the research group, the achievements of research group in the book are always impregnated with successive sweat and wisdom of doctoral students, graduate and undergraduate students, here, I would like to take the occasion of completing the book to give my heartfelt thanks to them.

The authors are aware that the expertise is limited and that there may be some errors in the book. If so, please do not hesitate to point them out.

Jinhui Peng
January 2015

Contents

Chapter I New Technology of Microwave Metallurgy

Chapter Ⅱ New Technology of Microwave Applications in Material and Chemical Engineering

Chapter III New Technology of Ultrasonic Metallurgy

Chapter IV New Technology of Microfluidic

About the Author

Jinhui Peng was born in December 1964 and got the doctor's degree from Kunming University of Science and Technology in 1992, then got to Germany and England for postdoctoral research from 1994 to 1996 and from 1999 to 2000, respectively. Now he is a tutor of Ph. D students in the field of non – ferrous metallurgy and chairman of the Key Laboratory of Unconventional Metallurgy, Ministry of Education, at the same time, he is foreign academician of Academy of Natural Sciences, Russia, and Institute of International Mineral Resources, meantime, he also enjoys the special allowance of the State Department.

Prof. Jinhui Peng has devoted himself to establishing the new technology of unconventional metallurgy, accomplished more than 70 projects of 973 Program, 863 Program, National Natural Science Foundation of China, International S&T Cooperation Program of China, Science Foundation of Ministry of Education of China, Key Enterprise Entrusted Brainstorm Project, etc. He has won many prizes of state technological invention award, science and technology innovation prize of Ho Leung Ho Lee foundation, award nomination for the top ten national excellent science and technology workers, outstanding contribution award for national science & technology during the "11th Five – Year Plan", etc. In addition, **Prof. Jinhui Peng** is one of the sixth discipline appraisal group members of the state department, a subject matter expert for 863 Program during the "12th Five – Year Plan", and one of reviewing expertise group members of post – doctoral research station. **Prof. Jinhui Peng** has published 4 books, more than 500 papers, and 160 patents.

As a project engineer, the research results achieved by **Prof. Jinhui Peng** are in the international advanced level. The group independently developed a series of new type of microwave high temperature device, and established some large – scale, continuous, and automatic microwave high temperature pilot lines, which are successfully used for material modification, waste comprehensive utilization, and metallurgical products, etc., achieving the aim of high efficiency, energy saving and environment friendly.

Chapter Ⅰ

New Technology of Microwave Metallurgy

Microwave Sensor for Measuring the Properties of a Liquid Drop❶

Ming Huang, Jingjing Yang, Jiaqiang Wang, Jinhui Peng

Abstract: A novel microwave sensor for measuring the properties of a liquid drop has been invented, its analytical theory established and a working prototype has been constructed and tested. It was also found that the theory based on the microwave sensor is in good agreement with the experimental results. Excellent linearity is achieved by optimizing the design, with an accuracy of distilled water drop volume measurement of approximately 0.5μL, and this microwave sensor has been used to measure surface tension, species concentration and the microwave absorption properties of a liquid drop simultaneously, which are the key parameters in the fields of physical chemistry and microwave chemistry.

Keywords: liquid drop; microwave sensor; surface tension; absorption properties

1 Introduction

The formation of drops is a phenomenon ubiquitous in daily life, science and technology[1]. It is found that a great deal of information on liquid properties is contained in the process of drop formation. This makes it possible to measure several physical parameters of a liquid by using dropanalysis. The development of a fibre drop multianalyser has been reported over the last 15 years[2]. It has proved to be a powerful analytical tool for determining the physical and chemical characteristics of liquids. More recently, capacitive tensiography has been reported[3,4]. It has been demonstrated that the capacitive transducer gives a direct measurement of the volume in the pendant liquid drop, with a resolution of 1μL.

It is well known that microwave and infrared form a continuous electromagnetic spectrum that extends from RF frequency to optical wave. It has been shown that the RF capacitive sensor and fibre sensor can measure the parameters of a liquid drop[2,3]. Therefore, it is possible to measure the parameters of a liquid drop by a microwave sensor. The objective of this paper is to apply the microwave sensor for the measurement of the parameters of a liquid drop. Preliminary experiments have been carried out and these show that the microwave sensor is capable of measuring drop volumes with an accuracy of down to 0.5μL. It can also measure microwave absorption properties, species concentration and surface tension simultaneously.

2 Theory

Microwave sensors based on cavity perturbation techniques have been studied by many research-

---

❶ This article was reviewed in "Nature CHINA" on 6 June 2007.

ers[5,6]. Measurements of a liquid drop are performed by inserting a small, appropriately shaped liquid drop into a cavity and determining the properties of the liquid drop from the resultant change in the resonant frequency and loaded quality factor which is given by[7]

$$f_0 - f_s = \frac{1}{2}(\varepsilon_r' - 1)f_s W_0^{-1}\int_{v_s} E \cdot E_0^* \, dv \tag{1}$$

$$Q_s^{-1} - Q_0^{-1} = \frac{1}{2}W_0^{-1}\varepsilon_r''\int_{v_s} E \cdot E_0^* \, dv \tag{2}$$

here $\varepsilon_r = \varepsilon_r' - j\varepsilon_r''$ is the complex permittivity of the liquid drop; ε_r' and ε_r'' are the real part and the imaginary part; E_0 is the field in the unperturbed cavity and E is the field in the interior of the liquid drop; v_s is the volume of the liquid drop; Q_0 and f_0 are the quality factor and resonance frequency of the cavity in the unperturbed condition respectively and Q_s, f_s the corresponding parameters of the cavity loaded with the liquid drop; W_0 is the total energy stored in the cavity.

Under the quasi – static approximation, the electric field within a liquid drop sphere placed in a uniform external electric field E_0 is given by[8]

$$E = \frac{3E_0}{\varepsilon_r' + 2} \tag{3}$$

Substitution of this expression into (1) yields the usual expression for the perturbation of the frequency by a small liquid drop sphere,

$$\Delta f = f_0 - f_s = \frac{3E_0^2(\varepsilon_r' - 1)f_s}{2W_0(\varepsilon_r' + 2)}V_s(t) \tag{4}$$

where $V_s(t)$ is the volume of the liquid drop which grows in the process of drop formation. Eq. (4) indicates that the resonant frequency change Δf of the cavity is directly proportional to $V_s(t)$.

The microwave cavity is a two – port network. The insertion loss and half power width of this network can be written as[9]

$$T = \frac{2\sqrt{\beta_1\beta_2}}{1 + \beta_1 + \beta_2} \tag{5}$$

$$Q_s = \frac{f_s}{B} \tag{6}$$

where T is the insertion loss of the network, and $T = (P_{in} - P_{out})/P_{in}$; P_{in} and P_{out} are the microwave input power and the microwave output power of the cavity respectively; β_1 and β_2 are the input coupling coefficient and the output coupling coefficient of the network respectively, and $\beta_1 = Y_{01}/n_1^2 G$, $\beta_2 = Y_{02}/n_2^2 G$; Y_{01} is the equivalent input admittance of the network; Y_{02} is the equivalent output admittance of the network; n_1 and n_2 are the turns ratio of the input ideal transformer and the turns ratio of the output ideal transformer, respectively; G is the equivalent conductance of the networks; B is the half power width of the network.

Suppose that n_1, n_2, Y_{01}, Y_{02} are constant, and $R = 1/G = k\varepsilon_r''V_s(t)$, where R is directly proportional to $V_s(t)$ with a coefficient k, $\beta_1 \ll 1$, $\beta_2 \ll 1$, then from Eq. (5), the following can be obtained

$$P_{out} = 1 - \frac{2\sqrt{\beta_1\beta_2}}{1 + \beta_1 + \beta_2} \cdot P_{in} \approx 1 - 2\sqrt{\beta_1\beta_2} \cdot P_{in}$$

$$= 1 - \frac{2P_{in}\varepsilon_r''k\sqrt{Y_{01}Y_{02}}}{n_1n_2} \cdot V_s(t) = 1 - \frac{2P_{in}k'\varepsilon_r''}{n_1n_2} \cdot V_s(t) \tag{7}$$

where $k' = k\sqrt{Y_{01}Y_{02}}$. Eq. (7) indicates that the larger the volume $V_s(t)$ of the liquid drop, the smaller the output power P_{out} of the cavity. Therefore, the smaller the P_{out}, the smaller the output voltage of the cavity, and the output voltage of the cavity is inversely proportional to $V_s(t)$.

Substituting Eq. (3) into Eq. (2) would yield

$$Q_s^{-1} = Q_0^{-1} + \frac{3\varepsilon_r''E_0^3}{2W_0(\varepsilon_r' + 2)} \cdot V_s(t) \tag{8}$$

Eq. (8) indicates that the larger $V_s(t)$ is, the smaller Q_s will be, and Q_s of the cavity is inversely proportional to $V_s(t)$.

3 Measuring equipment

The microwave sensor consists of a cavity, a microwave generator, an interface circuit, a detecting circuit and a computer. The cavity is a circular cylindrical E_{010} mode resonator, of which the resonance frequency is 2.45GHz, aperture diameter is 8 mm, the outside diameter of the liquid delivery tube is 3.5mm and the inside diameter of the liquid delivery tube is 1mm. The microwave generator is scan frequency with a resolution of 1MHz. The detecting circuit is composed of a linear detector, a low – pass filter and a 12 – bit high speed A/D converter. The output voltage accuracy of the microwave sensor is 1.22mV. The sensor system was controlled by the computer. Its software ran in a windowsXP environment. The control software was programmed by visual basic. A structure diagram of the microwave sensor system is schematically presented in Fig. 1. The computer controlled the microwave generator through the interface circuit. The microwave signals were transmitted into the cavity. The output signals of the cavity were picked up by the detecting circuit. The data processing of the microwave sensor system was based on the computer.

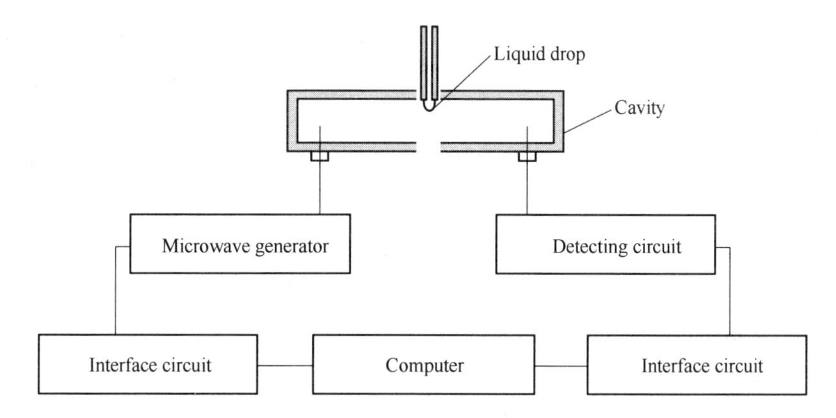

Fig. 1 Sketch of the microwave sensor for measuring the properties of a liquid drop

4 Results and discussion

Fig. 2 shows a 3D graph for constant – pressure delivery obtained from distilled water. Fig. 3 is obtained from Fig. 2. It can be seen from Fig. 3(a) that the relative frequency shift of the cavity is linear with time in the process of drop formation, which is in good agreement with Eq. (4). It is obvious that the output voltage of the cavity in Fig. 3(b) and the Q_s value in Fig. 3(c) are inversely proportional to time in the process of drop formation, which is in agreement with Eqs. (7) and (8), respectively. Therefore, the theory based on the microwave sensor is in good agreement with the experimental results. Based on the experimental result in Fig. 3(b), the difference of output voltage is 100mV in the process of distilled water drop formation, and the weight of a drop is 38.5mg, and the resolution of the measuring equipment for voltage is 1.22mV, therefore, the accuracy of drop volume measurement of the microwave sensor is approximately 0.5μL for distilled water. The different solutions and species concentration variations usually cause change in the physical properties of a solution, such as surface tension, concentration and mass density etc, which in turn will influence the drop size and its microwave absorption properties. Therefore, these combined effects can be measured by detecting the output signal of the microwave sensor in the process of drop formation.

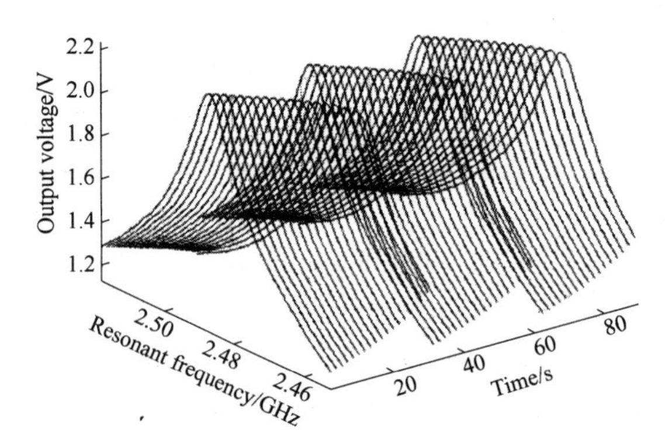

Fig. 2 3D graph detected by the microwave sensor in the process of the formation of three distilled water drops

Fig. 4 gives experimental results for glycerol, distilled water, and 10% NaCl – water solution. The fitting curves of Fig. 4 can be simulated as $y = at + b$, where t is the time of the formation of a liquid drop, y is equal to output voltage V, resonant frequency f_s, or half power width B of the microwave sensor for different models respectively. The fitting parameters aandbdisplayed in Table 1 are obtained by the least – squares fit.

The fitting equation $V = at + b$ in Table 1 shows that b decreasing monotonically from 2.2336 to 2.1434 indicates the imaginary part ε''_r of permittivity of the liquid drop increasing gradually from 10% NaCl – water to glycerol. Similarly, the fitting equation $f_s = at + b$ shows that the real part ε'_r of permittivity of the liquid drop decreases from glycerol to 10% NaCl – water, while b increases monotonically from glycerol to 10% NaCl – water. The fitting equation $B = at + b$ shows that the half power width of the microwave sensor increases monotonically from glycerol to 10% NaCl – water. Meanwhile, the fitting parameter a means the slope of the fitting curve $y = at + b$ is a dynamic characteristic in relation to the volume, surface tension, density and pressure, etc. in the process of drop formation. Therefore, the microwave absorption properties of the liquid contain a great deal of

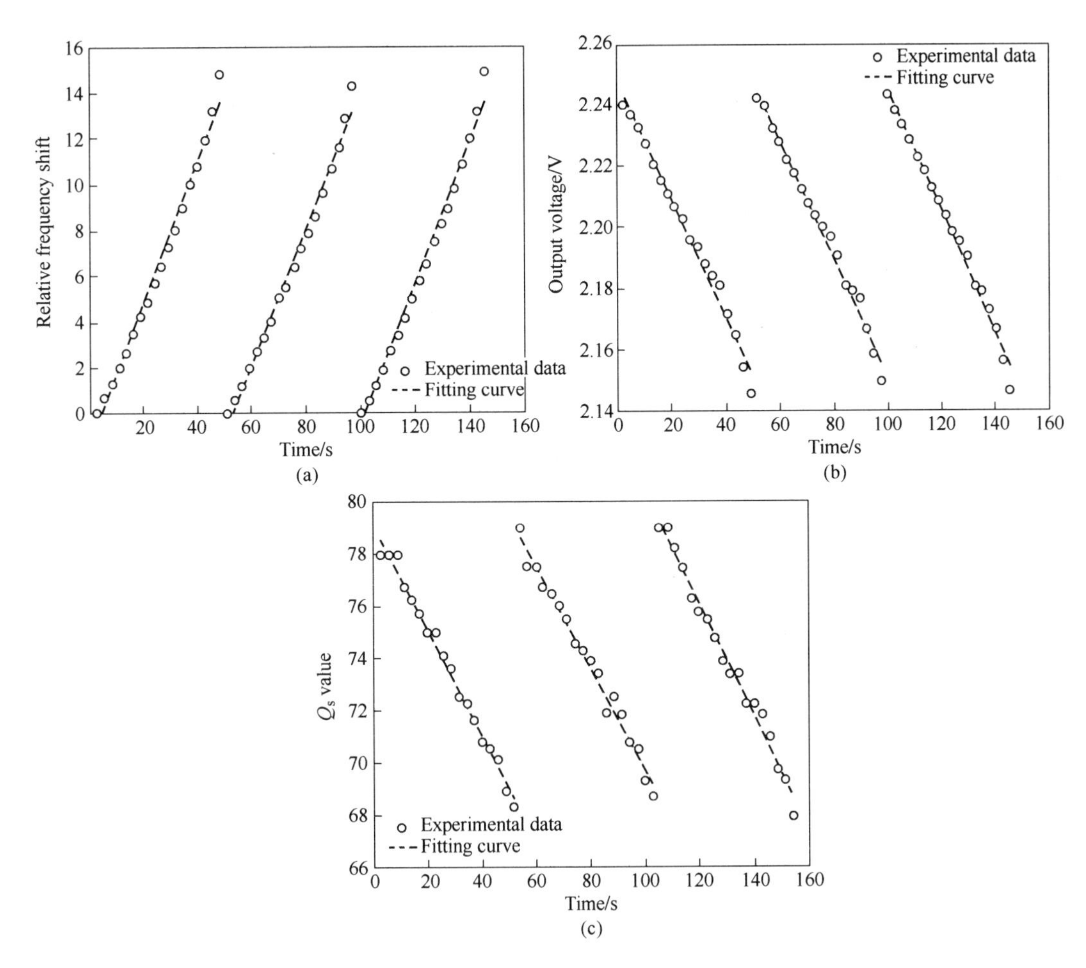

Fig. 3 The formation processes of three distilled water drops detected by the microwave sensor
(a) The relation between Δf and time; (b) The relation between output voltage and time;
(c) The relation between Q_s value and time
○ the experimental data;---- the least-squares fit to the experimental data
The final point lies away from the line of best fit. The main reason is the shape of the drop is significantly altered just before detachment

information in the process of drop formation. Just like the fibre drop multianalyser[2] capacitive tensiography[3], the microwave sensor has been demonstrated to be a new method to determine the physical and chemical characteristics of liquids in an experiment.

By measuring the static liquid in the capillary, the relations between the output signal of the microwave sensor and the concentration by weight of glycerol - water solutions are obtained, and shown in Figs. 5 and 6. Fig. 5 shows that the resonant frequency of the microwave sensor is approximately linearly dependent on the concentration of glycerol - water solutions, which is in agreement with experimental results in Fig. 7(a) of Ref. [3]. More interestingly, Fig. 6 shows that the output voltage and the half power width are not monotonically dependent on concentration, which would be difficult to explain by mixing formulae of dielectric material[10,11].

The surface tension of a liquid drop is a key parameter in the field of physical chemistry. How-

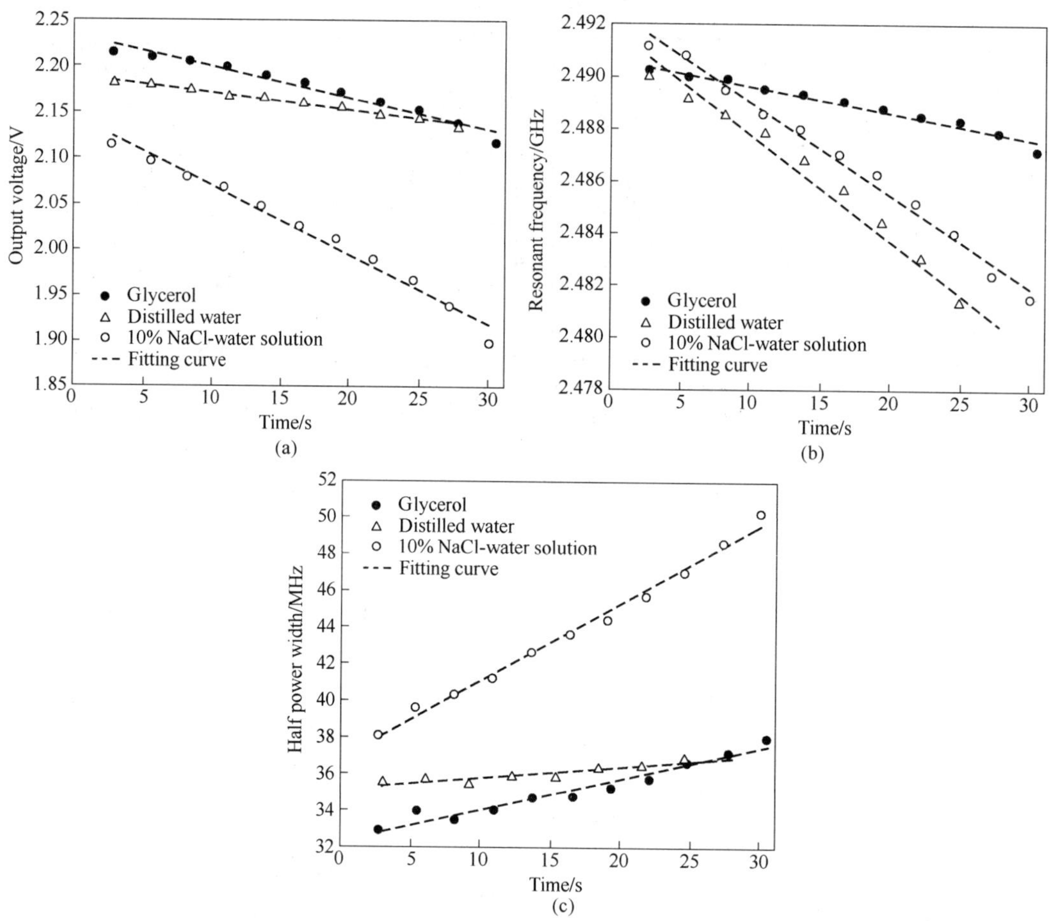

Fig. 4 The formation processes of glycerol, distilled water and 10% NaCl - water solution drops detected by the microwave sensor

(a) The relation between output voltage and time; (b) The relation between relative frequency shift and time; (c) The relation between half power width B and time

The dashed lines are the least - squares fit to the experimental data

Table 1 Parameters of the fitting curves in Fig. 4

Liquids	$V-t$		f_s-t		$B-t$	
	a	b	a	b	a	b
Glycerol	-0.0034	2.2336	-0.00010	1.4907	0.1709	32.2909
Distilled water	-0.0019	2.1905	-0.00041	2.4919	0.0639	35.0944
10% NaCl - water	-0.0075	2.1434	-0.00036	2.4926	0.4264	36.8127

ever, the surface tension of aqueous glycerol varies from 6.28mN/m to 7.20mN/m, while the concentration of glycerol - water changes from 0% to 100%, and the surface tension difference of 12.5% and 25% glycerol - water is 0.5mN/m in Table 2 of Ref. [3]. Therefore, it is difficult to measure the concentration of the solution by the surface tension of the liquid. Based on the experimental results in Figs. 5 and 6, the differences in the resonant frequency, output voltage and half

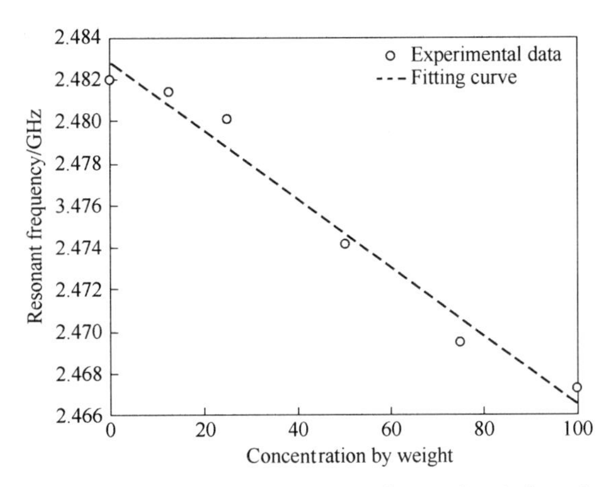

Fig. 5 The resonant frequency versus concentration by weight of glycerol – water solutions

power width are 2MHz, 111. 6mV, and 7. 2MHz, respectively, while the concentration of glycerol – water changes from 12. 5% to 25%. It shows that the microwave sensor for measuring the concentration of liquid is more accurate. The nonlinear phenomena found in our experiment (Fig. 6) are not understood and the physical/chemical origin of the interesting phenomena remains a subject for the future.

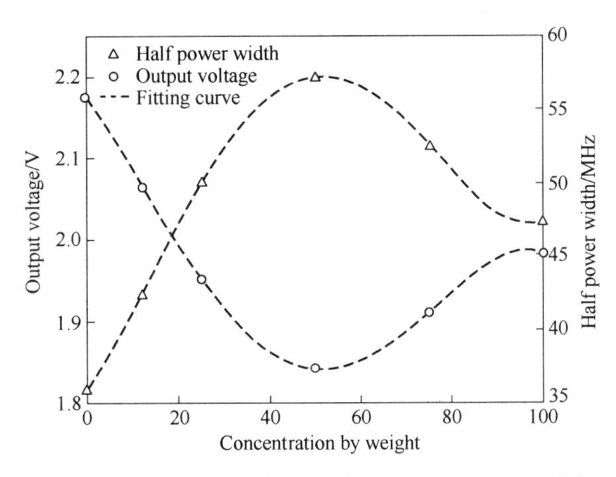

Fig. 6 Plot of output voltage and resonant frequency of the microwave sensor against concentration by weight of glycerol – water solutions

5 Conclusion

The theory of the microwave sensor has been established and a prototype has been constructed with a volume resolution of 0. 5μL. Besides surface tension and the microwave absorption properties of a pendant liquid drop, the concentration of liquid can also be measured more accurately by a microwave sensor.

Acknowledgements

The authors thank the National Natural Science Foundation of China (No. 20463003) and the Natural Science Foundation of Yunnan Province(No. 2004E0003Z) for financial support. Chinese Patent is pending.

References

[1]Eggers J. Nonlinear dynamics and breakup of free surface flows[J]. Reviews of Modern Physics,1997,69(3):69865 - 69930.

[2]Mcmillan N D,Finlayson O,Fortune F,et al. The fibre drop analyser:a new multianalyser instrument with applications in sugar processing and for the analysis of pure liquids[J]. Measurement Science and Technology, 1992,3(8):746 - 764.

[3]Wang C H,Augousti A T,Mason J,et al. The capacitive drop tensiometer—a novel multianalysing technique for measuring the properties of liquids[J]. Measurement Science and Technolog,10(1):19 - 24.

[4]Augousti A T,Mason J,Mcmillan N D,et al. Application of capacitive tensiography to investigation of pendant drop growth[J]. Measurement Science and Technology,2006,17(10):N48 - N52.

[5]Kupfer K,Kraszewski A. Sensors Update[M]. Oxford:Oxford University Press,2000.

[6]Huang M,Peng J H,Yang J J,et al. Microwave cavity perturbation technique for measuring the moisture content of minerals sulphide concentrate[J]. Minerals Engineering,2007,20(1):92 - 94.

[7]Metaxas A C,Meredith R J. Industrial Microwave Heating[M]. London:Peter Peregrinus,1983.

[8]Bleaney B I,Bleaney B. Electricity and Magnetism[M]. Oxford:Oxford University Press,1957.

[9]Bahl I,Bhartia P. Microwave Solid State Circuit Design(2nd)[M]. Beijing:Publishing House of Electronics Industry,2006.

[10]Sihvola A H,Kong J A. Effective permittivity of dielectric mixtures[J]. IEEE Transactions on Geoscience and Remote Sensing,1988,26(4):420 - 429.

[11]Brosseau C. Modelling and simulation of dielectric heterostructures:a physical survey from an historical perspective[J]. Journal of Physics D:Applied Physics,2005,39(7):1277 - 1294.

Microwave Sensor for Measuring the Properties of a Liquid Drop

The process of drop formation can reveal a large amount of information on the properties of liquids. Now, Ming Huang and co - workers at Yunnan University and Kunming University of Science and Technology have developed a sensor that uses microwaves to probe the physical and chemical make - up of a liquid drop. The sensor could prove useful for measuring important quantities in physical chemistry.

Research Highlights

Subject Categories: Chemistry Materials

Published online: 6 June 2007 | doi:10.1038/nchina.2007.99

Liquid drops: Wavy measurement

Tim Reid

A new technique uses microwaves to investigate the physical and chemical properties of liquid drops

Original article citation
Huang, M., Yang, J., Wang, J. & Peng, J. Microwave sensor for measuring the properties of a liquid drop. *Meas. Sci. Technol.* **18**, 1934-1938 (2007).

The process of drop formation can reveal a large amount of information on the properties of liquids. Now, Ming Huang and co-workers at Yunnan University and Kunming University of Science and Technology have developed a sensor that uses microwaves to probe the physical and chemical make-up of a liquid drop[1]. The sensor could prove useful for measuring important quantities in physical chemistry.

© (2007) Ming Huang

The researchers inserted a liquid delivery tube into the cavity of a cylindrical resonator and passed microwaves through the cavity. When drops were formed on the end of the tube, the resonant frequency of the cavity increased in proportion to the size of the drop. In addition, the output voltage of the cavity decreased owing to the absorption of microwaves by the drop. Both findings are consistent with theory, enabling the drop volume to be calculated to an accuracy of 0.5μl and providing information on the liquid's electrical permittivity, density, pressure and surface tension.

The sensor was tested using distilled water and both glycerol and salt solutions, and gave results consistent with previous measurements. The resonant frequency of the cavity was also found to be linearly dependent on the concentration of the glycerol solution, which provides a highly accurate method of measuring solution concentrations.

The authors of this work are from:
School of Information Science and Engineering, Yunnan University, Kunming, China; Department of Applied Chemistry, Yunnan University, Kunming, China; Faculty of Materials and Metallurgical Engineering, Kunming University of Science and Technology, Kunming, China.

References

1. Huang, M., Yang, J., Wang, J. & Peng, J. Microwave sensor for measuring the properties of a liquid drop. *Meas. Sci. Technol.* **18**, 1934-1938 (2007).

The researchers inserted a liquid delivery tube into the cavity of a cylindrical resonator and passed microwaves through the cavity, when drops were formed on the end of the tube, the resonant frequency of the cavity increased in proportion to the size of the drop. In addition, the output voltage of the cavity decreased owing to the absorption of microwaves by the drop. Both findings are consistent with theory, enabling the drop volume to be calculated to an accuracy of 0.5μL and providing information on the liquid' s electrical permittivity, density, pressure and surface tension.

The sensor was tested using distilled water and both glycerol and salt solutions, and gave results consistent with previous measurements. The resonant frequency of the cavity was also found to be linearly dependent on the concentration of the glycerol solution, which provides a highly accurate method of measuring solution concentrations.

* Microwave sensor for measuring the properties of a liquid drop[J]. Measurement Science and Technology, 2007, 18(7): 1934 – 1938.

A New Equation for the Description of Dielectric Losses under Microwave Irradiation

Ming Huang, Jinhui Peng, Jingjing Yang, Jiaqiang Wang

Abstract: A new equation was developed to describe quantitatively the dielectric loss of materials that is a key parameter in the fields of microwave chemistry and microwave processing of materials under microwave irradiation. This equation can be applied to explain the complicated thermal runaway phenomenon that has relations with dielectric relaxation and to simulate the dielectric loss under microwave irradiation. It was also found that the simulation results based on this equation are in good agreement with the Volge – Fulcher law. The new equation has potential application in the study of microwave interaction with solid materials, especially in materials' processing which involves microwave chemistry, synthesis, sintering, melting, joining, surface – modifications, quality improvements, etc and it is hoped that the dynamic process of material under microwave irradiation could be explained.

Keywords: microwave; dielectric losses; volge – fulcher law; thermal runaway

1 Introduction

Since the early 1980s, microwaves have been widely used in chemistry[1-3] and also used to process a wide variety of absorbing materials[4-8]. A few possible mechanisms were proposed to explain the microwave interaction with absorbing materials. An initial explanation was the rotation of polar molecules[9]. The molecular dipoles are induced by microwave to rotate. This rotation causes molecular collisions that generate heat. Unfortunately this mechanism does not explain the intensity of the observed microwave effects. Other possible mechanisms are superheating effects[10], formation of hot spots[11], the presence of ionic molecules[12], photochemical focus[13] and the induction losses caused by eddy current[14]. However, mechanistic details of these microwave – driven processes are still far from understood, particularly, since few theoretical calculation was conducted to simulate these dynamic processes. On the other hand, for polar molecules, the traditional approach[15] to the interpretation of the frequency dependence of the dielectric loss is based on the Debye relaxation mechanism for which the time domain response is given by

$$F(t) = \exp(-t/\tau) \tag{1}$$

where τ and t are relaxation time and observation time, respectively.

The dielectric loss does not obey Eq. (1) except in some liquid dielectrics and departs seriously from it in most solids[16-18]. It is shown that some relaxation processes, particularly, chemical reaction kinetics, would follow the time domain response law[19]:

$$F(t) = \exp\left[-\left(\frac{t}{\tau}\right)^{\alpha}\right] \quad (0 < \alpha < 1) \tag{2}$$

While α equals 0.5, $F(t)$ is called slow relaxation, the slow response[20] or non - Debye relaxation for which the relaxation mechanism is governed by the bound charges. Eq. (2) was obtained in the absence of microwave irradiation, and it is very important to develop an equation, which can be used to calutate the dielectric loss of non - Debye dielectric relaxation under microwave irradiation. However, so far there is no report on the equation that needs to find.

Here we report that an equation was deduced to calculate the dielectric loss mechanism for non - Debye dielectric relaxation. To the best of our knowledge, this seems to be the first equation used to describe the non - Debye relaxation phenomena in solids under microwave irradiation. Based on this equation and experimental results, which are the relation between the imaginary permittivites(ε'') and temperature of two different materials, we simulated the relation between relaxation time τ and temperature of these materials. More interestingly, the simulation results indicate that the relation between dielectric relaxation time τ and temperature of the materials is in good agreement with the Volge - Fulcher law[17].

The new equation successfully explained the thermal runaway phenomenon of material under microwave irradiation. Based on this equation, the higher the temperature of material is, the smaller the relaxation time τ will be, and more power will be absorbed by the material. The equation also has potential application in the study of microwave interaction with solid materials, especially in sintering of important ceramics, processing of metallic materials, microwave - assisted organic synthesis, nano - chemistry, etc. Moreover, the dynamic process of material under microwave irradiation could be explained using the equation.

2 Dielectric loss computation methods

2.1 Dielectric loss per unit volume

From Maxwell's differential equations, Ohm's loss in dielectric medium of materials is given by

$$\int_v \boldsymbol{E}\cdot\boldsymbol{J}\mathrm{d}v = \int_v \boldsymbol{E}\cdot\left(\nabla\times\boldsymbol{H}-\frac{\partial\boldsymbol{D}}{\partial t}\right)\mathrm{d}v = \int_v\left[\boldsymbol{H}\cdot(\nabla\times\boldsymbol{E})-\nabla\cdot(\boldsymbol{E}\times\boldsymbol{H})-\boldsymbol{E}\cdot\frac{\partial\boldsymbol{D}}{\partial t}\right]\mathrm{d}v$$

Taking time derivatives of $\boldsymbol{E}\cdot\boldsymbol{D}$ and $\boldsymbol{H}\cdot\boldsymbol{B}$, we have

$$\frac{\partial}{\partial t}(\boldsymbol{E}\cdot\boldsymbol{D}) = \boldsymbol{E}\cdot\frac{\partial\boldsymbol{D}}{\partial t}+\boldsymbol{B}\cdot\frac{\partial\boldsymbol{E}}{\partial t} \quad \frac{\partial}{\partial t}(\boldsymbol{H}\cdot\boldsymbol{B}) = \boldsymbol{H}\cdot\frac{\partial\boldsymbol{B}}{\partial t}+\boldsymbol{B}\cdot\frac{\partial\boldsymbol{H}}{\partial t}$$

Finally, we obtain

$$\begin{aligned}\oint_s(\boldsymbol{E}\cdot\boldsymbol{H})\mathrm{d}s &= \int_v\left[\boldsymbol{H}\cdot\frac{\partial\boldsymbol{B}}{\partial t}+\boldsymbol{E}\cdot\frac{\partial\boldsymbol{D}}{\partial t}+\boldsymbol{E}\cdot\boldsymbol{J}\right]\mathrm{d}v \\ &= \int_v\left[\frac{1}{2}\frac{\partial}{\partial t}(\boldsymbol{E}\cdot\boldsymbol{D})+\frac{1}{2}\frac{\partial}{\partial t}(\boldsymbol{H}\cdot\boldsymbol{B})\right]\mathrm{d}v+\int_v\boldsymbol{E}\cdot\boldsymbol{J}\mathrm{d}v+ \\ &\quad \int_v\left[\frac{1}{2}\left(\boldsymbol{E}\cdot\frac{\partial\boldsymbol{D}}{\partial t}-\boldsymbol{D}\cdot\frac{\partial\boldsymbol{E}}{\partial t}\right)\right]\mathrm{d}v+\int_v\left[\frac{1}{2}\left(\boldsymbol{H}\cdot\frac{\partial\boldsymbol{B}}{\partial t}-\boldsymbol{B}\cdot\frac{\partial\boldsymbol{H}}{\partial t}\right)\right]\mathrm{d}v\end{aligned}$$

The term on the left - hand side of this equation represents the power flowing on closed surface s. The first term on the righthand side represents the electric energy and magnetic energy stored within the volume v. The second term on the righthand side represents the power dissipated within the volume v. The third term on the right - hand side represents electric losses of material within the volume v. The fourth term on the right - hand side represents the magnetic losses of material within the volume v. Therefore, for electric dielectric material, the dielectric loss per unit volume is

$$P_d = \frac{1}{2}\left(\boldsymbol{E} \cdot \frac{\partial \boldsymbol{D}}{\partial t} - \boldsymbol{D} \cdot \frac{\partial \boldsymbol{E}}{\partial t}\right) \tag{3}$$

where $\boldsymbol{E}$ is electric field intensity vector, in V/m; $\boldsymbol{H}$ is magnetic field intensity vector, in A/m; $\boldsymbol{D}$ is electric flux density vector, in C/m^2; $\boldsymbol{B}$ is magnetic flux density vector, in Wb/m^2; $\boldsymbol{J}$ is volume current density vector, in A/m^2; the volume v is bounded by the surface s.

2.2 Debye dielectric relaxation

If Debye dielectric relaxation mechanism is followed and governed by Eq. (1) and the process occurring under an electric field with $E(t) = E\cos(\omega t)$ would still followthe linear superposition principle, we obtain

$$\begin{aligned} D(t) &= \varepsilon_h E(t) + \varepsilon_l \int_{-\infty}^{t} E(t') \frac{\partial}{\partial t}\left[-\exp\left(-\frac{t-t'}{\tau}\right)\right] dt' \\ &= \varepsilon'(\omega) E\cos(\omega t) - j\varepsilon''(t)\sin(\omega t) \\ P_d &= \frac{1}{2} E\cos(\omega t) \frac{\partial}{\partial t}[\varepsilon'(\omega) E\cos(\omega t) - j\varepsilon''(t) E\sin(\omega t)] \\ P_d &= \frac{1}{2}\varepsilon'' E^2 \omega \end{aligned} \tag{4}$$

where P_d is the dielectric loss per unit volume based on the Debye relaxation; ε_h is the high frequency permittivity of the material; ε_l is the low frequency permittivity or staticpermittivity of the material; $\varepsilon'(\omega)$ is the real part of the dielectric permittivity of the material and $\varepsilon' = \varepsilon_h + \varepsilon_l/(1+\omega^2\tau^2)$; $\varepsilon''(\omega)$ is the imaginary part of the dielectric permittivity of the material and $\varepsilon'' = \omega\tau\varepsilon_l/(1+\omega^2\tau^2)$; the electric field $E(t)$ is switched on at $t = t'$ and ω is the angular frequency of microwave.

Eq. (4) indicates that the dielectric loss per unitvolume is in agreement with the traditional law[8].

2.3 Non - Debye relaxation

If non - Debye dielectric relaxation mechanism is followed and governed by Eq. (2) and the other conditions are the same as 2.2, we obtain

$$\begin{aligned} D(t) &= \varepsilon_h E(t) + \varepsilon_l \int_{-\infty}^{t} E(t') \frac{\partial}{\partial t}\left[-\exp\left(-\sqrt{\frac{t-t'}{\tau}}\right)\right] dt' \\ &= \varepsilon_h E(t) + \frac{\varepsilon_l E}{2\tau}\int_{-\infty}^{t} \cos(\omega t')\left(\frac{t-t'}{\tau}\right)^{-1/2} \times \exp\left(-\sqrt{\frac{t-t'}{\tau}}\right) dt' \\ &= \varepsilon_t' E\cos(\omega t) + \varepsilon_t'' E\sin(\omega t) \\ P_t &= \frac{1}{2}E^2\omega\varepsilon_t'' + \frac{E^2}{2}\left[\cos^2(\omega t)\frac{\partial \varepsilon_t'}{\partial t} + \cos(\omega t)\sin(\omega t)\frac{\partial \varepsilon_t''}{\partial t}\right] \end{aligned} \tag{5}$$

where

$$P_t = \frac{1}{2}E^2\omega\varepsilon''_t + \frac{E^2}{2}\left[\cos^2(\omega t)\frac{\partial\varepsilon'_t}{\partial t} + \cos(\omega t)\sin(\omega t)\frac{\partial\varepsilon''_t}{\partial t}\right]$$

$$\varepsilon'_t = \varepsilon_h + \frac{\varepsilon_1}{2\tau}\int_0^{\infty}\left(\frac{t''}{\tau}\right)^{-1/2}\exp(-\sqrt{t'/\tau})\cos(\omega t'')\,dt''$$

$$\varepsilon''_t = \frac{\varepsilon_1}{2\tau}\int_0^{\infty}\left(\frac{t''}{\tau}\right)^{-1/2}\exp(-\sqrt{t'/\tau})\sin(\omega t'')\,dt''$$

$$t'' = t - t'$$

and P_t is the dielectric loss per unit volume based on the non - Debye relaxation.

Comparing Eqs. (4) and (5), the first term on the right - hand side of Eq. (5) is a little similar to Eq. (4), but ε''_t is different from ε'', and the second term on the right - hand side of Eq. (5) is a function of t and $\partial/\partial t$. Therefore, the dynamic processes of non - Debye dielectric under microwave irradiation could be simulated using Eq. (5).

3 Results and discussion

The dielectric loss per unit volume P_d in Eq. (4) is a time independent quantity, while P_t in Eq. (5) is a function of t and τ. The characteristic of material under power microwave irradiation is variable, which is the foundation of microwave chemistry and materials' processing. Therefore, Eq. (5) could be used to explain the dynamic processes under power microwave irradiation.

The frequency of microwave irradiation is 2460 MHz, the interaction between material and microwave is about 2. 46109 times per second and temperature rise of the material under microwave irradiation is less than 0. 1℃ per second[21]; therefore, the electric dielectric characteristic of the material is invariable within less than a second, and we can get that $\frac{\partial\varepsilon'_t}{\partial t}$ and $\frac{\partial\varepsilon''_t}{\partial t}$ is equal to zero, respectively.

Substituting $\frac{\partial\varepsilon'_t}{\partial t}=0$ and $\frac{\partial\varepsilon''_t}{\partial t}=0$ into Eq. (5) would yield

$$P_t = \frac{1}{2}E^2\omega\varepsilon''_t(\tau,t) \tag{6}$$

Using Eq. (6), the dependence of the power absorbed by material on time t at different $\tau(\tau_1 > \tau_2 > \cdots > \tau_i > \tau_{i+1})$ is shown in Fig. 1, which indicates that the smaller the relaxation time τ, the greater the power absorbed by the material.

Fig. 2 shows the imaginary permittivites(ε'') of nickeliferous limonitic laterite ores measured by Pickles[22] at frequencies of 2460MHz at temperature up to about 1000℃ using the cavity perturbation technique.

It can be seen from Fig. 2 that imaginary permittivites (ε'') of nickeliferous limonitic laterite increases while the temperature is increasing, and ε'' is the function of temperature, which can be written as $\varepsilon''(T)$. Therefore, it can be obtained from Eq. (4) that the power absorbed by material increases while the temperature is increasing. The temperature of nickeliferous limonitic laterite is a

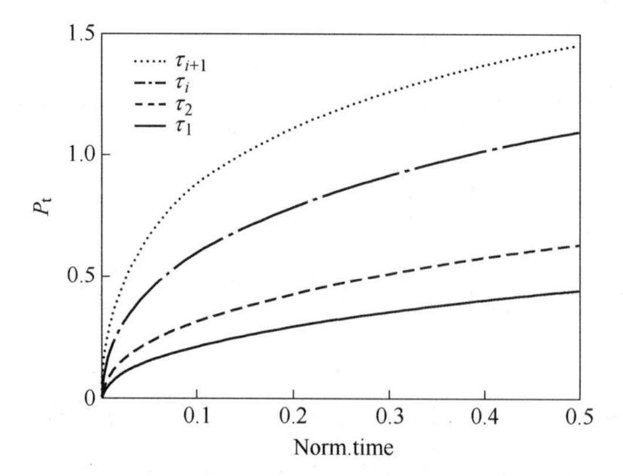

Fig. 1 The dependence of the power absorbed by dielectric on the normalized time

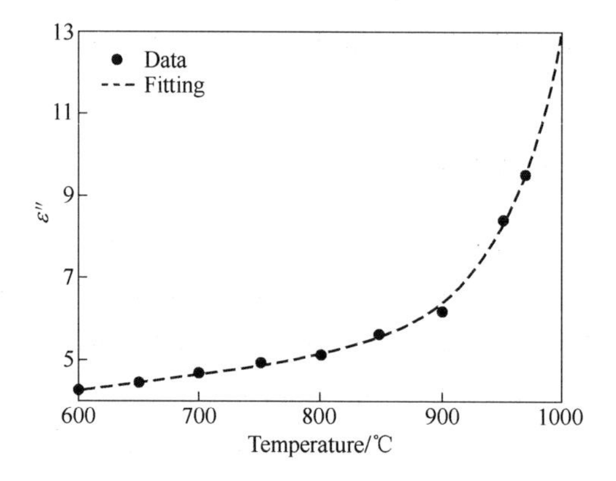

Fig. 2 The imaginary permittivities (ε'') of nickeliferous limonitic laterite as a function of temperature at 2460MHz
● data from Ref. [22]; ---- the fitting curve

function of irradiation time, and $T = g(t)$, then using Eq. (4) the microwave power absorbed by dielectric would be directly proportion to $\varepsilon''(T)$:

$$P_d = \frac{1}{2}\varepsilon''(T)E^2\omega \tag{7}$$

In order to use Eq. (6) to explain the power absorbed by nickeliferous limonitic laterite under microwave irradiation, we suppose that Eq. (6) is directly proportional to Eq. (7) with a coefficient k, and we have

$$\varepsilon''_t(\tau, t) = k\varepsilon''(T) \tag{8}$$

Supposing that time $t_1, t_2, \cdots, t_9$ is corresponding to 600℃, 650℃, ⋯, 1000℃, respectively, and dielectric relaxation time τ is invariable in the time interval $\Delta t(\Delta t \to 0)$, then from Eq. (8), the followings can be obtained:

$$\frac{\varepsilon_1}{2\tau_1}\int_{t_1}^{t_1+\Delta t}\left(\frac{t-t_1}{\tau_1}\right)^{-1/2}\exp(-\sqrt{(t-t_1)/\tau_1})\,dt = k\varepsilon''(600)$$

$$\frac{\varepsilon_1}{2\tau_2}\int_{t_2}^{t_2+\Delta t}\left(\frac{t-t_2}{\tau_2}\right)^{-1/2}\exp(-\sqrt{(t-t_2)/\tau_2})\,dt = k\varepsilon''(650)$$

$$\frac{\varepsilon_1}{2\tau_3}\int_{t_3}^{t_3+\Delta t}\left(\frac{t-t_3}{\tau_3}\right)^{-1/2}\exp(-\sqrt{(t-t_3)/\tau_3})\,dt = k\varepsilon''(1000)$$

where $\tau_1, \tau_2, \cdots, \tau_9$ is the relaxation time that is corresponding to microwave irradiation time t_1, $t_2, \cdots, t_9$, respectively.

Solving the preceding equations by any standard method, we obtain the relation between relaxation time τ and temperature T showed in Fig. 3.

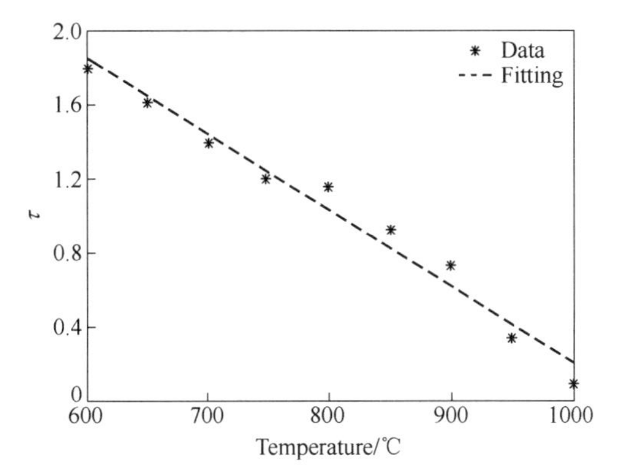

Fig. 3 Simulation result of relaxation time versus the temperature of nickeliferous limonitic laterite

* the value of relaxation time from simulation; ---- the fitting curve

Using the same simulation procedure and the experimental result of dielectric loss in alumina as a function of temperature that had been measured by Westphal and Sils[23], the relation between relaxation time and the temperature of alumina was also obtained and shown in Fig. 4.

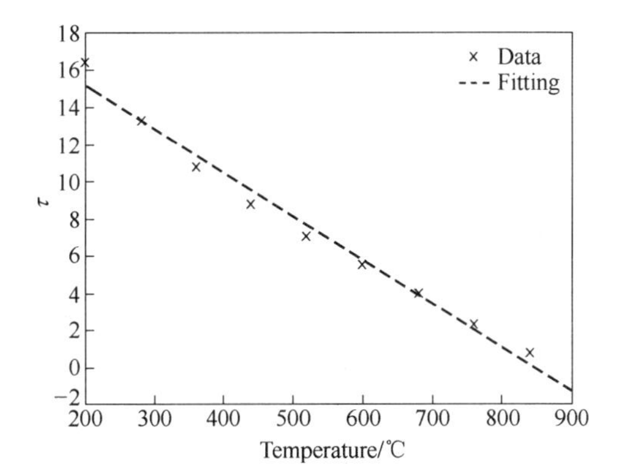

Fig. 4 Simulation result of relaxation time versus the temperature of alumina

× the value of relaxation time from simulation; ---- the fitting curve

In the two examples above, the simulation results show that the relaxation time τ is inversely proportional to temperature, which is in good agreement with the Vogle – Fulcher law, and is confirmed by the good linear fit.

4 Conclusions

(1) It was found that under sinusoidal microwave irradiation the dielectric response is not sinusoidal again, the response of non – Debye dielectric time dependent power absorbed by the dielec-

tric exists and the power absorbed by material between $0-\tau$ is variable.

(2) The results calculated using Eq. (6) are in good agreement with the Vogle - Fulcher law, which show that the relaxation time τ has an inverse ratio to the temperature of the material.

(3) The new equation has potential application in the study of the microwave interaction with solid materials, and it is hoped that the dynamic processes of material under microwave irradiation could be explained.

Acknowledgements

The authors thank the National Natural Science Foundation of China (No. 20463003) and the Natural Science Foundation of Yunnan Province (No. 2004E0003Z) for financial support.

References

[1] Adam D. Nature. 2003:421, 571 - 572.

[2] Nüchter M, Ondruschka B, Bonrath W, et al. Microwave assisted synthesis—a critical technology overview[J]. Green Chemistry, 2004, 6(3), 128 - 141.

[3] Vijayan Raghavan G S, Dai J. The 12th National Conf. on Microwave Power Application Symp. Proc. (Chengdu, China). 2005:10.

[4] Mingos D M P, Baghurst D R, Lecture T. Applications of microwave dielectric heating effects to synthetic problems in chemistry[J]. Chemical Society Reviews, 1991, 20(1):1 - 47.

[5] Clark D E, Sutton W H. Microwave processing of materials[J]. Annual Review of Materials Science, 1996, 26 (1):299 - 331.

[6] Roy R, Agrawal D, Cheng J, et al. Full sintering of powdered - metal bodies in a microwave field[J]. Nature, 1999, 399(6737):668 - 670.

[7] Risman P O, Celuch - Marcysiak. Microwave, Radar and Wireless Communications (MIKON - 2000: 13th Int. Conf. , Wroclaw, Poland). 2000, 3:22 - 24.

[8] Metaxas A C, Meredith R J. Industrial Microwave Heating[M]. London: Peter Peregrinus, 1983:72.

[9] Grant E H. Microwaves: Industrial, Scientific and Medical Applications[M]. London: Artech House, 1992:12.

[10] Baghurst D R, Mingos D M P. A new reaction vessel for accelerated syntheses using microwave dielectric super - heating effects[J]. Journal of the Chemical Society, Dalton Transactions, 1992(7):1151 - 1155.

[11] Rybakov K I, Semenov V E. 7th Int. Conf. on Microwave and High Frequency Heating (Valencia). 1999: 17 - 19.

[12] Senise J T, Jermolovicius L A. 2003 Proc. 2003SBMO/IEEE - S IMOC (Foz Do Iguacu, Brazil). 2003, 3: PD 01 - 06.

[13] Isaacs N. Physical Organic Chemistry (Belfast: Longman). 1987:96.

[14] Cherradi D G, Provost J, Raveau B. Electric and magnetic field contributions to the microwave sintering of ceramics[J]. Electroceramics Ⅳ, 1994, 2:1219 - 1222.

[15] Gaiduk V I. Dielectric Relaxation and Dynamics of Polar Molecules [M]. Singapore: World Scientific, 1999:16.

[16] Jonscher A K. Universal Relaxation Law[M]. London: Chelsea Dielectrics Press, 1996:11.

[17] Jonscher A K. Dielectric relaxation in solids[J]. Journal of Physics D: Applied Physics, 1999, 32(14):R57.

[18] Jonscher A K. Physical basis of dielectric loss[J]. Nature, 1975: 717 - 719.

[19] Williams G, Watts D C. Non - symmetrical dielectric relaxation behaviour arising from a simple empirical decay function[J]. Transactions of the Faraday Society, 1970, 66: 80 - 85.

[20] Li J, Chen M, Zheng F, Zhou Z. Sci. China, 1997, 40: 290 - 295.

[21] Hui Q W, Xhilin T, Hong Y. The 12th National Conf. on Microwave Power Application Symp. Proc. (Chengdu, China). 2005: 68 - 70.

[22] Pickles C A. Microwave heating behaviour of nickeliferous limonitic laterite ores[J]. Minerals Engineering, 2004, 17(6): 775 - 784.

[23] Westphal W, Sils A. Microwaves: Industrial, Scientific and Medical Applications. London: Artech House, 1992: 10.

Microwave Absorbing Properties of High Titanium Slag

Libo Zhang, Guo Chen, Jinhui Peng, Jin Chen, Shenghui Guo, Xinhui Duan

Abstract: Microwave absorbing properties of high titanium slag were investigated by using microwave cavity perturbation technique. High titanium slag containing more than 90% TiO_2 was prepared by carbothermal reduction of ilmenite. The temperature rise curve of high titanium slag in microwave heating process was obtained. Crystalline compounds of high titanium slag before and after microwave irradiation were obtained and characterized by X - ray diffractometry (XRD). Effects of particle size of high titanium slag and mixtures of high titanium slag with different mass fractions of V_2O_5 on microwave absorbing properties were investigated systematically. The results show that high titanium slag has good microwave absorption property; untreated high titanium slag mainly consists of crystalline compounds of anatase and iron titanium oxide, while the microwave - irradiation treated one is mainly composed of crystalline compounds of rutile and iron titanium oxide. Synthetic anatase is transformed completely into rutile at about 1050℃ for 20min under microwave irradiation. High frequency shift and low amplitude of voltage make high titanium slag an ideal microwave absorbent. 180μm of particle size and 10% mass fraction of V_2O_5 are found to be the optimum conditions for microwave absorption.

Keywords: high titanium slag; microwave absorbing; microwave cavity perturbation; microwave irradiation

1 Introduction

It is known that titanium dioxide (TiO_2) exists in three main polymorphic phases: rutile, anatase and brookite[1]. Anatase and brookite are metastable phases, and their exothermic and irreversible conversions to rutile at high temperatures have been widely investigated. The transformation of anatase to rutile in TiO_2 is influenced by several experimental conditions such as temperature, particle size and synthetic method of dioxide[2-4].

Titanium dioxide pigments are manufactured by the chloride or sulfate processes, in which natural or synthetic rutile is used as raw materials. However, these processes pollute the environment[5]. Since available resources of high grade natural rutile tend to diminish, the shortage of natural rutile has encouraged researchers to find an efficient method to convert anatase in high titanium slag to synthetic rutile. To explore new method to produce synthetic rutile with low energy consumption and less environment pollution is necessary[6-8]. Microwave can eliminate the environmental pollution from production source and realize sustainable development of titanium resources, and produce synthetic rutile from high titanium slag.

The microwave absorbing property of materials is an important physical indicator in the fields of microwave chemistry, microwave detection, and microwave processing[9-11]. Microwave cavity perturbation technique is widely adopted for microwave dielectric properties measurements[12]. A microwave experiment is based on the cavity perturbation method, employing the single resonant mode[13]. The change in the cavity characteristics in the presence of a sample is measured. In addition, the complex conductivity is directly evaluated. Huang, et al.[14] invented a novel microwave sensor for measuring the properties of a liquid drop, and the technique can easily be extended to microwave processing materials. Its analytical theory is established and a working prototype has been constructed and tested. It is also found that the theory based on the microwave sensor is in a good agreement with the experimental results. Huang, et al.[15] investigated the technique of microwave cavity perturbation in measuring the characteristics of microwave absorbing of the mixtures of carbonaceous reducer and ilmenite. Huang, et al.[16] also used the microwave cavity perturbation technique to measure the moisture content of a sulphide mineral concentrate.

In this work, a new method, based on microwave cavity perturbation technique and digital signal processing technique, was applied to detecting the microwave absorbing properties of high titanium slag. The effects of particle size of high titanium slag and the mixtures of high titanium slag with different mass fractions of V_2O_5 on microwave absorbing properties were investigated. The optimum conditions of producing synthetic rutile from high titanium slag by microwave irradiation were obtained.

2 Experimental

2.1 Materials

High titanium slag was obtained from Kunming City, Yunnan Province, China. High titanium slag sample was prepared from ilmenite by carbothermal reduction in an electric arc furnace. The chemical compositions of high titanium slag are shown in Table 1.

Table 1 Chemical composition of high titanium slag (mass fraction, %)

TiO_2	FeO	S	P	C	Al_2O_3	SiO_2	MnO	MgO	Bal.
90.120	5.260	0.049	0.014	0.049	2.750	2.570	1.040	2.300	1.158

2.2 Measuring principles and apparatus

The microwave sensor system was based on microwave cavity perturbation technique and digital signal processing technique. Derived from the theory of electric – magnetic field, the frequency shift and the output voltage of the microwave cavity are given by[16,17]

$$\frac{\Delta\omega}{\omega} = -\omega_0(\varepsilon'_r - 1)\int_{V_e} E_0^* E \mathrm{d}V / 4W \tag{1}$$

$$\frac{1}{Q} - \frac{1}{Q_0} = 2\varepsilon_0 \varepsilon''_r \int_{V_e} E_0^* E \mathrm{d}V / 4W \tag{2}$$

$$W = \int_V [(E_0^* D_0 + H_0^* B_0) + (E_0^* D_1 + H_0^* B_1)] \mathrm{d}V \tag{3}$$

where $\Delta\omega = \omega - \omega_0$, E_0^*, D_0, H_0^* and B_0 are the fields in the interior of the sample; V and V_e are the volumes of the cavity and the sample, respectively; dV is the elemental volume; Q_0 and ω_0 are the quality factor and resonance frequency of cavity in the unperturbed condition, respectively; Q and ω are the corresponding parameters of the cavity loaded with the sample; ε'_r and ε''_r are the real and the imaginary part of the complex permittivity of the sample, respectively; and W is the storage energy.

According to Eqs. (1) – (3), the data of microwave absorbing properties can be acquired by measuring the output voltage and the frequency shift of the microwave sensor. A structure diagram of the microwave sensor system is illustrated in Fig. 1. The computer controls the fast scanning microwave generator through multipurpose card. The microwave signals are transmitted into the microwave sensor. The output signals of the microwave sensor are picked up by the linear detector and Digital Signal Processor (DSP). Then they are fed into the low pass filter. After that, the output signals of the low pass filter are amplified and converted by the A/D converter. The data processing of the microwave sensor system is finished on the computer[18].

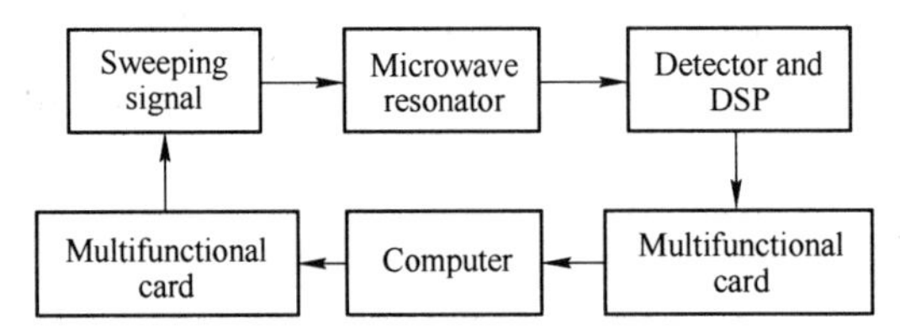

Fig. 1 Sketch of microwave sensor system for measuring microwave absorbing properties

Temperature rise curve of high titanium slag was obtained with microwave reactor. The schematic diagram of the microwave reactor is shown in Fig. 2. The microwave heating apparatus consists of a magnetron, a power controller, a matched load, a wave guide, and a cavity. A self – made microwave reactor has a multi – mode cavity and continuous controllable power capacity. The microwave power supply for the microwave reactor consists of two magnetrons, which is cooled by water circulation, at a frequency of 2.45GHz and a power of 1.5kW. A ceramic tube, 50mm (outer diameter) × 80mm (inner diameter) × 600mm (length), is positioned at the center of the microwave oven, by drilling holes on the side faces, with ends projecting on both sides. The temperature of the sample is monitored using an infrared pyrometer (Raytek, Marathon Series, USA) with the circular crosswire focusing on the sample cross – section. The temperature is also measured using a thermocouple (Shengyun Company, China) as a reference. XRD patterns were obtained using a Rigaku diffractometer (D/Max 2200, Japan) with CuK_α radiation and a Ni filter operated at 35kV, 20mA and a scanning rate of 0.25°/min.

2.3 Methods

The sample (100g) was placed in the self – made microwave heating equipment and heated to 1050℃ for 20min at 3kW, and then it was naturally cooled in the furnace to room temperature.

The high titanium slag with different mean particle sizes and the mixtures of different mass frac-

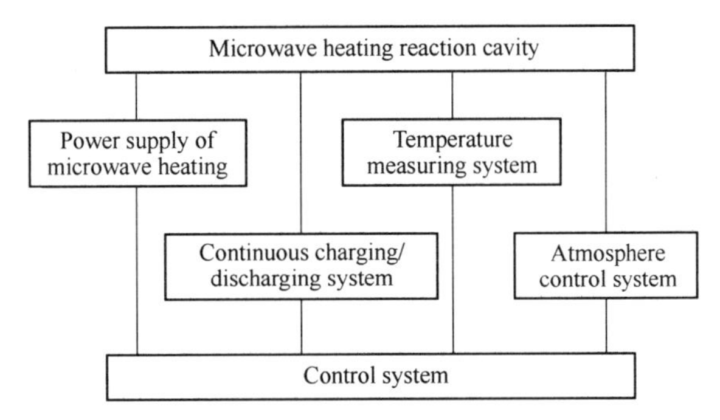

Fig. 2 Schematic diagram of microwave reactor

tions of V_2O_5 were prepared. Each sample(2g) was dried at 105℃ for 2h. Finally, the microwave absorbing properties of the samples were obtained by putting the samples into the microwave resonant sensor in turn.

3 Results and discussion

3.1 Temperature rise curve of high titanium slag

The relationship between temperature and time of the high titanium slag under microwave irradiation is shown in Fig. 3. It can be seen that the temperatures of the samples increase with increasing microwave irradiation time. The temperature reaches 1050℃ after 5min, and then the microwave power is adjusted to keep this temperature for 5min. Fig. 3 shows the temperature can in fact be kept at 1200℃, which indicates that the self - made microwave heating equipment can be well controlled, thus making it suitable for other studies. Furthermore, the results indicate the high titanium slag is a good microwave absorbent.

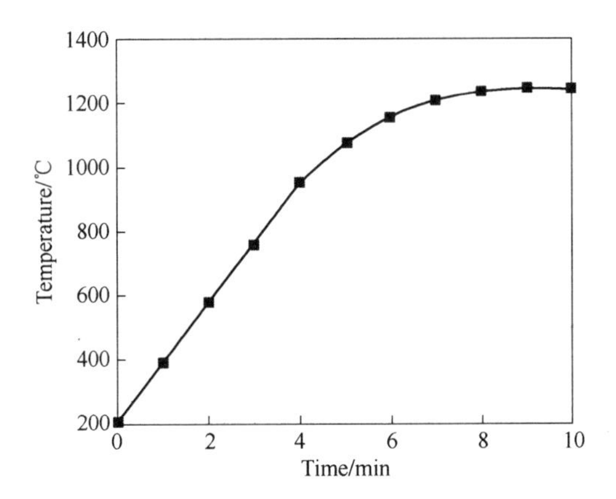

Fig. 3 Temperature rise curve of high titanium slag by microwave irradiation

3.2 Characterization of high titanium slag

The high titanium slag before and after microwave irradiation is characterized by XRD, and the results are shown in Fig. 4. It can be seen from Fig. 4(a) that anatase and iron titanium oxide are mainly crystalline compounds in the high titanium slag. In addition, a minor amount of rutile is present. The iron titanium oxide has the strongest diffraction peak at $2\theta = 25.22°$, which is close to the strongest diffraction peak of anatase at $2\theta = 25.28°$, so the two peaks are overlapped. High titanium slag after microwave irradiation at 1050℃ for 20min is characterized by XRD, which is shown in Fig. 4 (b). It can be found from Fig. 4 (b) that the diffraction peaks of rutile gradually are broadened and their intensities are increased under microwave irradiation. Rutile has the strongest diffraction peak at $2\theta = 25.44°$. All the X - ray diffraction peaks of samples well match with those of the standard XRD pattern of rutile phase. It can be inferred from Fig. 4 (b) that anatase is completely converted to synthetic rutile.

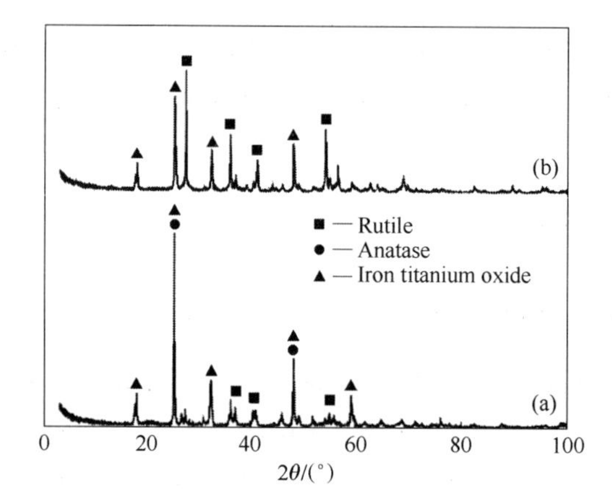

Fig. 4 XRD patterns of samples

(a) Untreated by microwave irradiation; (b) Treated by microwave irradiation at 1050℃ for 20min

3.3 Absorbing properties

Eq. (1) indicates that the real part of the permittivity is directly proportional to frequency shift. Eq. (2) indicates that the imaginary part of the permittivity is inversely proportional to amplitude of voltage. By analyzing of the changing behavior and calculating the amplitude of voltage and the frequency of the microwave spectrum's first wave crest with computer program, the effects particle size of high titanium slag and mass fraction of V_2O_5 on microwave absorbing properties are obtained[19,20].

3.3.1 Effect of particle size of high titanium slag on microwave absorbing properties

The microwave spectra of particle size of high titanium slag are illustrated in Fig. 5. The microwave sensor with empty chamber gives rise to the resonant curve with the highest resonant amplitude and the largest resonant frequency. The other resonant curves indicate lower resonant amplitude and smaller resonant frequency, resulted from the microwave sensor filled with different sizes of high titanium slag, respectively.

The effect of particle size of high titanium slag on the frequency shift is illustrated in Fig. 6. It can be seen that the frequency shift increases gradually from about 47kHz to 84kHz with increasing the particle size of high titanium slag from 120μm to 180μm, and then it decreases from 84kHz to

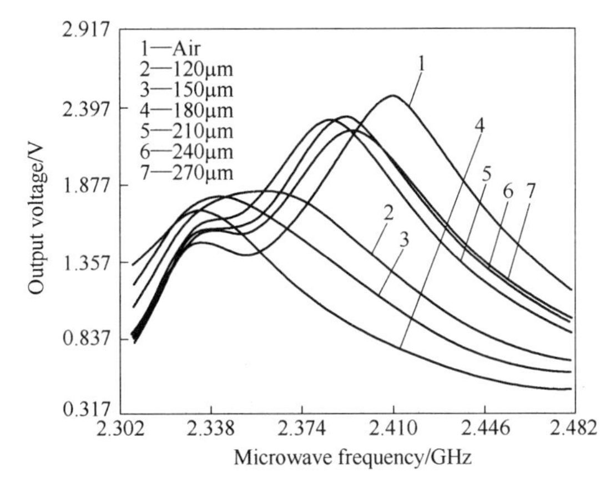

Fig. 5 Microwave spectra of different particle sizes of high titanium slag

14kHz with increasing the particle size of high titanium slag from 180μm to 270μm. Fig. 7 shows the relationship between the amplitude of voltage of microwave sensor and particle size of high titanium slag. Similarly, the amplitude of voltage also decreases from about 1.868V to 1.701V with increasing the particle size of high titanium slag from 120μm to 180μm; it increases from 1.701V to 2.317V with increasing the particle size of high titanium slag from 180μm to 240μm; and then it decreases from 2.317V to 2.217V with further increasing the particle size of high titanium slag from 240μm to 270μm. Therefore, 180μm of particle size is chosen as the optimum size of high titanium slag in microwave field.

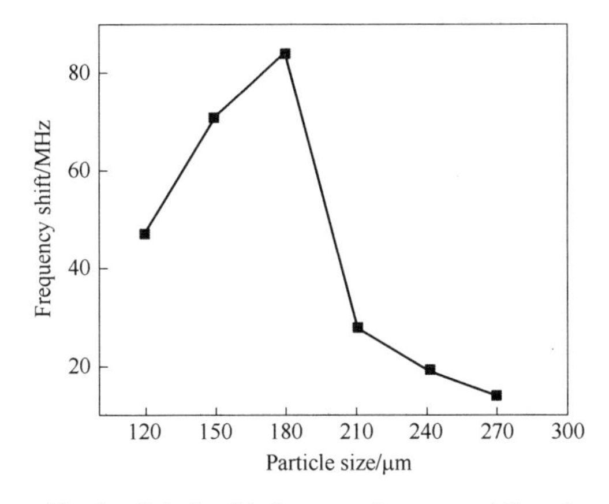

Fig. 6 Relationship between frequency shift and particle size of high titanium slag

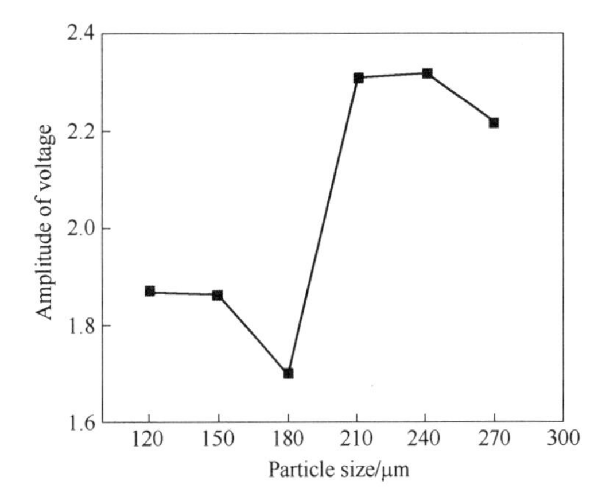

Fig. 7 Relationship between amplitude of voltage of microwave sensor and particle size of high titanium slag

The result of perturbation technique is that the presence of high titanium slag in the resonant cavity causes a shift of resonant frequency and changes in the amplitude of voltage of the resonant cavity. The changes in the amplitude of voltage of the cavity exist because of dielectric loss of the sample. The dielectric constant and loss tangent of high titanium slag can be calculated from the shift of frequency and amplitude of voltage.

3.3.2 Effect of mass fraction of V_2O_5 on microwave absorbing properties

Fig. 8 shows the microwave spectra of mixtures of high titanium slag with different mass fractions of V_2O_5. It is found that the microwave sensor with empty chamber gives rise to the resonant curve with the highest resonant amplitude and the biggest resonant frequency. The other curves show different results for mixtures of high titanium slag with different mass fractions of V_2O_5.

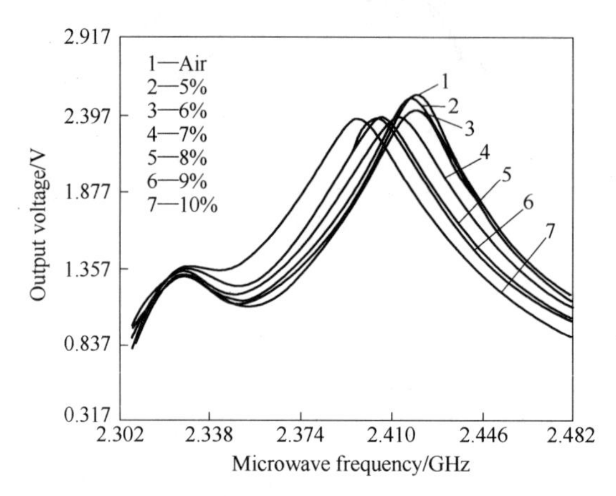

Fig. 8 Microwave spectra of mixtures of high titanium slag with different mass fractions of V_2O_5

Fig. 9 shows the relationship between the mass fraction of V_2O_5 and the frequency shift. It can be found that frequency shift decreases gradually from about 8MHz to 3MHz with increasing the mass fraction of V_2O_5 form 5% to 6% ,and then it increases from 3MHz to 41MHz when the mass fraction of V_2O_5 increases from 6% to 10%. Fig. 10 shows the relationship between the amplitude of voltage and the mass fraction of V_2O_5. It is clear that the amplitude of voltage decreases with increasing the mass fraction of V_2O_5 (it decreases abruptly after 5% and then decreases smoothly after 6%). The above discussion indicates that the mass fraction of V_2O_5 has a significant effect on the microwave absorbing properties of high titanium slag. Yu, et al. [21] investigated the optimum microwave absorbing a property of mixtures is acquired with the mass fractions of V_2O_5 less than

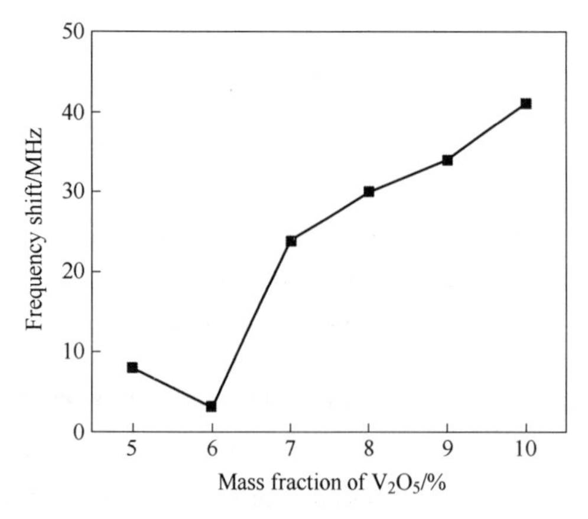

Fig. 9 Relationship between frequency shift and mass fraction of V_2O_5

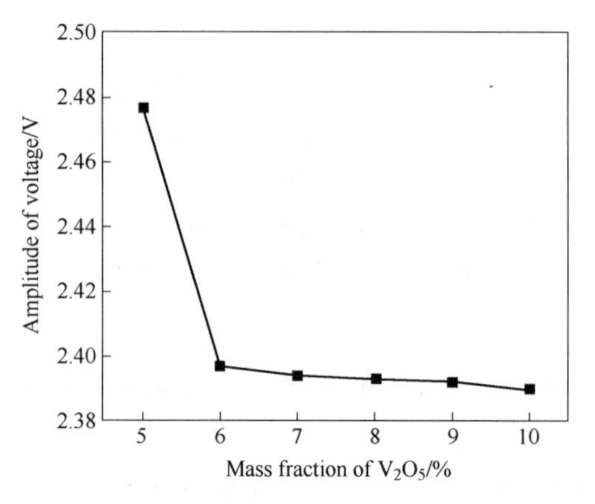

Fig. 10 Relationship between amplitude of voltage and mass fraction of V_2O_5

10%. The optimum content, 10% of the mass fraction of V_2O_5, is obtained. The phase transformation from anatase to rutile in high titanium slag is irreversible. V_2O_5 can enhance the phase transformation of TiO_2. It can also lower the phase transformation temperature of TiO_2 and increase the transformation ratio.

4 Conclusions

(1) The temperature of high titanium slag is raised up to 1050℃ after 5min by microwave heating at a microwave frequency of 2.45GHz and output microwave power of 3kW. When the temperature is kept at 1050℃ for 20min, the phase transformation, from anatase to rutile of high titanium slag in microwave field, is observed. High titanium slag has good microwave absorbing properties in the microwave field.

(2) The microwave absorbing properties of high titanium slag, which are measured by microwave cavity perturbation technique, are investigated in the frequency range of 2.302 - 2.482GHz. 180μm of particle size and 10% of the mass fraction of V_2O_5 are the optimum conditions for producing synthetic rutile from high titanium slag in microwave field.

(3) The microwave cavity perturbation technique is used to optimize the experimental parameters and will provide the guidance for the study of microwave heating of high titanium slag in the future.

References

[1] Wang J, Ma T, Zhang Z H, et al. Investigation on the transition crystal of ordinary rutile TiO_2 powder by microwave irradiation in hydrogen peroxide solution and its sonocatalytic activity[J]. Ultrasonics Sonochemistry, 2007, 14(5): 575 - 582.

[2] Dondi M, Cruciani G, Balboni E, et al. Titania slag as a ceramic pigment[J]. Dyes and Pigments, 2008, 77(3): 608 - 613.

[3] Villiers J, Verryn S, Fernandes M. Disintegration in high - grade titania slags: low temperature oxidation reactions of ferro - pseudobrookite[J]. Mineral Processing and Extractive Metallurgy, 2004, 113(8): 66 - 74.

[4] Wang M Y, Li L S, Zhang L, et al. Effect of oxidization on enrichment behavior of TiO_2 in titanium - bearing slag[J]. Rare Metals, 2006, 25(2): 106 - 110.

[5] Li C, Liang B, Guo L H. Dissolution of mechanically activated Panzhihua ilmenites in dilute solutions of sulphuric acid[J]. Hydrometallurgy, 2007, 89(1/2): 1 - 10.

[6] Samal S, Rao K K, Mukherjee P S, et al. Statistical modelling studies on leachability of titania - rich slag obtained from plasma melt separation of metallized ilmenite[J]. Chemical Engineering Research and Design, 2008, 86(2): 187 - 191.

[7] Bessinger D, Geldenhuis J M A, Pistorius P C, et al. The decrepitation of solidified high titania slags[J]. Journal of Non - Crystalline Solids, 2001, 282(1): 132 - 142.

[8] Pistorius P C, Motlhamme T. Oxidation of high - titanium slags in the presence of water vapour[J]. Minerals Engineering, 2006, 19(3): 232 - 236.

[9] Li W, Peng J H, Zhang L B, et al. Pilot - scale extraction of zinc from the spent catalyst of vinyl acetate synthesis by microwave irradiation[J]. Hydrometallurgy, 2008, 92(1/2): 79 - 85.

[10] Li W, Zhang L B, Peng J H, et al. Effects of microwave irradiation on the basic properties of wood - ceramics

made from carbonized tobacco stems impregnated with phenolic resin[J]. Industrial Crops and Products, 2008,28(2):143 - 154.

[11] Verma A, Saxena A K, Dube D C. Microwave permittivity and permeability of ferrite - polymer thick films[J]. Journal of Magnetism and Magnetic Materials, 2003, 263(1/2):228 - 234.

[12] Krasovitsky V, Terasawa D, Nakada K, et al. Microwave cavity perturbation technique for measurements of the quantum hall effect[J]. Cryogenics, 2004, 44(3):183 - 186.

[13] Sheen J. Measurements of microwave dielectric properties by an amended cavity perturbation technique[J]. Measurement, 2009, 42(1):57 - 61.

[14] Huang M, Yang J J, Wang J Q, et al. Microwave sensor for measuring the properties of a liquid drop[J]. Measurement Science and Technology, 2007, 18(7):1934 - 1938.

[15] Huang M Y, Peng J H, Huang M, et al. Microwave - absorbing characteristics of mixtures about different proportions of carbonaceous reducer and ilmenite in microwave field[J]. The Chinese Journal of Nonferrous Metals, 2007, 17(3):476 - 480(in Chinese).

[16] Huang M, Peng J H, Yang J J, et al. Microwave cavity perturbation technique for measuring the moisture content of sulphide minerals concentrates[J]. Minerals Engineering, 2007, 20(1):92 - 94.

[17] Carter R G. Accuracy of microwave cavity perturbation measurements[J]. IEEE Transactions on Microwave Theory and Techniques, 2001, 49(5):918 - 923.

[18] Huang M, Peng J H, Yang J J, et al. A new equation for the description of dielectric losses under microwave irradiation[J]. Journal of Physics D: Applied Physics, 2006, 39(10):2255 - 2258.

[19] Cristallo G, Roncari E, Rinaldo A, et al. Study of anatase - rutile transition phase in monolithic catalyst V_2O_5/TiO_2 and V_2O_5 - WO_3/TiO_2[J]. Applied Catalysis A: General, 2001, 209(1/2):249 - 256.

[20] Haber J, Nowak P. A catalysis related electrochemical study of the V_2O_5/TiO_2 (rutile) system[J]. Langmuir, 1995, 11(3):1024 - 1032.

[21] Yu X F, Wu N Z, Xie Y C, et al. The monolayer dispersion of V_2O_5 and its influence on the anatase - rutile transformation[J]. Journal of Materials Science Letters, 2001, 20(4):319 - 321.

Microwave Cavity Perturbation Technique for Measuring the Moisture Content of Sulphide Minerals Concentrates

Ming Huang, Jinhui Peng, Jingjing Yang, Jiaqiang Wang

Abstract: The moisture content of a sulphide mineral concentrate was measured by the microwave cavity perturbation technique. Comparative experiments were performed using this technique and by oven drying. It was found that the deviation in the measurement of the moisture content of the concentrate was less than 0.5%, indicating that this would be a suitable technique for moisture content determination of sulphide mineral concentrates.

Keywords: sulphide ores; on - line analysis; modelling; microwave resonator; moisture content

1 Introduction

Moisture determination has been a major problem in many branches of industry for many years. In the metallurgical industry, the exact moisture content of a specified material has to be determined in order to allow control of the water dosage, of the quality of the product, and of the reduction of applied energy.

Moisture determination by microwaves is applied in many branches of industry[1]. However, few applications[2,3] have been reported in the metallurgical industry. It has been shown that the penetration depth of microwaves is much greater than that of infrared radiation, and microwave methods can measure the volume moisture content of the materials. In addition, microwave methods are much safer and faster than ionizing radiation methods[4].

The objective of this paper is to apply the microwave cavity perturbation technique[5,6] for the rapid measurement of the moisture content of sulphide mineral concentrates containing sphalerite, chalcopyrite and nickel sulphide minerals.

2 Experimental techniques

2.1 Materials

Three different materials, sphalerite concentrate, chalcopyrite concentrate, and nickel sulphide concentrate were used in the microwave moisture determination experiments. These samples were obtained from a smelter in Yunnan (China). The composition of the sphalerite concentrate was: Fe 7.24%; SiO_2 3.48%; S 28.47%; Zn 50.18%; other elements 10.63%, respectively. The composition of the chalcopyrite concentrate was: S 31.13%; Fe 30.17%; Cu 20.10%; Zn 2.99%; SiO_2 5.28%; Ni 0.012%; Pb 0.63%; Bi 0.09%; Al_2O_3 0.68%; CaO 0.44%; MgO 0.39%; As

0.56%, respectively. The composition of the nickel sulphide concentrate was Ni 2.86%; Cu 2.36%, respectively.

2.2 Measuring equipment

The mechanism for this technique is the measurement of resonant frequency and the output voltage of the resonant sensor[7] unloaded and loaded with samples, as resonant frequency and output voltage of the resonant sensor are very sensitive to moisture content.

The main parts of the equipment include resonant sensor[8], sweeping signal, detector and DSP, interface circuit, and computer. The software control of the set-up was performed by Windows XP operating system, and programmed by Visual Basic 6.0.

2.3 Method

To obtain the samples with different moisture content, each sulphide concentrate (typically 5kg in weight) was dried at 105℃ for at least 2h, and the samples of the three different concentrates were each divided into 10 shares, every share weighing 0.5kg. Subsequently, samples with different moisture content were obtained by adding different proportions of water. Finally, the moisture contents of the samples were obtained by putting the samples into the resonant sensor one by one.

3 Results and discussion

A series of 10 tests were carried out with sphalerite concentrates of different moisture content, the testing time per sample being about 5s. The results are plotted as a function of moisture content in Fig. 1. It can be seen that the output voltage of the resonant sensor decreases as the moisture content increases. Interestingly, their relationship follows a nonlinear fitting equation: $V = -0.0007x^2 - 0.0385x + 2.0497$. The resonant frequency of the resonant sensor shows a gradually decrease as moisture content increases, and follows a nonlinear fitting equation: $f = -0.0004x^2 + 0.0010x + 2.4250$, where x is the moisture content of the samples. Based on the fitting equation, the measure-

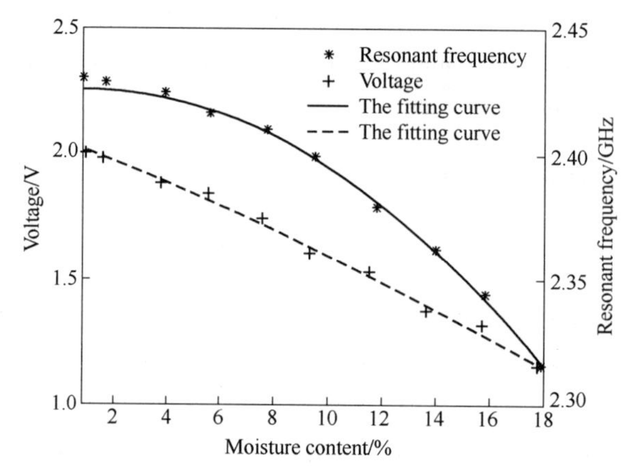

Fig. 1 Plot of output voltage and resonant frequency of the resonant sensor against moisture content of sphalerite concentrates

ment deviation of sphalerite concentrates is less than 0. 42%. Using the same method, the measurement deviations of chalcopyrite concentrates and nickel sulphide concentrates are less than 0. 37% and 0. 41%, respectively.

Microwave cavity perturbation measurement is an indirect technique, which must be calibrated against a reference method (e. g. oven drying) for different materials. The initial temperature of the material influences the measured moisture content of the material, so the temperature at set - up and the material needs to be compensated for. Generally, the natural inhomogeneity of the samples used (particularly at large particle sizes) influences the moisture content measurement, and it is difficult to assess accurately the moisture content of the sample[2]. However, for sulphide mineral concentrates used in our experiments, the particle sizes were smaller than 0. 07mm, so that this did not affect the measured results.

4 Conclusions

(1) Using the microwave cavity perturbation technique, it is possible to determine the moisture content of sulphide mineral concentrates.

(2) The measurement deviation is less than 0. 5% under laboratory conditions.

(3) The commercial prototype of the microwave cavity perturbation technique for measuring the moisture content of sulphide mineral concentrates is now under development[9] and patents are pending.

Acknowledgements

The authors would like to thank the National Natural Science Foundation of China (Project 20463003) for financial support. Also the authors thank Kunming Jin - Hui - Tong Wireless and Microwave Sensor Institute for their cooperation.

References

[1] Kraszewski A W. Microwave Aquametry[M]. New York: IEEE Press, 1996.

[2] Cutmore N G, Evans T G, Mcewan A J, et al. Low frequency microwave technique for on - line measurement of moisture[J]. Minerals Engineering, 2000, 13(14): 1615 - 1622.

[3] Klein A. Microwave determination of moisture in coal: comparison of attenuation and phase measurement[J]. Journal of Microwave Power, 1981, 16(3 - 4): 289 - 303.

[4] Kupfer K. Electromagmetic Aquametry[M]. New York: Springer - Verlag, 2005.

[5] Carter R G. Accuracy of microwave cavity perturbation measurements[J]. Ieee Transactions on Microwave Theory and Techniques, 2001, 49(5): 918 - 923.

[6] Harrington R F. Time - Harmonic Electromagmetic Fields[M]. New York: McGraw - Hill, 1961.

[7] Radmanesh M M. Radio Frequency and Microwave Electronics Illustrated[M]. Beijing: Publishing House of Electronics Industry, 2002.

[8] Huang M, et al. 2002, China patent 02125071. 5.

[9] Huang M, et al. 2005, Patent application number in China 200510048695. 1.

Effect of Temperature on Dielectric Property and Microwave Heating Behavior of Low Grade Panzhihua Ilmenite Ore

Chenhui Liu, Libo Zhang, Jinhui Peng, Bingguo Liu,
Hongying Xia, Xiaochun Gu, Yifeng Shi

Abstract: The permittivity of low grade Panzhihua ilmenite ore at 2.45GHz in the temperatures from 20℃ up to 100℃ was measured using the technology of open – ended coaxial sensor combined with theoretical computation. The results show that both the real (ε') and imaginary (ε'') part of complex permittivity ($\varepsilon' - j\varepsilon''$) of the ilmenite significantly increase with temperature. The loss tangent ($\tan\delta$) is a quadratic function of temperature, and the penetration depth of ilmenite decreases with temperature increase from 20℃ to 100℃. The increase of the sample temperature under microwave radiation displays a nonlinear relationship between the temperature (T) and microwave heating time (t). The positive feedback interaction between complex permittivity and sample temperature amplifies the interaction between ilmenite and the microwave radiation. The optimum dimensions for uniform heat deposition vary from 10cm to 5cm (about two power penetration depths) in a sample being irradiated from both sides in a 2.45GHz microwave field when temperature increases from room temperature to 100℃.

Keywords: dielectric properties; Panzhihua ilmenite low grade; microwave heating; temperature increase

1 Introduction

Due to the worldwide intensive consumption of rutile resource, ilmenite ore is becoming the major resource for the titanium industry[1]. Many processes are used to upgrade low grade ilmenite ore to produce synthetic rutile or high grade titanium slag for further produce titanium dioxide (TiO_2). These processes involve a combination of pyrometallurgy, hydrometallurgy and electrometallurgy. However, these processes can be highly energy intensive and expensive[2]. Panzhihua ilmenite ore accounts for more than 90% of titanium reserves of China, it contains low grade TiO_2 and high content of impurities (especially high content of MgO and CaO), which made it is difficult to upgrade the ilmenite ore[3]. Therefore, it is desirable to develop new technology for processing ilmenite ore.

The application of microwave heating in minerals processing and treatment has recently attracted high interests from metallurgical industry. Compared with conventional heating methods, the advantages of microwave heating include rapid and selective heating, uniform distribution, high efficiency, fast switch on and off as well as flexible and modular design[4]. As a result of these advantages, microwave heating has been applied in a variety of mineral processing and extractive metallurgy, such as microwave – assisted ore grinding, microwave – assisted carbothermic reduction and leaching of ilmenite, microwave drying, microwave – assisted roasting and smelting sulfide concentrate, microwave – assisted spent carbon regeneration and microwave – assisted waste management[5-8]. Though the application of microwave heating in the metallurgical industry is promising, further studies on the interaction between microwave radiation and minerals and microwave heating mechanism are highly desired to develop innovative techniques for minerals processing and ore treatment using microwave heating.

In the absence of few ores and minerals have magnetic properties, complex permittivity is the fundamental property that determine the microwave absorption of an ore[9]. Complex permittivity ε can be expressed as follows:

$$\varepsilon = \varepsilon' - j\varepsilon'' \tag{1}$$

where $j = \sqrt{-1}$. The real part (ε') is often referred as the dielectric constant, which reflects the ability of the material to store electromagnetic energy within its structure. The imaginary part (ε'') is the dielectric loss factor that characterizes the ability of the material to convert the stored electromagnetic energy into thermal energy. Another important parameter is loss tangent which describes how well the material dissipates stored energy into heat at a given frequency and temperature. Loss tangent ($\tan\delta$) can be expressed as follows:

$$\tan\delta = \frac{\varepsilon''}{\varepsilon'} \tag{2}$$

the permittivities are a function of both temperature and temperature.

Dielectric properties of and microwave heating behaviors of Panzhihua ilmenite and ilmenite concentrate have been reported by several previous researchers. Huang et al. studied the microwave absorbing characteristics and temperature increase characteristics of Panzhihua ilmenite concentrate, results shows that ilmenite concentrate is a good microwave absorbing mineral and the microwave characteristics are largely reduced after the ilmenite concentrate was oxidized[10]. Ouyang et al. studied the temperature increase characteristics of different grade of Panzhihua ilmenite ore, results indicate that the ilmenite increase from room temperature to 500℃ in 6min under microwave irradiation at 2.45GHz[11]. The dielectric constant and dielectric loss of synthetic $FeTiO_3$ were measured in the temperature range of 4.2K to room temperature in the frequency range of 0.1kHz to 100kHz in a previous study[12]. The results indicate that both the dielectric constant and dielectric loss increase with the increase in sample temperature and conductivity due to the electron hopping between Fe^{2+} and Fe^{3+} at higher temperatures. Seshadri and Viswanath studied the dielectric constant of ilmenite ore in the temperature range between 25℃ and 377℃. The plot of $\log\varepsilon'$ vs. $1/T$ exhibits two linear regimes and the slope changes around 270℃[13]. At this temperature, a phase transition was observed in the material. Chiteme and Mulaba – bafubiand investigated the dielectric properties of natural ilmenite treated with microwave radiation and found that the dielectric properties of the microwave – treated sample were similar to those of the sample treated by conventional heating techniques. Their experimental results showed a nonlinear relationship between dielectric loss and sample temperature[14].

To our best knowledge, no literature has been reported regarding on the temperature dependence of dielectric property of low grade Panzhihua ilmenite at the frequency of 2.45GHz. Therefore, the major objectives of this work are: (1) to measure the complex permittivity (dielectric constant ε' and dielectric loss ε'') and loss tangent ($\tan\delta$) of the ilmenite ore and their variation with sample temperature (from 20℃ to 100℃) at the frequency of 2450MHz; (2) to calculate the penetration depth of low grade Panzhihua ilmenite based on the permittivities; (3) to study the characteristics of temperature increase of the ilmenite ore in different time durations of microwave irradiation.

2 Experiment

2.1 Sample preparation

Samples of low grade ilmenite ore were provided by Panzhihua Steel Plant (located in Sichuan Province, China). The chemical compositions and grain sizes of the low grade ilmenite are shown in Tables 1 and 2.

Table 1 Chemical components of the low grade ilmenite ore (%)

TiO_2	TFe	Fe_2O_3	FeO	CaO	MgO	SiO_2	Al_2O_3
38. 36	27. 36	5. 48	30. 27	4. 32	6. 03	10. 02	1. 76

Table 2 Grain size of ilmenite concentrate

Size/μm	<250	250 - 100	100 - 75	75 - 48	<48
Concent/%	1. 59	10. 27	70. 88	16. 37	2. 48

As shown in Table 1, titanium dioxide and iron oxides are the main components of the ilmenite ore. Therefore, the dielectric properties of this ore are closely related to that of those components. The original sample was dried at the temperature of 105℃ for 24h in a desiccator.

2.2 Dielectric property measurement system

Several methods have been employed to measure dielectric properties of solid samples including cavity perturbation[15], open - ended coaxial probes[16], waveguide transmission line[17], and free - space[18]. Nevertheless, the open - ended coaxial probe method is currently the most widely used technique for measuring complex dielectric permittivity of materials. It is easy to use and has a flexible requirement for the sample shape. In addition, this method can take measurements in a large range with high accuracy[19].

In this work, a hybrid experimental/computational permittivity measuring system developed by the Institute of Applied Electromagnetics at Sichuan University, was used to determine the complex permittivity of the ilmenite at different sample temperatures. The schematic diagram of this system was shown in Fig. 1. In the experiment, sample powder was sealed in a resonant cavity (80mm inner diameter with 100mm length) made of stainless steel and heated by an electric furnace placed inside the holder cavity. An open ended coaxial sensor (see Fig. 2) connected to the Agilent PNA5230 Network Analyzer was used to measure the reflected signals. The probe was placed on the flat surface of a solid sample or inserted into the sample powder for a full contact. The electromagnetic fields around the probe change as the probe contacts with the specimen. The probe receives reflected signals from the sample which contains the information related to the complex dielectric permittivity of the measured sample. The reflected signals were measured by the probe and recorded by the analyzer. A thermocouple was used for temperature measurements. The FDTD method was employed to calculate the distribution of the electromagnetic field around the sensor and the reflection coefficients at different frequencies[20]. Based on the experimental/computation reflection coefficient, genetic algorithm was used to calculate the complex permittivity of measuring sample[21].

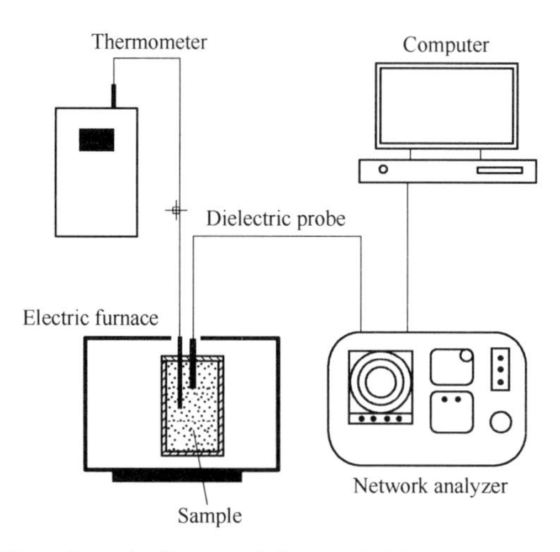

Fig. 1 The schematic diagram of the permittivity measurement system

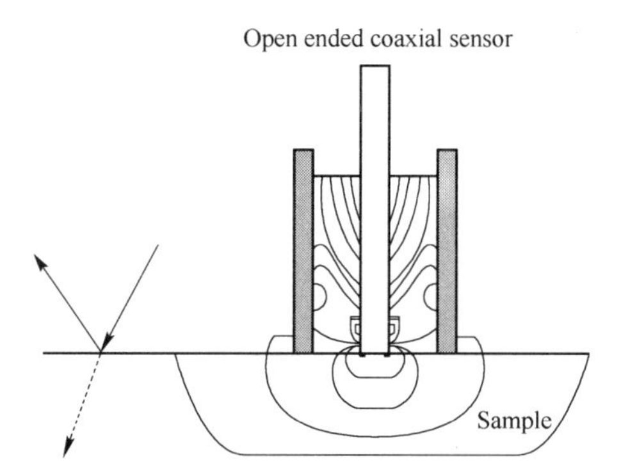

Fig. 2 The schematic diagram of the open – ended coaxial sensor

Before the measurements were taken, the network analyzer was warmed for at least 30min and then calibrated with a load of 50Ω. The error of the system was around 5% for high – loss materials if a standard calibration process was followed. Permittivity measurements were taken at the frequency of 2.45GHz. The complex permittivity of the ilmenite sample was measured at 20℃, 40℃, 60℃, 80℃ and 100℃.

2.3 Microwave heating equipment

Microwave heating equipment was made by the Key Laboratory of Unconventional Metallurgy, has the ability to alter power intensity in the range of 0 – 3000W at the frequency of 2.45GHz. The microwave system consists of two magnetrons, a waveguide and a multi – mode cavity. It is equipped with a water – cooled condenser and a temperature controller to adjust the microwave power level for a preset temperature. The schematic diagram of the microwave heating equipment was shown in Fig. 3. A ceramic crucible container with an inner diameter of 100mm and a length of 200mm was located at the center of the stainless steel oven. A thermocouple pyrometer was used to measure the sample temperature. The measurements were recorded by a designated computer.

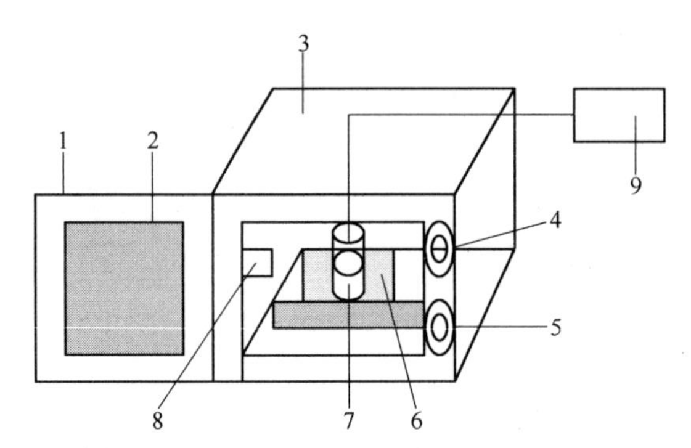

Fig. 3 Schematic diagram of the microwave heating system

1—oven door; 2—observation window; 3—microwave multi – mode cavity; 4—timer; 5—temperature controller; 6—fireproof materials; 7—raw materials; 8—ventilation hole; 9—temperature measurement system

To investigate the behavior of microwave heating, a sample of 500g ilmenite ore was dried at 105℃ for 2h and then was placed in the crucible. The thermocouple pyrometer was inserted into the center of the sample for taking temperature measurements. Microwave radiation was imposed on the sample. After a preset residence time, the microwave irradiation was stopped and temperature was recorded.

3 Results and discussions

3.1 Real and imaginary permittivity

The real permittivity of Panzhihua ilmenite ore variation with the temperature increase at 2.45GHz was shown in Fig. 4. The real permittivity increased with temperature increasing. The real permittivity increased slightly from room temperature to 40℃. However, it increased rapidly with the increase in sample temperature with a rate of 0.04/℃ when temperature was above 40℃. Above about 80℃, the increase rate of real permittivity with temperature became extremely high. Although the temperature limit in the measurement was 100℃, the real permittivity of ilmenite increase with increasing temperature had the same trend with the dielectric constant of nickeliferous laterite ores and kaolin ores[22]. Compared with other ores and minerals, the low grade ilmente ore has a higher value. Fig. 5 shows the imaginary permittivity of Panzhihua ilmenite ore variation with the temperature increase at 2.45GHz. The imaginary permittivity was lower than the real part in the range of tested temperatures, while the trend of temperature increase was similar with that of real permittivity. In the temperatures between 20℃ and 60℃, the imaginary permittivity increased slightly with temperature. This increase became more rapidly when temperature was above 60℃ and this quick increase was maintained in the temperature range of 60 – 100℃.

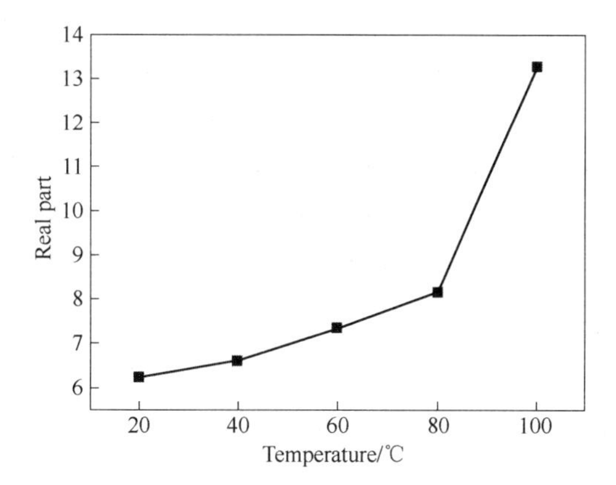

Fig. 4 Variation of real part of complex permittivity with sample temperature

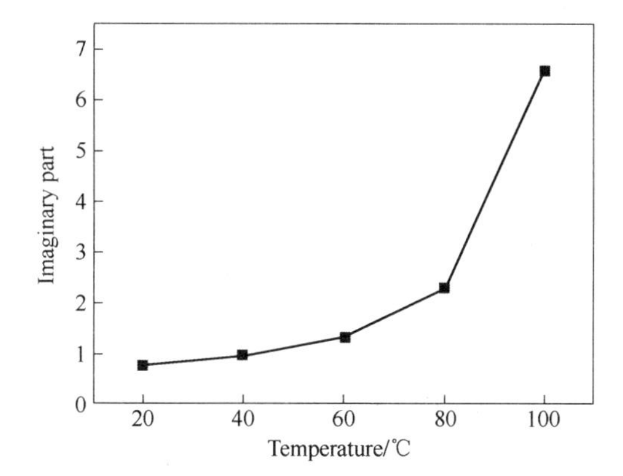

Fig. 5 Variation of real part of complex permittivity with sample temperature

The real part of anatase TiO_2 powder is 2.9 ± 0.29 and the imaginary part is $6.2 \times 10^{-3} \pm 2.1 \times 10^{-3}$ in the microwave frequency regime between 2.9GHz and 3.2GHz[23]. The low grade ilmenite has higher values of ε' and ε'' than anatase TiO_2 powder at room temperature, especially the values of ε'', which indicates that ilmenite ore is very good at absorbing microwave radiation. Dervos measured the dielectric properties of high temperature sintered pure TiO_2 powder at room temperature and between 20kHz and 1000Hz, result shows that dielectric constant of rutile has the similar trend with the Panzhihua ilmenite ore[23]. But the dielectric constant is easily affected by microwave frequency and is instability. The ilmenite in our study is an ion crystal and has complex minerals, the interfaces between various compounds exists a large number of ions and electrons, most ions and electrons will produce relaxation phenonmenon under microwave field. The imaginary of permittivity of a dielectric is directly proportional to the conductivity. The numbers of ions and electrons of ilmenite will increase with increasing temperature, making loss factor of ilmenite increase with temperature increase, ilmenite will absorb more electromagnetic energy and convert it into thermal energy, and thermal energy makes ilmenite generate temperature increase.

Pervious studies on ilmenite have shown that the increase in conductivity of ilmenite in higher temperatures is due to the increase in the number of pairs of (Fe^{2+}, Fe^{3+}) and (Ti^{3+}, Ti^{4+})[24]. When ilmenite is heated with microwave radiation at the frequency of 2.45GHz, orientation (dipolar) polarization is the dominant polarization and the associated relaxation phenomena constitute the loss mechanisms[25]. At room temperature, the Ti^{4+} ion has the minimum potential. Therefore, it is off-centered and gives rise to electric dipoles. The energy levels of Ti^{4+} increase rapidly with temperature. At a large scale, this feature makes the dielectric constant of ilmenite increase expeditiously with temperature, while it triggers a fast ionic relaxation process at a micro scale. Due to the relationship between ε'' and conductivity σ, more microwave energy is converted into thermal energy with the increase in ε'', which leads to a quick increase in ilmenite temperature[26].

3.2 Loss tangent and penetration depth

The values of loss tangent of the ilmenite ore at the frequency of 2.45GHz were calculated using Eq. (2) and the results are depicted in Fig. 6.

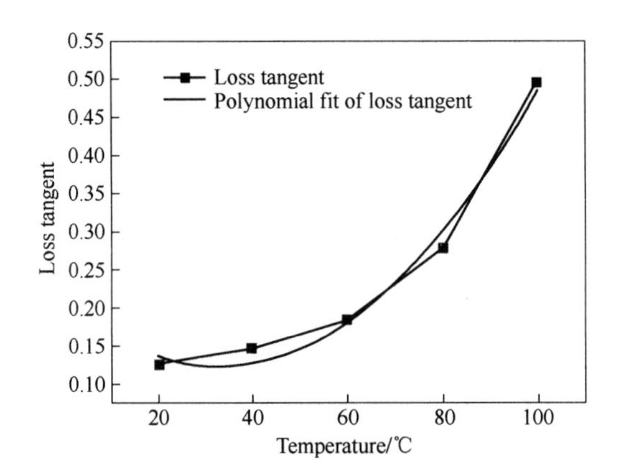

Fig. 6 Variation of the loss tangent with sample temperature

From Eq. (2) we know that an increase in $\tan\delta$ is a result of more rapid increase in ε'' than that in ε'. The experimental results reveal that the variation of $\tan\delta$ with sample temperature had a similar pattern with that of ε' and ε''. The values of $\tan\delta$ increased with a rate of 0.0014/℃ from room temperature to 60℃. The increase rate was about 0.01/℃ when temperature was above 60℃. At room temperature, $\tan\delta$ was only 0.12, but it increased rapidly to 0.5 when the sample was heated to 100℃. The variation of loss tangent at 2450MHz with temperature can be expressed as follows when the temperatures were between 20℃ and 100℃:

$$\tan\delta = 0.2104 - 0.00533T + 0.00008T^2 \tag{3}$$

The square of correlation coefficient (R^2) of Eq. (3) is 0.97. Values of loss tangent of ilmenite are much higher than pure TiO_2 powder, limonite ore and goethite ore. Panzhihua ilmenite not only has the similar high dielectric constant as pure TiO_2 powder but also has the high loss factor, indicating that Panzhihua ilmenite has strong ability to absorb microwave energy and transform the thermal energy.

The power penetration depth(D_p) is defined as the depth where the strength of microwave field is reduced to 1/e of its surface value and is expressed with the following Eq. (4):

$$D_p = \frac{\lambda_0}{2\sqrt{2}\pi\sqrt{\varepsilon'\left[\sqrt{1+\left(\frac{\varepsilon''}{\varepsilon'}\right)^2}-1\right]}} \tag{4}$$

where λ_0 is wavelength; λ_0 = 12.24cm in 2.45GHz; π is a constant. The penetration depth of Panzhihua ilmenite ore can be calculated by Eq. (4) based on the measured results of dielectric properties, Fig. 7 shows the results of penetration depth of various with temperature at 2.45GHz. Second order polynomial equation was used to fit the curve of temperature dependence of penetration depth, the result was shown in

$$D_p = 5.6806 - 0.00368T - 2.54643e^{-4}T^2 \tag{5}$$

The R - square(R^2) value of Eq. (5) is 0.99846. A high correlation value and low deviations indicate this equation can be used for prediction.

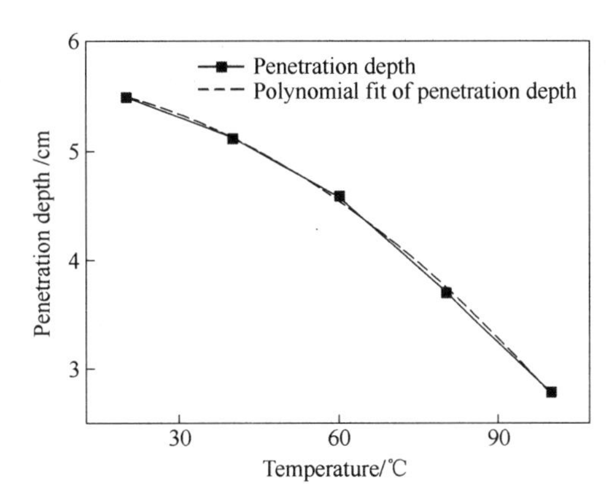

Fig. 7 The variation of penetration depth of the sample with temperature and a polynomial fitting curve

As can be seen from Fig. 7, the optimum dimensions for uniform heat deposition in a sample being irradiated from both sides in a 2.45GHz microwave field vary from 10cm to 5cm (about two power penetration depths) when temperature increase from room temperature to 100℃. The deposited microwave energy can be relatively uniformly distributed by double - sided irradiation according to the penetration depth, and rapidly relatively uniform temperature increases can be achieved. In a larger sample, there will be obvious temperature gradients. In a smaller sample, a central hot spot may be produced by surface cooling. The accurate determination of penetration depth helps to optimize the load size in the microwave applicator.

3.3 Microwave heating behavior

The temperature variation of the Panzhihua low grade ilmenite with microwave heating time was shown in Fig. 8. The sample temperature increased nonlinearly with heating time. It can be seen from the Fig. 8 that the sample temperature increased from room temperature to 1230℃ within

60min, which indicates that ilmenite ore has strong ability to absorb microwave energy and convert it into heat.

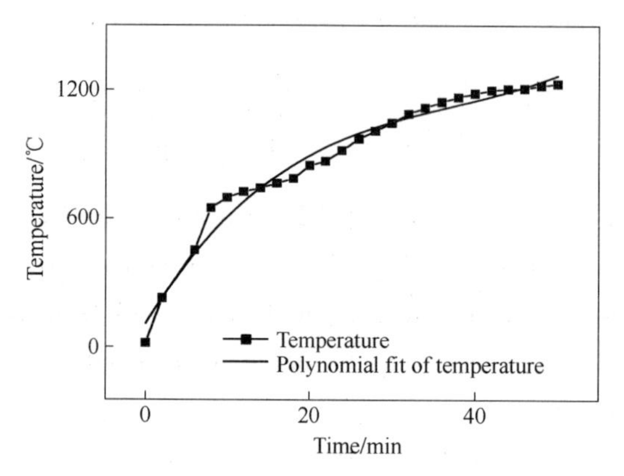

Fig. 8 The variation of sample temperature with microwave heating time

Temperature increase rate of the ilmenite under microwave irradiation between 25℃ and 700℃, 700℃ and 900℃, 900℃ and 1120℃ are 67.9℃/min, 15.8℃/min, 20℃/min. This result is in good agreement with the dielectric properties of the ilmenite. At lower temperatures, the ilmenite ore could absorb only a small fraction of microwave radiation energy imposed to it because of the low ε' and ε'' values. Consequently, only a small portion of the radiation energy was converted to thermal energy and the sample temperature increased slowly. When temperature was above 60℃, the values of ε' and ε'' increased swiftly that induced a fast increase in sample temperature in a short time. This positive feedback interaction between dielectric property and sample temperature explains the mechanism that the ilmenite ore can be heated to a very high temperature in a short time by the application of microwave radiation because the material can absorb a much larger fraction of radiation energy at higher temperatures. In the temperature varies from 700℃ up to 900℃, temperature increase rate of sample was lower than the rate of temperature increase in the temperature between 100℃ and 700℃. The sample kept a high temperature increase rate from 900℃ up to 1120℃, with the growing effect of conductivity of sample, the temperature increase rate of sample decreased with time, making the sample keep a constant temperature. This result suggests that one of the differences between conventional heating and microwave heating is that the process of microwave heating is nonlinear because the heating starts inside. Regression analysis shows that the relation between the sample temperature and microwave heating time can be expressed by the following equation:

$$T = 111.7214 + 62.32872t - 1.4069t^2 + 0.01247t^3 \tag{6}$$

With an R - square of 0.97357, Eq. (6) is a good representation of microwave heating behavior of the low grade ilmenite. This result agrees well with the previous studies of C. Chiteme[14]. This phenomenon also occurred in other minerals and ores such as nickeliferous limonitic laterite ores and oil shall[9]. Chen et al. investigated microwave absorbing properties of high titanium slag, results show that the temperature of high titanium slag can increase from room temperature to 1050℃ in 5min, indicating a higher temperature increase rate than Panzhihua ilmenite ore[27]. High titanium slag has more content of TiO_2 and less impurities, dielectric properties of high content of TiO_2 makes the high titanium slag has a quicker temperature increase rate. Different dielectric phases are commonly contained in minerals. When heated by microwave radiation, components of high dielectric constant and high dielectric loss can absorb more electromagnetic energy from microwave radiation and cause increase in temperature. This temperature increase induces the change in dielectric properties. On the other hand, components of low dielectric constant cannot be well heated by microwave radiation. In these materials, mass and heat transfer is the key factor for temperature changes. Some minerals have very low dielectric constants and hardly absorb microwave radiation at low temperatures. However, the dielectric constants of these minerals appear high at higher temperatures. In this case, materials of high dielectric constant can be added to those minerals for effective microwave heating. Since different components of an ore have different dielectric properties, stress discrepancy occurs on the surface of a lump sample of such ore, and such discrepancy generates cracks and crackles on the surface. Therefore, the grinding time of microwave - pretreated minerals can be significantly shorter than that without microwave pretreatment.

4 Conclusions

The real and imaginary parts of complex permittivity of low grade Panzhihua ilmenite vary significantly with temperature increase. Loss tangent is related to sample temperature by a quadratic function between 20℃ and 100℃:

$$\tan\delta = 0.2104 - 0.00533T + 0.00008T^2$$

With the temperature increase, the imaginary part increases rapidly, and therefore, the coupling of microwave radiation with the ilmenite ore is dramatically improved.

The optimum dimensions for uniform heat deposition in a sample being irradiated from both sides in a 2.45GHz microwave field vary from 10cm to 5cm (about two power penetration depths) when temperature increase from room temperature to 100℃.

The variation of sample temperature with microwave heating time can be expressed with the following equation:

$$T = 111.7214 + 62.32872t - 1.4069t^2 + 0.01247t^3$$

The increase in sample temperature triggers the increase in conductivity of some ions at grain boundaries, which leads to an increase in the dielectric loss factor, in turn, speeds up the increase in temperature. This positive feedback interaction between complex permittivity and sample temperature makes the ilmenite ore absorbs a much larger fraction of microwave radiation imposed to it and converts the electromagnetic energy into thermal energy more effectively.

References

[1] Chachula F, Liu Q. Upgrading a rutile concentrate produced from Athabasca oil sands tailings[J]. Fuel, 2003, 82(8): 929 - 942.

[2] Zhang W S, Zhu Z W, Chen C Y. A literature review of titanium metallurgical process[J]. Hydrometallurgy, 2011, 108(3 - 4): 177 - 188.

[3] Zhang L, Hu H P, Wei L P, Chen Q Y, Tan J. Hydrochloric acid leaching behaviour of mechanically activated Panxi ilmenite($FeTiO_3$) [J]. Separation and Purification Technology, 2010, 73(2): 173 – 178.

[4] Chen G, Chen J, Guo S H, Li J, Srinivasakannan C, Peng J H. Dissociation behavior and structural of ilmenite ore by microwave irradiation[J]. Applied Surface Science, 2012, 258(10): 4826 – 4829.

[5] Guo S H, Chen G, Peng J H, Chen J, Li D B, Liu L J. Microwave assisted grinding of ilmenite[J]. Transactions of Nonferrous Metals Society of China, 2011, 21(9): 2122 – 2126.

[6] Li Y, Lei Y, Zhang L B, Peng J H, Li C L. Microwave drying characteristics and kinetics of ilmenite[J]. Transactions of Nonferrous Metals Society of China, 2011, 21(1): 202 – 207.

[7] Peng J H, Yang J J, Huang M, Huang M Y. Microwave – assisted reduction and leaching process of ilmenite [C]//Antennas, Propagation and EM Theory, 2008. ISAPE. 8th International Symposium. Kunming, China, 2008: 1383 – 1386.

[8] Pickles C A. Microwave in extractive metallurgy: Part 2—A review of applications[J]. Minerals Engineering, 2009, 22(13): 1112 – 1118.

[9] Pickles C A. Microwave in extractive metallurgy: Part 1—Review of fundaments[J]. Minerals Engineering, 2009, 22(13): 1102 – 1111.

[10] Huang M Y, Peng J H, Lei Y, Huang M, Zhang S M. The temperature rise and microwave – absorbing characteristics of ilmenite concentrate in microwave field[J]. Journal of Sichuan University(Engineering Science Edition), 2007, 39(2): 111 – 115(in Chinese).

[11] Ouyang H Y, Yang Z, Xiong X L, Wang K H. Study on elevated temperature curve and fluidization leaching behaviour of illmenite in microwave field[J]. Mining and Metallurgical Engineering, 2010(2): 73 – 75(in Chinese).

[12] Iwauchi K, Kiyama M, Nakamura T. Dielectric properties of $FeTiO_3$[J]. Physica Status Solidi, 1991, 127(2): 567 – 575.

[13] Viswanth R P, Seshsdri A T. The ferroelectric characteristics in Fe – Ti – O system[J]. Solid State Communications, 1991, 92(10): 831 – 842.

[14] Chiteme C, Mulaba – bafubiandi A F. An investigation on electric properties of microwave treated natural ilmenite[J]. Journal of Materials Science, 2006, 41(8): 2365 – 2372.

[15] Matthew K T, Raveendranath U. Cavity perturbation techniques for measuring dielectric parameters of water and other allied liquids[J]. Sensors Update, 2000, 7(1): 185 – 210.

[16] Nelson S, Bartley P G. Measuring frequency and temperature – dependent dielectric properties of food materials [J]. IEEE Transactions on Instrumentation and Measurement, 2002, 51(4): 589 – 592.

[17] Deshpande M D, Reddy C J, Tiemsin P I, Cravey R. A new approach to estimate complex permittivity of dielectric materials at microwave frequencies using waveguide measurements[J]. IEEE Transactions on Microwave Theory and Technology, 1997, 45(3): 359 – 365.

[18] Seo I S, Chin W S, Lee D G. Characterization of electromagnetic properties of polymeric composite materials with free space method[J]. Composite Structures, 2004, 66(1 – 4): 533 – 542.

[19] Sheen N I, Woodhead I M. An open – ended coaxial probe for broad – band permittivity measurement of agricultural products[J]. Journal of Agricultural Engineering Research, 1999, 74(2): 193 – 202.

[20] Francois T, Bernard J. Complete FDTD analysis of microwave heating processes in frequency – dependent and temperature – dependent media[J]. IEEE Transactions on, Microwave Theory and Technology, 1997, 45(1): 108 – 117.

[21] Huang K M, Cao X J, Liu C J, Xu X B. Measurement/computation of effective permittivity of dilute solution in saponification reaction [J]. IEEE Transactions on Microwave Theory and Technology, 2003, 51(10):

2106 - 2111.

[22] Pickles C A. Microwave heating behaviour of nickeliferous limonitic laterite ores[J]. Minerals Engineering, 2004,17(6):775 - 784.

[23] Dervosc T, Thirios E, Novacovich J, Vassiliou P. Permittivity properties of thermally treated TiO_2[J]. Materials Letters, 2004,58(9):1502 - 1507.

[24] Jonscher A K. Dielectric relaxation in solids[J]. Journal of Physics D: Applied Physics, 1999,32(14):R57 - R70.

[25] Metaxas A C. Microwave heating[J]. Power Engineering Journal, 1991,5(5):237 - 247.

[26] Wright R A, Cocks F H, Vaniman D T, Blake R D, Meek T T. Thermal processing of ilmenite and titania - doped haematite using microwave energy[J]. Journal of Materials Science, 1989,24(4):1337 - 1342.

[27] Zhang L B, Chen G, Peng J H, Chen J, Guo S H, Duan X H. Microwave absorbing properties of high titanium slag[J]. Journal of Central South University of Technology, 2009,16(4):588 - 593.

Dielectric Properties and Temperature Increase Characteristics of Zinc Oxide Dust from a Fuming Furnace

Libo Zhang, Aiyuan Ma, Chenhui Liu, Wenwen Qu,
Jinhui Peng, Yongguang Luo, Yonggang Zuo

Abstract: Cavity perturbation method was used to determine the dielectric properties (ε', ε'' and $\tan\delta$) of zinc oxide dust in different apparent densities. The process was conducted to study the microwave – absorption properties of zinc oxide dust and the feasibility of microwave roasting zinc oxide dust to remove fluorine and chlorine. The dielectric constant, dielectric loss, and loss tangent were proportional to the apparent density of zinc oxide dust. The effects of sample mass and microwave power on the temperature increase characteristics in the microwave field were also studied. The results showed that the apparent heating rate of the zinc oxide dust increased with the increase in microwave roasting power and decreased with the increase in the sample mass. The heating rate of the samples could reach approximately 800℃ after microwave treatment for 8min, which indicates that zinc oxide dust has strong microwave – absorption ability.

Keywords: zinc oxide dust; apparent density; dielectric properties; microwave heating; temperature increase characteristics

1 Introduction

Zinc is an important metal that is used in the metallurgical, chemical and textile industries. It is extracted mainly from sulfide ore. A part of zinc is recycled from zinc – containing wastes such as zinc dust, zinc ash and zinc dross. The zinc industry is developing rapidly but its operations are constrained by limited resources, energy shortages, and ecological and environmental problems[1-3]. Large amounts of zinc oxide dust are produced during the smelting of zinc, lead and copper, as well as the scrap recycling process of galvanized steel. Zinc oxide dust has a significant recovery value and profit margins because of the high content of valuable Zn, Pb and In[4-6]. Zinc oxide dust obtained from a lead fuming furnace also contains relatively large amounts of fluorine and chlorine impurities during the zinc hydrometallurgy process; this condition results in equipment corrosion, high production costs, and low zinc recovery[7,8]. Thus, finding an effective method to solve these problems and reuse the secondary resource is necessary[9-11]. Currently, two main methods in fluorine and chlorine removal from zinc oxide dust are adopted: pretreatment removal and leach liquor removal[12,13]. However, both methods involve high energy consumption and management difficulties in the subsequent process. The microwave removal method is considered as an environmentally friendly and energy – saving alternative.

As an efficient and clean form of energy, microwave heating has been widely used in mineral processing and metallurgy. Compared with conventional heating, microwave heating has the following advantages: fast heating rate, volume heating, selective heating, easy automatic control and absence of environmental pollution. It can also reduce the reaction temperature, shorten the reaction time and promote energy conservation[14 – 20]. With these advantages, microwave heating is introduced to remove fluorine and chlorine from zinc oxide dust. Microwave heating is employed in the microwave field dielectric loss of metallurgical material to ensure the overall heating of the material. The heating is determined by the material dielectric parameters (complex permittivity), which are key physical parameters to describe the interaction mechanism between microwave and dielectric materials. However, microwave heating is selective with regard to materials of different dielectric properties. Microwave selectively heats high – loss substances and low – loss substances that have no obvious absorption[21 – 24]. To date, no new study on the process of zinc oxide dust treatment to remove fluorine and chlorine via microwave heating has been published in China and elsewhere.

The dielectric properties (ε', ε'' and $\tan\delta$) of material are employed to describe the microwave absorption characteristics in the microwave heating process[25 – 28]. Dielectric properties can be defined in terms of relative permittivity (ε) composed of a real part (ε') and an imaginary part (ε''). The real part of the relative permittivity is known as dielectric constant and the imaginary part as loss factor. These properties can be expressed using the equation: $\varepsilon = \varepsilon' - j\varepsilon''$.

Loss tangent ($\tan\delta$) is a parameter employed to describe how well a material absorbs microwave energy. It represents the ratio of the dielectric loss factor to the dielectric constant ($\tan\delta = \varepsilon''/\varepsilon'$). The dielectric constant and loss tangent are functions of measurement frequency, material homogeneity and anisotropy, moisture, and material temperature.

The effects of various factors such as microwave frequency, relative density and particle size of powder samples on the permittivity and permeability were investigated. The ε' values of powder SiO_2 and Fe_3O_4 samples with the relative density below 1 are smaller than the values estimated using the linear relation between ε' and the relative density, and larger than those estimated using the Lichtenecker's logarithm mixed law[29].

In this study, the cavity perturbation method was adopted to measure the dielectric properties of zinc oxide dust with different apparent densities. The zinc oxide dust temperature increase characteristics through microwave heating were also investigated to provide a theoretical basis for the removal of fluorine and chlorine from zinc oxide dust.

2 Experimental

2.1 Experimental materials

The zinc oxide dust used in this study was obtained from a lead and zinc smelting plant in Yunnan province. The main chemical composition of the zinc oxide dust sample is shown in Table 1, and its

XRD spectra are presented in Fig. 1.

Table 1 Chemical compositions of zinc oxide dust sample

Component	Zn	Pb	Ge	Cd	Fe	Sb
Content(mass fraction)/%	53.17	22.38	0.048	0.21	0.38	0.23
Component	S	As	F	Cl	SiO_2	CaO
Content(mass fraction)/%	3.84	1.04	0.0874	0.0783	0.65	0.096

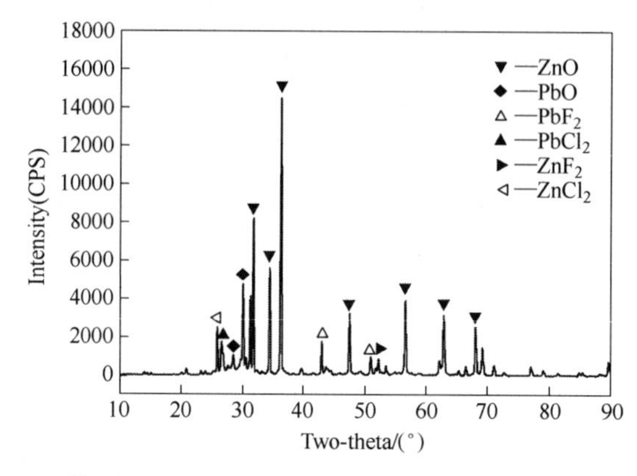

Fig. 1 XRD patterns of zinc oxide dust sample

As shown in Table 1 and Fig. 1, the zinc oxide dust sample contained a significant amount of lead and zinc, mainly in the form of lead and zinc oxides. Fig. 2 shows three different morphologies of zinc oxide dust: cubic crystal structure, floc structure, and spherical structure. Zn mainly existed in the ZnO phase, Pb mainly existed in the PbO phases, while F and Cl mainly existed in zinc and lead halide. A certain amount of fluorine and chlorine compounds were also found. The fluorine and chlorine compounds were dissolved in leaching solution during the leaching process. The excess content of fluorine and chlorine elements resulted in equipment corrosion, high production costs, and low zinc recovery in the subsequent zinc electrolysis process. Microwave was directly applied to roast the zinc oxide fumes according to the different microwave absorbance among halides, lead, and zinc oxides in combination with the advantages of selective microwave heating, for the purpose to strengthen the separation of halides as volatile components.

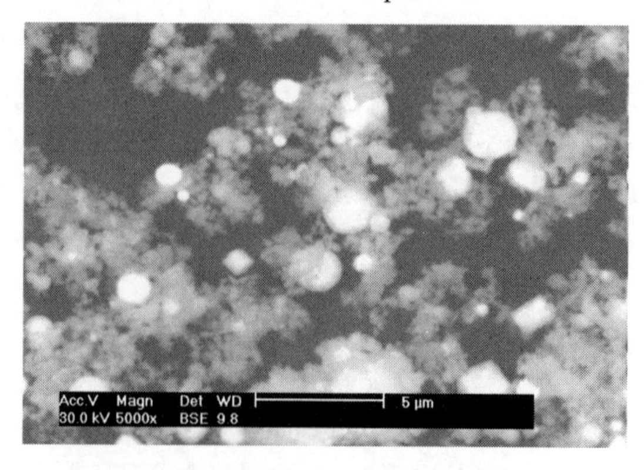

Fig. 2 SEM images of zinc oxide dust sample

2.2 Microwave heating system to measure temperature increase characteristics

The experiment equipment included a 3kW box – type microwave metallurgical reactor developed by the Key Laboratory of Unconventional Metallurgy, Kunming University of Science and Technology, China. A diagram of the microwave heating system is shown in Fig. 3.

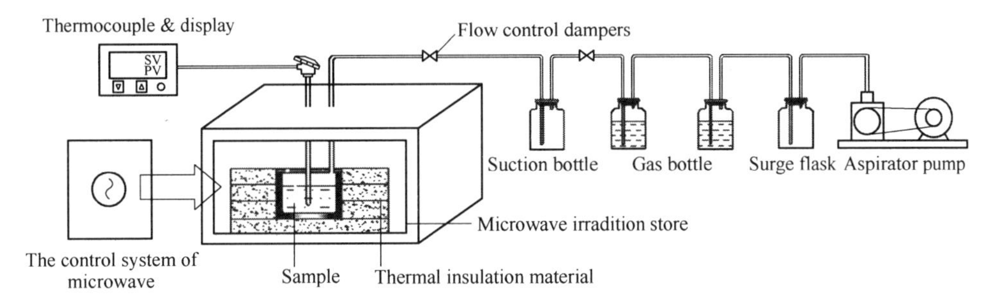

Fig. 3 Diagram of the microwave heating system

The microwave device had automatic temperature control with microwave heating frequency of 2450MHz and continuous adjustable power of 0kW to 3kW. A thermocouple with a shielded sleeve was used to measure the temperature of the samples, which ranged from 0℃ to 1300℃. The zinc oxide dust samples were placed in a mullite crucible with an inner diameter of 90mm and a height of 120mm. Thereafter, the samples were placed into the microwave cavity resonator sensor. The mullite crucible showed good properties, such as wave transparency and heat shock resistance. The smoke soot absorption system was composed of a dust collection bottle, two water bottles, an alkali absorption bottle, a buffer bottle, and a miniature pump. This system was able to collect and absorb the flue gas and flue dust generated during the experimental process.

2.3 Test principle of dielectric characteristics and system

2.3.1 Cavity perturbation test principle

The resonant cavity perturbation method[30] has been widely adopted because of its high measurement accuracy and wide sphere of application. The dielectric property measurement method is based on the cavity perturbation method that employs a single resonant mode. The change in the cavity characteristics in the presence of a sample has been measured. Based on a sample in the resonant cavity, the principles can be expressed as follows[31]:

$$\frac{\Delta\omega}{\omega} = -\omega_0(\varepsilon_r' - 1)\int_{V_e} E_0^* \cdot E\mathrm{d}v/(4W) \tag{1}$$

$$\frac{1}{Q} - \frac{1}{Q_0} = 2\varepsilon_0\varepsilon_r''\int_{V_e} E_0^* \cdot E\mathrm{d}v/(4W) \tag{2}$$

$$W = \int_V [(E_0^* \cdot D_0 + H_0^* \cdot B_0) + (E_0^* \cdot D_1 + H_0^* \cdot B_1)]\mathrm{d}v \tag{3}$$

where $\Delta\omega = \omega - \omega_0$, $\Delta\omega$ is the angular frequency deviation; ω_0 is the resonance frequency of the cavity in the unperturbed condition; ω represents the corresponding parameters of the cavity loaded with the sample; ε_r' and ε_r'' are the real and imaginary parts of the complex permittivity of the sample, respectively; V and V_e are the volumes of the cavity and the sample, respectively; E_0^*, D_0, H_0^*, and B_0 are the interior fields of the sample; E is the field strength of the cavity; dv is the elemental volume; Q_0 is the resonance frequency of the cavity in the unperturbed condition; Q represents the corresponding parameters of the cavity loaded with the sample; W is the storage energy; D_1 and B_1 are the added values of the electric displacement and magnetic induction intensity, respectively.

The measurement principle of the cavity perturbation is illustrated in Fig. 4. Both the reflection coefficient and resonant frequency were recorded by the microwave sensor connected to the software. The dielectric constant and loss factor were calculated based on the reflection coefficient and resonance frequency variation under the empty sensor and filled with the sample, respectively.

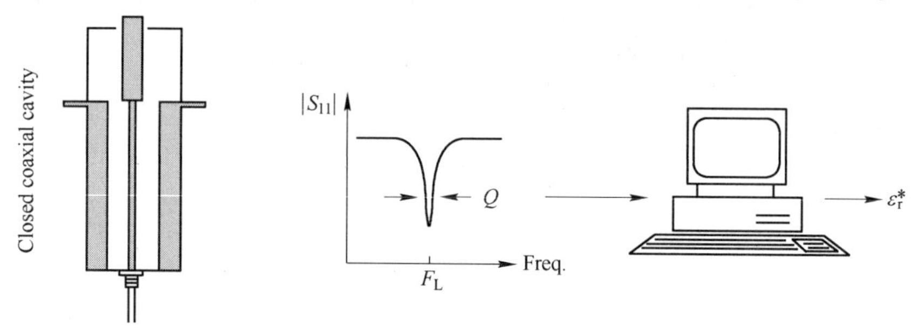

Fig. 4 Schematic of measurement principle of the cavity perturbation

2.3.2 Test equipment and process

The dielectric parameter measuring device for zinc oxide dust is shown in Fig. 5. The dielectric property testing equipment was a dielectric parameter tester (dielectric kit for vials). The device consisted of a microwave power source, a microwave receiver, and a cavity resonator. The microwave signal receiver of AD – 8320 integrated circuit could detect the signal amplitude and phase. A resonator was also used in the cavity. A test control unit was connected to the computer test software via a USB data cable. A test sample was first placed in a small bottle with almost the same size as the cavity resonator. The bottle was placed into the cavity resonator through the opening holes. The dielectric parameters could be calculated using the cavity perturbation theory by comparing the resonance frequency and quality factor before and after the test.

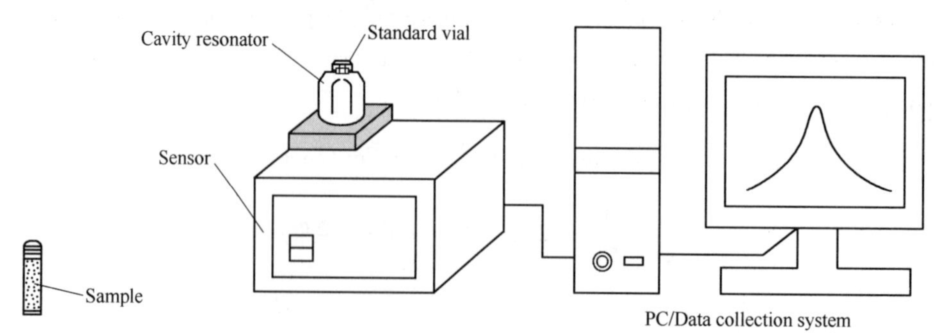

Fig. 5 Schematic diagram of zinc oxide dust dielectric constant measurement device

The dried zinc oxide dust was divided into 11 parts that were weighted and placed in sealed quartz standard tubes of known volume at 21℃ ± 1℃ and frequency of 2.45GHz. Each sample was subsequently pressed to obtain 11 different density values. The apparent density of the zinc oxide dust was calculated using the sample mass and volume in the standard pipe. The system error on the real part of the dielectric coefficient was controlled at 3% to 5% [26] during the perturbation method tests. Deionized water(the real part of the dielectric constant was 78) and polytetrafluoroethylene(the real part of the dielectric coefficient was 2.08) as standards, results were 76.79 and 2.04, respectively. The real and measured values had differences of 1.55% and 1.92%, respectively, which would indicate that the measurement results of small volume perturbation was made credible by using the perturbation method.

3 Results and discussion

3.1 Dielectric properties of zinc oxide dust

The measured material of the dielectric properties is related to a number of factors, such as density, humidity and frequency. This study mainly discusses the effect of apparent density on the dielectric properties of zinc oxide dust.

3.1.1 Effect of apparent density on dielectric properties

The effect of apparent density on the dielectric constant, dielectric loss, and loss tangent are presented in Fig. 6(a) to (c), respectively.

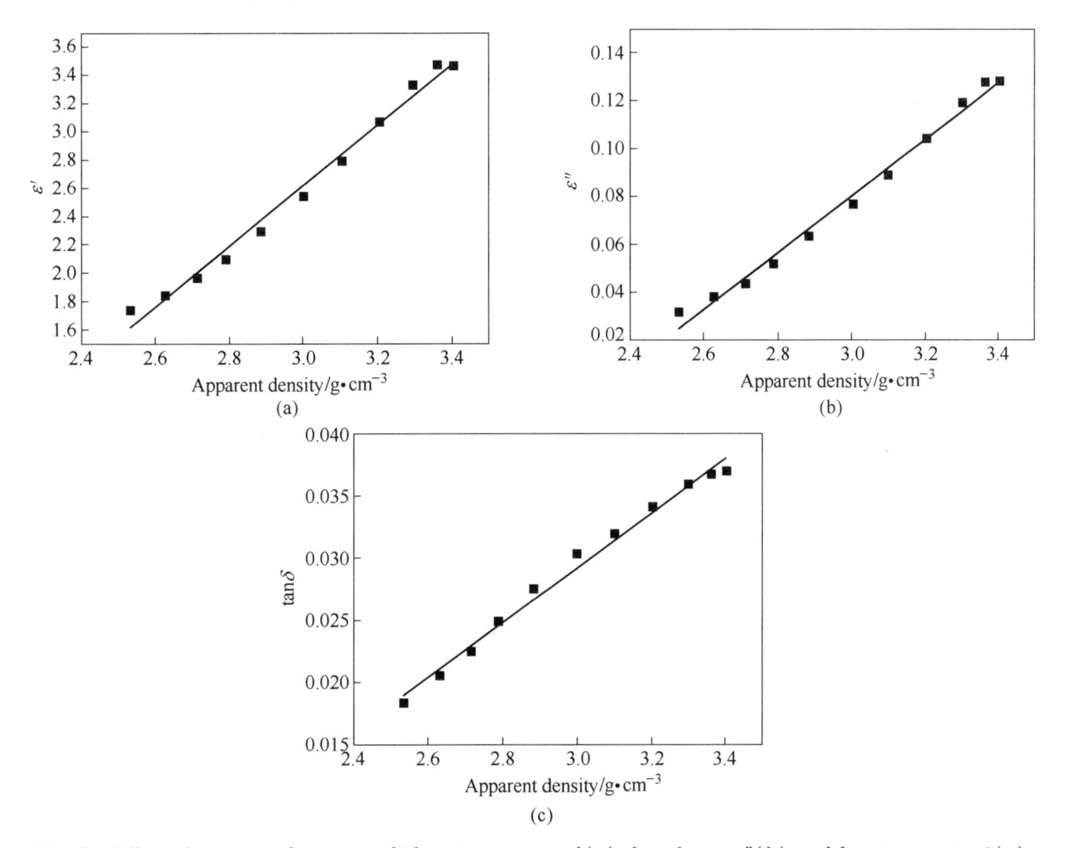

Fig. 6 Effect of apparent density on dielectric constant ε'(a), loss factor ε''(b) and loss tangent $\tan\delta$(c)

As shown in Fig. 6, the dielectric properties(ε', ε'' and $\tan\delta$) of the zinc oxide dust has a good linear relationship with the apparent density. Table 2 shows the linear regression equation and the correlation coefficient R^2.

Fig. 6 showed that apparent density is small($<2.8\text{g/cm}^3$), dielectric constant(ε'), loss factor (ε'') and loss tangent($\tan\delta$) stay at a relatively low level and the change trend is not obvious, because of the zinc oxide dust apparent density relatively is small, standard tube for testing has large number of voids between material particles where full of air. And with the increase of apparent density, the air which is between material particles is continuous discharge, dielectric characteristics value of zinc oxide dust(ε', ε'') changes significantly.

When the apparent density is low($<3.2\text{g/cm}^3$), the value of the dielectric properties(ε', ε'' and $\tan\delta$) increases linearly with the apparent density. Thus, the change trend is not obvious. The increase in the apparent density increases the dielectric properties, thereby increasing the microwave absorption ability. More heat energy converted from microwave energy may be the reason for the increase in the dielectric properties.

3.1.2 Apparent density effect on microwave penetration depth

The microwave penetration depth was measured as the distance from the surface to the inner part of the material, where the microwave field intensity was reduced to 1/e of the original field intensity. This value can be expressed as follows[32-34]:

$$D_p = \frac{\lambda_0}{2\sqrt{2}\pi\sqrt{\varepsilon'\left[\sqrt{1+\left(\frac{\varepsilon''}{\varepsilon'}\right)^2}-1\right]}} \tag{4}$$

In Eq. (4), D_p determines the microwave heating uniformity of the material at 2.45GHz and $\lambda_0 = 12.24\text{cm}$.

The microwave penetration depth of the zinc oxide dust was calculated at different apparent densities. The effect of apparent density on the microwave penetration depth is illustrated in Fig. 7. The results showed that D_p decreased when the apparent density increased. Based on the curves and the regression equations reported in Table 2, with the increase in the apparent density, the material absorbed more microwave energy. The energy was converted into heat along with the decay of the microwave field strength and power, which determined the microwave penetration depth. This result is consistent with the conclusion presented by Peng and Yang[22]. When the material penetration depth was greater than the size of the heated sample, the influence was negligible. By contrast, the microwave energy penetration was limited when the sample size was larger than the penetration depth. Non - uniformity heating also occurred during the microwave heating. Thus, the suitable apparent density was essential to fully employ the volumetric microwave heating process.

Table 2 Regression equations on zinc oxide dust material dielectric property in different densities

(x = apparent density)

Name	Linear regression equation	R^2	Name	Linear regression equation	R^2
ε'	$\varepsilon' = 2.14119x - 3.81184$	$R^2 = 0.9892$	$\tan\delta$	$\tan\delta = 0.02186x - 0.03641$	$R^2 = 0.98909$
ε''	$\varepsilon'' = 0.11806x - 0.27402$	$R^2 = 0.98927$	D_p	$D_p = 49.8097x^2 - 353.8286x + 655.0319$	$R^2 = 0.99934$

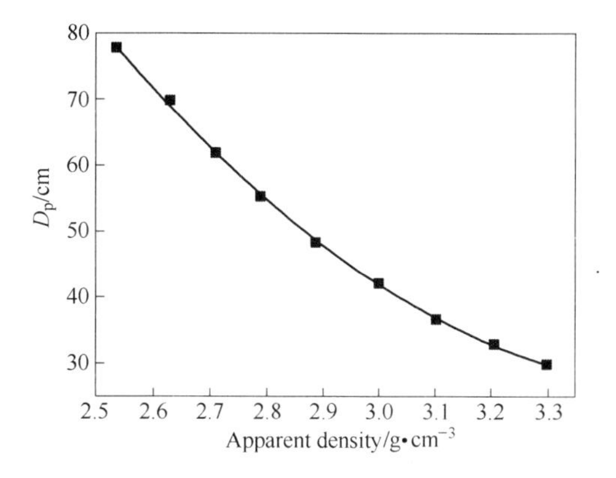

Fig. 7 Apparent density effect on microwave penetration depth

3.2 Microwave heating characteristics of zinc oxide dust

3.2.1 Effect of sample mass on temperature increase characteristics

The temperature increase characteristics and the sample mass in the microwave field are closely related to each other. In the microwave field, the sample mass of the zinc oxide dust affected the heating behavior under a microwave power setting of 900W, as shown in Fig. 8.

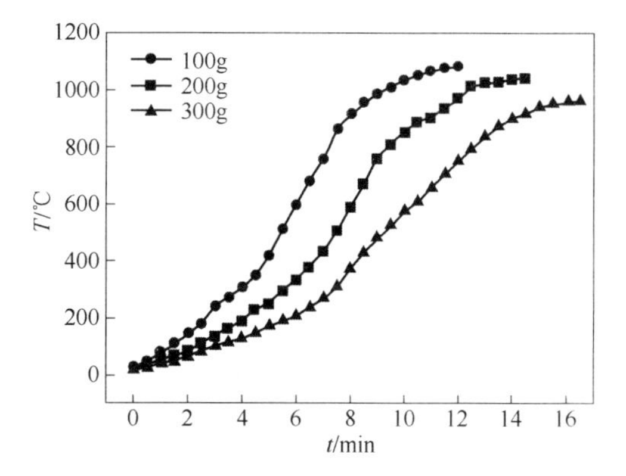

Fig. 8 Zinc oxide dust heating rate curves at different qualities in the microwave field

The relationship between the temperature of the zinc oxide dust with different sample temperatures (T_m) for sample mass of 100g, 200g and 300g is illustrated in Fig. 8. The time and empirical equations are shown in Eqs. (5) to (7):

$$T_m = 54.455 + 17.309t + 28.073t^2 - 1.635t^3 \quad (R^2 = 0.9950) \tag{5}$$

$$T_m = 66.289 + 40.393t + 20.787t^2 - 0.923t^3 \quad (R^2 = 0.9942) \tag{6}$$

$$T_m = 62.586 + 33.955t + 13.119t^2 - 0.465t^3 \quad (R^2 = 0.9960) \tag{7}$$

The results show that the zinc oxide fume average heating rates were 90℃/min, 72℃/min and 58℃/min, respectively. Thus, a smaller the sample mass indicates a faster apparent heating rate.

In the microwave field, the effect of the sample mass of zinc dust on the heating rate can be calculated as follows[35]:

$$\frac{dT}{d\tau}=\frac{T-T_0}{\tau}=\frac{2\pi f\varepsilon_0\varepsilon''E^2}{\rho VC_p}=\frac{2\pi f\varepsilon_0\varepsilon''E^2}{mC_p} \tag{8}$$

where T is the material heating temperature; T_0 is the material initial temperature; τ is the time; m is the material quality; C_p is the specific heat capacity material; f is the frequency of the microwave; ε_0 is the vacuum permittivity; ε'' is the dielectric loss factor; E is the electric field strength.

As Eq. (8) shows, the greater the dust mass is, the smaller the heating rate is. This observation is in accordance with the experimental results. On the one hand, the sample masses of the materials increased as the microwave power density per unit sample mass decreased. The increasing sample mass also increased the contact areas of the sample and resulted in increasing heat dissipation to the external environment. On the other hand, when microwave power was constant, a larger amount of zinc oxide dust indicated a thicker sample and the need for more microwave power. During the experimental condition range, the power density of the sample decreased with the increase in the sample mass, which resulted in a slower rate of temperature increase.

3.2.2 Effect of microwave power on temperature increase characteristics

The heating curves of 300g zinc oxide dust at microwave power settings of 900W, 1200W and 1800W are presented in Fig. 9. The empirical formula of T_m and time are shown in Eqs. (9) to (11):

$$T_m = 62.586 + 33.955t + 13.119t^2 - 0.465t^3 \quad (R^2 = 0.9960) \tag{9}$$

$$T_m = 65.370 + 37.373t + 19.040t^2 - 0.813t^3 \quad (R^2 = 0.9963) \tag{10}$$

$$T_m = 48.632 + 11.821t + 26.286t^2 - 1.519t^3 \quad (R^2 = 0.9965) \tag{11}$$

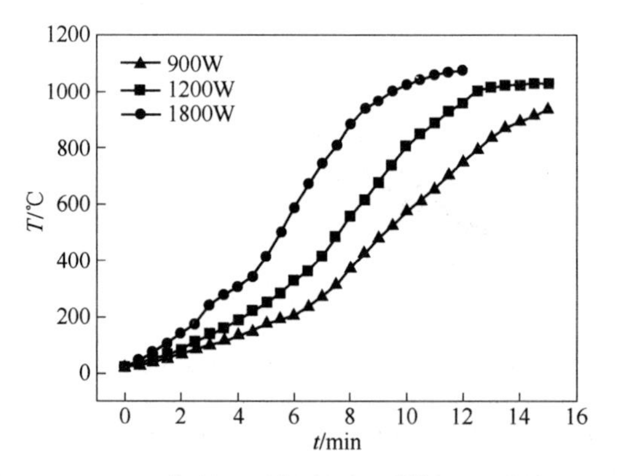

Fig. 9 Heating rate curves of zinc oxide dust at different microwave power settings

As illustrated in Fig. 9, microwave power had obvious effects on the temperature increase rate under the same conditions. The microwave power increased the apparent average heating rate of zinc oxide dust, and higher microwave power required shorter heating time to reach the same temperature. The experimental result is similar to reported results for microwave heating of low - grade nickel oxide ore[36]. The reason for this observation is that the temperature of the material increased with the increasing microwave output power within a certain range.

The per unit volume of the zinc oxide dust also absorbed microwave power or the microwave energy dissipated power in the dust. This value can be expressed as the following equation[37]:

$$P = 2\pi f\varepsilon'' E^2 \tag{12}$$

where f is the microwave frequency; ε'' is the dielectric loss factor that is a function of temperature; E is the electric field strength.

In Eq. (12), the increasing microwave power indicated that the increasing electric field strength in other conditions was unchanged. Microwaves could penetrate the interior of the material and would result in heating uniformity. The zinc oxide dust absorbed more microwave energy with the increase in E, which resulted in temperature increase. Therefore, increasing the microwave heating power properly could reduce the heating time and improve the apparent average heating rate of the zinc oxide dust.

In the process of roasting, microwave heating take full advantage of dielectric loss of the zinc oxide dust and selectively heat molecules or atoms which have good microwave absorbance. The zinc oxide dust in microwave field reaches 800℃ within 8min, which provides a well thermodynamic condition for defluorination and dechlorination by microwave roasting. The microwave absorbance of chloride and sulfide is strong, while that of Zn and Pb oxide is weak in the zinc oxide fume. Capitalizing on the selective heating property of microwaves can strengthen the separation of fluorides and chlorides in volatile components. The successful outcome of this research will provide the theoretical basis for effectively separating fluorine and chlorine as well as efficiently producing zinc with low energy consumption from the zinc oxide dust.

4 Conclusions

The cavity perturbation method was adopted to study the dielectric properties and temperature increase characteristics of zinc oxide dust under microwave radiation. The results are summarized as follows:

(1) The dielectric properties (ε', ε'' and $\tan\delta$) of zinc oxide dust had a good linear relationship with the apparent density. The dielectric properties increased when the apparent density increased. D_p decreased when the apparent density increased.

(2) The material heating rate was affected by the sample mass at the output power of 900W in the microwave field. As the zinc oxide dust mass increased, the heating rate slowed down. When mass was kept constant, the temperature increase rate of zinc oxide dust was affected by the microwave power. Increasing the microwave heating power appropriately could reduce the heating time and improve the apparent average heating rate of the zinc oxide dust.

(3) Zinc oxide dust had good absorption properties and could be heated quickly in the microwave field. This condition provides a theoretical to remove fluorine and chlorine at 800℃ in 8min by microwave, which could provide a clean and energy - efficient method to treat zinc - containing wastes. The results could also provide a theoretical basis for exploring new technology to remove fluorine and chlorine in zinc oxide dust.

Acknowledgement

The authors are grateful for financial supports by the National Natural Science Foundation

(No. 51104073), the National Technology Research and Development Program of China (No. 2013AA064003), the Applied Basic Research Project of Yunnan Province(2011FZ038), the Yunnan Province Young Academic Technology Leader Reserve Talents(2012HB008).

References

[1] Ma H J, He H J, Li H M. Comprehensive utilization of restoring volatilizable secondary zinc oxide[J]. Nonferrous Metals, 2010(4): 16, 17, 28(in Chinese).

[2] Li Y H, Liu Z H, Zhao Z W, Li Q H, Liu Z Y, Zeng L. Determination of arsenic speciation in secondary zinc oxide and arsenic leachability [J]. Transactions of Nonferrous Metals Society of China, 2012, 22(5): 1209 - 1216.

[3] Ruiz O, Clemente C, Alonso M, Alguacil F J. Recycling of an electric arc furnace flue dust to obtain high grade ZnO[J]. Journal of Hazardous Materials, 2007, 141(1): 33 - 36.

[4] Liu Q, Yang S H, Chen Y M, He J, Xue H T. Selective recovery of lead from zinc oxide dust with alkaline Na(2)EDTA solution[J]. Transactions of Nonferrous Metals Society of China, 2014, 24(4): 1179 - 1186.

[5] Li X H, Zhang Y J, Qin Q L, Yang J, Wei Y S. Indium recovery from zinc oxide flue dust by oxidative pressure leaching[J]. Transactions of Nonferrous Metals Society of China, 2010, 20(supl): S141 - S145.

[6] Peng J, Peng B, Yu D, Tang M T, Lobel J, Kozinski J A. Volatilization of zinc and lead in direct recycling of stainless steel making dust[J]. Transactions of Nonferrous Metals Society of China, 2004, 14(2): 392 - 396.

[7] Gresin N, Topkaya Y A. Dechlorination of a zinc dross[J]. Hydrometallurgy, 1998, 49(1 - 2): 179 - 187.

[8] Şahin F Ç, Derin B, Yücel O. Chloride removal from zinc ash[J]. Scandinavian Journal of Metallurgy, 2000, 29(5): 224 - 230.

[9] Ma H J, Shi W G, Zheng Y Q. Research on a new technology of producing electrolytic zinc by complex - componential sub - zinc oxide[J]. Journal of Non - ferrous Metallurgy, 2010(3): 52 - 55(in Chinese).

[10] Li D, Wang J H, Guo X H, Hu Y M. Proctical study on producing electrolytic zinc from secondary zinc oxide [J]. Non - ferrous Mining and Metallurgy, 2011, 27(3): 33 - 37(in Chinese).

[11] Jha M K, Kumar V, Singh R J. Review of hydrometallurgical recovery of zinc from industrial wastes[J]. Resources Conservation and Recycling, 2001, 33(1): 1 - 22.

[12] Mason C R S, Harlamovs J R, Dreisinger D B, Grinbaum B. Solvent extraction of a halide from a aqueous sulphate solution: US, 7037482 B2[P]. 2006 - 52.

[13] Bodson F J J. Process for the elimination of chloride from zinc sulphate solution: US, 4005174[P]. 1977 - 1 - 25.

[14] Jin Q H, Dai S S, Huang K M. Microwave chemistry[M]. Beijing: Science Press, 1999(in Chinese).

[15] Tong Z F, Bi S W, Yang Y H. Present situation of study on microwave heating application in metallurgy[J]. Journal of Materials and Metallurgy, 2004, 3(2): 117 - 120(in Chinese).

[16] Cai W Q, Li H Q, Zhang Y. Recent development of microwave radiation application in metallurgical processes [J]. The Chinese Journal of Process Engineering, 2005, 5(2): 228 - 232(in Chinese).

[17] Zheng Y, Niu Y Q, Niu X J, Wu P S. The application of microwave for processing minerals[J]. Uranium Mining and Metallurgy, 2002, 21(3): 151 - 153(in Chinese).

[18] Kingman S W, Rowson N A. Microwave treatment of minerals—a review[J]. Minerals Engineering, 1998, 11(11): 1081 - 1087.

[19] Haque K E. Microwave energy for mineral treatment processes—a brief review[J]. International Journal of Mineral Processing, 1999, 57(1): 1 - 24.

[20] Bao R, Yi J H, Peng Y D, Zhang H Z. Effects of microwave sintering temperature and soaking time on microstructure of WC - 8Co[J]. Transactions of Nonferrous Metals Society of China, 2013, 23(2): 372 - 376.

[21] Bergese P. Specific heat, polarization and heat conduction in microwave heating systems: A nonequilibrium thermodynamic point of view[J]. Acta Materialia, 2006, 54(7): 1843 - 1849.

[22] Peng J H, Yang X W. The New Applications of Microwave Power[M]. Kun Ming: Yunnan Science and Technology Press, 1997 (in Chinese).

[23] Metaxas A C, Meredith R J. Industrial Microwave Heating[M]. London: Peter Peregrinus, 1983.

[24] Clark D E, Folz D C, West J K. Processing materials with microwave energy[J]. Materials Science and Engineering: A, 2000, 287(2): 153 - 158.

[25] Lin W G. Microwave Theory and Technology[M]. Beijing: Science Press, 1979 (in Chinese).

[26] Dong S Y. Microwave Measuring[M]. Beijing: Beijing University of Science and Technology Press, 1991 (in Chinese).

[27] Büyüköztürk O, Yu T Y, Ortega J A. A methodology for determining complex permittivity of construction materials based on transmission - only coherent, wide - bandwidth free - space measurements[J]. Cement and Concrete Composites, 2006, 28(4): 349 - 359.

[28] Ding D H, Zhou W C, Luo F, Zhu D M. Influence of pyrolytic carbon coatings on complex permittivity and microwave absorbing properties of Al_2O_3 fiber woven fabrics[J]. Transactions of Nonferrous Metals Society of China, 2012, 22(2): 354 - 359.

[29] Hotta M, Hayashi M, Nishikata A, Nagata K. Complex permittivity and permeability of SiO_2 and Fe_3O_4 powders in microwave frequency range between 0.2 and 13.5GHz[J]. ISIJ International, 2009, 49(9): 1443 - 1448.

[30] Huang M, Peng J H, Zhang S M, Zhang L B, Xia H Y, Yang J J. Measuring methods of the material permittivity and it's applications [C]//The 12th National Microwave Can Application Academic Conference Proceedings. Chengdu: China Microwave Society, 2005: 79 - 91.

[31] Carter R G. Accuracy of microwave cavity perturbation measurements[J]. IEEE Transaction on Microwave Theory and Techniques, 2001, 49(5): 918 - 923.

[32] Guo W C, Wang S J, Gopal T, Judy A J, Tang J M. Temperature and moisture dependent dielectric properties of legume flour associated with dielectric heating[J]. LWT - Food Science and Technology, 2010, 43(2): 196 - 201.

[33] Kumar P, Coronel P, Simunovic J, Truong V D, Sandeep K P. Measurement of dielectric properties of pumpable food materials under static and continuous flow conditions[J]. Journal of Food Science, 2007, 72(4): E177 - E183.

[34] Peng Z W, Hwang J Y, Mouris J, Hutcheon R, Huang X D. Microwave penetration depth in materials with non - zero magnetic susceptibility[J]. ISIJ International, 2010, 50(11): 1590 - 1596.

[35] Chen J, Lin W M, Zhao J. The Coking Coal Metallurgy Technology[M]. Beijing: Chemical Industry Press, 2007 (in Chinese).

[36] Hua Y X, Tan C E, Xie A J, Lv H. Microwave - aided chloridizing of nickel - bearing garnierite ore with $FeCl_3$ [J]. Nonferrous Metals, 52(1): 59, 60 (in Chinese).

[37] Thostenson E T, Chou T W. Microwave processing: fundamentals and applications[J]. Composites Part A: Applied Science and Manufacturing, 1999, 30(9): 1055 - 1071.

Temperature and Moisture Dependence of the Dielectric Property of Silica Sand

Chenhui Liu, Libo Zhang, Jinhui Peng, C. Srinivasakannan, Bingguo Liu, Hongying Xia, Junwen Zhou, Xu Lei

Abstract: The major objective of this work was to investigate the effects of temperature and moisture content on the dielectric properties of silica sand. The dielectric properties of moist silica sand at five temperatures between 20℃ to 100℃, covering different moisture content levels at a frequency of 2.45GHz, were measured with an open - ended coaxial probe dielectric measurement system. The wave penetration depth was calculated based on the measured dielectric data. The results show moisture content to be the major influencing factor for the variation of dielectric properties. Dielectric constant, loss factor and loss tangent all increase linearly with increasing moisture content. Three predictive empirical models were developed to relate the dielectric constant, loss factor, loss tangent of silica sand as a linear function of moisture content. An increase in temperature between 20℃ to 100℃ was found to increase the dielectric constant and loss factor. The penetration depth decreased with increase in moisture content and temperature. Variation in penetration depth was found to vary linearly with decrease in moisture content. An predictive empirical model was developed to calculate penetration depth for silica sand.

This study offers useful information on dielectric properties of silica sand for developing microwave drying applications in mineral processing towards designing better microwave sensors for measuring silica sand moisture content.

Keywords: dielectric properties; penetration depth; silica sand; moisture content; temperature increase; microwave drying; temperature dependence

1 Introduction

Silica sand is one of the silicate minerals. It is hard, anti - friction and less reactive, with the major mineral component being SiO_2. High purity quartz sand is the main raw material to manufacture silica film, quartz glass, semiconductor silicon, fiber optics and metallurgical fluxes[1]. The fine silica sand is commonly obtained from mineral dressing processes such as scrubbing, magnetic separation and flotation. Silica sand contains 8% - 10% water that needs to be reduced to less than 1%, for being utilized as ingredient in the glass manufacturing process.

The most common method for drying silica sand in metallurgical industry are hot air convective drying, fluidized bed drying and rotary kiln drying. Although these methods are simple, the drawbacks include inability to handle large quantities and to achieve consistent quality standards, contamination, long duration, low energy efficiency and high cost[2]. Microwave drying has established itself to be energy - efficient than conventional drying[3]. The major advantages of microwave drying include rapid heat transfer, volumetric and selective heating, compactness of equipment, quick start/stop and pollution – free since nothing is burned[4]. Microwave drying has been recently applied in processing of foods, woods, minerals, pharmaceutical materials and other industrial raw materials[5-8].

Material processing with microwave radiation involves several complicated physical processes, including the absorption of electromagnetic energy that transforms into heat energy and transportation of generated heat[9]. A better understanding of the interaction between microwave radiation and the material is essential for using it more efficiently and effectively. The dielectric properties of a material are closely related to the interaction between the material and microwave radiation. Complex permittivity is one of the widely studied dielectric parameter of a material, and it is defined as detailed below.

$$\varepsilon^* = \varepsilon' - j\varepsilon'' \tag{1}$$

The complex permittivity (ε^*) is related to the ability of a material to couple with electric energy from microwave fields. The dielectric constant (ε') reflects the ability of the material to store electromagnetic energy within its structure, and the loss factor (ε'') characterizes the ability of the material to convert the stored electromagnetic energy into thermal energy. Loss tangent ($\tan\delta$) is another dielectric parameter of a material and is defined as:

$$\tan\delta = \frac{\varepsilon''}{\varepsilon'} \tag{2}$$

where, $\tan\delta$ demonstrates the ability of a material to convert stored energy into heat. In microwave drying, the dielectric constant and loss factor of a material are needed to estimate penetration depth of the material and the generated heat. Both dielectric constant and loss factor are a function of microwave frequency and sample moisture and temperature. Mathematical models that predict the microwave drying of silica sand needs equations for the dielectric properties as a function of moisture and temperature[10].

The dielectric properties of foods, pharmaceutical powders and woods have been reported for a wide range of moisture contents[11-13]. However, metallurgical materials, especially on the silica sand mixed with water is seldom reported. To the knowledge of the authors, the dielectric properties of wet silica sand particles that are utilized as industrial raw material, have not been reported in open literature.

The objectives of the present work are: (1) to assess the effects of temperature and moisture content on dielectric properties of silica sand at 2.45GHz; (2) to assess the effects of moisture content and temperature on dielectric constant, loss factor and loss tangent; (3) to calculate the power penetration depth of silica sand particles under different conditions in order to estimate the optimal thickness of the drying bed for the microwave drying system.

2 Materials and methods

2.1 Materials and sample preparation

The flotation treated silica sand was supplied by a metallurgical plant in Yunnan province, China. The chemical composition of silica sand(Fig. 1) used in the present work is presented in Table 1. It consists mainly of quartz with small quantities of aluminum oxide, hematite and rutile.

Fig. 1 SEM image of silica sand

Table 1 Chemical component of high purity silica sand

Component	SiO_2	Al_2O_3	Fe_2O_3	CaO	MgO	TiO_2	K_2O
Content(mass fraction)/%	98.67	0.58	0.30	0.03	0.02	0.02	0.05

The grain size distribution of the sample is shown in Table 2. Silica sand is mostly free composition water, but moisture could be up to 8% - 10%.

Table 2 Grain size of high purity silica sand

Grain size/μm	>210	210 - 150	150 - 100	<100
Content/%	21.4	60.1	18.5	1.4

2.2 Open - ended coaxial - line dielectric measurement system

A hybrid experimental/computational permittivity measuring system(Fig. 2), which was developed by the Institute of Applied Electromagnetics at Sichuan University, was used to determine the complex permittivity of the silica sand[14]. The sample powder was sealed in a stainless steel holder (80mm inner diameter with 100mm length) and placed in the resonant cavity, heated by an electric furnace. An open ended coaxial sensor(Fig. 3) connected to the Agilent PNA5230 Network Analyzer was used to measure the reflected signals. The probe was inserted into sample powder for full contact. The electromagnetic field around the probe changed as the probe contacted the specimen. The probe receives reflected signals from the sample which provides the information related to the complex dielectric permittivity of the measuring sample. The reflected signals were measured by the probe and recorded by the analyzer. A thermocouple was used for measuring the temperature. The method of genetic algorithm, which was based on the experimental/computation reflection coefficient, was used to calculate the complex permittivity of the sample[10,15,16].

Permittivity measurements were made at the frequency of 2.45GHz. The complex permittivity measurement of the silica sand sample with different moisture content(0 - 10%) at 20℃ and a sample with 5% water content is measured at 20℃, 40℃, 60℃, 80℃ and 100℃.

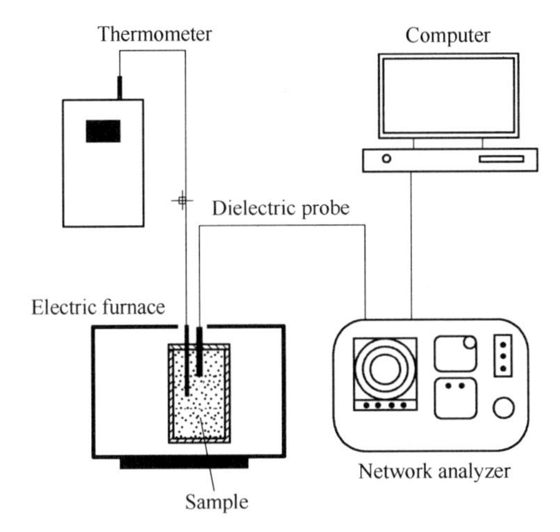

Fig. 2 The scheme of permittivity measurement system

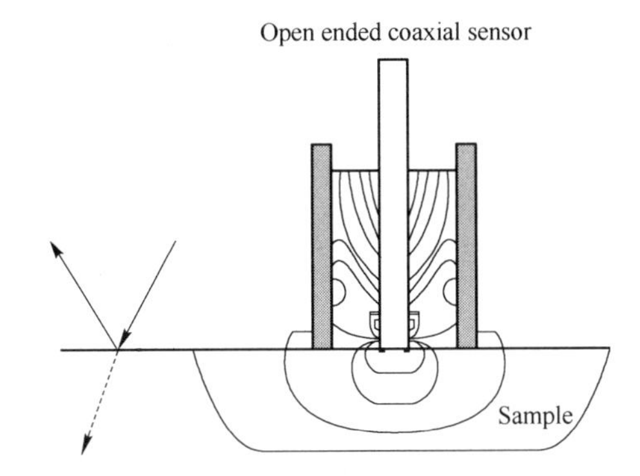

Fig. 3 The schematic of the open – ended coaxial sensor

3 Results and discussions

3.1 Effects of moisture on dielectric properties

Table 3 as well as the Figs. 4 – 6, show the linear increase in the dielectric constant, loss factor and loss tangent of the testing silica sand particles with increase in moisture content at 20℃.

Table 3 Regression equations on moisture contents and dielectric parameters at 2.45GHz at 25℃

Dielectric properties	Regression equations	Correlation efficient(R^2)
Dielectric constant	$\varepsilon' = 1.96208 + 0.4212M$	0.95647
Loss factor	$\varepsilon'' = 0.11162 + 0.04897M$	0.94858
Loss tangent	$\tan\delta = 0.06862 + 0.00313M$	0.99665
Penetration depth	$D_p = 17.7549 - 1.08141M$	0.98738

At a given frequency, increasing moisture content increases the dielectric constant of silica sand sample (Fig. 4), water is a dipolar compound that couples microwave energy at microwave frequen-

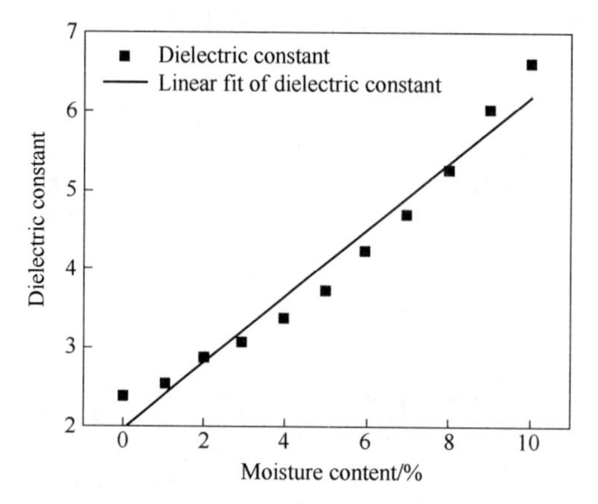

Fig. 4 Variation of dielectric constant of silica sand particles as a linear function of moisture content at 2.45GHz

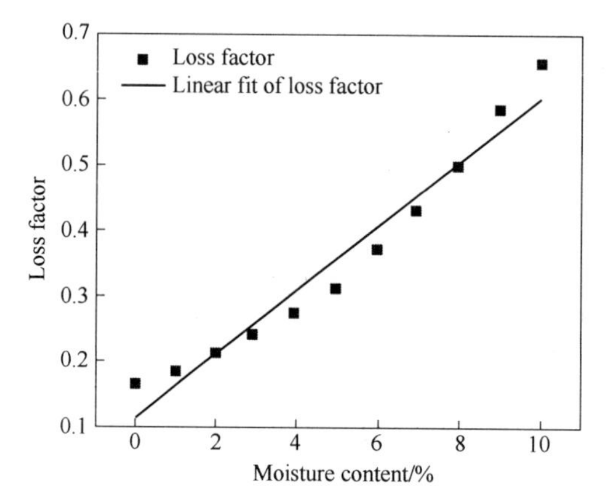

Fig. 5 Variation of relative loss factor ε_r'' of quartz sand as a linear function of moisture content at 2.45GHz

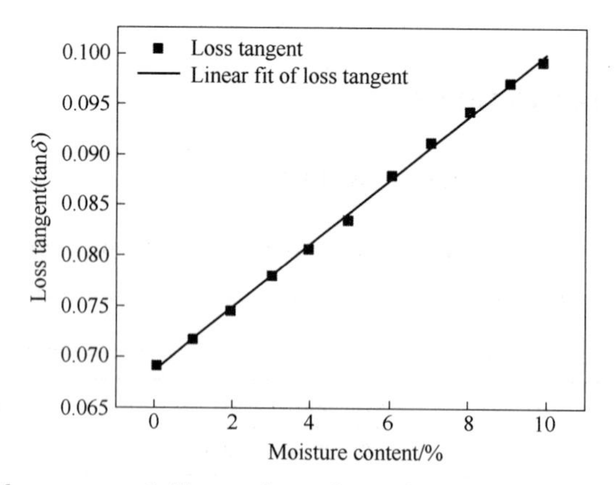

Fig. 6 Variation of loss tangent of silica sand as a linear function of moisture content at 2.45GHz

cies more efficiently than most other components of minerals. The dielectric constant was found to be highest at the maximum moisture content(10%), while it is the lowest for the dry sample. This result agrees well with previous studies on clays and soils[5,17-19].

The variation in dielectric constant with moisture content was found to be linear and of regression coefficients are provided in Table 3 along with the coefficient of determination R^2. The high R^2 value justifies the linear relationship, while the regressed equation can be utilized to estimate the dielectric constant of the silica particles with moisture content.

As shown in Fig. 5, the loss factor increase linearly with moisture content at 2.45GHz at 20℃. For minerals that contain abundant free water, the loss factor may either stay constant or decreases with moisture content[11,20,21]. Although water in silica sand exists as free water, it was observed that at low moisture content, loss factor increases with increase in moisture content.

The variation in of loss factor with moisture content was found to be linear and the regression coefficients are provided in Table 3 (along with the R^2). A high R^2 and low deviations ensure utilization of the regressed equation for determination of the loss factor of silica sand with moisture content.

Similar to the dielectric constant, a linear variation loss factor with moisture content has been reported for material such as clays and soils[22,23]. Differences of dielectric properties between different materials were embodied by various slopes and intercepts on the linear lines[14,24,25].

The loss tangent was calculated from Eq. (2), as shown in Fig. 6. As expected, since the dielectric constant and loss factor were found to exhibit a linear increase with increase in moisture content, the ratio of which the loss tangent is also expected to be linear as shown in Fig. 6. A high correlation value and low deviations indicate that this equation can be used to determine the loss tangent of silica sand at given moisture content.

Based on the linear relationship between dielectric property of a material and its moisture content, several new technologies have been developed for measuring moisture content of materials. The time domain reflectometry (TDR) has been widely used in measuring soil moisture due to its advantages of high accuracy and stability[5,12,26,27].

The moisture content in sand - like mixed materials is commonly categorized into three layers: the hygroscopic water layer, the capillary water layer and the free water layer[12]. The dielectric property of water within mixtures is not the same as that of pure water because the ion content of water within mixtures is different. The exposure water to silica sand particles, promotes the ions transfer to water, which form an ionic halo around the particles. These ions contribute to the electrical conduction which return back to silica sand particle when the water is removed[28,29]. Interaction may occur between the solid and the solvent, which causes the dielectric property of the mixture to be lower or higher than the sum of the free water moisture plus solid.

In microwave drying, the microwave energy is converted to thermal energy in dielectric materials according to Eq. (3):

$$P_{abs} = 2\pi f \varepsilon'' E_{in}^2 \tag{3}$$

where P_{abs} is the amount of absorbed energy through dielectric dispersion per unit volume; E_{in} is the electric field intensity within dielectric. The absorbed energy is mainly used to evaporate water promoting drying. The ability of absorbing microwave energy of silica sand decreases with decrease in moisture content at 2.45GHz. Hence the drying rate of silica sand is faster at the beginning of drying than at the ending of microwave dying, due to the reduction in moisture content during microwave drying process. The advantage of microwave drying system is attributed to the uniform moisture content of the entire sample[30,31].

3.2 Effects of temperature on dielectric properties

The temperature dependence of dielectric properties of silica with 5% moisture content over the temperature range from 20℃ to 100℃ are presented in Figs. 7 – 9.

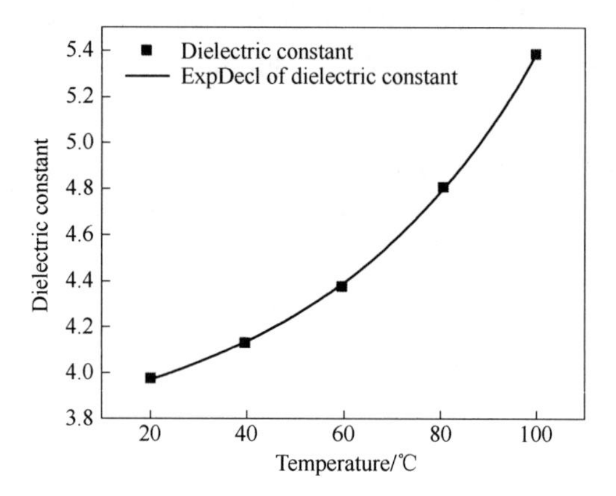

Fig. 7 Variation of dielectric constant of silica sand with 5% moisture content with temperature at 2.45GHz

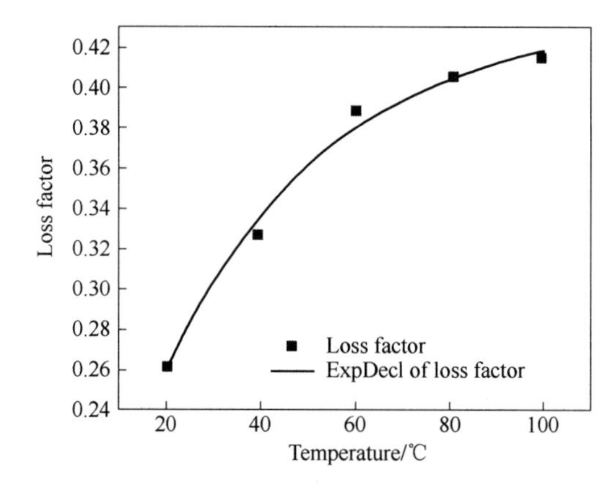

Fig. 8 Variation of loss factor of silica sand with 5% moisture content with temperature at 2.45GHz

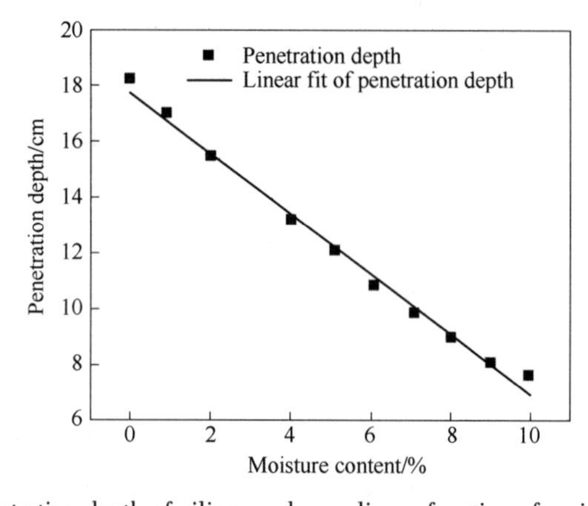

Fig. 9 Variation of penetration depth of silica sand as a linear function of moisture content at 2.45GHz

Fig. 7 shows that, the dielectric constant of silica sand increase almost linearly with temperature from 20℃ to 100℃, at 5% moisture content of the sample. At low temperatures (below 40℃), change in dielectric constant is very minimal, while rapid increase in dielectric constant was observed at temperature above 40℃. The relationship between the dielectric constant and the temperature could be established through the fit of an exponential function and equation parameters are shown in Table 4. An increase in dielectric constant of minerals with temperature was also reported for food products[27,32,33]. The temperature dependence of water in terms of ε' and ε'' has both been reported to decrease with increase in temperature from 20℃ to 100℃[34]. However, the temperature dependence behavior of dielectric constant with 5% water content is not similar to that of water at 20 – 100℃. Contrary to pure water the dielectric constant of wet silica sand keeps increasing, which can be attributed to the variations of dielectric polarization of free water which is much more than bound water.

Table 4 Regression equations on temperature and dielectric parameters at 2.45GHz at 25℃

Dielectric properties	Regression equations	Correlation efficient(R^2)
Dielectric constant/$F \cdot m^{-1}$	$\varepsilon' = 3.66 + 0.2e^{(T/46.6)}$	0.9985
Loss factor/$F \cdot m^{-1}$	$\varepsilon'' = 0.44 - 0.313e^{(-T/35.05)}$	0.97453

The effect of temperature on loss factor of silica sand with 5% moisture content is shown in Fig. 8. The relationship between the loss factor and the temperature could be established through the fit of an exponential function and equation parameters are shown in Table 4. The mixture at higher temperatures has lager loss factors than at lower temperatures. Water will absorb much more energy, and reach a higher temperature than silica sand under the same treatment time[35]. This enables water to reach boiling point while silica sand still at lower temperatures. That is the advantages of selective heating with microwave drying.

For pure silica sand, the loss factor increase with increasing temperature, while for pure water, the loss factor decrease with increasing temperature. At 5% moisture content, most water is tightly bound to solids with little mobility at room temperature to respond to microwave field at 2.45GHz. Raising temperature increases the mobility of water molecules and ionic conductivity, as indicated by an increase in loss factor in silica sand with temperature[36]. The loss factor is the sum of two components: ionic, ε''_σ, and dipole, ε''_d loss. These two components respond oppositely to temperature. The dipole loss decreases and ionic loss increases with temperature. The increasing trend of ε'' with temperature can be attributed to the possibility of runaway effects at lower moisture ranges[37].

Heating efficiency and non – uniformity are important factors in developing microwave drying techniques of minerals and ores[38]. The measured results obtained in this study show both dielectric constant and loss factor of silica sand increase with temperature and moisture content at 2.45GHz. Temperature rise leads to higher loss factor which in turn increases the rate of temperature rise. When heated in a multimode microwave cavity, moist silica sand may experience thermal runaway and hot spots effects.

The dielectric behavior of the silica sand illustrated in this paper is highly influenced by its water content compared with temperature in microwave drying process.

3.3 Relationship between penetration depth and moisture content and temperature

The power penetration depth is defined as the depth where the strength of microwave field is reduced to 1/e of its surface value and was expressed with the following Eq. (4):

$$D_p = \frac{\lambda_0}{2\sqrt{2}\pi\sqrt{\varepsilon'\left[\sqrt{1+\left(\frac{\varepsilon''}{\varepsilon'}\right)^2}-1\right]}} \tag{4}$$

where, D_p is the penetration depth, cm; λ_0 is the wavelength (12.24cm) at 2.45GHz[39].

Changes in penetration depth of the silica sand with respect to different moisture content are presented in Fig. 9 and Table 3.

Fig. 9 depicts the effects of moisture content on penetration depth at 2.45GHz. From literature, it is well known that the power penetration depth of pure water at room temperature is 1.68cm at 2.45GHz and that of dry silica sand particles is 18.28cm. The penetration depth of silica sand at any moisture content is smaller than that of dry silica sand and larger than that of pure water. At moisture content ranging from 0% to 10%, the higher the moisture, the shorter is the penetration depth. According to Eq. (4), materials with high loss factor, microwave energy does not penetrate deeply[40]. Shorter penetration depths indicate more ready absorption of microwave power. Water has higher loss factor and lower penetration depth compared with dry material, and therefore, water will be selectively heated in moist minerals. In general, the moisture content in wet silica sand is less than 10%, which is much lower than that in foods such as apples and breads[12,41]. The marginal increase in penetration depth with the reduction in moisture content for silica sand mixture, is different from that of the food material such as apple which exhibit a sharp change at the moisture content of 30%. Regression analysis reveals that the relationship between the power penetration depth of wet quartz sand particles and moisture content can be expressed by the second polynomial function and details of which are shown in Table 3.

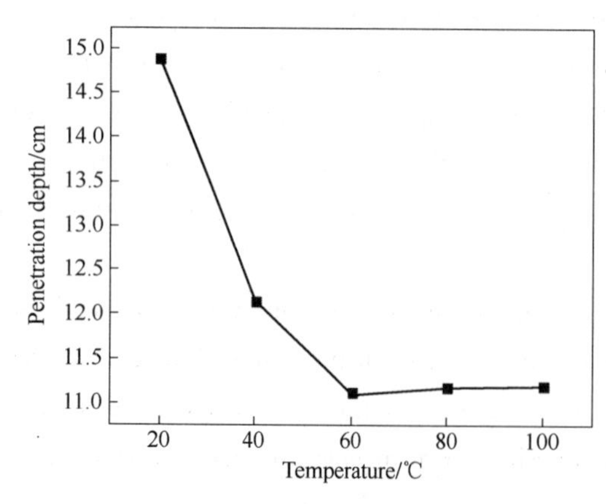

Fig. 10 Variation of penetration depth of silica sand (5% moisture) with temperature increase at 2.45GHz

Fig. 10 depicts changes in penetration depth of silica sand with 5% moisture content from 20℃ to 100℃. An increase in temperature results in decrease of penetration depth between room temperature and boiling point of water. The penetration depth reduced from 14.88cm to 11.10cm corresponding to the increase in temperature from 20℃ to 100℃. The reduction in penetration depth is significant only until a temperature of 60℃. Limited penetration depth at high moisture content could result in non - uniform heating of silica sand restricting the thickness of the bed.

For uniform drying with dielectric heating, the thickness of silica sand should not be more than two or three times the penetration depth[42]. Moisture content has more significant effects on dielectric properties of the sample than temperature. Considering penetration depth of silica sand, thickness for microwave drying could be between 30cm and 50cm at 2.45GHz.

4 Conclusions

The effects of moisture content and temperature on the dielectric property of silica sand were investigated by using openended senor dielectric measurement system at the frequency of 2.45GHz.

Moisture content is the major influence factor contributing to the variation of dielectric properties. Dielectric constant, loss factor and loss tangent all increase linearly with increase in moisture content. Three empirical predictive models were developed to relate the dielectric constant, loss factor and loss tangent of silica sand with the moisture content.

Temperatures between 20℃ to 100℃ were found to have a positive effect on dielectric constant and loss factor. A positive influence on loss factor contribute to the non - uniformity in microwave heating.

The penetration depth decreased with increase in moisture content and temperature. The penetration depth was found to reduce with an increase in the moisture content. A predictive model was developed to calculate penetration depth for silica sand with moisture content.

The dielectric properties model could be applied to predict microwave drying characteristics as a continuous function of changing moisture content.

Acknowledgements

The authors acknowledge support from foundations of the National Technology Research and Development Program of China (863Program) (No. 2013AA064003), the National Natural Science Foundation of China Project (No. 5114703), the International S&T Cooperation Program of China (No. 2012DFA70570) and a grant of dielectric properties measurements supported by Professor Huang at Sichuan University.

References

[1] Mostaghel S, Samuelsson C. Metallurgical use of glass fractions from waste electric and electronic equipment (WEEE) [J]. Waste Management, 2010, 30(1): 140 - 144.

[2] Therdthai N, Zhou W, Pattanapa K. Microwave vacuum drying of osmotically dehydrated mandarin cv. (Sai - Namphaung) [J]. International Journal of Food Science & Technology, 2011, 46(11): 2401 - 2407.

[3] Sharma G, Prasad S. Drying of garlic (allium sativum) cloves by microwave - hot air combination [J]. Journal of Food Engineering, 2001, 50(2): 99 - 105.

[4] Pickles C. Microwaves in extractive metallurgy: Part 2—A review of applications [J]. Minerals Engineering, 2009, 22(13): 1112 - 1118.

[5] Guo W, Wang S, Tiwari G, Johnson J A, Tang J. Temperature and moisture dependent dielectric properties of

legume flours associated with dielectric heating[C]. In American Society of Agricultural and Biological Engineers Annual International Meeting, American Society of Agricultural and Biological Engineers, June 24, 2009, Reno, NV, United states. 1599 - 1608.

[6] Haque K E. Microwave energy for mineral treatment processes—a brief review[J]. International Journal of Mineral Processing, 1999, 57(1): 1 - 24.

[7] Nelson S O. Agricultural applications of dielectric measurements[J]. IEEE Transactions on Dielectrics and Electrical Insulation, 2006, 13(4): 688 - 702.

[8] Wang W, Chen G. Theoretical study on microwave freeze - drying of an aqueous pharmaceutical excipient with the aid of dielectric material[J]. Drying Technology, 2005, 23(9 - 11): 2147 - 2168.

[9] Al - Harahsheh M, Kingman S, Saeid A, et al. Dielectric properties of Jordanian oil shales[J]. Fuel Processing Technology, 2009, 90(10): 1259 - 1264.

[10] Dev S R S, Gariepy Y, Orsat V, Raghavan G S V. FDTD modeling and simulation of microwave heating of in - shell eggs[J]. Progress in Electromagnetics Research M, 2010, 13: 229 - 243.

[11] McLoughlin C M, McMinn W A M, Magee T R A. Physical and dielectric properties of pharmaceutical powders [J]. Powder Technology, 2003, 134(1 - 2): 40 - 51.

[12] Nelson S O, Trabeisi S. Influence of water content on RF and microwave dielectric behavior of foods[J]. Journal of Microwave Power and Electromagnetic Energy, 2009, 43(2): 65 - 70.

[13] Wang S, Monzon M, Gazit Y, et al. Temperature - dependent dielectric properties of selected subtropical and tropical fruits and associated insect pests[J]. Transactions of the American Society of Agricultural Engineers, 2005, 48(5): 1873 - 1881.

[14] Hu L, Toyoda K, Ihara I. Dielectric properties of edible oils and fatty acids as a function of frequency, temperature, moisture and composition[J]. Journal of Food Engineering, 2008, 88(2): 151 - 158.

[15] Muley P D, Boldor D. Investigation of microwave dielectric properties of biodiesel components[J]. Bioresource Technology, 2013, 127(0): 165 - 174.

[16] Torres F, Jecko B. Complete FDTD analysis of microwave heating processes in frequency - dependent and temperature - dependent media[J]. Microwave Theory and Techniques, IEEE Transactions, 1997, 45(1): 108 - 117.

[17] Kraus M, Kopinke F D, Roland U. Influence of moisture content and temperature on the dielectric permittivity of zeolite NaY[J]. Physical Chemistry Chemical Physics, 2011, 13(9): 4119 - 4125.

[18] Nelson S O. Measurement and use of dielectric properties of agricultural products[J]. IEEE Instrumentation and Measurement Technology Conference, 1991, 14 - 16: 636 - 640.

[19] Nelson S O, Trabelsi S. Dielectric properties of agricultural products and applications[C]. American Society of Agricultural and Biological Engineers Annual International Meeting, June 21 - 24, 2009: 2901 - 2919.

[20] Martín - Esparza M E, Martínez - Navarrete N, Chiralt A, Fito P. Dielectric behavior of apple (var. Granny Smith) at different moisture contents[J]. Journal of Food Engineering, 2006, 77(1): 51 - 56.

[21] Navarrete A, Mato R B, Cocero M J. A predictive approach in modeling and simulation of heat and mass transfer during microwave heating[J]. Application to SFME of Essential Oil of Lavandin Super. Chemical Engineering Science, 2012, 68(1): 192 - 201.

[22] Okiror G P, Jones C L. Effect of temperature on the dielectric properties of low acyl gellan gel[J]. Journal of Food Engineering, 2012, 113(1): 151 - 155.

[23] Ratanadecho P, Aoki K, Akahori M. A numerical and experimental investigation of the modeling of microwave heating for liquid layers using a rectangular wave guide (effects of natural convection and dielectric properties) [J]. Applied Mathematical Modelling, 2002, 26(3): 449 - 472.

[24] Jiao S, Johnson J A, Tang J, Tiwari G, Wang S. Dielectric properties of cowpea weevil, black - eyed peas and mung beans with respect to the development of radio frequency heat treatments[J]. Biosystems Engineering, 2011, 108(3): 280 - 291.

[25] Liao X, Raghavan G S V, Yaylayan V A. Dielectric properties of aqueous solutions of - D - glucose at 915MHz[J]. Journal of Molecular Liquids, 2002, 100(3): 199 - 205.

[26] Nelson S O. Measurement and applications of dielectric properties of agricultural products[J]. IEEE Transactions on Instrumentation and Measurement, 1992, 41(1): 116 - 122.

[27] Thomas Z M, Zahn M. Dielectrometry measurements of moisture diffusion and temperature dynamics in oil impregnated paper insulated electric power cables[C]. Conference Record of the 2008 IEEE International Symposium on Electrical Insulation, 2008, 1 - 2: 539 - 542.

[28] Liu Y, Tang J, Mao Z. Analysis of bread dielectric properties using mixture equations[J]. Journal of Food Engineering, 2009, 93(1): 72 - 79.

[29] Mangalaraja R V, Ananthakumar S, Manohar P, et al. Microwave - flash combustion synthesis of $Ni_{0.8}Zn_{0.2}$ - Fe_2O_4 and its dielectric characterization[J]. Materials Letters, 2004, 58(10): 1593 - 1596.

[30] Hussain A, Li Z, Ramanah D R, et al. Microwave drying of ginger by online aroma monitoring[J]. Drying Technology, 2010, 28(1): 42 - 48.

[31] Therdthai N, Zhou W. Characterization of microwave vacuum drying and hot air drying of mint leaves(Mentha cordifolia Opiz ex Fresen)[J]. Journal of Food Engineering, 2009, 91(3): 482 - 489.

[32] Durance T, Yaghmaee P. Microwave Dehydration of Food and Food Ingredients[M]. In Comprehensive Biotechnology(Second Edition) Academic Press, Burlington. 2011: 617 - 628.

[33] Nelson S O. Dielectric spectroscopy for agricultural applications[C]//Proceedings of the 21st IEEE Instrumentation and Measurement Technology Conference, IMTC/04, May 18 - 20, Institute of Electrical and Electronics Engineers Inc., Como, Italy. 2004: 752 - 754.

[34] Idris A, Khalid K, Omar W. Drying of silica sludge using microwave heating[J]. Applied Thermal Engineering, 2004, 24(5 - 6): 905 - 918.

[35] Li X, Zhang W, Fu X. Epidemiological characteristics of varicella cases reported in Kunming Yanan Hospital from 2007 to 2009[J]. The Chinese Journal of Dermatovenereology, 2011, 3: 016.

[36] Andrés A, Bilbao C, Fito P. Drying kinetics of apple cylinders under combined hot air - microwave dehydration [J]. Journal of Food Engineering, 2004, 63(1): 71 - 78.

[37] Metaxas A, Meredith R J. Industrial microwave heating[M]. Inst of Engineering & Technology, 1983.

[38] Clark D E, Folz D C, West J K. Processing materials with microwave energy[J]. Materials Science and Engineering: A, 2000, 287(2): 153 - 158.

[39] Perelaer J, de Gans B J, Schubert U S. Ink - jet printing and microwave sintering of conductive silver tracks [J]. Advanced materials, 2006, 18(16): 2101 - 2104.

[40] Decareau R V, Mudgett R E. Microwaves in the food processing industry[M]. Food Science and Technology, USA, 1985.

[41] Liu Y, Tang J, Mao Z. Analysis of bread loss factor using modified Debye equations[J]. Journal of Food Engineering, 2009, 93(4): 453 - 459.

[42] Schiffmann R F. Commercializing microwave systems: paths to success or failure[J]. Ceramic Transactions, 1995, 59: 7 - 16.

Dielectric Properties and Optimization of Parameters for Microwave Drying of Petroleum Coke Using Response Surface Methodology

Chenhui Liu, Libo Zhang, C. Srinivasakannan, Jinhui Peng, Bingguo Liu, Hongying Xia

Abstract: The dielectric properties of petroleum coke at five temperatures between 20℃ to 100℃, covering different moisture content levels at 2.45GHz was measured using an open - ended coaxial probe dielectric measurement system. The effects of drying temperature, drying time and sample mass on the moisture content and the dehydration rate of petroleum coke was studied by the response surface methodology. Results show that dielectric constant, loss factor and loss tangent all increase nearly linearly with increasing moisture content. Three predictive empirical models were developed to relate loss factor, loss tangent depth of petroleum coke as a linear function of moisture content from 3% - 10%. An increase in temperature between 20℃ to 100℃ was found to increase the dielectric properties. The penetration depth increases with decrease in moisture content and temperature. Variation in penetration depth was found to vary linearly with a decrease in moisture content range of 3% to 10%. A predictive empirical model was developed to calculate penetration depth for petroleum coke. Two mathematical models established and analysed by RSM were adequate to describe the relationship between the microwave drying conditions and the responses of moisture content and dehydration rate. Statistical analysis with response surface regression shows that microwave drying temperature, drying time and sample mass were significantly correlated with moisture content and dehydration rate. Based on the RSM analysis, the optimum values of the process variables were obtained as: microwave drying temperature 75℃, drying time 10s, and sample mass 60g, at these conditions, the moisture content was 0.34, dehydration rate was 2.94.

Keywords: dielectric properties; microwave drying; response surface methodology; petroleum coke

1 Introduction

In the titanium industry, high titanium slag mainly as the raw material in fluidized chlorination process to prepare titanium tetrachloride ($TiCl_4$), further to produce titanium dioxide pigment and titanium sponge[1]. The fluidized chloride process involves high titanium slag powders are mixed with petroleum coke to produce titanium tetrachloride ($TiCl_4$) at high temperatures[2]. Petroleum coke in the process of storage picks up moisture from the atmosphere, as a result having moisture content in the range of 0.8% - 2%. Moisture content in the petroleum coke should not be more than 0.5%, since water tends to react with chlorine forming hydrogen chloride during the chlorination process. Petroleum coke must be dried to prevent severe corrosion of equipments and hence it is desirable to reduce the moisture content less than 0.3% [3].

The most common methods for drying petroleum coke in metallurgical industries are hot air convective drying, fluidized bed drying and rotary kiln drying[4]. Although these methods are simple, the drawbacks include inability to handle large quantities to achieve consistent quality standards, contamination, long duration, low energy efficiency and high cost[5]. Microwave drying has established itself to be an energy - efficient process compared with conventional drying[6]. The major advantages of microwave drying include rapid heat transfer, volumetric and selective heating, compactness of equipment, quick start/stop and pollution - free[7]. Microwave drying has been recently applied in processing of iron oxide pelletes and nickeliferous limonitic laterite ores, ilmenite concentrate, coals and other minerals[8].

Though the application of microwave drying in metallurgical industry is promising, it demands a clear understanding of the microwave - mineral interaction and its mechanism, in order to design and develop industrial scale equipments for minerals processing and ore treatment using microwave energy[9].

Microwave drying involves several complicated physical processes, including the absorption of electromagnetic energy that transforms into heat energy and transportation of the heat thus generated. A better understanding of the interaction between microwave and the material is essential for using it more efficiently and effectively. The most important material property that governs microwave heating is the permittivity of the material, $\varepsilon = \varepsilon' - i\varepsilon''$, where the dielectric constant ε', and loss factor ε'', are the dielectric properties[10]. The dielectric constant is a measure of its ability to store electric energy, while loss factor is a measure of the ability to convert electric energy into heat.

Dielectric property is function of temperature and moisture content, which is an integral part in simulating the temperature distribution and absorbance of microwave power density inside the minerals[11]. Dielectric properties of a material also determine penetration of microwave power and local power absorption rates. The change in dielectric properties during microwave drying alters the penetration of radiation, causing uneven temperature distribution. In order to understand these characteristics and to provide guidelines for the design of microwave drying systems, it is mandatory to understand the influence of temperature and moisture content in the complex permittivity of high titanium slag, as these parameters continuously change during the drying process.

The dielectric properties of foods, pharmaceutical powders and coals have been reported for a wide range of moisture contents[12,13]. However, the effect of variation in the moisture content on the microwave absorbance characteristics of petroleum coke is seldom reported. To the knowledge of authors, the dielectric properties of wet petroleum coke that are utilized as industrial raw material, has not been reported in open literature.

For a microwave drying system, there is a need to optimize the drying parameters such as drying temperature, drying time, mass weight for minimizing the moisture content and maximizing the dehydration rate. A statistical experimental design based on response surface methodology (RSM) can be an effective method for optimizing microwave drying parameters[14,15]. The RSM combines mathematics with statistics to model and analyze, where several factors control a response of interest[16]. Several previous researchers have proved that RSM was a powerful statistical tool in optimizing drying parameters of products such as foods[17], woods[18], agricultural products[19] and pharmaceutical powder[20]. However, literature reports pertaining to the optimization of microwave drying of petroleum coke using the RSM approach is not available.

The present work, attempts to assess and optimize the effect of process parameters such as the drying temperature, duration of drying and sample mass on the final moisture content and the drying rate. A two parameter optimization was performed as it is desirable to have low final moisture content along with high drying rate. The drying rate is known to decrease with decrease in moisture content and hence it is necessary to identify the optimal combination dielectric properties of petroleum coke. A three - level, three variables Box - Behnken Design (BBD) was established to optimize the three variables on the two responses.

2 Materials and methods

2.1 Materials

Petroleum coke was obtained from a titanium factory in Yunnan province of China. The moisture content of the sample was measured using a standard hot - air oven at 105℃ by retaining the samples for 24h.

The moisture content and dehydration rate of petroleum coke were calculated using the following equations:

$$W_t = \frac{m_t - m_g}{m_t} \times 100\% \tag{1}$$

where, W_t is the moisture content at the time of t, min; m_t is the weight of sample at the time of t, g; m_g is the weight of dried sample, g.

$$V = \frac{\Delta m}{\Delta t} \tag{2}$$

where, V is dehydration rate, g/min; Δm is water loss amount of sample during Δt time, g; Δt is microwave drying time, min.

2.2 Dielectric property measurement system

In this work, a hybrid experimental/computational permittivity measuring system developed by the Institute of Applied Electromagnetics at Sichuan University, was used to determine the complex permittivity of the petroleum coke at different temperatures[21]. The schematic of this system is shown in Fig. 1.

In the experiment, sample powder was sealed in a resonant cavity (80mm inner diameter with 100mm length) made of stainless steel (sample holder) and heated by an electric heater placed inside the holder cavity. An open ended coaxial sensor (see Fig. 2) connected to the Agilent PNA5230 Network Analyzer was used to measure the reflected signals. The probe is placed on the flat surface of a solid sample or inserted into the sample powder for a full contact. The electromagnetic fields around the probe change as the probe contacts with the specimen. The probe receives reflected signals from the sample which contains the information related to the complex dielectric

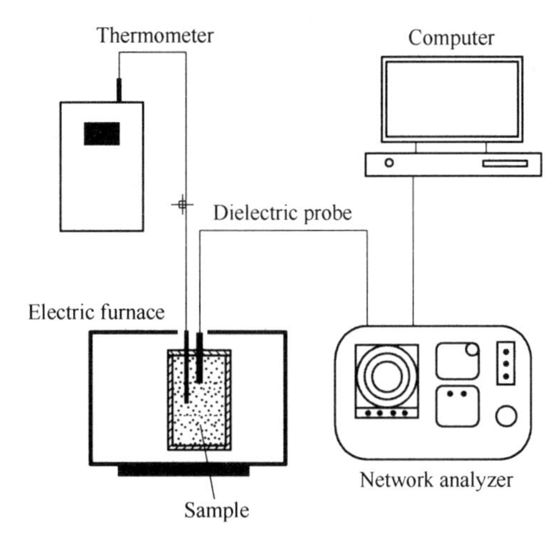

Fig. 1 The schematic of permittivity measurement system

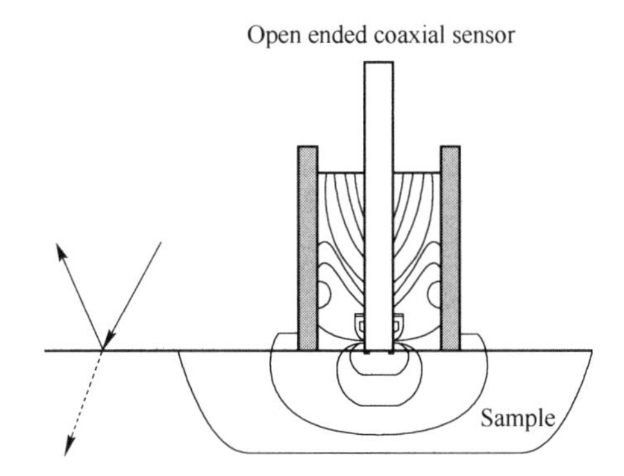

Fig. 2 The schematic of the open – ended coaxial sensor

permittivity of the measured sample, the reflected signals were measured by the probe and recorded by the analyzer. A thermocouple was used for temperature measurements. The method of genetic algorithm (GA) – based inverse – calculation, was based on the experimental/computation reflection coefficient, was used to calculate the complex permittivity of measuring sample[22].

Permittivity measurements were made at the frequency of 2.45GHz. The complex permittivity of the petroleum coke sample with different moisture content (0 – 10%) was measured at 20℃, 40℃, 60℃, 80℃ and 100℃.

2.3 Microwave drying equipment and drying experiment

Microwave drying equipment was made by the Key Laboratory of Unconventional Metallurgy, Ministry of Education, Kunming University of Science and Technology. It has the ability to alter the power intensity in the range of 0 – 3000W at the frequency of 2.45GHz (the wavelength of 12.24cm). The dimensions of microwave cavity are 215mm by 350mm by 330mm. A fan located in the oven could be used to entrain water vapor out of the chamber during the process of drying. The two magnetrons were equipped with a water – cooled condenser.

The schematic diagram of the microwave heating equipment is shown in Fig. 3. Weight measurements were made by the means of an electronic balance attached to the top of the drying cavity. This balance was connected to a computer to record the measurements continuously. A thermocouple pyrometer was used to monitor the temperature. Experiments were conducted by placing the sample in a glass plate at the center of the oven. Variance in the experimental measurements were based on the repeat runs as suggested by the software at the center point of all the three variables.

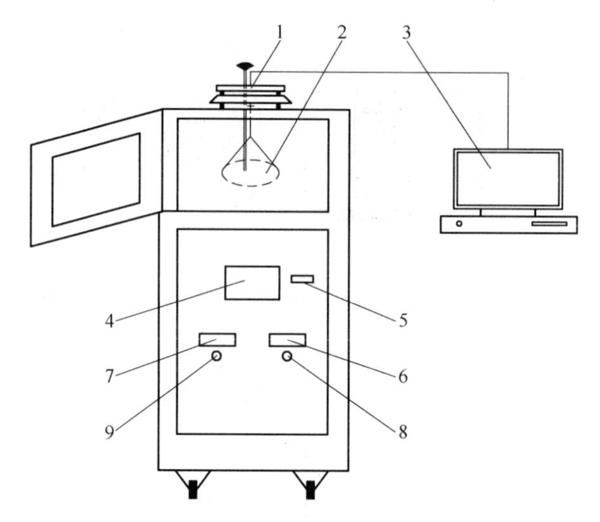

Fig. 3 The schematic of microwave drying system

1—electronic balance;2—container;3—computer system;4—temperature display instrument;
5—power controller;6—ampere meter;7—voltmeter;8—start swith;9—stop swith

2.4 Design of response surface methodology

Design Expert Software(version 7.1.5,from Stat - Ease Inc.,USA) was used for the experimental design,model building and data analysis. Microwave drying temperature(χ_1),time(χ_2) and sample mass(χ_3) at three levels,were chosen as three independent variables and the response variables were final moisture content(Y_1) and the dehydration rate(Y_2). A three - level experimental design proposed by BBD was applied for response surface fitting,with a design matrix of 15 trials including three replicates at the central point(Table 1).

Table 1 Independent variables and their levels used for Box - Behnken design

Variables	Code	Coded levels		
		-1	0	1
Microwave drying temperature/℃	χ_1	50	70	90
Time/s	χ_2	10	30	50
Sample mass/g	χ_3	30	50	70

The quadratic equation for predicting the optimal conditions can be expressed according to Eq. (3)[23]:

$$Y = \beta_0 + \sum_{i=1}^{n} \beta_i \chi_i + \sum_{i=1}^{n} \beta_{ii} \chi_i^2 + \sum_{i<j} \beta_{ij} \chi_i \chi_j \qquad (3)$$

where Y is the predict response values; β_0 is a constant coefficient; β_i, β_{ii}, β_{ij} are the linear, quadratic and ij th interaction coefficients, respectively; χ_i, χ_j are independent variables. The analyses of the present work, attempts to assess and optimize the effect of process parameters such as the drying temperature, duration of drying and sample mass on the final moisture content and the drying rate. A two parameter optimization was performed as it is desirable to have low final moisture content along with high drying rate. The drying rate is known to decrease with decrease in moisture content and hence it is necessary to identify the optimal combination dielectric properties of petroleum coke. A three - level, three variables Box - Behnken Design (BBD) was established to optimize the three variables on the two responses.

Variance (ANOVA) and response surfaces were performed using the Design Expert Software. Optimization of microwave drying conditions for petroleum coke was obtained using the software's numerical and graphical optimization functions.

3 Results and discussions

3.1 The effects of temperature and moisture content on dielectric properties of petrol coke

3.1.1 The effects of moisture content on dielectric properties

The effects of moisture content on dielectric properties of petroleum coke are depicted in Figs. 4 - 6.

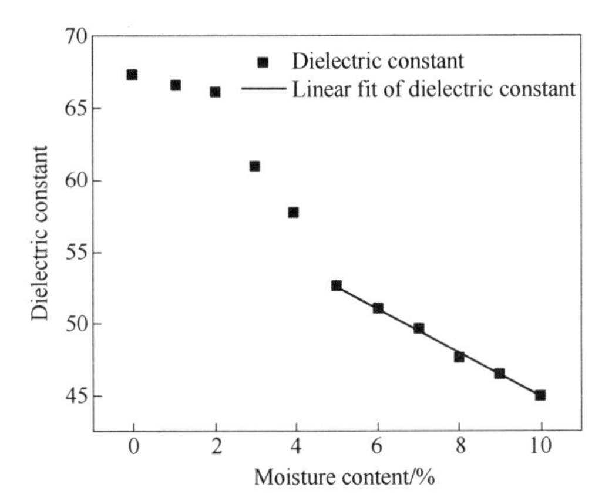

Fig. 4 Variation of dielectric constant of petroleum coke with moisture content

At room temperature, the dielectric constant was found to decrease from 67.38 to 44.915 for a corresponding increase in moisture content from 0 to 10%. The minimum value corresponds to bone dry sample while the maximum value corresponds to the sample with 10% moisture content. These results has the opposite trend with previous reported trends for materials such as foods, woods, clays and soils[24-27]. The variation in dielectric constant with moisture content from 5% to 10% was found to be linear and of regression coefficients are provided in Table 2 along with the coefficients of determination R^2. The high R^2 value justifies the linear relationship, while the regressed equation can be utilized to estimate the dielectric constant of the petroleum coke with moisture content from 5% to 10%.

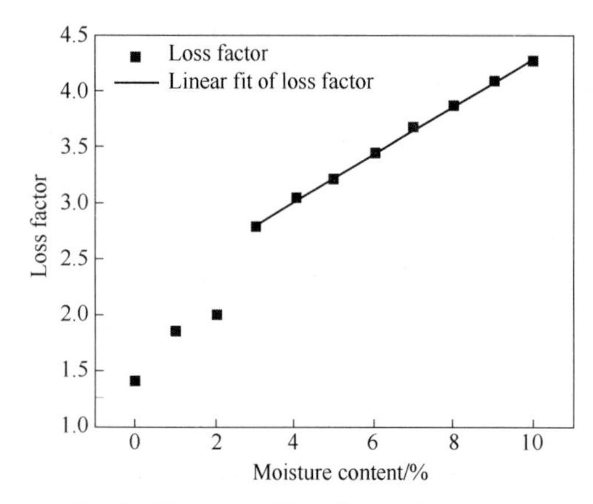

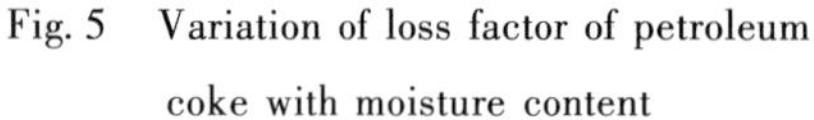
Fig. 5 Variation of loss factor of petroleum coke with moisture content

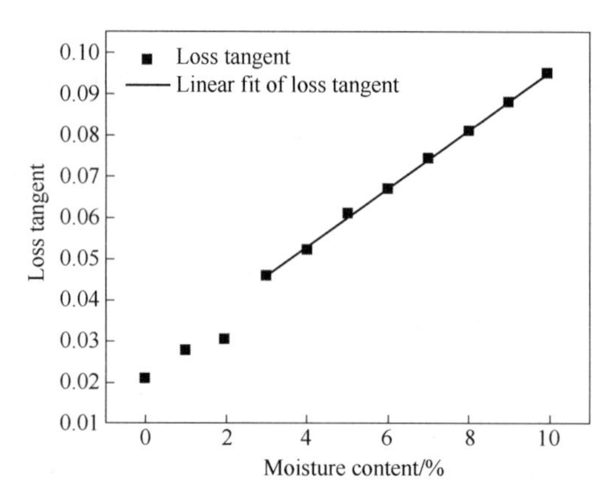

Fig. 6 Variation of dielectric loss tangent $\tan\delta$ of petroleum coke as a linear function of moisture content

Table 2 Regression equations on moisture contents and dielectric properties of petroleum coke

Dielectric properties	Regression equations	Moisture range	Correlation efficient(R^2)
Dielectric constant	$\varepsilon' = 60.30818 - 1.54733M$	5% - 10%	0.9962
Loss factor	$\varepsilon'' = 65.64591 - 2.19815M$	3% - 10%	0.9302
Loss tangent	$\tan\delta = 0.02493 + 0.0073M$	3% - 10%	0.99929
Penetration depth	$D_p = 8.21928 - 0.3347M$	3% - 10%	0.95882

Figs. 5 and 6 illustrate the loss factor and loss tangent of petroleum coke vary with moisture content. The loss factor and loss tangent both show the quasilinear dependence on moisture content from 3% to 10% at 2.45GHz. The increases imply that the loss factor and loss tangent present an enhanced response to increasing moisture content.

The correlation coefficients of linear regression(R^2) were 0.930 and 0.999, respectively (Table 2). These high R^2 - values imply that these linear models can be used for moisture content prediction of petroleum coke between 3% - 10% moisture content. The results of these nearly linear variations with moisture content agree well with previous studies on tomato and pomace[28]. Earlier researches on clays, soils and foods have also shown a linear relationship between moisture contents and dielectric parameters. Differences of dielectric properties between different materials were embodied by various slopes and intercepts on the linear lines[29,30].

Based on the relationship between dielectric property of a material and its moisture content, several new technologies have been developed for measuring moisture content of materials. The time domain reflectometry (TDR) has been widely used in measuring soil moisture due to its advantages of high accuracy and stability[31]. Moschler and Hanson[32] developed parallel resonant sensors to measure moisture content of hardwood lumbers, and the results show that the sensor working at 4.5GHz to 6GHz displays a linear response to moisture content over a range of 6% - 100%. A complete understanding of dielectric properties of materials will be useful in developing new sensors for measuring moisture contents of the material more accurately in a wider range of moisture contents.

The dielectric property of water within mixtures is not the same as that of pure water because the ion content of water within mixtures is different. The exposure water to petroleum coke, promotes the ions contribute to water, which form an ionic halo around the particles. These ions contribute to the electrical conduction and turn back to the petroleum coke particle when the water is removed[33]. Interaction may occur between the solid and the solvent, which causes the dielectric property of the mixture to be lower or higher than the sum of the free moisture plus solid[21].

In microwave drying, the absorbed energy is proportional to the loss factor, and is mainly used to evaporate water promoting drying. The ability of absorbing energy is of petroleum coke decreases with decrease in moisture at 2.45GHz. Hence the drying rate of petroleum coke is faster at the beginning than at the ending of drying period, due to reduction in moisture during microwave drying process. The advantage of microwave drying system is attributed to the uniform moisture content of the entire sample[5].

3.1.2 The effects of temperature on dielectric properties

The temperature dependence of dielectric properties of petroleum coke with 10% moisture content over the temperature range from 20℃ to 100℃ are presented in Figs. 7 – 9.

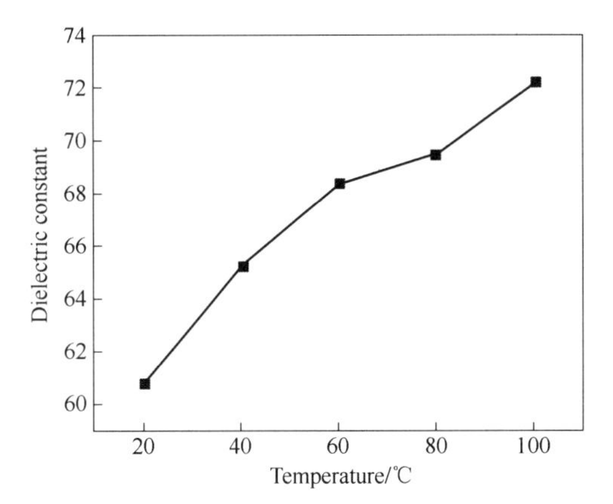

Fig. 7 Dielectric constant of petroleum coke with 10% moisture varies with temperature

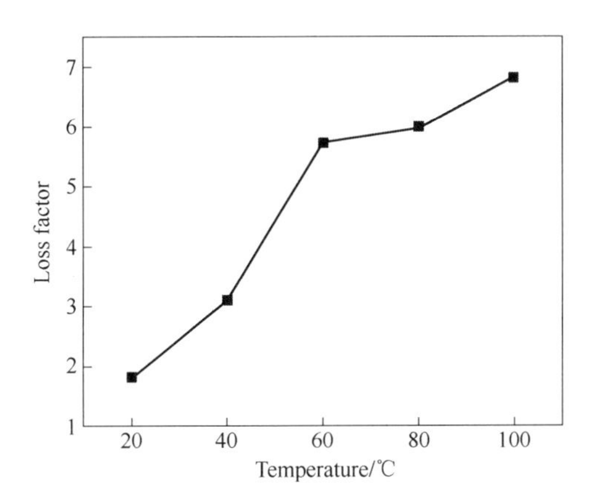

Fig. 8 Loss factor of petroleum coke with 10% moisture varies with temperature

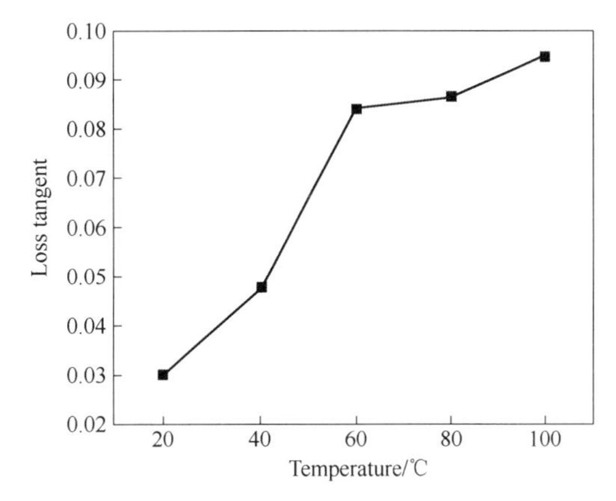

Fig. 9 Loss tangent of petroleum coke with 10% moisture varies with temperature

Fig. 7 shows that, the dielectric constant of petroleum coke increases almost linearly with temperature from 20℃ to 100℃, at 10% moisture content of the sample. The dielectric constant was found to increase from 60.76 to 72.15 at the rate of 0.11/℃ in the temperature range covered in the present work. The result shows that the effect of temperature on dielectric constant is as dominant as moisture content. An increase in dielectric constant of minerals with temperature was also reported for food products[34,35]. The temperature dependence of water in terms of ε' and ε'' has both been reported to decrease with increase in temperature from 20℃ to 100℃[36]. However, the temperature dependence behavior of dielectric constant of petroleum coke with 10% water content is not similar to that of water at 20 – 100℃. Contrary to pure water the dielectric constant of petroleum coke keeps increasing, which can be attributed to the variation of dielectric polarization of petroleum coke which is much more than water.

The effects of temperature on loss factor and loss tangent of petroleum coke with 10% moisture content are shown in Figs. 8 and 9. The mixture at higher temperatures has larger loss factors and loss tangents than at lower temperatures. Water will absorb much more energy, and reach a higher temperature than the mixture under the same treatment time due to water has a larger loss factor compared with the mixture at the same temperature. This enables water to reach boiling point while petroleum coke still at lower temperatures. That is the advantages of selective heating with microwave drying.

For pure petroleum coke, the loss factor increase with increasing temperature, while for pure water, the loss factor decrease with increasing temperature. At 10% moisture content, most water is free water with big mobility at room temperature to respond to microwave field at 2.45GHz. Raising temperature increases the mobility of water molecules and ionic conductivity, as indicated by an increase in loss factor in petroleum coke with temperature[37]. The loss factor is the sum of two components: ionic, ε''_{σ}, and dipole, ε''_{d} loss. These two components respond oppositely to temperature. The dipole loss decreases and ionic loss increases with temperature. The increasing trend of ε'' with temperature can be attributed to the possibility of runaway effects at lower moisture ranges[38].

Heating efficiency and non – uniformity are important factors in developing microwave drying techniques of minerals and ores[39]. The measured results obtained in this study show both dielectric constant and loss factor of petroleum coke increase with temperature and moisture content at 2.45GHz. Temperature rise leads to higher loss factor which in turn increasing the rate of temperature rise. When heated in a multimode microwave cavity, moist petroleum coke may experience thermal runaway and hot spots effects.

The dielectric behavior of the petroleum coke illustrated in this paper is highly influenced by its water content and temperature in microwave drying process.

3.1.3 Relationship between penetration depth and moisture content

The power penetration depth is defined as the depth at which the strength of microwave field is reduced to 1/e of its surface value and is expressed with the following equation[4]:

$$D_p = \frac{\lambda_0}{2\sqrt{2}\pi\sqrt{\varepsilon'\left[\sqrt{1+\left(\frac{\varepsilon''}{\varepsilon'}\right)^2}-1\right]}} \tag{4}$$

where, D_p is the penetration depth; λ_0 is the free space wavelength; ε' is the dielectric constant; ε'' is the loss factor; D_p is a important parameter in characterizing temperature distribution of materials in microwave heating process.

The power penetration depths of the petroleum coke particles at different moisture contents were computed using Eq. (4). Fig. 10 depicts the effects of moisture content on power penetration depth at 2.45GHz. The penetration depth of petroleum coke is found to vary from 11.4cm to 3.06cm at 2.45GHz. From previous research we know that the power penetration depth of pure water at room temperature is 1.68cm at 2.45GHz[40] and that of dry petroleum coke particles is 11.4cm. The penetration depth of wet petroleum coke particles is smaller than that of dry petroleum coke and lager than that of pure water. The power penetration depth of petroleum coke particles increases nearly linearly with the decrease in moisture content from 10% to 3% while it has a sharp increase from 3% moisture content to dried petroleum coke. This slow increase with the reduction in moisture is different from that in the case of apple in which a sharp change happens at the moisture content of 30%[41]. Regression analysis reveals that the relationship between the power penetration depth of wet petroleum coke particles and moisture content between 3% to 10% can be expressed by a linear equation in Table 2. The R – square(R^2) value is 0.9589. A high correlation value and low deviations indicate that this equation can be used for predicting penetration depth of petroleum coke with moisture content from 3% to 10%.

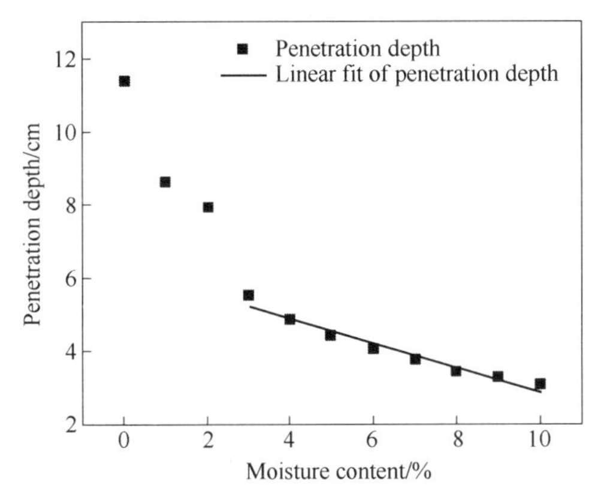

Fig. 10 Effects of moisture content on power penetration depth of petroleum coke at 2.45GHz

Fig. 11 depicts variations of penetration depth of with 10% moisture content from 20℃ to 100℃. An increase in temperature from 20℃ to 100℃ contributes to a reduction in the penetration from 8.44cm to 2.42cm. Limited penetration depth led to non – uniform heating characteristics in high titanium slag with high moisture content and large thickness. It is favorable to have low moisture content and low temperature drying in order to have high penetration depth, which will facilitate larger drying bed thickness. Bed thickness larger than the penetration depth will result in non uniform heating and hence the bed thickness needs to be limited less than the penetration depth of the microwave.

Considering moisture content and temperature dependence of dielectric properties of petroleum

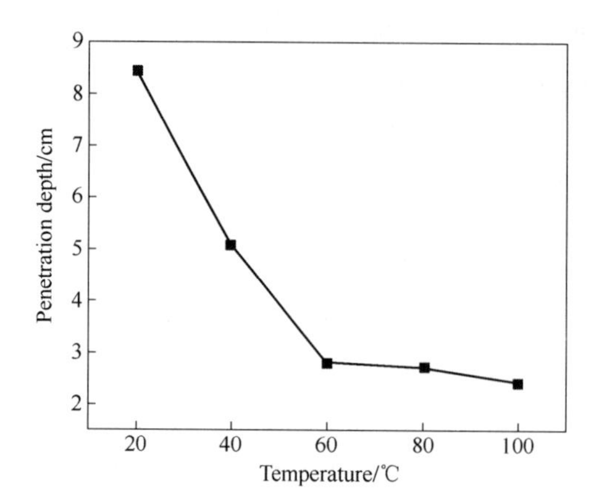

Fig. 11 Effects of temperature on power penetration depth of petroleum coke at 2.45GHz

coke at 2.45GHz, to obtain effective microwave drying, the material thickness should be within the penetration depth of 3cm, for an effective microwave drying.

3.2 Optimization of microwave drying parameters

The experimental data of various responses during MWD of petrol coke are presented in Table 3.

Table 3 Experimental design martrix and results

Number	Variables			Moisture content	Dehydration rate
	Drying temperature χ_1/℃	Drying time χ_2/s	Sample mass χ_3/g	Y_1/%	Y_2/g · min^{-1}
1	50	10	50	0.49	2.01
2	90	10	50	0.19	2.91
3	50	50	50	0.4	0.46
4	90	50	50	0.11	0.63
5	50	30	30	0.31	0.51
6	90	30	30	0.05	0.67
7	50	30	70	0.63	0.74
8	90	30	70	0.28	1.23
9	70	10	30	0.21	1.71
10	70	50	30	0.12	0.37
11	70	10	70	0.45	2.98
12	70	50	70	0.35	0.68
13	70	30	50	0.27	0.89
14	70	30	50	0.24	0.92
15	70	30	50	0.30	0.86

3.2.1 Development of regression model

The design together with the response values from the experiments are shown in Table 3. Runs 13 – 15 at the center point are repeated to determine the experimental error. The moisture content and dehydration rate are utilized in the quadratic model suggested by the software. The polynomial regression equation is developed using Box – Behnken. The final empirical models in terms of coded factors for final moisture content(Y_1) and dehydration rate(Y_2) are shown in Eqs. (5) and (6), respectively.

$$Y_1 = 0.29 - 0.15\chi_1 - 0.045\chi_2 + 0.13\chi_3 \quad (5)$$

$$Y_2 = 0.89 + 0.21\chi_1 - 0.93\chi_2 + 0.30\chi_3 - 0.18\chi_1\chi_2 + 0.083\chi_1\chi_3 - 0.24\chi_2\chi_3 - 0.018\chi_1^2 + 0.63\chi_2^2 - 0.085\chi_3^2 \quad (6)$$

The suitability of model equation is evaluated using the correlation coefficients(R^2), which are 0.97 and 0.99, respectively. The proximity of R^2 value to unity, indicate the suitability of the model equation. Both of the R^2 values of final moisture content and dehydration rate are close to unity, indicating that there is a good agreement between the experimental and the predicated final moisture content as well as dehydration rate.

The results of the response surface models(Eq. (5)) in the form of ANOVA are given in Table 4. The Model F value of 128.86 and Prob > F less than 0.0001 prove that the model is significant. The value of model terms Prob > F less than 0.05 indicates that the model terms are significant. In this case, microwave drying temperature(χ_1), drying time(χ_2) and sample mass(χ_3) are significant model terms. Based on the F values(Table 4), χ_1 shows the largest F value of 213.29, indicating that it has the most significant effect on the final moisture content, compared to χ_2 and χ_3. The effect of microwave drying temperature on final moisture content is more significant than drying time and sample mass, with F value of 128.86 and 19.20, respectively.

Table 4 Analysis of variance(ANOVA) for response surface liner model for final moisture content

Source	Sum of squares squar	df	Mean Square	F value	Prob > F
Model	0.33	3	0.11	128.86	<0.0001
χ_1	0.18	1	0.18	213.29	<0.0001
χ_2	0.016	1	0.016	19.20	0.0011
χ_3	0.13	1	0.13	154.10	<0.0001
Residual	9.28×10^{-3}	11	8.44×10^{-4}	—	—

Note: $R^2 = 0.97$; $R^2_{adj} = 0.96$; adequate precision = 37.00(>4).

The ANOVA of the quadratic model of dehydration rate is shown in Table 5. An F value of 52.13 and Prob > F of 0.0002 prove that the model is significant. The three drying variables χ_1, χ_2, χ_3, and interaction parameters ($\chi_2\chi_3$, χ_2^2) are significant whereas the interactions trems($\chi_1\chi_2$, $\chi_1\chi_3$, χ_1^2, χ_3^2) are insignificant to the response.

Table 5 Analysis of variance(ANOVA) for response surface quadratic model for drying rate

Source	Sum of squares squar	df	Mean Square	F value	Prob > F
Model	9.98	9	1.11	52.13	0.0002
χ_1	0.37	1	0.37	17.38	0.0087
χ_2	6.98	1	6.98	327.85	<0.0001

Continues Table 5

Source	Sum of squares squar	df	Mean Square	F value	Prob > F
X_3	0.70	1	0.70	33.00	0.0022
X_1X_2	0.13	1	0.13	6.26	0.0543
X_1X_3	0.03	1	0.03	1.28	0.3093
X_2X_3	0.23	1	0.23	10.83	0.0217
X_1^2	1.13×10^{-3}	1	1.13×10^{-3}	0.05	0.8268
X_2^2	1.47	1	1.47	68.88	0.0004
X_3^2	0.03	1	0.03	1.25	0.3137
Residual	0.11	5	0.02	—	—

Note: $R^2 = 0.99$; $R^2_{adj} = 0.97$; adequate precision = 21.38 (>4).

3.2.2 Moisture content

Fig. 12 shows the comparison of the predicted values versus experimental final moisture content of petroleum coke. Actual values are the measured response data for a particular run, and the predicted values were evaluated from the model and generated by using the approximating functions[42]. As can be seen, the predicted values match well with the experimental values, indicating the ability of the model to successfully capture the correlation between the drying variables and the final moisture content. From the statistical results obtained, it is shown that the model is adequate to predict the moisture content within the reasonable range of the variables studied.

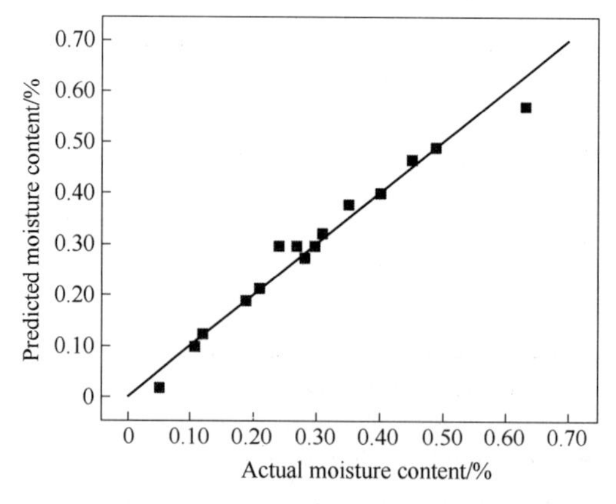

Fig. 12 Predicted values versus experimental values of final moisture content

Fig. 13 shows the three - dimensional response of the combined effects of microwave drying time and drying temperature on the moisture content of petroleum coke after drying, at a sample mass of 50g. As can be seen, the moisture content decreases with an increase in the drying temperature and drying time. The linear effect of both terms is clearly observed. When drying time is 10s at the sample weight of 50g, the moisture content is 0.49% at drying temperature of 50℃ while it is reduced to 0.19% at drying temperature of 90℃. The higher of the drying temperature the less moisture content of petroleum coke. At a certain drying temperature, the longer drying time, the more moisture was removed. Drying temperature has a more significant effect on moisture content compared with drying time during microwave drying petroleum coke.

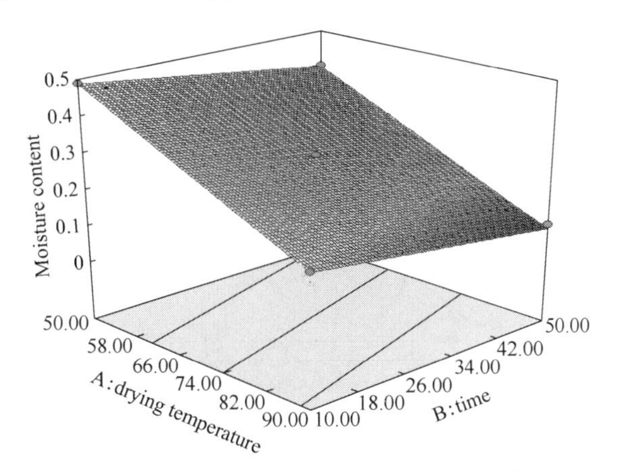

Fig. 13 Three - dimensional response surface plot of moisture content: effect of drying temperature and drying time on the moisture content(mass = 50g)

Fig. 14 reveals the interaction between sample weight and drying temperature on the moisture content of petroleum coke after drying. The linear effect of both terms is clearly observed. It can be noticed that by increasing the weight of the sample, the moisture content increases. This can be attributed to the higher amount of moisture trapped in the sample which needs to be removed. The higher the amount of moisture in petroleum the more energy and time are required to evaporate it.

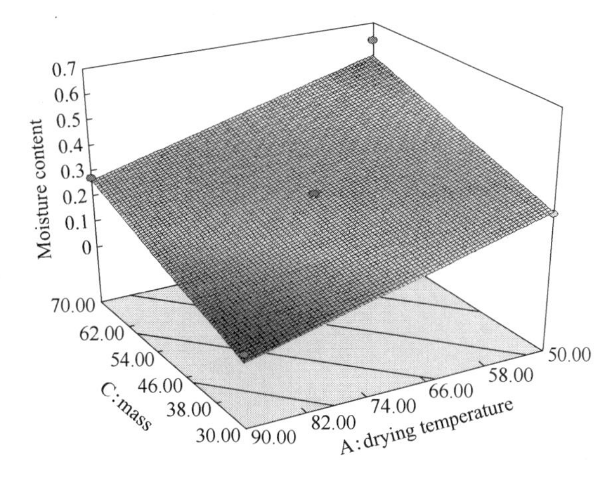

Fig. 14 Three - dimensional response surface plot of moisture content: effect of drying temperature and mass on the moisture content(drying time = 30s)

3. 2. 3 Dehydration rate

Fig. 15 shows the comparison of the predicted values versus dehydration rate of petroleum coke. As can be seen, the predicted values match well with the experimental values, indicating the ability of the model to successfully capture the correlation between the drying variables and the dehydration rate.

In a word, from the statistical results obtained, it is shown that the model is adequate to predict the dehydration rate within the reasonable range of the variables studied.

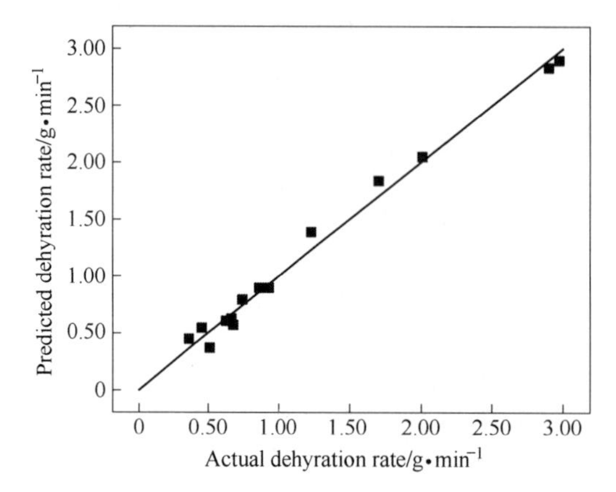

Fig. 15 Predicted values versus experimental values of dehydration rate

The 3D interactions between terms drying temperature and drying time on the dehydration rate are plotted in Fig. 16. In this figure, the second order effect of drying time and linear effect are clearly observed. Increasing the drying time led to the decrease in the dehydration rate while the increase in microwave drying temperature led to little increase in dehydration rate. The dehydration rate decreases with the increase of drying time while drying tempeature is insignificant. Under the same drying conditions of drying temperature (70℃) and sample mass (70g), the dehydration rate is 0.68g/min when the drying time is 50s while it reaches 2.98g/min when the drying time is 10s, the latter is 4.38 times of the former. Petroleun coke and water both have the strong microwave absorbing ability according to the dielectric property measurements. In the temperature increasing period of microwave drying process, sharply increase in temperature of sample lead to rapidly evaporation of water molecules. Dehydration rate has larger values at the temperature increase period of drying process. With drying time increasing, moisture content of petroleum coke decrease to a certain extent and water loss in per unit time decreases, leading to a decrease in dehydration rate.

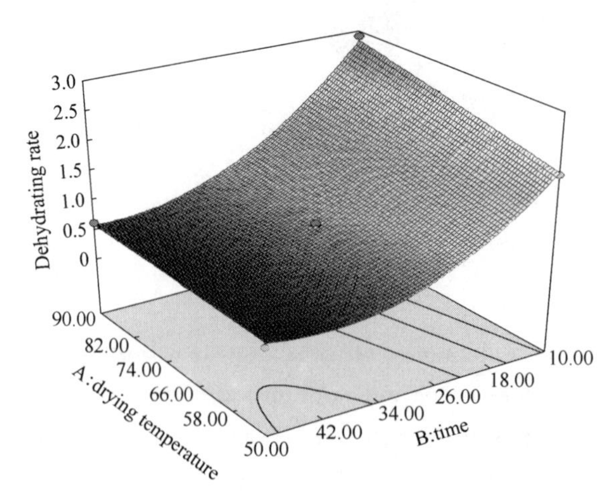

Fig. 16 Three - dimensional response plot of drying rate
(effect of dring temperature and time, weight = 50g)

Fig. 17 shows the three – dimensional response of the combined effects of mass and drying temperature on the dehydration rate of petroleum coke during microwave drying. As can be seen, the dehydration rate decreases with increase in the drying temperature and drying time. The nearly linear effects of both terms are clearly observed. When drying time is 10s at the drying temperature of 70℃, the dehydration rate is 0.45% at the mass weight of 70g, while it is reduced to 0.21% at mass weight of 30g. The higher of the mass weight the higher of dehydration rate of petroleum coke. Mass weight has more significant effect on dehydration rate compared with drying temperature during microwave drying petroleum coke.

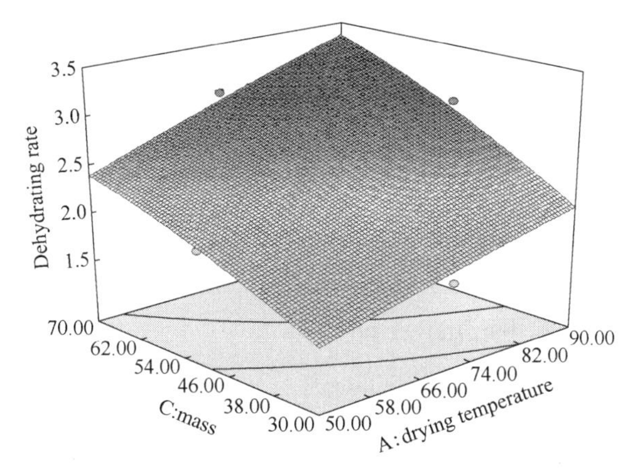

Fig. 17 Three – dimensional response plot of dehydration rate
(effect of drying temperature and mass, drying time = 10s)

3.2.4 Optimization by response surface modeling

In order to obtain the desired final moisture content, as well as the maximum dehydration rate of petroleum coke, the drying parameters were optimized by the Design Expert software. The most desirable experimental condition suggested by the software was selected to be verified. The experimental condition of microwave drying petroleum coke was given in Table 6. The predicted moisture content and dehydration rate are 0.36% and 2.81g/min while the experimental data are 0.34% and 2.94g/min, respectively. This experimental result agrees well with the prediction from the response surface, which indicates the success of the optimization process.

Table 6 Predicted and experimental values of the responses at optimum conditions

Drying temperature χ_1/℃	Time χ_2/s	Sample mass χ_3/g	Moisture Content/%		Dehydration rate/g · min^{-1}	
			Predicted	Experimental	Predicted	Experimental
75	10	60	0.36	0.34	2.81	2.94

4 Conclusions

The dielectric properties of petroleum coke at five temperatures between 20℃ to 100℃, covering different moisture content levels at 2.45GHz was measured with an open – ended coaxial probe dielectric measurement system. The results shows that dielectric constant, loss factor and loss tangent all increase nearly linearly with increasing moisture content. Three predictive empirical models were developed to relate loss factor, loss tangent depth of petroleum coke as a linear function of moisture content from 3% – 10%. An increase in temperature between 20℃ to 100℃ was found to increase the dielectric properties. The penetration depth increased with decrease in moisture content and temperature. Variation in penetration depth was found to vary linearly with decrease in moisture content. A predictive empirical model was developed to calculate penetration depth for petroleum coke.

The effects of drying temperature, drying time and sample mass on the moisture content and dehydration rate of petroleum coke were studied by the response surface methodology. Two mathematical models established and analyzed by RSM were adequate to describe the relationship between the microwave drying conditions and the response of moisture content and dehydration rate. Statistical analysis with response surface regression showed that microwave drying temperature, drying time and sample mass were significantly correlated with moisture content and dehydration rate. Based on the RSM analysis, the optimum values of the process variables were obtained as: microwave drying temperature 75℃, drying time 10s, and sample mass 60g, at these conditions, the moisture content was 0.34, dehydration rate was 2.94.

Acknowledgements

We thank for the Key National Basic Research Program of China (51090385 and 5114703), Project 2013AA064003 supported by 863 Project and Project (5114703) supported by National Natural Science Foundation of China; Project supported by the International S&T Cooperation Program of China (No. 2012DFA70570), and the Yunnan Provincial International Cooperative Program (No. 2011IA004) and Applied Basic Research Project of Yunnan Province (KKSA201152054). Thank for the dielectric properties measurements supported by Professor Huang at Sichuan University.

References

[1] Yuan Z F, Zhu Y Q, Xi L, Xiong S F, Xu B S. Preparation of $TiCl_4$ with multistage series combined fluidized bed [J]. Transactions of Nonferrous Metals Society of China, 2013, 23(1): 283 – 288.

[2] Zhang W S, Zhu Z W, Cheng C Y. A literature review of titanium metallurgical processes [J]. Hydrometallurgy, 2011, 108(3): 177 – 188.

[3] Peng J H, Zhang L B, Yang G, Ma X, Liu B G, Xia H Y, Guo S H. A method of microwave drying high titanium [P], China, CN101545709A. 2009 – 9 – 30.

[4] Li Y, Lei Y, Zhang L B, Peng J H, Li C L. Microwave drying characteristics and kinetics of ilmenite [J]. Transactions of Nonferrous Metals Society of China, 2011, 21(1): 202 – 207.

[5] Nantawan T, Zhou W B. Characterization of microwave vacuum drying and hot air drying of mint leaves (Mentha Cordifolia Opiz Ex Fresen) [J]. Journal of Food Engineering, 2009, 91(3): 482 – 489.

[6] Vongpradubchai S, Rattanadecho P. Microwave and hot air drying of wood using a rectangular waveguide [J]. Drying Technology, 2011, 29(4): 451 – 460.

[7] Gökçe D, Dilek K A, Özbek B. Microwave drying kinetics of okra [J]. Drying Technology, 2007, 25 (5):

917 - 924.

[8] Pickles C A. Microwaves in extractive metallurgy: Part 2—a review of applications[J]. Minerals Engineering, 2009, 22(13): 1112 - 1118.

[9] Pickles C A. Microwaves in extractive metallurgy: Part 1—a review of fundamentals[J]. Minerals Engineering, 2009, 22(13): 1102 - 1111.

[10] Bulatovic S, Wyslouzil D M. Process development for treatment of complex perovskite, ilmenite and rutile ores [J]. Minerals Engineering, 1999, 12(12): 1407 - 1417.

[11] Feng H, Tang J, Cavalieri R P, Plumb O A. Heat and mass transport in microwave drying of porous materials in a spouted bed[J]. AIChE Journal, 2004, 47(7): 1499 - 1512.

[12] Koubaa A, Perré P, Hutcheon R M, Julie L. Complex dielectric properties of the sapwood of aspen, white birch, yellow birch, and sugar maple[J]. Drying Technology, 2008, 26(5): 568 - 578.

[13] Mcloughlin C M, Mcminn W A M, Magee T R A. Physical and dielectric properties of pharmaceutical powders [J]. Powder Technology, 2003, 134(1): 40 - 51.

[14] Madamba P S. The response surface methodology: an application to optimize dehydration operations of selected agricultural crops[J]. LWT, 2002, 35(7): 584 - 592.

[15] Giri S K, Prasad S. Optimization of microwave - vacuum drying of button mushrooms using response - surface methodology[J]. Drying Technology, 2007, 25(5): 901 - 911.

[16] Pe'rez - Francisco J M, Cerecero - Enrı'quez R, Andrade - Gonzalez I, et al. Optimization of vegetal pear drying using response surface methodology[J]. Drying Technology, 2008, 26(11): 1401 - 1405.

[17] Sutar P, Prasad S. Optimization of osmotic dehydration of carrots under atmospheric and pulsed microwave vacuum conditions[J]. Drying Technology, 2011, 29(3): 371 - 380.

[18] Theppaya T, Prasertsan S. Optimization of rubber wood drying by response surface method and multiple contour plots[J]. Drying Technology, 2004, 22(7): 1637 - 1660.

[19] Madamba P S. Optimization of the drying process: An application to the drying of garlic[J]. Drying Technology, 1997, 15(1): 117 - 136.

[20] Jangam S V, Thorat B N. Optimization of spray drying of ginger extract[J]. Drying Technology, 2010, 28(12): 1426 - 1434.

[21] Kaderka M. Dielectric properties of nickel - doped sodium chloride crystals[J]. Czech. Journal of Physics B, 1969, 39(8): 530 - 536.

[22] Barnett R N, Landman U. Water adsorption and reactions on small sodium chloride clusters[J]. Journal of Physical Chemistry, 1996, 100(3): 13950 - 13958.

[23] Martins R M, Siqueira S, Freitas L A P. Spray congealing of pharmaceuticals: study on production of solid dispersions using Box - Behnken design[J]. Drying Technology, 2012, 30(9): 935 - 945.

[24] Sutar P P, Prasad S. Modeling microwave vacuum drying kinetics and moisture diffusivity of carrot slices[J]. Drying Technology, 2007, 25(10): 1695 - 1702.

[25] Vongpradubchai S, Rattanadecho P. The microwave processing of wood using a continuous microwave belt drier [J]. Chemical Engineering and Processing: Process Intensification, 2009, 48(5): 997 - 1003.

[26] Haque K E. Microwave energy for mineral treatment processes - a brief review[J]. International Journal of Mineral Processing, 1999, 57(1): 1 - 24.

[27] Saarenketo T. Electric properties of water in clay and silty soils[J]. Journal of Applied Geophysics, 1998, 40 (1 - 3): 73 - 88.

[28] Al - Harahsheh M, Al - Muhtase A H, Magee T R A. Microwave drying kinetics of tomato pomace: effect of osmotic dehydration[J]. Chemical Engineering and Processing: Process Intensification, 2009, 48(1): 524 - 531.

[29] Koubaa A, Perre P, Hutcheon R M, Lessard J. Complex dielectric properties of the sapwood of aspen, white brich, yellow brich, sugar maple[J]. Drying Technology, 2008, 26(5): 568 - 578.

[30] Guo W C, Zhu X H, Liu Yi, Zhuang H. Sugar and water contents of honey with dielectric property sensing[J]. Journal of Food Engineering, 2010, 97(2): 275 - 281.

[31] Wang K D, Wang Y M. Soil moisture measurement method based on TDR using phase detecting technology [C]//Proceedings of 2008 International Conference on Informationization Automation and Electrification in Agriculture, Zhenjiang, November, 28, 2008.

[32] Willian W M, Gregory R H. Microwave moisture measurement system for hardwood lumber drying[J]. Drying Technology, 2008, 26(9): 1155 - 1159.

[33] Liu Y, Tang J, Mao Z. Analysis of bread loss factor using modified Debye equations[J]. Journal of Food Engineering, 2009, 93(4): 453 - 459.

[34] Nelson S O, Trabelsi S. Influence of water content on RF and microwave dielectric behavior of foods[J]. Journal of microwave power and electromagnetic energy 2009, 43(2): 13 - 23.

[35] Thomas Z M, Zahn M. Dielectrometry measurements of moisture diffusion and temperature dynamics in oil impregnated paper insulated electric power cables[C]//Electrical Insulation, 2008. ISEI 2008. Conference Record of the 2008 IEEE International Symposium on, IEEE: 2008: 539 - 542.

[36] Idris A, Khalid K, Omar W. Drying of silia sludge using microwave drying[J]. Applied Thermal Engineering, 2004, 24(5 - 6): 905 - 918.

[37] Andrés A, Bilbao C, Fito P. Drying kinetics of apple cylinders under combined hot air - microwave dehydration [J]. Journal of Food Engineering, 2004, 63(1): 71 - 78.

[38] Metaxas A C, Meredith R. Industrial Microwave Heating[M]. London: Peter Peregrinus Ltd., 1983.

[39] Clark D E, Folz D C, West J K. Processing materials with microwave energy[J]. Materials Science and Engineering: A, 2000, 287(2): 153 - 158.

[40] Bows J R. A classification system for microwave heating of food[J]. International Journal of Food Science and Technology, 2000, 35(4): 417 - 430.

[41] Tang J, Feng H, Lau M. Microwave heating in food processing[J]. In Advances in Agricultural Engineering, Singapore, 2002.

[42] Zhang Z, Peng J, Qu W, Zhang L, Zhang Z, Li W, Wan R. Regeneration of high - performance activated carbon from spent catalyst: optimization using response surface methodology[J]. Journal of the Taiwan Institute of Chemical Engineers, 2009, 40(5): 541 - 548.

Dimension Optimization for Silica Sand Based on the Analysis of Dynamic Absorption Efficiency in Microwave Drying

Xiaobiao Shang, Junruo Chen, Weifeng Zhang, Jinhui Peng,
Hua Chen, Shenghui Guo, Guo Chen

Abstract: We propose a guide for achieving maximum microwave absorption in microwave drying by optimizing the thickness based on the analysis of reflection loss (*RL*) of a silica sand layer. The microwave reflection loss (*RL*) of the silica sand layer is studied in the moisture content range from 1% to 5% at 20℃ and the silica sand (5% moisture content) in the temperature range from 20℃ to 100℃ at 2.45GHz. The calculated reflection losses for various moisture contents and temperatures show that the reflection loss sensitively depends on the thickness of the silica sand, the *RL* of silica sand appear microwave absorption peaks with increasing thickness, the microwave absorption peak shifts towards a smaller thickness side as the moisture content and temperature of the silica sand increase, and the intensity of microwave absorption peaks in the *RL* patterns of silica sand decrease with decreasing moisture content, and achieving the highest absorption at 60℃.

Keywords: silica sand; microwave drying; reflection loss; absorption; moisture content

1 Introduction

Microwave drying has distinguishing characteristics such as material - selective and volumetric heating compared with conventional methods, leading to extremely broad applications in materials drying[1-5]. The efficiency of microwave drying was found to be dependent on various factors, such as permittivity[6], moisture content[7], material dimension[8], microwave frequency[9], etc. Among these parameters, the material dimension plays an important role in microwave drying of materials.

Although many researchers have demonstrate the importance of dimension optimization of materials in microwave drying by experimental work, general rules from a theoretical perspective are not available. To reduce the number of tests and achieve maximum microwave absorption in the entire drying process, a study for obtaining optimal dimensions of materials becomes very important due to a strong material dimension dependence of microwave power absorption. Recent study indicates that the analysis of reflection loss (*RL*) of materials under microwave radiation may be very helpful for obtaining optimal dimensions of materials throughout microwave drying by simultaneously considering the effects of various factors mentioned above.

For obtaining dimension optimization of materials in microwave drying, there is no detailed study considering the moisture content and temperature dependences of microwave absorption properties. A valid dimension optimization rule for materials is still highly demanded. To solve this issue, this work is devoted to achieve the maximum microwave absorption in microwave drying of materials by calculating the reflection losses of materials. The microwave reflection loss (RL) of silica sand is studied in the moisture content range from 0% to 5% at 20℃ and silica sand (5% moisture content) in the temperature range from 20℃ to 100℃ at 2.45 GHz using formulas that derived from the transmission line theory. The results provide a general rule for dimension optimization of materials in microwave drying, which could provide insight on achieving the maximum absorption during microwave drying of materials.

2 Theory of reflection loss

Fig. 1 shows a material slab backed by a metallic convey belt under microwave irradiation. Reflection loss (RL) quantifies the amount of microwave power reflected from the surface of a material slab backed by metal under microwave irradiation. The smaller RL there is for a sample, the better (larger) the microwave absorption in the material. According to the transmission line theory[10], the reflection loss (RL) of the materials slab can be determined by the following equation:

$$RL = 20\lg \frac{\left| \sqrt{\frac{\mu_r}{\varepsilon_r}} \tanh\left(j \frac{2\pi f}{c} \sqrt{\mu_r \varepsilon_r} d - 1 \right) \right|}{\left| \sqrt{\frac{\mu_r}{\varepsilon_r}} \tanh\left(j \frac{2\pi f}{c} \sqrt{\mu_r \varepsilon_r} d + 1 \right) \right|} \tag{1}$$

where μ_r and ε_r are the complex relative permeability ($\mu_r = \mu'_r - j\mu''_r$) and permittivity ($\varepsilon_r = \varepsilon'_r - j\varepsilon''_r$), respectively, of the slab; j is the imaginary unit; c is the velocity of microwave in free space; d is the thickness of the slab.

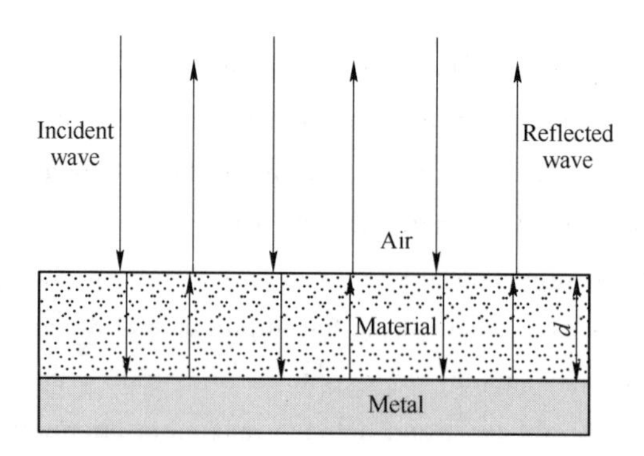

Fig. 1 Schematic of a single - layer absorber under microwave irradiation

As indicated by Eq. (1), reflection loss of the material slab depends on the microwave dielectric properties of silica sand layer at a given frequency, moisture content and temperature. In our previous study, these properties of silica sand in the temperature range from 20℃ to 100℃ and the moisture content range from 1% to 5% at 20℃, 2.45GHz were measured with an open - ended coaxial probe dielectric measurement system, as shown in Figs. 2 and 3[11], respectively. The moisture content, temperature and thickness dependences of reflection loss of the silica sand can be determined based on the reported parameters. The values of μ_r of silica sand are assumed to be 1 in the *RL* calculations($\mu' = 1$ and $\mu'' = 0$).

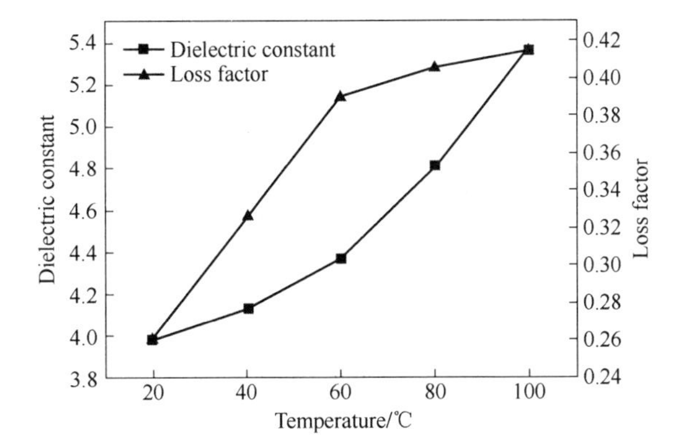

Fig. 2 Variation of complex permittivity of silica sand particles (5% moisture content) with temperature at 2.45GHz

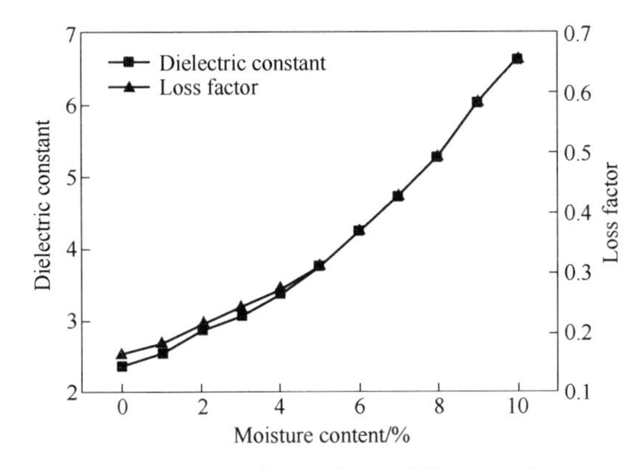

Fig. 3 The complex permittivity of silica sand particles at different moisture contents at 20℃, 2.45GHz

3 Results and discussion

The calculated results of *RL* versus thickness for silica sand with increasing moisture content (from 1% to 5%) at 20℃, 2.45GHz are shown in Fig. 4. As shown by Fig. 4, there are 4 microwave absorption peaks in the *RL* patterns, it was observed that there is a matching thickness for matching moisture content at which microwave absorption is maximum. Fig. 5 shows the matching thickness in the reflection loss of the silica sand at various moisture contents. Peak 1 at 5% moisture content indicates that the silica sand layer exhibits the maximum microwave absorption ($RL = -5.7$dB). However, as moisture content increases, a shift of peak position is observed, and the matching thickness decreases almost linearly with the increase of the moisture content of silica sand, For instance, the peak shifts from 0.025m to 0.018m as the moisture content increases from 1% to 5% as shown in Fig. 5. A moisture content dependence of sample thickness corresponding to the maximum microwave absorption is also observed in other peaks. This phenomenon can be attributed to the increased microwave phase constant and, therefore, a shorter microwave wavelength in the silica sand(λ_d) as the moisture content increases(Table 1). The value of λ_d is determined by[12]

$$\lambda_d = \sqrt{2}\lambda\{\varepsilon'\mu' - \varepsilon''\mu'' + [(\varepsilon'\mu')^2 + (\varepsilon''\mu'')^2 + (\varepsilon'\mu'')^2 + (\mu'\varepsilon'')^2]^{1/2}\}^{-1/2} \tag{2}$$

where λ is the microwave length in free space.

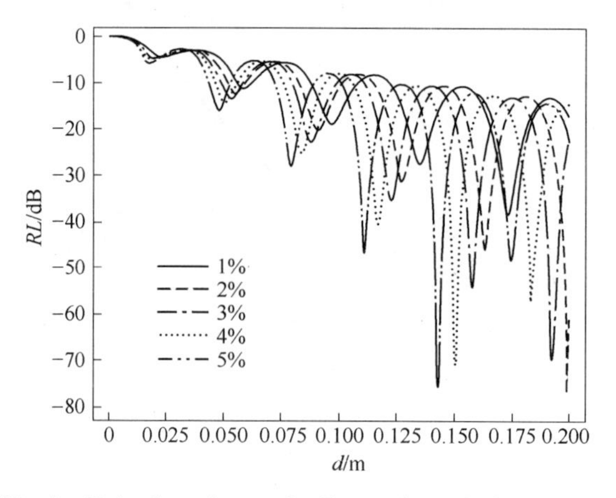

Fig. 4 Moist dependence of reflection loss of silica sand as the thickness varies from 0 to 0. 2m at 20℃ ;1% ,2% ,3% ,4% and 5%

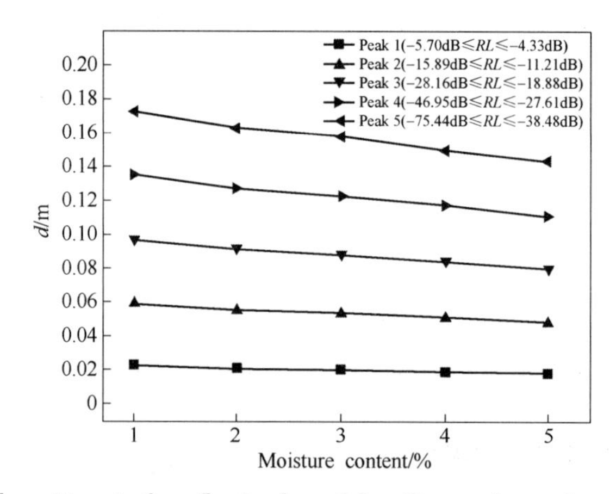

Fig. 5 Absorption peak positions in the reflection loss of the silica sand at various moisture content(20℃)

Table 1 The microwave length in the silica sand at temperatures between 20℃ and 100℃ (5% moisture content) and in the moisture content range of 1% to 5% at 20℃, 2. 45GHz

Parameter	Temperature/℃					Moisture content/%				
	20	40	60	80	100	1	2	3	4	5
λ_m/m	0. 0633	0. 0602	0. 0585	0. 0558	0. 0528	0. 0764	0. 0722	0. 0698	0. 0665	0. 0632

In Fig. 4 it can, further, be noticed that for all absorption peaks (Peaks 1 − 4) in the *RL* patterns, their intensity increases (*RL* decreases) with increasing moisture content. For example, *RL* of peak 1 increases from −4. 33dB to −5. 70dB as the moisture content increase from 1% to 5%. This can be explained by the fact that microwave penetration depth of the silica sand decreases with increasing moisture content[11].

These suggest that the highest drying efficiency can not be achieved throughout the microwave drying in materials with a fixed thickness because of the moisture content dependence of the variation of peak intensity and the shit of microwave absorption peaks. However, we can anticipate that a suitable material thickness should be limited in the range indicated by the positions of the microwave absorption peaks in the moisture content range(i. e. $d_{peak,5\%} \leqslant d \leqslant d_{peak,1\%}$). This suggest that four thicknesses for the silica sand corresponding to the four peaks in the *RL* patterns as shown in Fig. 4 should be chosen for achieving high drying efficiency throughout the microwave drying process.

It is also seen from Fig. 5 that the slope on the line of peak 1 tends to 0 and the slope becomes larger with the number of peak corresponding larger thickness. This suggests that the shift of peak is slight at small thicknesses and become larger with the increasing thickness, therefore it is impossible achieving the highest power absorption with larger shift of absorption peak. If one wants to achieve high absorption throughout the microwave drying at larger shift, he has to adjust the thickness of materials. Peng has found that high microwave absorption during the microwave heating can only be achieved in a sample with a small thickness in which a slight absorption peak shift(less than one eighth – wavelength)[13].

Figs. 6 and 7 show the temperature dependence *RL* of silica sand and absorption peak positions, respectively. It is seem that the variation trend of reflection loss with the temperature is similar to that with the moisture content. There are 5 absorption peaks in the *RL* patterns were shown in Fig. 6. peak 1 at 60℃ indicates that the silica sand layer exhibits the maximum microwave absorption($RL = -6.53$dB). The peak shifts from 0.018m to 0.014m as the moisture content increases from 20℃ to 40℃. A temperature dependence of sample thickness corresponding to the maximum absorption is also observed in other peaks. The matching thickness also decreases almost linearly with the increase of the temperature.

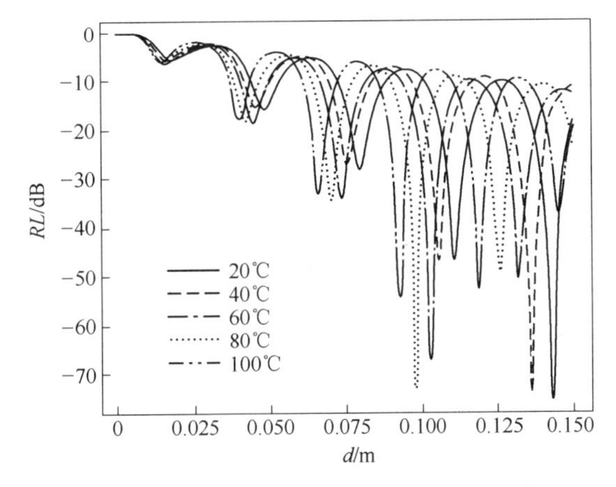

Fig. 6 Temperature dependence of reflection loss of silica sand(5% moisture content) as the thickness varies from 0 to 0.15m: 20℃, 40℃, 60℃, 80℃, 100℃

The *RL* value of < -20 is comparable to the 99% of microwave absorption and thus "*RL* <

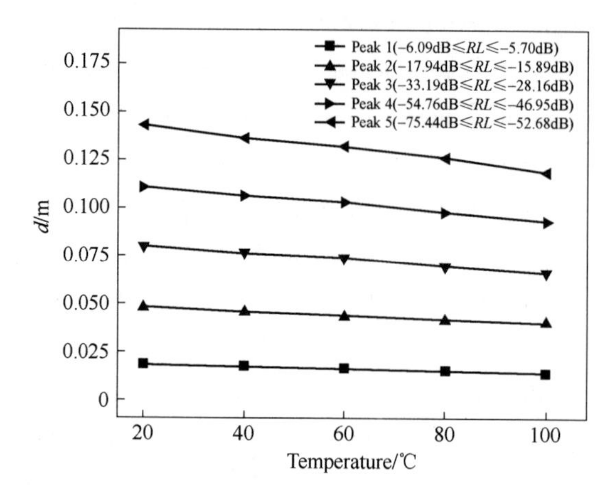

Fig. 7 Absorption peak positions in the reflection loss of silica sand(5% moisture content) at various temperature

−20dB" is considered as an adequate microwave absorption[14,15]. The *RL* values less than −20dB were recorded at the 2.45GHz, as shown in Table 2.

Table 2 The thickness ranges of the *RL* values less than −20dB at temperatures between 20℃ and 100℃(5% moisture content) and in the moisture content range of 1% to 5% at 20℃, 2.45GHz

Peak No.	Temperature/℃				
	20Thickness/m	40Thickness/m	60Thickness/m	80Thickness/m	100Thickness/m
1					
2	0.077 ~ 0.082	0.073 ~ 0.078	0.071 ~ 0.077	0.068 ~ 0.073	0.064 ~ 0.068
3	0.107 ~ 0.116	0.102 ~ 0.11	0.099 ~ 0.107	0.095 ~ 0.102	0.09 ~ 0.096
4	0.137 ~ 0.149	0.131 ~ 0.141	0.127 ~ 0.137	0.122 ~ 0.13	0.116 ~ 0.123
5				0.149 ~ 0.15	0.142 ~ 0.149
6					
Peak No.	Moisture content/℃				
	1Thickness/m	2Thickness/m	3Thickness/m	4Thickness/m	5Thickness/m
1					
2			0.086 ~ 0.09	0.081 ~ 0.086	0.077 ~ 0.082
3		0.123 ~ 0.132	0.118 ~ 0.128	0.113 ~ 0.122	0.107 ~ 0.116
4	0.13 ~ 0.139	0.157 ~ 0.17	0.151 ~ 0.165	0.144 ~ 0.157	0.137 ~ 0.149
5	0.166 ~ 0.180	0.193 ~ 0.2	0.185 ~ 0.2	0.177 ~ 0.191	0.168 ~ 0.181
6					0.0199 ~ 0.2

4 Conclusions

A guide for achieving maximum microwave absorption in microwave drying by optimizing the thickness based on the analysis of reflection loss(*RL*) of a silica sand layer has been proposed. The calculated reflection losses for various moisture contents and temperatures show that the reflection loss sensitively depends on the thickness of the silica sand, the *RL* of silica sand appear microwave absorption peaks with increasing thickness, the microwave absorption peak shifts towards a smaller thickness side as the moisture content and temperature of the silica sand increases, and the intensity of microwave absorption peaks in the *RL* patterns of silica sand decrease with decreasing moisture content, and achieving the highest absorption at 60℃. The variation of *RL* of the silica sand suggests that a thickness of silica corresponding to a slight shift is required to achieve the highest absorption at high moisture content and temperature.

Acknowledgements

This work was supported by the National Technology Research and Development Program of China (863 Program, No. 2013AA064003), the International S&T Cooperation Program of China (No. 2012DFA70570), the Yunnan Provincial International Cooperative Program (No. 2011IA004), and Project (51304097) supported by the National Natural Science Foundation of China.

References

[1] Chandrasekaran S, Ramanathan S, Basak T. Microwave material processing—a review[J]. AIChE Journal, 2012, 58(2): 330 - 363.

[2] Li Z Y, Wang R F, Kudra T. Uniformity issue in microwave drying[J]. Drying Technology, 2011, 29(6): 652 - 660.

[3] Dadalı G, Kılıç Apar D, Özbek B. Microwave drying kinetics of okra[J]. Drying Technology, 2007, 25(5): 917 - 924.

[4] Li C L, Peng J H, Zhang L B, et al. Optimization of microwave drying process for ammonia sulfate with response surface methodology[J]. Chemical Engineering (China), 2011, 39(3): 1 - 2.

[5] Vongpradubchai S, Rattanadecho P. The microwave processing of wood using a continuous microwave belt drier [J]. Chemical Engineering and Processing: Process Intensification, 2009, 48(5): 997 - 1003.

[6] Feng H, Yin Y, Tang J. Microwave drying of food and agricultural materials: basics and heat and mass transfer modeling[J]. Food Engineering Reviews, 2012, 4(2): 89 - 106.

[7] Heng P W, Loh Z H, Liew C V, et al. Dielectric properties of pharmaceutical materials relevant to microwave processing: effects of field frequency, material density, and moisture content[J]. Journal of Pharmaceutical Sciences, 2010, 99(2): 941 - 957.

[8] Zuo Y G, Zhang L B, Liu B G, et al. Optimization of microwave drying of CuCl residue using response surface methodology[J]. Advanced Materials Research, 2013, 803: 3 - 8.

[9] Al - Muhtaseb A H, Hararah M A, Megahey E K, et al. Dielectric properties of microwave - baked cake and its constituents over a frequency range of 0.915 - 2.450GHz[J]. Journal of Food Engineering, 2010, 98(1): 84 - 92.

[10] Shen G, Xu Z, Li Y. Absorbing properties and structural design of microwave absorbers based on W - type La - doped ferrite and carbon fiber composites[J]. Journal of Magnetism and Magnetic Materials, 2006, 301 (2): 325 - 330.

[11] Liu C, Zhang L, Peng J, et al. Temperature and moisture dependence of the dielectric properties of silica sand [J]. Journal of Microwave Power and Electromagnetic Energy, 2013, 47(3): 199 - 209.

[12] Peng Z, Hwang J, Andriese M. Microwave power absorption characteristics of ferrites[J]. Magnetics, IEEE Transactions on, 2013, 49(3): 1163 - 1166.

[13] Peng Z, Hwang J, Andriese M. Design of double - layer ceramic absorbers for microwave heating[J]. Ceramics International, 2013, 39(6): 6721 - 6725.

[14] Wei J, Zhao R, Liu X. Only Ku - band microwave absorption by Fe_3O_4/ferrocenyl - CuPc hybrid nanospheres [J]. Journal of Magnetism and Magnetic Materials, 2012, 324(20): 3323 - 3327.

[15] Dosoudil R, Franek J, Slama J, et al. Electromagnetic wave absorption performances of metal alloy/spinel ferrite/polymer composites[J]. Magnetics, IEEE Transactions on, 2012, 48(4): 1524 - 1527.

Optimization of Processing Parameters for Microwave Drying of Selenium – rich Slag Using Incremental Improved Back – propagation Neural Network and Response Surface Methodology

Yingwei Li, Jinhui Peng, Guian Liang, Wei Li, Shimin Zhang

Abstract: In the non – linear microwave drying process, the incremental improved back – propagation (BP) neural network and response surface methodology (RSM) were used to build a predictive model of the combined effects of independent variables (the microwave power, the acting time and the rotational frequency) for microwave drying of selenium – rich slag. The optimum operating conditions obtained from the quadratic form of the RSM are: the microwave power of 14.97kW, the acting time of 89.58min, the rotational frequency of 10.94Hz, and the temperature of 136.407℃. The relative dehydration rate of 97.1895% is obtained. Under the optimum operating conditions, the incremental improved BP neural network prediction model can predict the drying process results and different effects on the results of the independent variables. The verification experiments demonstrate the prediction accuracy of the network, and the mean squared error is 0.16. The optimized results indicate that RSM can optimize the experimental conditions within much more broad range by considering the combination of factors and the neural network model can predict the results effectively and provide the theoretical guidance for the follow – up production process.

Keywords: microwave drying; response surface methodology; optimization; incremental improved back – propagation neural network; prediction

1 Introduction

Microwave technology, with the characteristics of instantaneity, integrity, efficiency, safety, non – pollution and selective heating to the polar water molecules, is widely applied in the drying. The microwave magnetron can generate electromagnetic energy which can be transformed to internal energy in the interior of the dielectric materials[1-4].

In the microwave drying process, the influencing factors of microwave drying including the microwave input power, the acting time, the initial moisture content, the average material mass, the average material surface area and the rotational frequency have different affecting degrees in the drying process, which causes the longer testing cycle and the larger testing quantity. The parameters are difficult to be optimized.

In the present study, the experimental conditions are optimized by using central composite design (CCD) in response surface methodology (RSM), which is an empirical statistical modeling technique employed for multiple regression analysis using quantitative data obtained from properly designed experiments to solve multivariate equations simultaneously, obtaining the optimal process conditions through building up RSM optimization model[5-12].

And the back-propagation (BP) neural network, which has the ability of non-linear mapping, is chosen to build up the simulation model to predict the experimental process. The traditional BP neural network needs long convergent time and sometimes the convergent results cannot be obtained because of local minimum areas. The improved BP neural network based on the Levenberg-Marquardt (L-M) algorithm overcomes these limitations[13-17]. In the process of training the network, there are many problems. Much training data are probably offered by the way of increment batch and the limitation of the system memory can make the training data infeasible when the sample scale is large. Then, the incremental improved BP neural network is put forward[18-23].

2 Incremental improved back-propagation neural network

The BP neural network, as shown in Fig. 1, has a massively interconnected network structure consisting of many simple processing neurons capable of performing parallel computation for data processing. It is made of an input layer, an output layer and a number of hidden layers in which neurons are connected to each other with modifiable weighted interconnections.

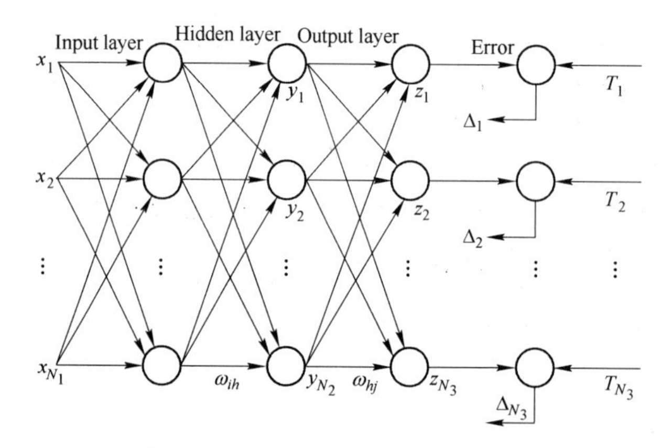

Fig. 1 BP neural network configuration

In Fig. 1, x_i is the input; y_h is the output of one node in the hidden layers; z_j is the output of one node in the output layer; T_j is the target output; ω_{ih} is the weight between the i-th node in the input layer and the h-th node in the hidden layers; ω_{hj} is the weight between the h-th node in the hidden layers and the j-th node in the output layer; N_1 is the number of node in the input layer; N_2 is the number of node in the hidden layers; N_3 is the number of node in the output layer.

The performed functions are expressed as the following equations:

The function of one node in hidden layers:

$$y_h = f\left(\sum_{i=1}^{N_1} \omega_{ih} \cdot x_i + \theta_h\right) \tag{1}$$

The function of one node in output layer:

$$z_j = f\left(\sum_{h=1}^{N_2} \omega_{hj} \cdot y_h + \gamma_j\right) = f\left[\sum_{h=1}^{N_2} \omega_{hj} \cdot f\left(\sum_{i=1}^{N_1} \omega_{ih} \cdot x_i + \theta_h\right) + \gamma_j\right] \tag{2}$$

The error function:

$$E = \frac{1}{2}\sum_{j=1}^{N_3} (T_j - z_j)^2 \tag{3}$$

The sigmoid transfer function:

$$f(x) = \frac{1}{1 + e^{-x}} \tag{4}$$

where θ_h is the threshold value of the h – th node; γ_j is the threshold value of the j – th node.

The L – M algorithm[13-17] is designed to approach the second – order training speed without having to compute the Hessian matrix. The L – M algorithm applies this approximation to the Hessian matrix in the following Newton – like update:

$$x_{i+1} = x_i - (J^T \cdot J + \mu \cdot I)^{-1} \cdot J^T \cdot E \tag{5}$$

where J is the Jacobian matrix that contains the first derivatives of the network errors with respect to the weights and biases; E is a vector of network errors. The Jacobian matrix can be computed through a standard back – propagation technique that is much less complex than that by computing the Hessian matrix. μ is a scalar. When μ is zero, it is just Newton's method, using the approximate Hessian matrix. When μ is large, it becomes gradient descent with a small step size.

The incremental learning is implemented by adjusting the weights of the BP neural network[18-23]. The effective extent of knowledge is settled based on the prior knowledge; the weight vector can be changed in the effective extent when the accuracy of the learned knowledge is unchanged. The weight is adjusted through the fixed network structure when the new sample is provided. This can make the indication value approach to the target value, and then the new sample knowledge is learned. So, the network can not only learn the new sample knowledge but also hold the original knowledge.

With the incremental learning, a scaling factor, s, which scales down all weight adjustments, is introduced, so all the weights are within bounds. The learning rule is

$$\Delta\omega_{ab}(k) = s(k) \cdot \eta \cdot \delta_b(k) \cdot O_a(k) \tag{6}$$

where $\Delta\omega_{ab}$ is the weight between the a – th node and the b – th node of all the network layers; η $(0 < \eta < 1)$ is a trial – independent learning rate; δ_b is the error gradient at the b – th node; O_a is the activation level at the a – th node; the parameter k is the k – th iteration.

3 Response surface methodology

The response surface methodology (RSM) is a collection of mathematical and statistical techniques that are useful for the modeling and analysis of problems in which a response of interest is influenced by several quantifiable variables or factors, with the objective of optimizing the response. Central composite design (CCD) method is suitable for fitting a quadratic surface and it helps to optimize the effective parameters with a minimum number of experiments, as well as to analyze the interaction between the parameters[5].

A full second - order polynomial model is fitted to the experimental data and the coefficients of the model equation are determined. The result in an empirical model related to the response is

$$y = \alpha_0 + \sum_{i=1}^{n} \alpha_i \cdot x_i + \sum_{i=1}^{n} \alpha_{ii} \cdot x_i^2 + \sum_{i=1}^{n-1} \sum_{j=i+1}^{n} \alpha_{ij} \cdot x_i \cdot x_j \tag{7}$$

where y is the predicted response; n is the number of factors; α_0 is the constant coefficient; α_i is the linear coefficient; α_{ii} is the quadratic coefficient; α_{ij} is the interaction coefficient.

4 RSM optimization model

4.1 Data preprocessing

In the experiment, the water content of seleniumrich slag was 38% (mass fraction) after pretreatment, the average mass of selenium - rich slag for each layer was 12kg and the average material surface area was 0.15m^2. Three input variables were used: the acting time x_1, the rotational frequency x_2 and the microwave power x_3; two output variables were used: the relative dehydration rate d and the material temperature t.

4.2 RSM optimization model

By analyzing the data in Table 1, the multiple quadratic regression equations of the temperature, the relative dehydration rate on the time, the rotational frequency and the microwave power is obtained:

Table 1 Experimental results designed by RSM

Run	Time/min	Rotational frequency/Hz	Power/kW	Temperature/℃	Relative dehydration rate /%
1	30.00	8.00	8.00	87	21.3
2	90.00	8.00	8.00	121	65.8
3	30.00	12.00	8.00	90	24.2
4	90.00	12.00	8.00	115	62.9
5	30.00	8.00	15.00	100	34.5
6	90.00	8.00	15.00	136	89.7
7	30.00	12.00	15.00	96	32.5
8	90.00	12.00	15.00	125	95.8
9	9.55	10.00	11.50	70	6.8
10	110.45	10.00	11.50	130	97.1
11	60.00	6.64	11.50	128	67.6
12	60.00	13.36	11.50	116	65.8
13	60.00	10.00	5.61	95	33.7
14	60.00	10.00	17.39	124	79.5
15	60.00	10.00	11.50	121	68.5
16	60.00	10.00	11.50	121	68.5
17	60.00	10.00	11.50	121	68.5
18	60.00	10.00	11.50	121	68.5
19	60.00	10.00	11.50	121	68.5
20	60.00	10.00	11.50	121	68.5

$$t = -34.05832 + 1.82141 \cdot x_1 + 2.47305 \cdot x_2 + 11.73004 \cdot x_3 - 0.033333 \cdot x_1 \cdot x_2 + 0.00714286 \cdot x_1 \cdot x_3 - 0.21429 \cdot x_2 \cdot x_3 - 0.00851056 \cdot x_1^2 + 0.029667 \cdot x_2^2 - 0.35108 \cdot x_3^2 \tag{8}$$

$$d = -102.86034 + 1.23766 \cdot x_1 + 6.80555 \cdot x_2 + 9.66546 \cdot x_3 + 0.00479167 \cdot x_1 \cdot x_2 + 0.042024 \cdot x_1 \cdot x_3 + 0.073214 \cdot x_2 \cdot x_3 - 0.00754896 \cdot x_1^2 - 0.39479 \cdot x_2^2 - 0.42041 \cdot x_3^2 \tag{9}$$

The variance analyses of the temperature and the relative dehydration rate quadratic model are listed in Tables 2 and 3, respectively.

Table 2 Variance analysis of temperature quadratic model

Source	Sum of squares	Degree of freedom	Mean square	F value	Prob > F
Model	5544.33	9	616.04	101.63	<0.0001
A - time	3703.88	1	3703.88	611.02	<0.0001
B - rotational frequency	106.75	1	106.75	17.61	0.0018
C - power	630.21	1	630.21	103.96	<0.0001
AB	32.00	1	32.00	5.28	0.0444
AC	4.50	1	4.50	0.74	0.4091
BC	18.00	1	18.00	2.97	0.1156
A^2	845.48	1	845.48	139.48	<0.0001
B^2	0.20	1	0.20	0.033	0.8585
C^2	266.56	1	266.56	43.97	<0.0001
Residual	60.62	10	6.06		

Table 3 Variance analysis of relative dehydration rate quadratic model

Source	Sum of squares	Degree of freedom	Mean square	F value	Prob > F
Model	12039.18	9	1337.69	89.80	<0.0001
A - time	9153.56	1	9153.56	614.51	<0.0001
B - rotational frequency	0.084	1	0.084	0.00566	0.9415
C - power	1766.60	1	1766.60	118.60	<0.0001
AB	0.66	1	0.66	0.044	0.8374
AC	155.76	1	155.76	10.46	0.0090
BC	2.10	1	2.10	0.14	0.7151
A^2	665.21	1	665.21	44.66	<0.0001
B^2	35.94	1	35.94	2.41	0.1514
C^2	382.23	1	382.23	25.66	0.0005
Residual	148.96	10	14.90		

According to Tables 2 and 3, the conclusions are drawn as follows. The respective Model F - Values of 101.63 and 89.80 imply that the models are significant. There is only a 0.01% chance that a "Model F - Value" could occur due to noise. Values of "Probability > F" are less than 0.05, indicating that model terms are significant. The respective values of "R - squared" are 0.9892 and 0.9878, indicating that the models are considered relatively higher as the value is close to unity, showing that there is a good agreement between the experimental and the predicted temperature results and relative dehydration rate results from respective model.

Fig. 2 shows the three - dimensional response surface diagrams. When the irradiation time and

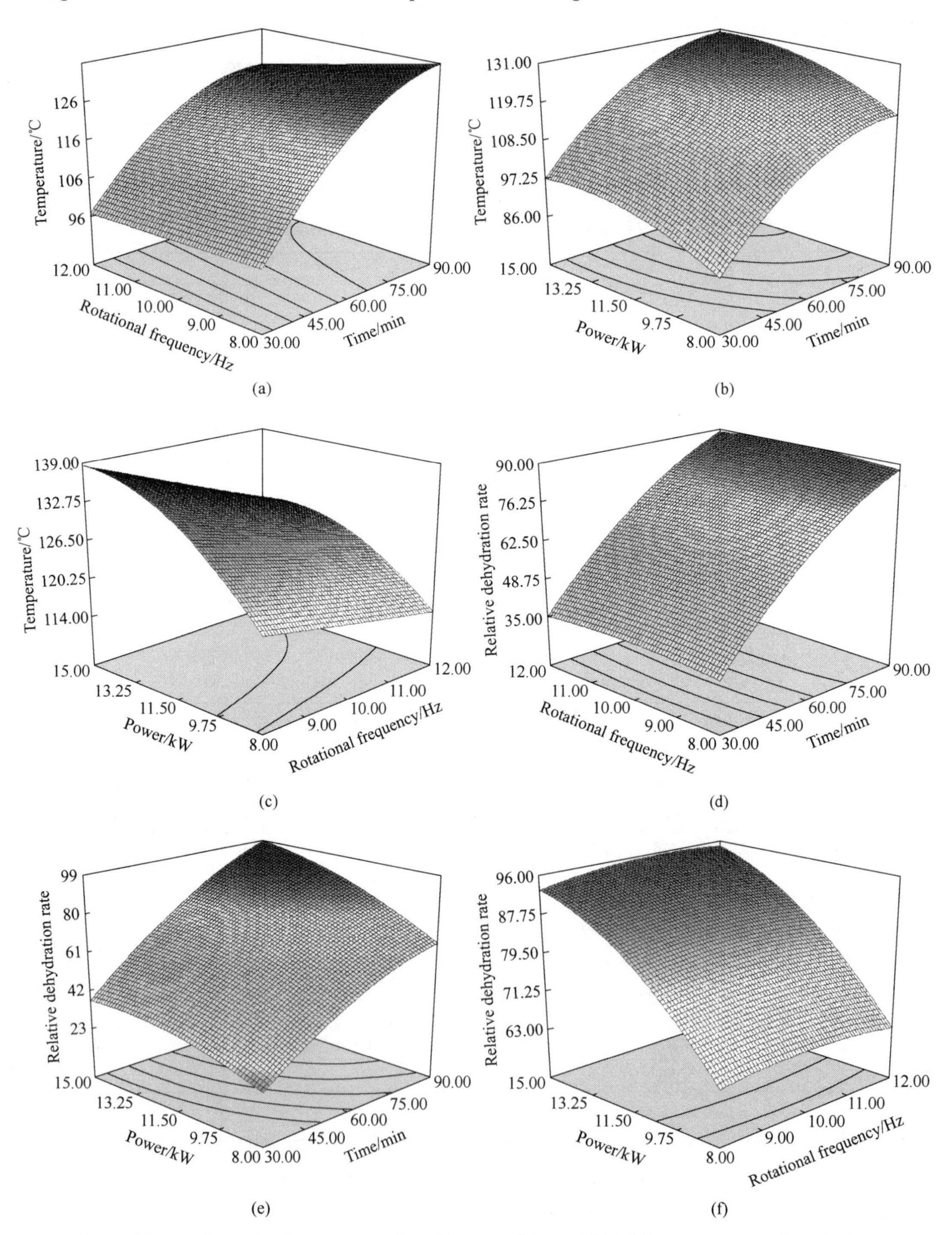

Fig. 2 Three - dimensional response surface diagrams: (a) and (d) Microwave power of 12.00kW; (b) and (e) Rotational frequency of 11.00Hz; (c) and (f) Irradiation time of 85min

the rotational frequency are held constant, the temperature and the relative dehydration rate increase with the increase of microwave power. Under constant irradiation time and microwave power, the temperature and the relative dehydration rate change only a little with the increase of rotational frequency. Under constant microwave power and the rotational frequency, the temperature and the relative dehydration rate increase with the increase of irradiation time. The microwave power and the irradiation time have the significant effect on the temperature and the relative dehydration rate, and the rotational frequency has no obvious effect on the temperature and the relative dehydration rate.

The optimal experimental conditions are listed in Table 4. In the optimum operating conditions, the incremental improved BP neural network prediction model can predict the drying process results and the different effects on the results of the independent variables, as listed in Table 5. After three verification experiments, the average acting time is 90min, the average rotational frequency is 10.5Hz, the average microwave power is 14.8kW, the average temperature is 131℃ and the average relative dehydration rate is 96.51%. These demonstrate that the optimal experimental conditions of the RSM and the prediction results of the incremental improved BP neural network are feasible.

Table 4 Optimal conditions and results using RSM

Time/min	Rotational frequency/Hz	Power/kW	Temperature/℃	Relative dehydration rate/%
89.58	10.94	14.97	136.407	97.1895

Table 5 Predicted results using incremental improved BP neural network

Time/min	Rotational frequency/Hz	Power/kW	Temperature/℃	Relative dehydration rate/%
90.00	11.0	15.2	132	96.95
90.00	10.5	14.7	131	96.68
90.00	10.0	14.5	130	95.89

5 Incremental improved BP neural network prediction model

In order to train the network conveniently and reflect the interrelations of the various factors preferably, the sample data must be pre-treated. In the training process, the number of layers of the hidden layer is determined and the network weights and threshold values are obtained through computing the minimum value of the error function. The optimum number of units of the hidden layers is 5 after training the network. When the network training is finished, the training network is tested, the non-training sample data are read to predict the results, and the output data are calculated and compared with the measured data. If the error is within the regulated scope, the neural network model is available and the simulation can be carried on.

The optimal experimental conditions and results designed by RSM are predicted by the incremental improved BP neural network, and the set target convergence accuracy is reached by comparing and analyzing the output and measured data of training and prediction and 166 times of iteration of the network.

Fig. 3 shows the predicted impacts of the acting time, the rotational frequency and the microwave power on the temperature and the relative dehydration rate.

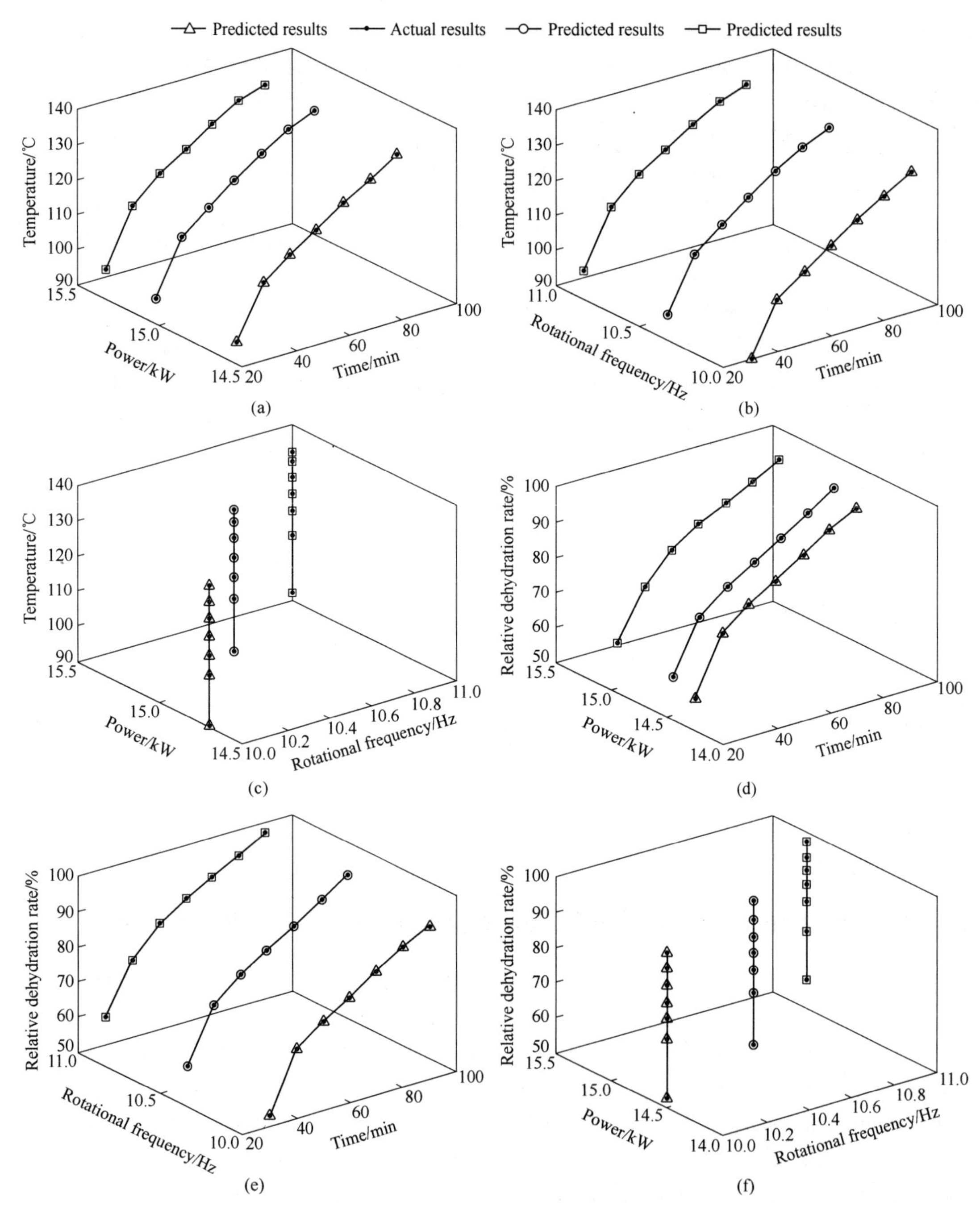

Fig. 3 Three – dimensional diagrams using incremental improved BP neural network

(a) Effect of time and power on temperature; (b) Effect of time and rotation frequency on temperature; (c) Effect of rotational frequency and power on temperature; (d) Effect of time and power on relative dehydration rate; (e) Effect of time and rotational frequency on temperature; (f) Effect of rotational frequency and power on temperature

△—Time: 10 – 90min, frequency: 10Hz, power: 14.5kW; ○—Time: 10 – 90min; frequency: 10.5Hz, power: 14.7kW; □—Time: 10 – 90min, frequency: 11Hz, power: 15.2kW

When the irradiation time and the rotational frequency are held constant, the temperature and the relative dehydration rate increase with the increase of microwave power. Under constant irradiation time and microwave power, the temperature and the relative dehydration rate change only a little with the increase of rotational frequency. And under constant microwave power and rotational frequency, the temperature and the relative dehydration rate increase with the increase of irradiation time. The microwave power and the irradiation time have the significant effect on the temperature and the relative dehydration rate, and the rotational frequency has no obvious effect on the temperature and the relative dehydration rate. The predicted values of the incremental improved BP neural network fit well to the actual values, and the mean square error is 0.16. The results of the irradiation time, the rotational frequency and the microwave power impact on the temperature and the relative dehydration rate are the same as the analysis results obtained from RSM. According to the verification experiments, the incremental improved BP neural network industrial prediction model can predict the drying process results and the different effects on the results of the independent variables can verify the accuracy of the optimum operating results obtained from the quadratic form of the RSM. The combination of the RSM industrial optimization model and the incremental improved BP neural network industrial prediction model can provide the basis for the production practice.

6 Conclusions

(1) The optimum operating conditions using RSM optimization model are: the microwave power of 14.97kW, the acting time of 89.58min, the rotational frequency of 10.94Hz, the temperature of 136.407℃, and the relative dehydration rate of 97.1895%. With the optimization model, the temperature and the relative dehydration rate of correlation coefficient are 0.9892 and 0.9878, which indicate that RSM can optimize the experimental conditions within much more broad range by considering the combination of factors.

(2) The predicted results with the optimum operating conditions using the incremental improved BP neural network are: the microwave power of 14.5 – 15.2kW, the acting time of 90min, the rotational frequency of 10.0 – 11.0Hz, the temperature of 130 – 132℃, and the relative dehydration rate of 95.89% – 96.95%, and the mean square error of the prediction model is 0.16, which indicate that the model can predict the results effectively and provide the theoretical guidance for the follow – up production process.

References

[1] Fito P, Chiralt A. Food matrix engineering: the use of water – structure – functionality ensemble in dried food product development[J]. Food Science and Technology International, 2003, 9(3): 151 – 156.

[2] Aguilera J M, Chiralt A, Fito P. Food dehydration and product structure[J]. Trends in Food Science & Technology, 2003, 14(10): 432 – 437.

[3] Peng J H, Yang X W. The New Applications of Microwave Power[M]. Yunnan: Yunnan Science & Technology Press, 1997 (in Chinese).

[4] Fito P, Chiralt A, Barat J M. Vacuum Impregnation for development of new dehydrated products[J]. Journal of Food Engineering, 2001, 49(4): 297 – 302.

[5] Jagannadha R K, Kim C H, Rhee S K. Statistical optimization of medium for the production of recombinant hiru-

din from Saccharomyces cerevisiae using response surface methodology[J]. Process Biochemistry, 2000, 35(7):639 - 647.

[6] Aktas N. Optimization of biopolymerization rate by response surface methodology(RSM)[J]. Enzyme and Microbial Technology, 2005, 37(4):441 - 447.

[7] Peng Z B, Li J, Peng W X. Application analysis of slope reliability based on Bishop analytical method[J]. Journal of Central South University: Science and Technology, 2010, 41(2):668 - 672(in Chinese).

[8] Zhong M, Huang K L, Zeng J G, Li S, Zhang L. Determination of contents of eight alkaloids in fruits of Macleaya cordata(Willd) R. Br. From different habitats and antioxidant activities of extracts[J]. Journal of Central South University of Technology, 2010, 17(3):472 - 479.

[9] Zainudin N F, Lee K T, Kamaruddin A H, Bhatia S, Mohamed A R. Study of absorbent prepared from oil palm ash(OPA) for flue gas desulfurization[J]. Separation and Purification Technology, 2005, 45(1):50 - 60.

[10] Azargohar R, Dalai A K. Production of activated carbon from Luscar char, experimental and modelling studies [J]. Microporous and Mesoporous Materials, 2005, 85(3):219 - 225.

[11] Kalil S J, Maugeri F, Rodrigues M. I. Response surface analysis and simulation as a tool for bioprocess design and optimization[J]. Process Biochemistry, 2000, 35(6):539 - 550.

[12] Roux W J, Stander N, Haftka R T. Response surface approximations for structural optimization[J]. International Journal for Numerical Methods in Engineering, 1998, 42(3):517 - 534.

[13] Kermani B G, Schiffman S S, Nagle H T. Performance of the Levenberg - Marquardt neural network training method in electronic nose applications[J]. Sensors and Actuators B - Chemical, 2005, 110(1):13 - 22.

[14] Singh V, Indra G, Gupta H O. ANN - based estimator for distillation using Levenberg - Marquardt approach [J]. Engineering Applications of Artificial Intelligence, 2007, 20(2):249 - 259.

[15] Lera G, Pinzolas M. Neighborhood based Levenberg - Marquardt algorithm for neural network training[J]. IEEE Transactions on Neural Networks, 2002, 13(5):1200 - 1203.

[16] Adeloye A J, Munari A D. Artificial neural network based generalized storage - yield - reliability models using the Levenberg - Marquardt algorithm[J]. Journal of Hydrology, 2006, 362(1 - 4):215 - 230.

[17] Mirzaee H. Long - term prediction of chaotic time series with multi - step prediction horizons by a neural network with Levenberg - Marquardt learning algorithm [J]. Chaos, Solitons & Fractals, 2009, 41(4): 1975 - 1979.

[18] Li Y W, Yu Z T, Meng X Y, Che W G, Mao C L. Question classification based on incremental modified bayes [J]. Proceedings of the 2008 2nd International Conference on Future Generation Communication and Networking, FGCN 2008, 2:149 - 152.

[19] Fu L M, Hsu H H, Principe J C. Incremental back - propagation learning network[J]. IEEE Transactions on Neural Networks, 1996, 7(3):757 - 761.

[20] Karayiannis N B, Mi G W Q. Growing radial basis neural networks: merging supervised and unsupervised learning with network growth techniques [J]. IEEE Transactions on Neural Networks, 1997, 8(6): 1492 - 1506.

[21] Ghosh J, Nag A C. Knowledge enhancement and reuse with radial basis function networks[J]. Proceeding of the 2002 International Joint Conference on Neural Networks, 2002, 1 - 3:1322 - 1327.

[22] Parekh R, Yang J H, Honavar V. Constructive neural network learning algorithms for pattern classification[J]. IEEE Transactions on Neural Networks, 2000, 11(2):436 - 451.

[23] Zhang J, Morris A J. A sequential learning approach for single hidden layer neural networks[J]. Neural Networks, 1998, 11(1):65 - 80.

Dielectric Properties and Microwave Heating Characteristics of Sodium Chloride at 2.45GHz

Chenhui Liu, Libo Zhang, Jinhui Peng, Wenwen Qu, Bingguo Liu, Hongying Xia, Junwen Zhou

Abstract: The effects of moisture content and temperature on the dielectric property of sodium chloride were investigated by using open – ended senor dielectric measurement system at the frequency of 2.45GHz. Moisture content is a major influencing factor in the variation of dielectric properties. Dielectric constant, loss factor and loss tangent all increase linearly with moisture content increasing. Three predictive models were developed to obtain dielectric constant, loss factor, loss tangent and of sodium chloride as linear functions of moisture content. Temperature between 20℃ and 100℃ has a positive effect on dielectric constant and loss factor. Penetration depth decreased nonlinearly with moisture and temperature increasing. A predictive model was developed to calculate penetration depth for sodium chloride as a fifth function of moisture content. In addition, the measurements indicate that the particles temperature increases linearly with microwaving heating time at different power levels. The knowledge gained from these results is useful in developing more effective applications of microwave drying and designing better sensors for measuring moisture content of sodium chloride.

Keywords: dielectric properties; sodium chloride; moisture content; microwave drying; temperature increase

1 Introduction

Sodium chloride is a major raw material for the chemical industry to produce sodium chlorate, metallic sodium, and other products. At the same time, it is also used in oil and gas exploration, textiles and dyeing, water treatment, rubber manufacture and metal processing. In metallurgical industry, sodium chloride is typically used in metallic titanium production. Because of its feature of hygroscopicity, the raw sodium chloride material needs to be dried before using it in the industry.

Currently, the most common methods for drying raw materials in the metallurgical industry are hot air drying, fluid bed drying and rotary kiln drying[1-3]. But these methods have some disadvantages: inability to handle large quantities and to achieve consistent quality standards low – energy efficiency, long drying time and high energy costs[4]. Microwave drying is more energy – efficient than conventional drying. The major advantages of microwave drying include rapid heat transfer, volumetric and selective heating, compactness of equipment, quick start/stop and pollution – free since nothing is burned[5]. Microwave drying has been recently applied in processing foods, woods, minerals, pharmaceutical materials and other industrial raw materials[6-9].

Material processing with microwave radiation involves several complicated physical processes, including the absorption of electromagnetic energy and transport of generated heat[10]. A better understanding of the interaction between microwave radiation and the material is essential for using it more efficiently and effectively. The dielectric properties of a material are closely related to the interaction between the material and microwave radiation. Complex permittivity is one of the widely studied dielectric parameters of a material, and it is defined as follows:

$$\varepsilon^{*} = \varepsilon' - j\varepsilon'' \tag{1}$$

The complex permittivity(ε^{*}) is related to the ability of a material to be coupled with electric energy from microwave fields. The dielectric constant(ε') reflects the ability of the material to store electromagnetic energy within its structure, and the loss factor(ε'') characterizes the ability of the material to convert the stored electromagnetic energy into thermal energy.

Loss tangent($\tan\delta$) is another dielectric parameter of a material and is defined as follows:

$$\tan\delta = \frac{\varepsilon''}{\varepsilon'} \tag{2}$$

It demonstrates how well a material convert stored energy into heat.

In microwave drying, the dielectric constant and loss factor is needed to estimate the penetration depth of the material and the generated heat. Both dielectric constant and loss factor are a function of microwave frequency and sample moisture and temperature. Measurements of dielectric properties(dielectric constant, loss factor and loss tangent) are required for understanding, explaining interactions between material and microwave energy[11]. Dielectric properties vary with the composition, moisture content and temperature of mineral materials and the frequency of the electric field[12]. Information about salt water is limited. Modeling changes in dielectric properties of sodium chloride with moisture content and temperature will allow predicting the same at any prescribed moisture content and temperature thereby facilitating microwave drying equipment and online moisture content sensor design[13].

The dielectric properties of foods, pharmaceutical powders and woods have been studied in a wide range of moisture contents[14,15]. However, research on metallurgical materials, especially on the sodium chloride particles mixed with water is limited. Kaderka[16] has studied the influence of nickel on dielectric losses in Sodium chloride crystals. Water adsorption and chemical reaction to small sodium chloride clusters have been observed by Barnet, et al. [17] The effects of sodium chloride on a lipid bi-layer were researched by Bokmann, et al[18]. We have not seen a work on the dielectric properties of wet sodium chloride particles as industrial raw materials at 2.45GHz.

The objectives of this study were: (1) to measure the dielectric properties of wet sodium chloride particles at ten moisture content levels and the material from ambient temperature to 100℃ at 20℃ intervals at the frequency of 2.45 GHz; (2) to formulate the effects of moisture content and temperature on dielectric constant, loss factor and loss tangent and calculated penetration depth; (3) to study temperature increasing characteristics of sodium chloride particles for given power levels in the auto-made multimode microwave oven at 2.45GHz.

2 Materials and methods

2.1 Materials and sample preparation

Wet sodium chloride particles were obtained from a titanium plant based at Kunming, Yunnan province, China. The compositions of the sample were shown in Table 1. The initial moisture content of the sample was 0.05kg/kg (wet basis) which was determined by drying the sample in an oven for 24h at the temperature of 105℃.

Table 1 Chemical composition of sodium chloride

Composition	NaCl	Mg	Ca	K	SO_4
%	99.800	0.003	0.007	0.020	0.050

2.2 Open ended coaxial line dielectric measurement system

Currently, several techniques have been adopted to determine the dielectric properties of solid samples including those of resonator cavity[19], transmission line[20,21] and free - space[22]. Open - ended coaxial - line probes have been used successfully for convenient dielectric measurements on liquid and semisolid materials, which includes most food samples[23]. Accuracies are poorer on very low loss materials, and solids must have accurately machined plane surfaces to avoid errors caused by any appreciable air gaps. It has been used to provide dielectric information on granular and pulverized materials when sample bulk densities were established[24]. It is easy to use and does not require special sample preparation[25]. To take measurements, the probe is placed on the flat surface of a solid sample or buried in the sample powder for a full contact. The electromagnetic fields at the probe's end fringe into the material and change as they come into contact, and the probe translates the changes in the permittivity of the testing sample into variations in the input reflection coefficient of the probe. This reflection coefficient is then recorded by the network analyzer, and the permittivity can be computed from the measured reflection signals.

In this work, a hybrid experimental/computation permittivity measuring system, which is developed by the Applied Radio physics Research Center at Sichuan University[26], was used to determine the complex permittivity of the powder samples at room temperature. The schematic diagram of the measuring system is depicted in Fig. 1.

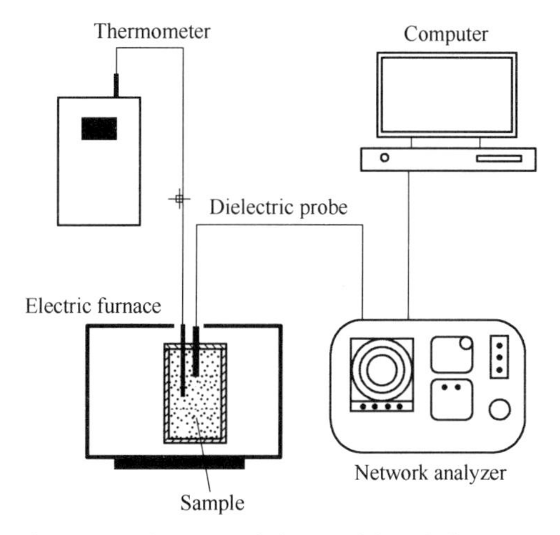

Fig. 1 Schematic diagram of the pen ended coaxial line dielectric measurement system

To make measurements, sample powder was filled in a resonant cavity (80mm inner diameter with a length of 100mm) of stainless steel heated by an electric furnace under the cavity. An open – ended coaxial sensor (Fig. 2) connected with the Agilent PNA5230 Network Analyzer was utilized to measure the reflection coefficients. When the sensor was inserted into the sample powder in the cavity, the reflection signals were recorded by the network analyzer. A thermocouple pyrometer was placed in the sample powder in the cavity for taking temperature measurements, and the readings were taken from a digital temperature indicator which was connected to the thermocouple. An FDTD method was employed to calculate the distribution of electromagnetic field surrounding the sensor and the reflection coefficients at different frequencies. The complex permittivity of the testing sample was then calculated from the measurements using an inversion algorithm developed by Huang, et al.[27] Bulk density also is a very important factor for accurate dielectric measurement, the bulk density of the sample in the sample container can be determined from sample mass and container dimensions. Temperature and humidity are the main factors in the research, and sample mass should keep the same at each measurement. The bulk density of the sample at each measurement is 1.417 kg/cm^3.

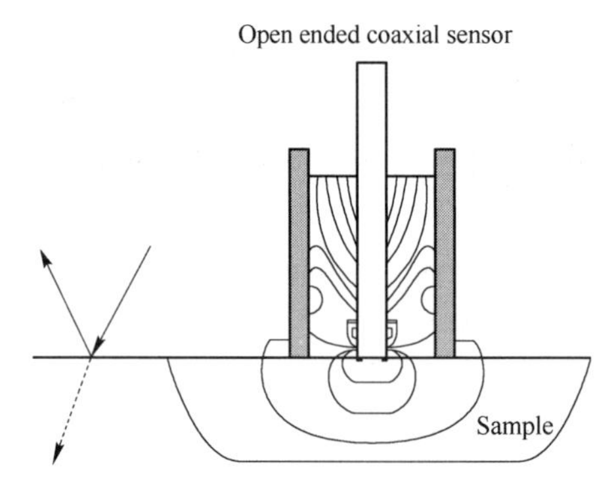

Fig. 2 The schematic diagram of the open – ended coaxial sensor

2.3 Microwave drying equipment and drying experiment

Microwave drying equipment was made by the Key Laboratory of Unconventional Metallurgy, Ministry of Education, Kunming University of Science and Technology. It has the ability to alter the power intensity in the range of 0 – 3000W at the frequency of 2.45GHz (the wavelength of 12.24cm). The dimensions of the microwave cavity were 215mm × 350mm × 330mm. A fan located in the oven could be used to flow water vapor out of the chamber in the drying process. The two magnetrons were equipped with a water – cooled condenser.

The schematic diagram of the microwave heating equipment is shown in Fig. 3. Weight measurements were made by the means of an electronic balance attached to the top of the drying cavity. This balance was connected to a computer to record the measurements continuously. A thermocouple pyrometer was used to monitor the temperature. In the experiment, each sample was put in a glass plate and placed at the center of the oven. There were three replications for each experiment and their averages were used for analysis.

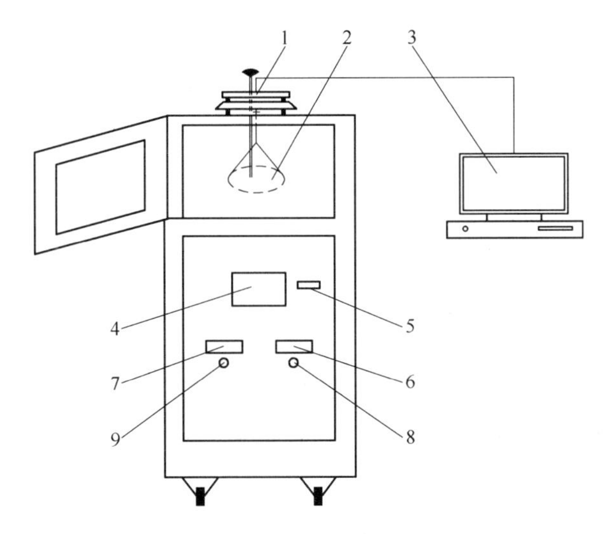

Fig. 3 Schematic diagram of microwave heating system

1—electronic balance; 2—container; 3—computer system; 4—temperature display instrument; 5—power controller; 6—ampere meter; 7—voltmeter; 8—startswitch; 9—stopswitch

3 Results and discussions

3.1 Effects of moisture on dielectric properties

Dielectric constant, loss factor and loss tangent of the testing sodium chloride particles increase steadily with moisture content. The results are given in Table 2 and depicted in Figs. 4 – 6.

Table 2 The complex permittivity and power penetration depths of sodium chloride particles at different moisture contents at 2.45GHz

Moisture content (wet basis)/%	Dielectric constant	Loss factor	Loss tangent	Penetration depth/cm
0	2.4181	0.20844	0.0862	14.55
0.5	3.1187	0.30345	0.0973	11.36
1	3.3806	0.45063	0.1333	7.97
1.5	3.722	0.68187	0.1832	5.53
2	3.9943	0.90551	0.2267	4.32
2.5	4.3735	1.21321	0.2774	3.39
3	4.5591	1.40694	0.3086	2.99
3.5	4.6201	1.68449	0.3646	2.53
4	4.6619	1.76733	0.3791	2.42
4.5	4.9521	2.14228	0.4326	2.06
5	5.1951	2.55651	0.4921	1.79

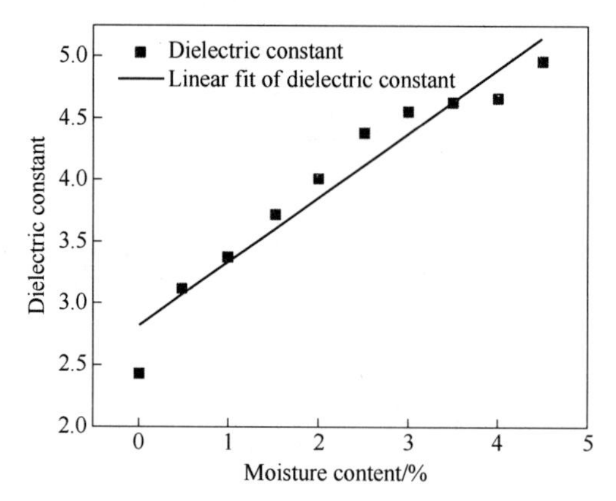

Fig. 4 Variation of dielectric constant of sodium chloride as a linear function of moisture content at 2.45GHz

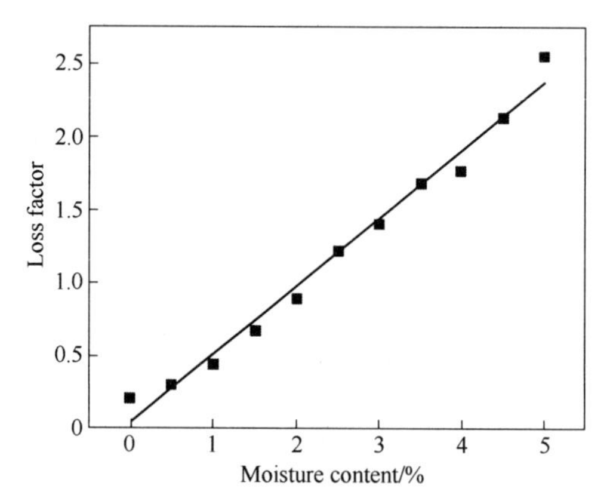

Fig. 5 Variation of loss factor of sodium chloride as a linear function of moisture content at 2.45GHz

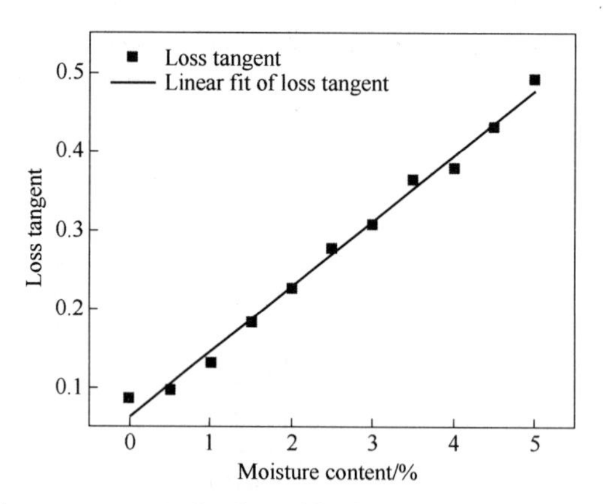

Fig. 6 Variation of dielectric loss tangent $\tan\delta$ of sodium chloride as a linear function of moisture content at 2.45GHz

At a given frequency, increasing moisture content increases the dielectric constant of sodium chloride sample (Fig. 4). Pure water is a dipolar compound that couples microwave energy at microwave frequencies more efficiently than most other components of minerals. Moist sodium chloride has a minimum value when the sample is dried and a maximum value when the sample contains 4% of water with wet basis in this study. This result agrees well with previous studies on clays and soils[28,29]. For the sodium chloride, the water between rough surface and pore structure is mainly in free form. The dielectric polarization attributable to free water is much more than bound water. The mobility of water molecules in a densely packed crystal is lower than in solution, and the mixture shows lower dielectric constants at lower moisture content. Mobility obviously increases when more water is present between the particle interfaces, and dielectric constant will increase with moisture content increasing.

The variation in dielectric constant with moisture content was found to be linear and a linear equation is given in Table 3 to describe the variation of dielectric constant of sodium chloride with moisture content. High R^2 – value implies that the linear models could be used for predicting moisture content at any given moisture content.

Table 3 Regression equations on moisture contents and dielectric parameters at 2.45GHz at 25℃

Dielectric properties	Regression equations	Correlation efficient(R^2)
Dielectric constant	$\varepsilon' = 2.81559 + 0.51753M$	0.92526
Loss factor	$\varepsilon'' = 0.04749 + 0.46459M$	0.982
Loss tangent	$\tan\delta = 0.06406 + 0.08278M$	0.99169
Penetration depth	$D_p = 13.95958 - 6.16704M + 0.77893M^2$	0.97489
Loss factor	$\varepsilon'' = 3.11217 - 0.0889T + 0.00132T^2$	0.971

As shown in Figs. 5 and 6, respectively, the loss factor and loss tangent both increase linearly with moisture content increasing at 2.45GHz at room temperature. The correlation coefficients of linear regression (R^2) were 0.982 and 0.991, respectively (Table 3). These high R^2 – values imply that these linear models can be used for predicting loss factor and loss tangent at any given moisture content. Earlier researches on clays, soils and foods have also shown a linear relationship between moisture contents and dielectric parameters[30,31,12]. Differences of dielectric properties between different materials were embodied by various slopes and intercepts on the linear lines.

Based on the relationship between dielectric property of a material and its moisture content, several new technologies have been developed for online measuring moisture content of materials. The time domain reflectometry (TDR) has been widely used in measuring soil moisture due to its advantages of high accuracy and stability[32]. Moschler and Hanson[33] developed parallel resonant sensors to measure moisture content of hardwood lumbers, and the results show that the sensor working at 4.5GHz to 6GHz displays a linear response to moisture content over a range of 6% – 100%. Dielectric properties data of sodium chloride will be useful in developing new sensors for measuring moisture contents of the material more accurately in a wider range of moisture contents.

Water in moisture and sand – like mixed materials is commonly categorized into three layers: the hygroscopic water layer, the capillary water layer and the free water layer[34]. The dielectric property of water within mixtures is not the same as that of pure water because the ion content of water within mixtures is different. A sodium particle is an ionic crystal with specific solubility. When solid sodium chloride particles are exposed to water, ions, mostly Na^+ and Cl^-, go into the solution and form an ionic halo around the particles. These ions contribute to the electrical conduction and turn back to the sodium chloride particle when the water is removed[35]. Interaction may occur between the solid and the solvent, which causes the dielectric property of the mixture to be lower or higher than the sum of the free moisture plus solid.

3.2 Effects of temperature on dielectric properties

The temperature dependence of dielectric properties of sodium chloride with 4% moisture content over the temperature range from 20℃ to 100℃ are presented in Figs. 7 and 8.

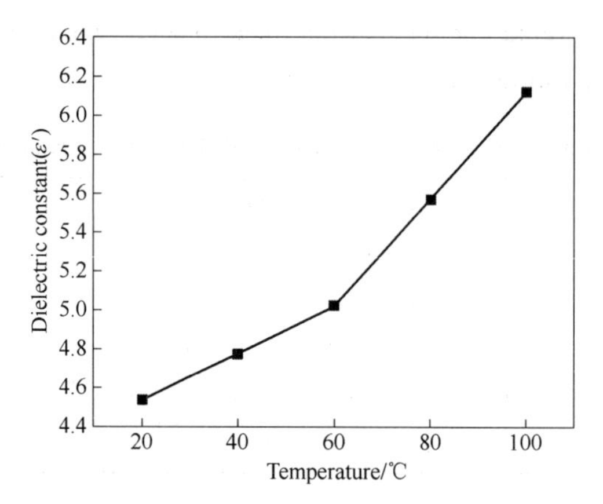

Fig. 7 Variation of dielectric constant of sodium chloride (4% moisture content) with temperature at 2.45GHz

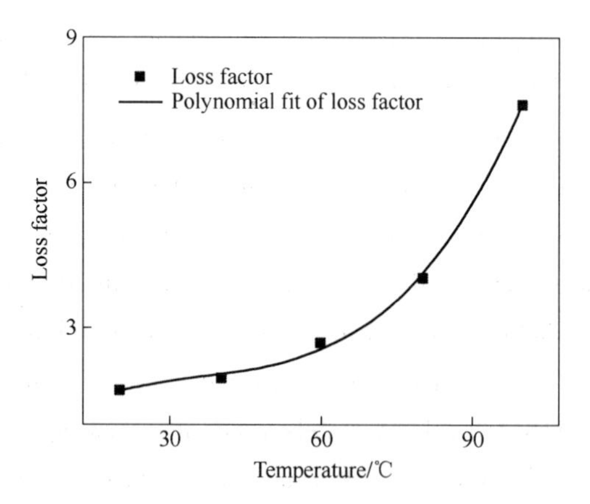

Fig. 8 Variation of loss factor of sodium chloride (4% moisture content) with temperature at 2.45GHz

Fig. 7 shows that at a given moisture, dielectric constant of sodium chloride increase almost linearly with temperature increasing from 20℃ to 100℃. At low temperatures (below 60℃), change in dielectric constant is very small. A rapid increase in dielectric constant was observed when the temperature was raised above 60℃. There is a positive linear relationship between the dielectric constant and temperature. An increase in dielectric constant with temperature increasing was noticed in food products[36,37]. Water in sodium chloride exists as free water and temperature dependence of the dielectric properties of water has been reported, ε' and ε'' both decrease with increasing temperature from 20℃ to 100℃[38]. However, the temperature dependent behavior of dielectric constant with 4% water content is not similar to that water at 20 – 100℃. Compared with pure water keeps decreasing with temperature, dielectric constant of wet sodium chloride keep increasing can be attributed to the varying of dielectric constant of sodium chloride with temperature. The effect of temperature on the loss factor of sodium chloride with 4% moisture content is shown in Fig. 8. Polynomial equations of 3rd order were used to adequately describe the variation of with temperature ranges of 20 – 100℃. The R^2 value of the model for loss factor is 0.99.

The mixture at higher temperatures has larger loss factors than at lower temperatures. The higher the loss factor, the more energy absorbed by dielectric. Water will absorb much more energy, and reaches a higher temperature than pure sodium chloride particles under the same treatment time[14]. This enables water to reach boiling point while sodium chloride is still at lower temperatures. That is the advantages of selective heating in microwave drying.

For pure sodium chloride particles, the loss factor increases with increasing temperature, while for pure water, the loss factor decreases with increasing temperature. At 4% moisture content, most water is tightly bound to solids with little mobility at room temperature to respond to microwave field at 2.45GHz. Researchers have found that the dielectric properties are related to the mobility of water and ions in solution, so an increase in mobility of the molecules in the mixture causes an increase in the dielectric properties of the system[12]. Raising temperature increases the mobility of water molecules and ionic conductivity, as indicated by an increase in loss factor in sodium chloride with temperature[18]. The loss factor is the sum of two components: ionic, ε''_σ, and dipole, ε''_d loss. These two components respond oppositely to temperature. The dipole loss decreases and ionic loss increases with temperature. The increasing trend of ε'' with temperature explains the possibility of runaway effects at lower moisture ranges[38].

Heating efficiency and non – uniformity are important factors in developing microwave drying techniques of minerals and ores[14]. The measured results obtained in this study showed both dielectric constant and loss factor of sodium chloride increase with temperature and moisture content increasing at 2.45GHz. Temperature rise leads to higher loss factor which in turn increases the temperature rising rate. When heated in a multimode microwave cavity, moist sodium chloride may experience thermal runaway and hot spots effects. The compressed sodium chlorides and water mixture sample was measured as a solid volume in dielectric measurement. But the air voids between particles and water may also result in non – uniform microwave heating.

The dielectric behavior of the sodium chloride illustrated in this paper is highly influenced by its water content compared with temperature in microwave dying process.

3.3 Relationship between penetration depth and moisture content

The power penetration depth is defined as the depth where the strength of microwave power is reduced to 1/e of its surface value and is expressed with the following equation[3]:

$$D_p = \frac{\lambda_0}{2\sqrt{2}\pi\sqrt{\varepsilon'\left[\sqrt{1+\left(\frac{\varepsilon''}{\varepsilon'}\right)^2}-1\right]}} \tag{3}$$

where λ_0 is the wavelength, which is a frequency – dependent parameter.

Penetration depth provides information related to the effective depth of microwave power disposition[39]. The power penetration depths of the sodium chloride particles at different moisture contents were computed using Eq. (3) and the results are shown in Table 2. Fig. 9 illustrates the effects of moisture content on power penetration depth at 2.45GHz. From previous research we know that the power penetration depth of pure water at room temperature is 1.04cm at 2450MHz and that of dry sodium chloride particles is 14.45cm. This depth of wet sodium chloride particles at any moisture content is smaller than that of dry sodium chloride particles and larger than that of pure water. In general, the moisture content in wet sodium chloride particles is less than 5%, which is much lower than that in foods such as apples and breads. Therefore, the power penetration depth of sodium chloride particles increases gradually with the decrease in moisture content. This slow increase with the reduction in moisture is different from that in the case of apple in which a sharp change happens at the moisture content of 30%[40]. Regression analysis reveals that the relationship between the power penetration depth of wet sodium chloride particles and moisture content can be expressed by a fifth degree polynomial function (Table 3).

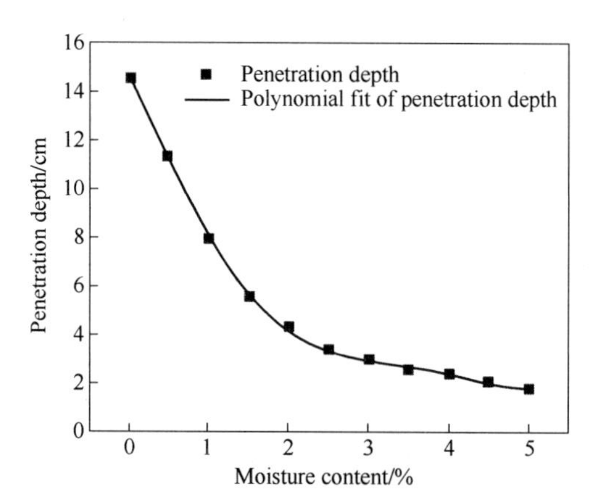

Fig. 9 Effects of moisture content on power penetration depth of sodium chloride at 2.45GHz

The R – square (R^2) value of Eq. (3) function is 0.998. A high correlation value and low deviations indicate that this equation can be used for prediction.

Fig. 10 depicts changes in penetration depth of sodium chloride with 5% moisture content from 20℃ to 100℃. An increase in temperature resulted in a decreased penetration between room temperature and boiling point of water. When the temperature increased from 20℃ to 100℃, the penetration depth was reduced from 14.88cm to 11.10cm. The effect of temperature became small above 60℃ where penetration depth keep same as at 60℃. Limited penetration depth led to non – uniform heating in sodium chloride with high moisture content and large thickness, because of the penetration depths.

For uniform drying with dielectric heating, the thickness of sodium chloride should not be more than two or three times the penetration depth[41]. Moisture content has more significant effects on dielectric properties of the sample than temperature. Considering the penetration depth of pure so-

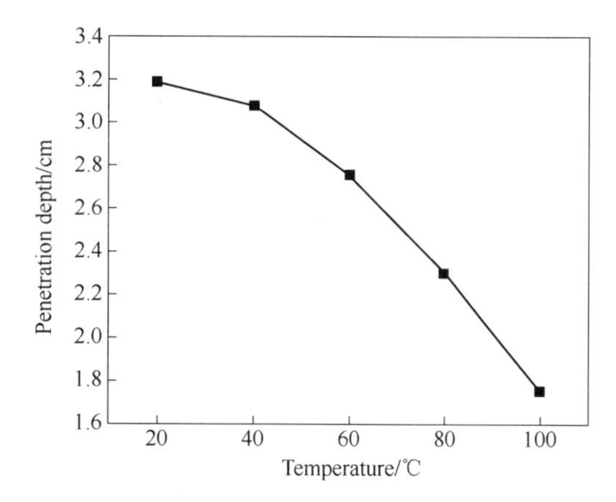

Fig. 10 Effects of temperature on power penetration depth of sodium chloride at 2. 45GHz

dium chloride, thickness for microwave dying would be between 3cm and 5cm at 2. 45GHz.

3. 4 Patterns of temperature increase at different power levels

Samples of 200g of wet sodium chloride particles with 5% of moisture were heated by microwave radiation at three power levels of 300W, 450W and 600W. Sample temperatures were measured during heating process. The results are depicted in Fig. 11. Because of their high dielectric property values, the mixture sample heated up fast to the water boiling point. Heating rates for the power of 600W were significantly different from that both 450W and 300W. At the power level of 300W, it takes 450s to heat the sample from a room temperature to a boiling temperature, while it took only 100s at the power level of 600W. But higher power densities may lead to non – uniformity in energy distribution within the sample. At all of the three power levels, sample temperatures increased linearly with microwave heating time. The results of regression analysis are depicted in Fig. 11 as fitted curves and the regression equations are given in Table 4. From the figure we can see that the sample temperatures increased from the inside of the particles immediately after microwave radiation was imposed. With the progress of this heating process, temperature graduates were established at the sample surfaces. These graduates could be attributed to the decrease in moisture content that led to a weaker coupling of microwave radiation with water in the portion of low moisture content[18]. In addition, the penetration depth and loss of the microwave energy could have contributed to the higher heating rate of the sodium chloride.

Table 4 Regression equations on sample temperature and microwave heating time

Microwave power/W	Regression equations	Correlation efficient(R^2)
300	$T = 35.04960 + 0.13829t$	0. 97379
450	$T = 27.7778 + 0.31697t$	0. 99667
600	$T = 25.18769 + 0.64954t$	0. 95782

Three stages can be identified on the curve of temperature increase in a microwave drying process. However, only the preheating stage was investigated in this study. The constant – rate dr-

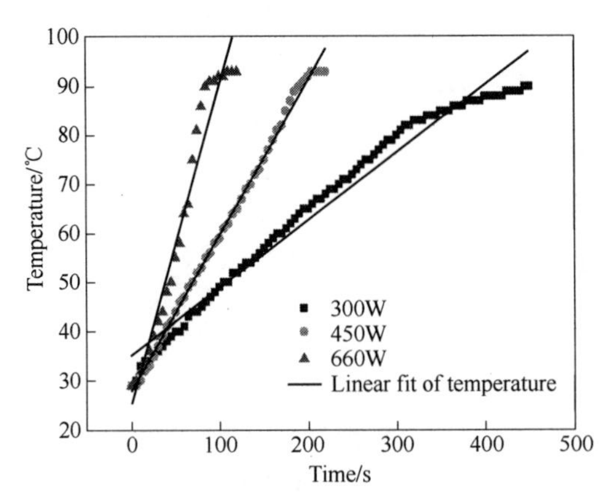

Fig. 11 Increase in sample temperature with microwave heating time when heated at power levels of 300W, 450W and 600W

ying and falling – rate drying stages were not examined in this work. Compared with dried and pure materials, samples with moisture have higher dielectric loss factor. At the beginning of microwave heating in the preheating stage, a large fraction of imposed microwave radiation was absorbed by the material because of high moisture content. The absorbed microwave energy can be converted into thermal energy in a short time and lead to increase in the temperature of the material and water, which in turn makes the vapor pressure inside the material greater than that in outside. Consequently, the water contained in the material evaporates to the surrounding air and the material was dried out gradually. This evaporation decreases with the reduction in moisture content of the material. Researchers have investigated the absorption of characteristic energy and power efficiency in the process of microwave drying[42]. The results show that energy consumption can be reduced in the process of drying when microwave drying is combined with other conventional drying techniques.

4 Conclusions

The effects of moisture content and temperature on the dielectric property of sodium chloride were investigated by using open – ended senor dielectric measurement system at the frequency of 2.45GHz.

Moisture content is a major influencing factor in the variation of dielectric properties. Dielectric constant, loss factor and loss tangent all increase linearly with moisture content increasing. Three predictive models were developed to obtain dielectric constant, loss factor, loss tangent and of sodium chloride as linear functions of moisture content. Temperature between 20℃ and 100℃ has a positive effect on dielectric constant and loss factor. A positive influence on loss factor will cause non – uniformity microwave heating. Penetration depth decreased with moisture and temperature increasing. Variation in penetration depth was found to vary nonlinearly with moisture content decreasing. A predictive model was developed to calculate penetration depth for sodium chloride as a fifth function of moisture content.

Such three dielectric properties model could be applied to predict microwave drying characteristics as a continuous function of changing moisture content under a drying process. The sample temperature increased linearly with microwave heating time in the preheating stage when the sample was heated at the power levels of 300W, 450W and 600W.

Acknowledgements

We thank for the Key National Basic Research Program of China (No. 51090385) and Project 2013AA064003 supported by 863 Project, Project 5114703 supported by National Natural Science Foundation of China; Project supported by the International S&T Cooperation Program of China (No. 2012DFA70570), and the Yunnan Provincial International Cooperative Program (No. 2011IA004) and Applied Basic Research Project of Yunnan Province (No. KKSA201152054). Thank for the dielectric properties measurements supported by Professor Huang at Sichuan University.

References

[1] Mujumdar A S. Research and development in drying: recent trends and future prospects[J]. Drying Technology, 2004, 22(1-2): 1-26.

[2] Groenewold H, Tsotsas E. A new model for fluid bed drying[J]. Drying Technology, 1997, 15(6-8): 1687-1698.

[3] Sass A. Simulation of heat-transfer phenomena in a rotary kiln[J]. Industrial & Engineering Chemistry Process Design and Development, 1967, 6(4): 532-535.

[4] Dadalı G, Kılıç Apar D, Özbek B. Microwave drying kinetics of okra[J]. Drying Technology, 2007, 25(5): 917-924.

[5] Zhang M, Tang J, Mujumdar A S, et al. Trends in microwave-related drying of fruits and vegetables[J]. Trends in Food Science & Technology, 2006, 17(10): 524-534.

[6] Zhang M, Jiang H, Lim R X. Recent developments in microwave-assisted drying of vegetables, fruits, and aquatic products—drying kinetics and quality considerations[J]. Drying Technology, 2010, 28(11): 1307-1316.

[7] Zhou B, Avramidis S. On the loss factor of wood during radio frequency heating[J]. Wood Science and Technology, 1999, 33(4): 299-310.

[8] Wang W, Chen G. Theoretical study on microwave freeze-drying of an aqueous pharmaceutical excipient with the aid of dielectric material[J]. Drying Technology, 2005, 23(9-11): 2147-2168.

[9] Appleton T J, Colder R I, Kingman S W, et al. Microwave technology for energy-efficient processing of waste[J]. Applied Energy, 2005, 81(1): 85-113.

[10] Bulatovic S, Wyslouzil D M. Process development for treatment of complex perovskite, ilmenite and rutile ores[J]. Minerals Engineering, 1999, 12(12): 1407-1417.

[11] Venkatesh M S, Raghavan G S V. An overview of dielectric properties measuring techniques[J]. Canadian Biosystems Engineering, 2005, 47(7): 15-30.

[12] Dev S R S, Raghavan G S V, Gariepy Y. Dielectric properties of egg components and microwave heating for in-shell pasteurization of eggs[J]. Journal of Food Engineering, 2008, 86(2): 207-214.

[13] Moschler W W, Hanson G R. Microwave moisture measurement system for hardwood lumber drying[J]. Drying Technology, 2008, 26(9): 1155-1159.

[14] Koubaa A, Perré P, Hutcheon R M, et al. Complex dielectric properties of the sapwood of aspen, white birch, yellow birch, and sugar maple[J]. Drying Technology, 2008, 26(5): 568 – 578.

[15] McLoughlin C M, McMinn W A M, Magee T R A. Physical and dielectric properties of pharmaceutical powders [J]. Powder Technology, 2003, 134(1): 40 – 51.

[16] Kaderka M. Dielectric properties of nickel – doped NaCl crystals [J]. Czechoslovak Journal of Physics B, 1969, 19(4): 530 – 536.

[17] Barnett R N, Landman U. Water adsorption and reactions on small sodium chloride clusters[J]. The Journal of Physical Chemistry, 1996, 100(33): 13950 – 13958.

[18] Böckmann R A, Hac A, Heimburg T, et al. Effect of sodium chloride on a lipid bilayer[J]. Biophysical Journal, 2003, 85(3): 1647 – 1655.

[19] Kraszewski A W, Nelson S O, You T S. Use of a microwave cavity for sensing dielectric properties of arbitrarily shaped biological objects [J]. Microwave Theory and Techniques, IEEE Transactions on, 1990, 38 (7): 858 – 863.

[20] Arai M, Binner J G P, Cross T E. Estimating errors due to sample surface roughness in microwave complex permittivity measurements obtained using a coaxial probe[J]. Electronics Letters, 1995, 31(2): 115 – 117.

[21] Deshpande M D, Tiemsin P I, Cravey R. A new approach to estimate complex permittivity of dielectric materials at microwave frequencies using waveguide measurements [J]. Microwave Theory and Techniques, IEEE Transactions on, 1997, 45(3): 359 – 366.

[22] Büyüköztürk O, Yu T Y, Ortega J A. A methodology for determining complex permittivity of construction materials based on transmission – only coherent, wide – bandwidth free – space measurements [J]. Cement and Concrete Composites, 2006, 28(4): 349 – 359.

[23] Seo I S, Chin W S, Lee D G. Characterization of electromagnetic properties of polymeric composite materials with free space method[J]. Composite Structures, 2004, 66(1): 533 – 542.

[24] Venkatesh M S, Raghavan G S V. An overview of microwave processing and dielectric properties of agri – food materials[J]. Biosystems Engineering, 2004, 88(1): 1 – 18.

[25] Nelson S O, Bartley P G. Open – ended coaxial – line permittivity measurements on pulverized materials[J]. Instrumentation and Measurement, IEEE Transactions on, 1998, 47(1): 133 – 137.

[26] Yan L P, Huang K M, Liu C J. A noninvasive method for determining dielectric properties of layered tissues on human back[J]. Journal of Electromagnetic Waves and Applications, 2007, 21(13): 1829 – 1843.

[27] Huang K, Cao X, Liu C, et al. Measurement/computation of effective permittivity of dilute solution in saponification reaction[J]. Microwave Theory and Techniques, IEEE Transactions on, 2003, 51(10): 2106 – 2111.

[28] Robinson D A. Measurement of the solid dielectric permittivity of clay minerals and granular samples using a time domain reflectometry immersion method[J]. Vadose Zone Journal, 2004, 3(2): 705 – 713.

[29] Hallikainen M T, Ulaby F T, Dobson M C, et al. Microwave dielectric behavior of wet soil – part 1: empirical models and experimental observations[J]. Geoscience and Remote Sensing, IEEE Transactions on, 1985(1): 25 – 34.

[30] Ahmed J, Ramaswamy H S, Raghavan V G S. Dielectric properties of butter in the MW frequency range as affected by salt and temperature[J]. Journal of food engineering, 2007, 82(3): 351 – 358.

[31] Saarenketo T. Electrical properties of water in clay and silty soils[J]. Journal of Applied Geophysics, 1998, 40 (1): 73 – 88.

[32] Whalley W R. Considerations on the use of time - domain reflectometry (TDR) for measuring soil water content[J]. Journal of Soil Science, 1993, 44(1): 1 – 9.

[33] Moschler W W, Hanson G R. Microwave moisture measurement system for hardwood lumber drying[J]. Drying

Technology, 2008, 26(9): 1155 - 1159.

[34] Liu Y, Tang J, Mao Z. Analysis of bread loss factor using modified Debye equations[J]. Journal of Food Engineering, 2009, 93(4): 453 - 459.

[35] Tulasidas T N, Raghavan G S V, Van de Voort F, et al. Dielectric properties of grapes and sugar solutions at 2.45GHz[J]. Journal of Microwave Power and Electromagnetic Energy, 1995, 30(2): 117 - 123.

[36] Okiror G P, Jones C L. Effect of temperature on the dielectric properties of low acyl gellan gel[J]. Journal of Food Engineering, 2012, 113(1): 151 - 155.

[37] Guo W, Liu Y, Zhu X, et al. Temperature - dependent dielectric properties of honey associated with dielectric heating[J]. Journal of Food Engineering, 2011, 102(3): 209 - 216.

[38] McMinn W A M, McLoughlin C M, Magee T R A. Temperature characteristics of pharmaceutical powders during microwave drying[J]. Drying Technology, 2006, 24(5): 571 - 580.

[39] Vongpradubchai S, Rattanadecho P. Microwave and hot air drying of wood using a rectangular waveguide[J]. Drying Technology, 2011, 29(4): 451 - 460.

[40] Guo W, Zhu X, Nelson S O, et al. Maturity effects on dielectric properties of apples from 10 to 4500MHz[J]. LWT - Food Science and Technology, 2011, 44(1): 224 - 230.

[41] Zhu X, Guo W, Wu X. Frequency - and temperature - dependent dielectric properties of fruit juices associated with pasteurization by dielectric heating[J]. Journal of Food Engineering, 2012, 109(2): 258 - 266.

[42] Zhang F, Zhang M, Mujumdar A S. Drying characteristics and quality of restructured wild cabbage chips processed using different drying methods[J]. Drying Technology, 2011, 29(6): 682 - 688.

Preparation of Reduced Iron Powders from Mill Scale with Microwave Heating: Optimization Using Response Surface Methodology

Qianxu Ye, Hongbo Zhu, Jinhui Peng, C. Srinivasakannan,
Jian Chen, Linqing Dai, Peng Liu

Abstract: Preparation of the reduced iron powder has been attempted with mill scale as the iron - bearing material and with wood charcoal as the reducing agent through microwave heating. The response surface methodology (RSM) is used to optimize the process conditions, with wood charcoal, process temperature, and holding time being the three process parameters. The regressed model equation eliminating the insignificant parameters through an analysis of variance (ANOVA) was used to optimize the process conditions. The optimum process parameters for the preparation of reduced iron powders have been identified to be the wood charcoal of 13.8%, a process temperature of 1391K (1118℃), and a holding time of 43min. The optimum conditions resulted in reduced iron powders with a total iron content of 98.60% and a metallization ratio of 98.71%. X - ray fluorescence (XRF) was used to estimate the elemental contents of the reduced iron powder, which meets the specification of the HY100.23 first - class iron powder standard. Additionally X - ray diffraction (XRD), energy - dispersive spectroscopy (EDS), and scanning electron microscopy (SEM) analysis were performed and the results are compiled.
Keywords: microwave heating; reduced iron powder; Mill scale; wood charcoal; RSM

1 Introduction

According to the different production methods and history, iron powders are classified in four types, reduced iron powder[1], atomized iron powder[2,3], and electrolytic iron powder[4], carbonyl iron powder[5], depending on the production method, and are used in various applications, taking advantage of their respective properties.

Reduced iron powders are one of the most essential materials in modern society. They have a very wide range of applications in different industries, including powder metallurgy, magnetic material[6-8], cutting and welding, coating[9-13], sewage treatment[14-16] as well as medical and food[17-20]. The demand for reduced iron powder is ever increasing due to the rapid development of car industry and overall due to betterment of living standards in developing countries. The oxidised, Mill scale[21] is formed in the process of rough machining or by further oxidizing. The output of Mill scale is just 1.0% of rough steel consumption. China output of rough steel is approximately 0.68 billion tons with the proportional output of Mill scale being 6.8 million tons in 2011. The global production of mill scale is reported to be about 13.5 million tons in 2008[22]. Mill scale is reported to be most suitable material for the preparation of high value magnetic materials and reduced iron powder owing to low impurities[23].

Traditional heating methods through conventional heat transfer modes lead to non – uniform heating, with large variation in temperature on the surface of the material to its interior, hindering the process of gas removal leading to long production duration and high energy consumption. Microwave heating is successfully applied in different scientific application due to its special characteristics such as selective, uniform, fast heating, no pollution and low equipment cost. The mode of heat transfer is through dipole polarization, ionic conduction and ferromagnetic resonance. When exposed to high frequency voltage, high frequency variation takes place in the direction of electromagnetic field, resulting in high frequency rotation of the polar molecule and dipole which generates heat quickly[24].

This paper deals with the preparation of reduced iron powder using wood charcoal as reducing agent, Mill scale as iron – bearing materials by microwave heating. Response Surface Methodology (RSM) is utilized to optimize the process parameters such as wood charcoal ratio, the activation temperature, holding time.

2 Materials and methods

2.1 Materials

The compositions of the Mill scale and wood charcoal are shown in the following Tables 1 and 2. The particle size distributions of the material and wood charcoal used is represented in Fig. 1. All these materials are milled by ball grinding mill. The percentages shown in full paper are weight percentages.

Table 1 Compositions of the Mill scale (total iron 74.25percentage) (%)

FeO	Fe_2O_3	Fe_3O_4	SiO_2	MnO	P	S	CuO	SnO_2	CaO	Cr_2O_3
61.31	36.36	1.61	0.096	0.30	0.014	0.017	0.12	0.09	0.032	0.029

Table 2 Compositions of wood charcoal (%)

Fixed carbon	Volatile organic matter	H_2O	CaO	FeO	MgO	S	K_2O	Na_2O
72.58	22.26	4.67	0.40	0.01	0.014	0.0028	0.069	0.0005

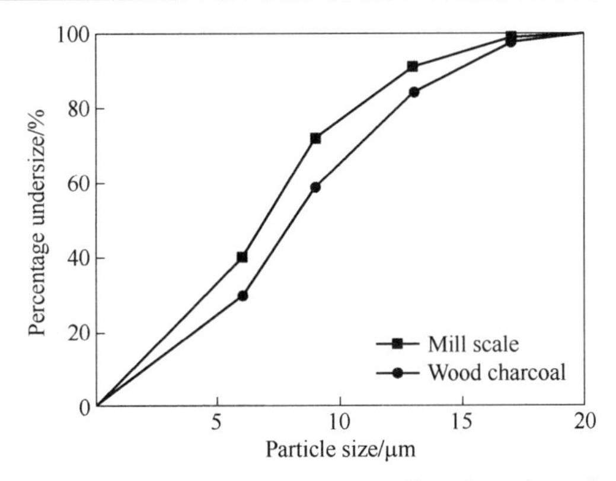

Fig. 1 Particle size distributions of the Mill scale and wood charcoal

2.2 Methods

A self - made microwave furnace, which utilizes a single - mode continuous controllable power is utilized for all experiments and is as shown in Fig. 2. The microwave frequency is 2.45GHz, while the output power is controlled within the maximum of 3000W. The activation temperature is controlled by varying the input microwave power. The activation temperature is measured by nickel chrome - nickel silicon armor type thermocouple which is in contact with the material. The thermocouple has dimension of length of 450mm, 8mm diameter, with the temperature range of 273K to 1523K (0℃ to 1250℃), and a measurement precision of ±0.5K.

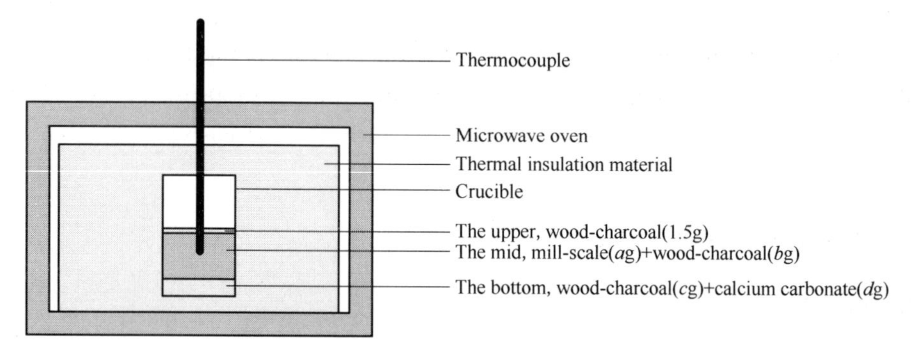

Fig. 2 Microwave cavity filling diagram

Feed proportion: the materials are divided into three layers. At the bottom layer is the mixture of wood charcoal and calcium carbonate, and at the mid - layer is the mixture of wood charcoal and Mill scale while at the upper layer is wood charcoal.

$$a + b = 100 \tag{1}$$

$$\frac{b}{a + b} \times 100\% = X_1\% \tag{2}$$

$$\frac{b + c}{a + b + c} \times 100\% = 20\% \tag{3}$$

$$\frac{d}{c + d} = 40\% \tag{4}$$

If $c = 0g$, $d = 0g$ and the reductant is C, the stoichiometric content of reductant required for reduced completely of mill scale is 15.78%.

2.3 Design of experiments

RSM is a collection of statistical and mathematical techniques useful for developing, improving, and optimizing processes in which a response of interest is influenced by several variables and the objective is to optimize this response[25]. RSM has important application in the design; development and formulation of new products, as well as in the improvement of existing product design[26-29].

The experimental conditions were generated using a Central Composite Design (CCD) with the three dependable variables being the wood charcoal (X_1), process temperature (X_2), holding time

(X_3). The total number of experiments using a full factorial CCD for 3 variables, with 8 factorial point, 6 axial points and 6 replicates at the center points can be calculated from the following equation[30]:

$$N = 2^n + 2n + n_c = 2^3 + 2 \times 3 + 6 = 20 \quad (5)$$

where N is the total number of experiments, while n is the number of dependent variables. Table 3 provides the experimental design matrix.

Table 3 Experimental design matrix and results

S. No.	X_1	X_2	X_3	Y_1	Y_2
1	10	1323K	20	92.18	79.8
2	15	1323K	20	97.53	94.02
3	10	1423K	20	96.27	89.51
4	15	1423K	20	97.32	93.5
5	10	1323K	60	95.51	86.66
6	15	1323K	60	97.61	94.6
7	10	1423K	60	96.75	91.02
8	15	1423K	60	96.98	92
9	8.3	1373K	40	93.34	78.94
10	16.7	1373K	40	98.11	96.42
11	12.5	1289K	40	96.85	92.1
12	12.5	1457K	40	98.42	97.23
13	12.5	1373K	7	94.49	84.75
14	12.5	1373K	74	97.34	93.38
15	12.5	1373K	40	98.18	96.77
16	12.5	1373K	40	97.95	95.96
17	12.5	1373K	40	98.07	96.55
18	12.5	1373K	40	98.27	96.02
19	12.5	1373K	40	97.66	94.7
20	12.5	1373K	40	98.03	96.52

The experimental data were analyzed using statistical software Design Expert software version 7.1.5 (STAT-EASE Inc, Minneapolis, USA) for regression analysis and to evaluate statistical significance of the equation.

3 Results and discussion

The experimental design matrix (Table 3) provides the experimental conditions and the results in terms of total iron (Y_1) and metallization ratio (Y_2).

$$\text{Total iron} = \frac{M_1}{M_0} \times 100\% \quad (6)$$

$$\text{Metallization ratio} = \frac{M_2}{M_1} \times 100\% \quad (7)$$

where M_0 is the mass of a sample; M_1 is the mass of Fe; M_2 is the mass of zero valent iron. Total iron is measured by the following steps. First, the Fe contained in the sample is oxidized to Fe^{3+} by strong acid, then the Fe^{3+} is reduced to Fe^{2+} by $TiCl_3$, finally the Fe^{2+} is titrated to Fe^{3+} by standard liquid of potassium dichromate. The measurement process of metallization is similar; the zero valent iron in the sample is first oxidized to Fe^{2+} by $FeCl_3$, and then titrated to Fe^{3+} by potassium dichromate.

The ranges of dependable variables covered in the present experiments were: fixed wood charcoal (10% to 15%); process temperature of 1323K to 1423K (1050℃ to 1150℃); holding time of 20min to 60min. The chemical and microstructure analysis were performed with the sample having the highest total iron content.

3.1 Total iron

The primary index to assess the quality of reduction is based on the estimate of total iron, which primarily depends on the metallization ratio and the presence of impurity. Fig. 3 shows the three - dimensional response surface plot of the process temperature and pct wood charcoal on the total iron, at a holding time of 40min. Increase in both the parameters are found to increase the total iron content, however wood charcoal having more significant effect as compared to the process temperature. Either an increase in the process temperature or the wood charcoal content facilitate an increased rate of metallization reaction contributing to an increased total iron content. A maximum of 98% total iron content was observed at the highest of the wood charcoal content and at the lowest of temperature, indicating the near complete conversion. Any further increase in the temperature would contribute to secondary reactions, which could have possibly reduced the total iron content. However, at low wood charcoal content, only an increase in the total iron content with the process temperature was observed.

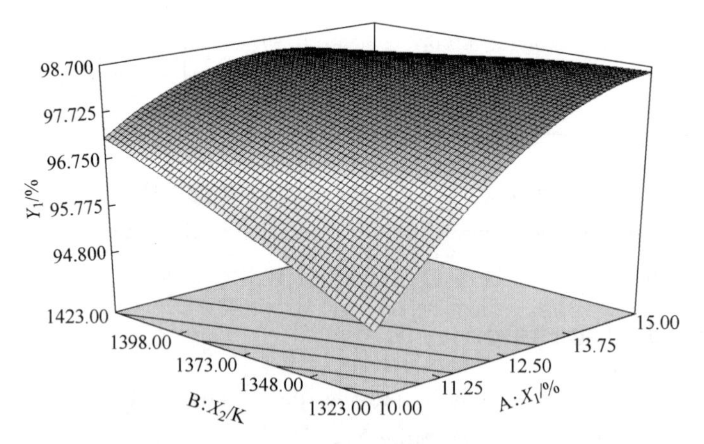

Fig. 3 Three - dimensional response surface plot of total iron: effect of reduced temperature and Wood charcoal ratio on the total iron (holding time: 40min)

Fig. 4 shows the three - dimensional response surface plot of the process temperature and holding time on the total iron at a wood charcoal content of 12.5%. An increase in both the process temperature is found to have significant effect on the total iron content, however the effect of holding time was more significant than the process temperature. According to Arrhenius equation, an increase in temperature contributes to the increased rate of reaction, while an increase in the holding time will increase the extent of reaction. However an optimum was observed at a combination corresponding to maximum holding time and temperature. The reduction in iron content beyond the optimum combination could be due to the secondary reactions as discussed in the earlier section.

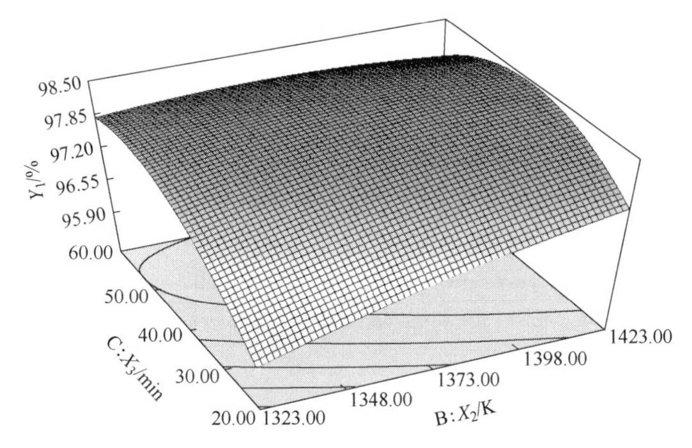

Fig. 4 Three - dimensional response surface plot of total iron: effect of reduced temperature and holding time on the total iron (Wood charcoal ratio: 12.5%)

3.2 Metallization ratio

Table 3 presents metallization ratio of reduced iron powder corresponding to the each experimental condition. As defined earlier, metallization ratio is yet another index, which helps to assess the quality of the reduction process. Fig. 5 is three - dimensional response surface plot of the reduced temperature and wood charcoal pct on the metallization ratio, while Fig. 6 is three - dimensional response surface plot of the reduced temperature and holding time on the metallization ratio. It can be observed from the figures that the trend very closely resembles to the Figs. 3 and 4, indicating that the effects are similar, however with a different magnitude.

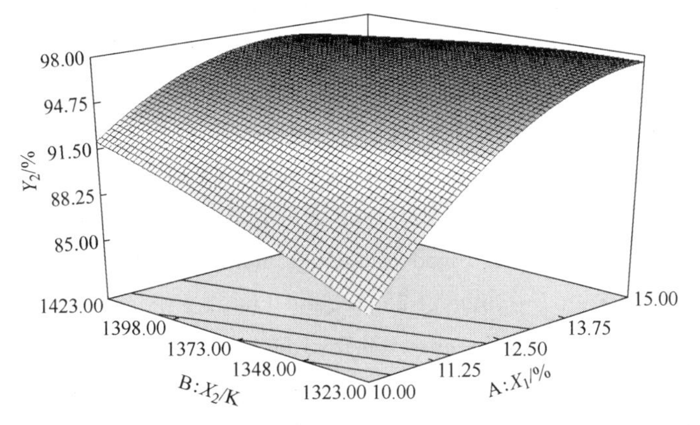

Fig. 5 Three - dimensional response surface plot of metallization ratio: effect of reduced temperature and Wood charcoal ratio on metallization ratio (Holding time: 40min)

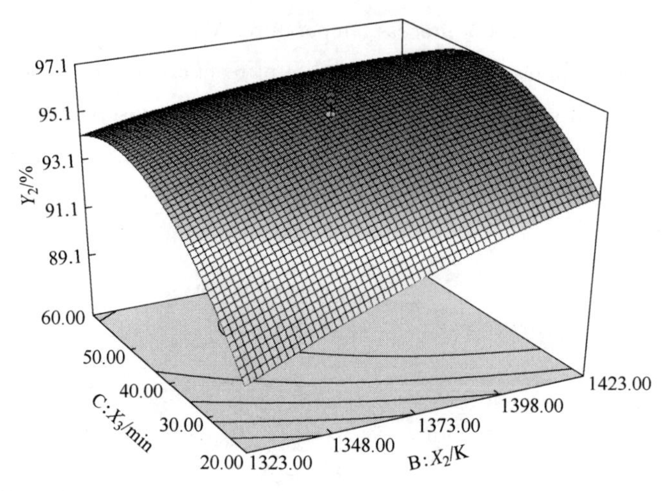

Fig. 6 Three – dimensional response surface plot of metallization ratio; effect of reduced temperature and holding time on the metallization ratio (Wood charcoal ratio: 12.5%)

3.3 Development of regression model

The dependable variables Y_1 and Y_2 are related to the independent variables through the regression analysis, with the help of the software and are presented as Eqs. (8) and (9). The model equation only accounts the variables that are significant. It can be observed that all the main effects as well interaction effects were found to be significant except the square term of the X_2 parameter. The significance of each of the parameters can be assessed from the low p values shown in Tables 4 and 5.

$$Y_1 = -125.47239 + 10.98959X_1 + 0.23675X_2 + 0.75917X_3 - 6.17E^{-3}X_1X_2 - 0.010175X_1X_3 - 4.0875E^{-4}X_2X_3 - 0.1322X_1^2 - 1.89764E^{-3}X_3^2 \quad (8)$$

$$Y_2 = -574.69049 + 33.3745X_1 + 0.72126X_2 + 1.88846X_3 - 0.01719X_1X_2 - 0.023225X_1X_3 - 9.2875E^{-4}X_2X_3 - 0.47525X_1^2 - 6.20135E^{-3}X_3^2 \quad (9)$$

The appropriateness of model equation is judged by correlation coefficient R^2, which is 0.9661 for Eq. (8), and 0.9583 for Eq. (9). The high R^2 values indicate the goodness of the fit between the independent and the dependent variables, which can further be ensured based on Figs. 7 and 8, both indicating good proximity between the experimental value and the predicted value using the model equation.

The rationality of the model can further be assessed based on the Analysis of Variance (ANOVA). The ANOVA for the response surface quadratic model for total iron is shown in Table 4. The value of model Prob > F smaller than 0.0001 and F – value of 31.65, both indicate the significance and appropriates of the proposed model equation. Similarly Table 5 presents the ANOVA for the metallization ratio. Both an F – value of 25.52, and Prob > F is smaller than 0.0006, indicate the significance of the model. It can be observed from both the ANOVA table that all the parameters are significant except the square term of the X_2 variable. A_p value of less than 0.05 shows the probability for the coefficient to be zero, indicating that the coefficients cannot be ignored in developing the model equation.

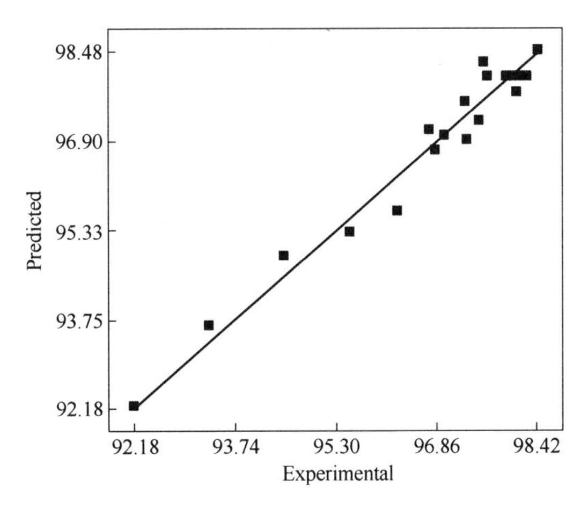

Fig. 7 Predicted as a function of experimental total iron

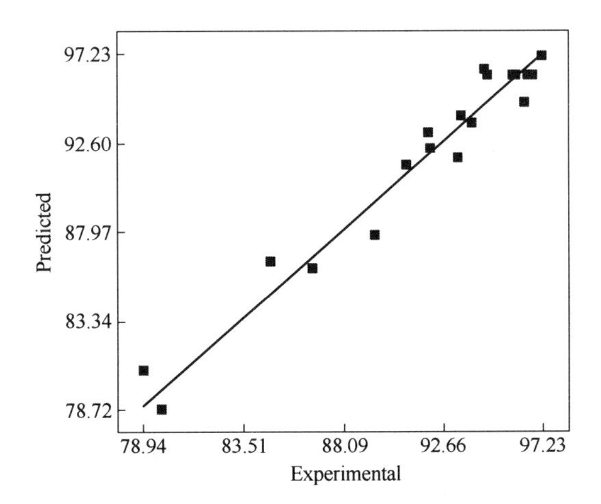

Fig. 8 Predicted as a function of experimental metallization ratio

Table 4 Analysis of variance (ANOVA) for response surface quadratic model for total iron

Source	Sum of squares	Degree of freedom	Mean square	F - value	Prob > F
Model	54.04	9	6.01	31.65	<0.0001
x_1	20.55	1	20.55	108.29	<0.0001
x_2	3.72	1	3.72	19.62	0.0013
x_3	5.10	1	5.10	26.86	0.0004
x_1x_2	4.76	1	4.76	25.08	0.0005
x_1x_3	2.07	1	2.07	10.91	0.0080
x_2x_3	1.34	1	1.34	7.04	0.0241
x_1^2	9.84	1	9.84	51.85	<0.0001
x_2^2	0.33	1	0.33	1.73	0.2177
x_3^2	8.30	1	8.30	43.76	<0.0001

Table 5 Analysis of variance(ANOVA) for response surface quadratic model for metallization ratio

Source	Sum of squares	Degree of freedom	Mean square	F - value	Prob > F
Model	548. 66	9	60. 97	25. 5217	<0. 0001
x_1	233. 98	1	233. 98	97. 95	<0. 0001
x_2	28. 07	1	28. 07	11. 75	0. 0065
x_3	35. 32	1	35. 32	14. 79	0. 0032
x_1x_2	36. 94	1	36. 94	15. 46	0. 0028
x_1x_3	10. 79	1	10. 79	4. 52	0. 0595
x_2x_3	6. 90	1	6. 90	2. 89	0. 1200
x_1^2	127. 14	1	127. 14	53. 23	<0. 0001
x_2^2	3. 61	1	3. 61	1. 51	0. 2470
x_3^2	88. 67	1	88. 67	37. 12	0. 0001

3. 4 Process optimization

The model equations were utilized to optimize the process conditions to maximize the total iron content using the Design Expert software. The optimum conditions were identified to be a wood charcoal percentage of 13. 8% , process temperature of 1391K(1118℃), holding time of 43min, with the total iron content being 98. 48% and metallization ratio of 97. 53%. Experiments are repeated to ensure the appropriateness of the optimized process conditions. The repeat experimental runs have resulted in a total iron average of 98. 61% and a metallization ratio of 98. 71% , with the relative pct error being 0. 13% and 1. 2% , which proves the effectiveness of the process optimization exercise.

3. 5 Chemical composition analysis

The commercial market demands not only high total iron, but also absence of impurities such as C, Si, Mn, P and S in reduced iron powder. XRF – 1800 of Shimadzu(Shimadzu, XRF – 1800 Sequential WDXRF) was utilized to assess the presence of various components in the sample and the results are listed in Table 6. It is obvious that Fe content is 99. 13% , while the contents of Si, Mn, P and S are low, which meets the standard of the first grade of Broad FHY100 · 23 iron powder. However, Mn content is high which is characteristic of using Mill scale reduced iron powder. The contents of other impurities such as Ca and Al were also observed to be high. The impurities could be either from the iron scale or from the ash content of wood charcoal.

Table 6 Compositions of reduced iron powders (%)

Fe	Mn	Si	P	S	Cu	Sn	Ca	Cr	K
99. 13	0. 3117	0. 0603	0. 0141	0. 0171	0. 1307	0. 0945	0. 0774	0. 0284	0. 0091

Table 7 shows the conversion of Table 6, which presents the state of components of reduced iron powder. It is based on X – ray fluorescence(XRF) which is not able to recognize light symbols such as O.

Fig. 9 is the X - ray diffraction(XRD) of reduced iron powder. Iron elements can be only recognized by XRD. It shows that content of impurity in iron powder is below the detection range of XRD.

Table 7 Compositions of reduced iron powders containing O (%)

Fe	MnO	SiO_2	P_2O_5	SO_3	CuO	SnO_2	CaO	Cr_2O_3	K_2O
98.60	0.3984	0.1275	0.0319	0.0436	0.1593	0.1195	0.1078	0.0385	0.0023

Note: FeO is not listed on Table 7.

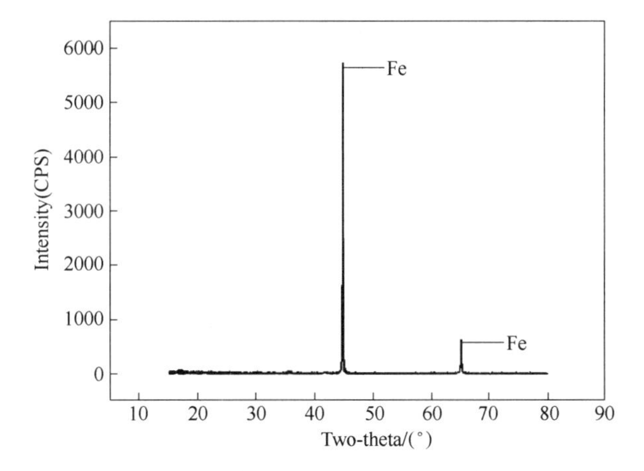

Fig. 9 XRD of reduced iron powder

3.6 Scanning electron microscopy(SEM) - energy - dispersive X - Ray spectroscopy(EDS) analysis

Fig. 10 shows a SEM of the mixture of 13.8% wood charcoal along with mill scale. Wood charcoal corresponds to the darker color, with smooth surface and clear profile, while the rest of the portion corresponds to Mill scale. It is obvious that mix of materials is uniform, and particle size of Mill scale distributes uniformly, with the size range from 2μm to 3μm to 10μm. The particle size of wood charcoal is about 5μm, and much more uniform.

Fig. 11 is SEM image of reduced iron powder. It shows a highly developed network structure. It is because that the wood charcoal is used up, and the Mill scale shrinks since oxygen(O) is released due to the reaction. The reduced iron particles fuse and bond each other, finally forming a network structure. There are some wide space cavities, left by big particles of coal consumed. There are some large network nodes, which were due to big particles of reduced Mill scale. It is clear that each particle is smooth and rod - like, with a diameter of about 2μm and length of 15μm. Some of them have irregular bright spots on the top.

Fig. 12 is EDS image of reduced iron powder. It shows that iron particles in principal part are Fe (Fig. 12(a)). It shows that irregular bright points on the top of iron particles are impurities, which include unreduced iron content. Most of elemental impurities being Ca, Mn, Si, P, C, possibly from

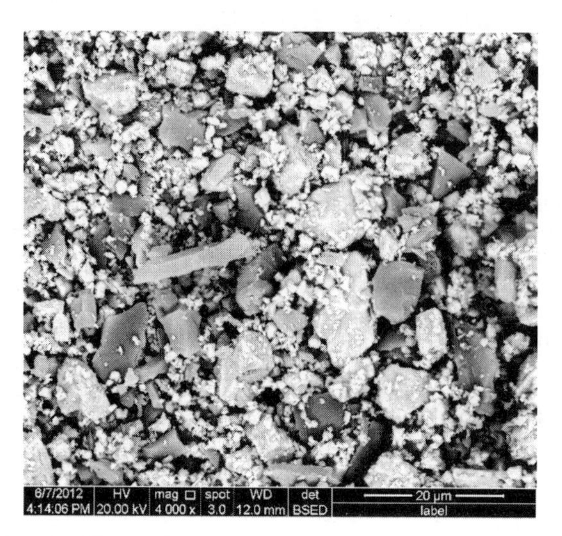

Fig. 10 SEM of the mixture of wood charcoal and Mill scale

Fig. 11 SEM of reduced iron powder

the unconsumed carbon along with ash content of wood charcoal and Mn from Mill scale. If the liberation of them from iron substrate is possible by crushing, and separation through magnetic separator, the content of impurity can further be reduced with proportionate increase in the total iron.

4 Conclusions

The RSM is used to optimize the process conditions, with wood charcoal, process temperature, and holding time being the three process parameters. The optimum process parameters for the preparation of reduced iron powders have been identified to be the rate of wood charcoal in the scale of 13.8%, process temperature of 1391K(1118℃), and holding time of 43min. The optimum conditions result in a reduced iron powers with total iron content of 98.60% and metallization ratio of 98.71%, which are complete agreement with the model equation predicted. XRF was used to esti-

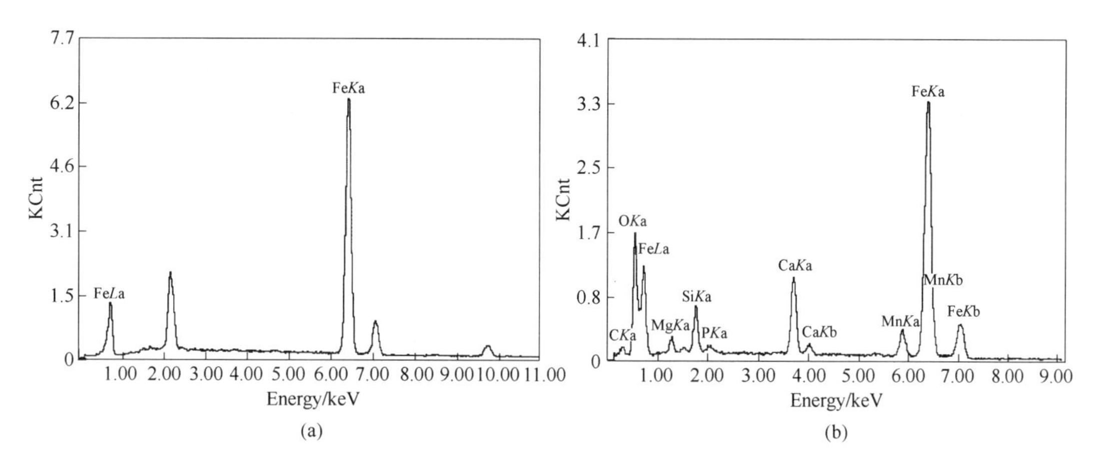

Fig. 12 EDS of the trunk of reduced iron powder(a) and EDS of the irregular bring points on the top of iron particles(b)

mate the elemental contents of the reduced iron powder, which meets the specification of the HY100. 23 first – class iron powder standard.

The results indicate the current process can produce high – quality reduced iron powders well in comparison with the popular Hogonas process, with a shorter process duration rendering the process economically viable.

Acknowledgements

Financial supports from the National Natural Science Foundation of China(No. 51090385), and the National Basic Research Program of China(No. 2007CB613606) were sincerely acknowledged.

References

[1] Yu X T. Spong iron and reduced iron powder produced by Höganäs progress[J]. Powder Metallurgy Industry. 1997, 7(3): 31 – 39.

[2] Han F L. Production of atomized iron and steel powders at QMP CO[J]. Powder Metallurgy Industry, 2003, 13 (2): 16 – 21.

[3] A S M. Metals Handbook[M]. Powder Metal Technologies and Applications, 1998, 7: 117.

[4] Chen Z D. Production of iron powder by electrolysis process[J]. Powder Metallurgy Industry, 1994, 20(2): 66 – 68.

[5] Jin Z H. Powder Metallurgy Industry. 1995, 5: 169 – 173.

[6] Bose A. Development of low – coercivity magnets from sponge iron powder[J]. Metal Powder Report, 2002, 57 (7): 88.

[7] Sidhu R K. Influence of particle size of iron powder on the microstructure of Nd – Fe – B alloy powder prepared by reduction – diffusion[J]. Journal of Alloys and Compounds, 2002, 346(1): 250 – 254.

[8] Lefebvre L P, Pelletier S, Gelinas C. Effect of electrical resistivity on core losses in soft magnetic iron powder materials[J]. Journal of Magnetism and Magnetic Materials, 1997, 176(2): L93 – L96.

[9] Bibikov S B, Kulikovskij E I, Kuznetsov A M, et al. Proceedings of the Second International Workshop on Ultra-wide Band Ultrashort Impulse Signal[C]. UWBUSIS, 2004: 129.

[10]Jana P B,Mallick A K,De K. IEEE Trans Electromagnetic Compatibility. 1992,34(4):478 -481.

[11]Soloman M A,Kurian P,Anantharaman M R,et al. Evaluation of the magnetic and mechanical properties of rubber ferrite composites containing strontium ferrite[J]. Polymer - Plastics Technology and Engineering, 2004,43(4):1013 -1028.

[12]Y R G. Electromagnetic properties and microwave absorption properties of $BaTiO_3$ - carbonyl iron composite in S and C bands[J]. Journal of Magnetism and Magnetic Materials,2011,323(13):1805 -1810.

[13] Liu L, Duan Y, Ma L, et al. Microwave absorption properties of a wave - absorbing coating employing carbonyl - iron powder and carbon black[J]. Applied Surface Science,2010,257(3):842 -846.

[14]Volpe A, Lopez A, Mascolo G, et al. Chlorinated herbicide (triallate) dehalogenation by iron powder[J]. Chemosphere,2004,57(7):579 -586.

[15] Ghauch A, Rima J, Amine C, et al. Rapid treatment of water contamined with atrazine and parathion with zero - valent iron[J]. Chemosphere,1999,39(8):1309 -1315.

[16]Ghauch A. Rapid removal of flutriafol in water by zero - valent iron powder[J]. Chemosphere,2008,71(5): 816 -826.

[17]Guan J,Zhou L,Nie S,et al. A novel gastric - resident osmotic pump tablet: in vitro and in vivo evaluation [J]. International Journal of Pharmaceutics,2010,383(1):30 -36.

[18]INACG. Guide lines for the eradication of iron deficiency anemia[J]. A Report of the International Utritional Anemia Consultative Group(INACG). Washington DC7 Nutrition Foundation,1977:1 -29.

[19]Arredondo M,Salvat V,Pizarro F,et al. Smaller iron particle size improves bioavailability of hydrogen - reduced iron - fortified bread[J]. Nutrition Research,2006,26(5):235 -239.

[20]Yeung C K,Miller D D,Cheng Z,et al. Bioavailability of elemental iron powders in bread assessed with an in vitro digestion/Caco -2 cell culture model[J]. Journal of Food Science,2005,70(3):S199 - S203.

[21]Umadevi T,Sampath Kumar M G,Mahapatra P C,et al. Recycling of steel plant mill scale via iron ore pelletisation process[J]. Ironmaking & Steelmaking,2009,36(6):409 -415.

[22]Cho S,Lee J. Metal recovery from stainless steel mill scale by microwave heating[J]. Metals and Materials International,2008,14(2):193 -196.

[23]Beatty R L,Sutton W H,Iskander M F. Microwave Processing of Materials Ⅲ:Symposium Held April 27 - May 1,1992,San Francisco,California,USA[M]. Materials Research Society,1992:3 -20.

[24]Venkatesh M S,Raghavan G S V. An overview of microwave processing and dielectric properties of agri - food materials[J]. Biosystems Engineering,2004,88(1):1 -18.

[25]Baş D,Boyacı İ H. Modeling and optimization I:usability of response surface methodology[J]. Journal of Food Engineering,2007,78(3):836 -845.

[26]Liyana - Pathirana C,Shahidi F. Optimization of extraction of phenolic compounds from wheat using response surface methodology[J]. Food Chemistry,2005,93(1):47 -56.

[27]Anupam K,Dutta S,Bhattacharjee C,et al. Adsorptive removal of chromium(Ⅵ) from aqueous solution over powdered activated carbon:optimisation through response surface methodology[J]. Chemical Engineering Journal,2011,173(1):135 -143.

[28]Erbay Z,Icier F. Optimization of hot air drying of olive leaves using response surface methodology[J]. Journal of Food Engineering,2009,91(4):533 -541.

[29]King V A E,Zall R R. A response surface methodology approach to the optimization of controlled low - temperature vacuum dehydration[J]. Food Research International,1992,25(1):1 -8.

[30]Azargohar R,Dalai A K. Production of activated carbon from Luscar char:experimental and modeling studies [J]. Microporous and Mesoporous Materials,2005,85(3):219 -225.

Carbothermal Reduction of Low - grade Pyrolusite by Microwave Heating

Qianxu Ye, Hongbo Zhu, Libo Zhang, Peng Liu, Guo Chen, Jinhui Peng

Abstract: Pyrolusite was carbothermally reduced using coal by microwave heating, and the crystal structures and microstructures of the samples were characterized after microwave heating using X - ray diffraction(XRD), scanning electron microscopy (SEM) and energy - dispersive spectroscopy (EDS). When the reductants proportion was of 10%, the reduction temperature was of 800℃ and the holding time was of 40min, the reduction ratio of MnO_2 to MnO was 97.2%, and Fe_2O_3 was almost completely transformed to Fe_3O_4 and there was no Fe(Ⅱ) produced. Moreover, it was found that the low - grade pyrolusite was carbothermally reduced using microwave heating with lower temperature and shorter processing time. These results show that microwave heating can be applied effectively and efficiently to the carbothermal reduction processes of low - grade pyrolusite.

Keywords: pyrolusite; microwave heating; coal; reduction

1 Introduction

With the extensive exploitation and consumption of high grade manganese ore (total Mn > 35%)[1], considerable attention is focused on the development and utilization of the low grade (total Mn < 35%) manganese ore, which has a low business value[2,3].

Pyrolusite is usually formed by MnO_2 and other oxides such as SiO_2 and Fe_2O_3[4,5]. It is granular, fibrous or columnar - like, which are good properties for absorbing microwave energy[6]. Pyrolusite is also a significant strategic resource and a key material for the production of MnO_2 and Mn. MnO_2 has many polymorphic forms such as a - , b - , g - and 3 - type[7]. It is widely used in the fields of catalysis, ion sieve, electrode materials of Li/MnO_2 and semiconductors[8-14]. Manganese is essential to iron and steel production because of its sulfur - fixing, deoxidizing, and alloying properties. Manganese from these ores can be extracted selectively using hydrometallurgical techniques.

One thing is to pre - reduce pyrolusite at high temperature (1000 - 1350℃)[15-18]. The reductants are used as natural gas and methane gas, which contains H_2, CO, CH_4 and carbonaceous materials like coal, wood - charcoal and graphite[19-27]. Firstly, MnO_2 in the pyrolusite is reduced to MnO, and then leached using a hot acid solution, and Mn^{2+} is obtained from the leaching solution. The leaching solution containing Mn^{2+} and Fe^{2+} needs to be purified and electrolyzed, and then the high quality electrolytic MnO_2 is acquired. Although this method is ell established, it is time - consuming and requires a large amount of energy (2 - 4h)[28]. Secondly, the current research hotspot is the direct reductive leaching method in a water solution. The reductants are oxalic acid[29], pyrite[30], aqueous sulfur dioxide[31], iron powder[32], iron(Ⅱ) sulfate[33], hydrogen eroxide[34], organic biomass reductants and bio - battery[35-37]. To each pyrolusite, pyrolusite and another mineral are mixed in an cid solution, by which the leaching solution containing Mn^{2+} and Fe^{2+} is obtained. The advantage of this method is the short process path, low energy - consumption and short processing cycle, but purification of the leaching solution is very difficult because of its complex chemical constituents.

Microwave heating has many characteristics, such as selective, uniform and fast heating, no pollution, low equipment cost, very fast reaction speed and high product yields[38]. It can successfully avoid the disadvantages of traditional heating methods, such as large temperature gradient, long processing period, low heating efficiency, high energy consumption and high pollution industries[39,40].

In this study, we try to make use of the advantage of microwave heating to solve the current problems (high temperature, high energy consumption and high reductants ratio) in the prereduction process of pyrolusite. The influence of reaction temperature and holding time on the reduction of pyrolusite were systematically investigated, and the aim is to determine a low temperature and a short period process for the reduction of pyrolusite.

2 Experimental

2.1 Materials

The chemical composition of the low - grade pyrolusite is presented in Table 1, and the compositions of the coal are shown in Table 2. The particle size distribution of the pyrolusite and coal used are presented in Fig. 1, and all these materials were milled by a ball grinding mill.

Table 1 Compositions of the pyrolusite (Total manganese 27.51%, mass, %)

MnO_2	Mn_3O_4	MnO	Fe_2O_3	Fe_3O_4	SiO_2	Al_2O_3	K_2O	CaO	BaO
41.00	1.67	0.51	11.97	1.15	36.73	3.66	0.86	0.82	0.38
P_2O_5	MgO	TiO_2	SO_3	Co_2O_3	NiO	ZnO	SrO	CuO	Y_2O_3
0.38	0.36	0.16	0.11	0.06	0.05	0.05	0.04	0.02	0.01

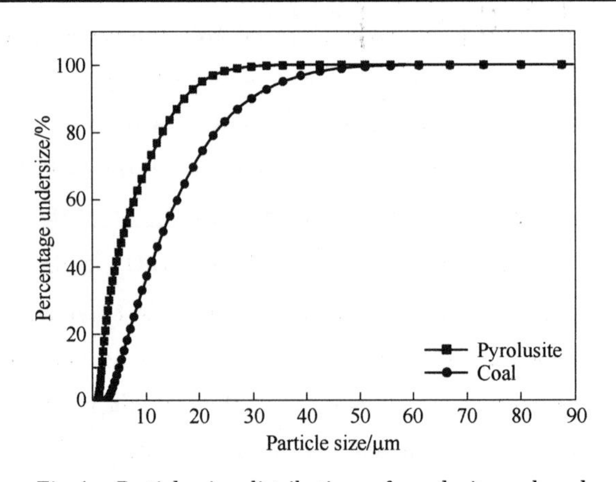

Fig. 1 Particle size distributions of pyrolusite and coal

Table 2 Compositions of the coal (mass, %)

Fixed carbon	Volatile organic matter	H_2O	SiO_2	Al_2O_3	SO_3	Fe_2O_3	TiO_2
67.58	10.89	4.67	7.19	3.55	2.55	1.83	0.76
CaO	K_2O	MgO	MnO	ZrO_2	P_2O_5	Cr_2O_3	SrO
0.63	0.19	0.08	0.02	0.02	0.02	0.01	0.01

2.2 Instruments

A self-made microwave tube furnace, which utilizes a single mode continuous controllable power was utilized for all experiments and are shown in Fig. 2. The microwave frequency was 2.45GHz, whereas the output power was controlled within the maximum of 3000W. The activation temperature was controlled by varying the input microwave power and measured by a nickel chrome - nickel silicon armor type thermocouple, which was in contact with the materials. The dimensions of the thermocouple were 1000mm in length, 3mm in diameter with the temperature range of 0 - 1250℃, and a measurement precision up to 0.5℃.

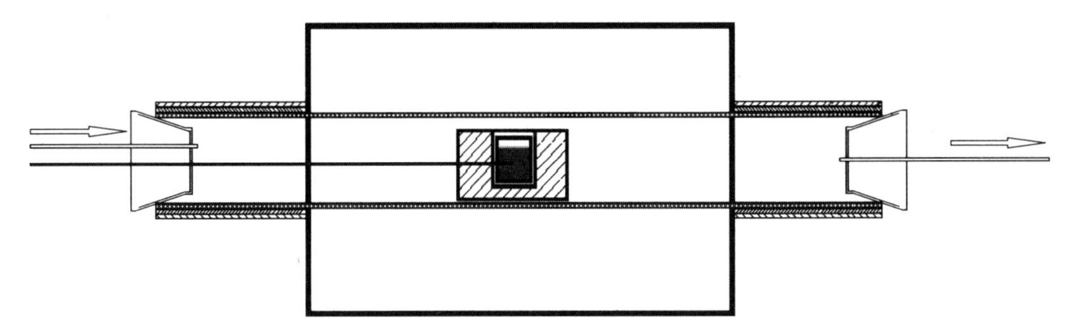

Fig. 2 Diagram of microwave tube furnace

2.3 Methods

The pyrolusite powder and the coal were thoroughly mixed in an agate mortar and placed into the corundum crucible, which has a temperature test hole at the waist, and the mixed material surface was covered by a layer of coal powder (about 1g), and the reductants proportion was 10%. The mixed materials were placed inside a microwave heating reactor. The schematic of the microwave reactor is shown Fig. 2. The cavity of the microwave heating reactor was filled with nitrogen by a gas cylinder for 10 min. Then, the carbothermal reduction experiments were started according to the process. Reduction ratios of Mn and the valence state of iron were chosen as independent research factors. At the end of the experiment, the reduction ratio of Mn was calculated based on the following equation:

$$\text{Degree of reduction} = \eta_{Mn} = \frac{M_2}{M_1} \times 100\% \qquad (1)$$

where M_1 is the mass of Mn; M_2 is the mass of Mn(Ⅱ); the determination of degree of reduction using ferrous ammonium sulfate as redox indicators.

2.4 Characterization

After microwave heating, the phase transitions of the pyrolusite were identified using XRD technology (D/Max 2200, Rigaku, Japan). XRD patterns were recorded using Rigaku diffractometer with CuK_α radiation and a Ni filter operated at the voltage of 35kV, anode current of 20mA and a scanning rate of 0.25°/min, respectively. The microstructure morphology of the pyrolusite and the microwave heat treated samples were investigated by scanning electron microscopy (SEM). The SEM instrument (XL30ESEM – TMP, Philips, Holland) was operated at 20kV in a low vacuum while the energy dispersion scanner spectrometer (EDX, USA) attached to the SEM was used for semi – quantitative chemical analysis.

3 Results and discussion

Fig. 3 shows the thermodynamics graph of the directly reduced manganese and iron oxide. The main chemical reductions can be calculated by the following equations:

$$2MnO_2 + C = Mn_2O_3 + CO(g) \tag{2}$$

$$3Mn_2O_3 + C = 2Mn_3O_4 + CO(g) \tag{3}$$

$$Mn_3O_4 + C = 3MnO + CO(g) \tag{4}$$

$$MnO + C = Mn + CO(g) \tag{5}$$

$$2MnO_2 + C = Mn_2O_3 + CO(g) \tag{6}$$

$$3Fe_2O_3 + C = 2Fe_3O_4 + CO(g) \tag{7}$$

$$Fe_3O_4 + C = 3FeO + CO(g) \tag{8}$$

$$FeO + C = Fe + CO(g) \tag{9}$$

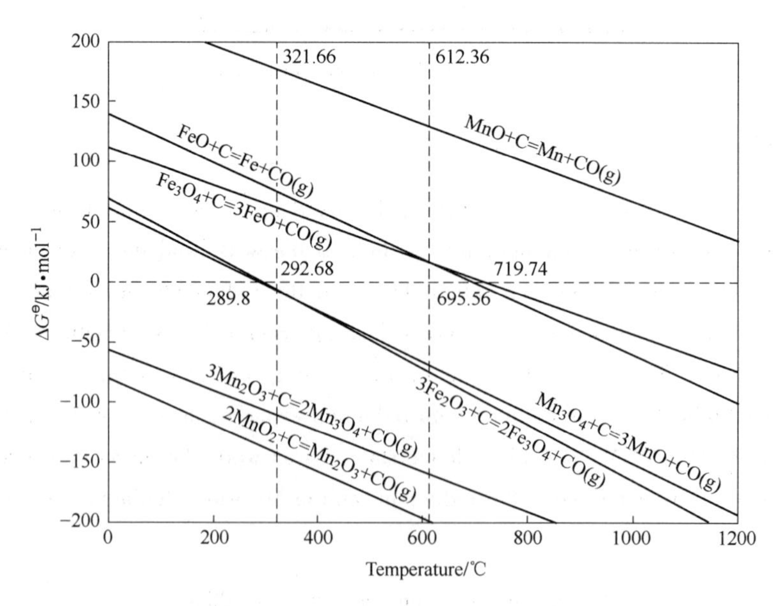

Fig. 3 Thermodynamics graph of direct reducing manganese and iron oxide

It can be seen in Fig. 3 that the reactions of MnO_2, Mn_2O_3, Mn_3O_4 (Eqs. (2) and (3)) can be performed at room temperature and the starting reduction temperatures of the reactions, 289.8 – 719.74℃, were used further in the research in order to transform Mn_xO_y to MnO. Mainly carbothermal reductions of the pyrolusite are shown in the following equation, which can be observed from the thermodynamic graph of the manganese and iron oxide by indirect reduction, as shown in Fig. 4.

$$MnO_2 + CO(g) = MnO + CO_2(g) \quad (10)$$

$$3Mn_2O_3 + CO(g) = 2Mn_3O_4 + CO_2(g) \quad (11)$$

$$Mn_3O_4 + CO(g) = 3MnO + CO_2(g) \quad (12)$$

$$MnO + CO(g) = Mn + CO_2(g) \quad (13)$$

$$3Fe_2O_3 + CO(g) = 2Fe_3O_4 + CO_2(g) \quad (14)$$

$$Fe_3O_4 + CO(g) = 3FeO + CO_2(g) \quad (15)$$

$$FeO + CO(g) = Fe + CO_2(g) \quad (16)$$

$$2C + O_2(g) = 2CO(g) \quad (17)$$

$$C + CO_2(g) = 2CO(g) \quad (18)$$

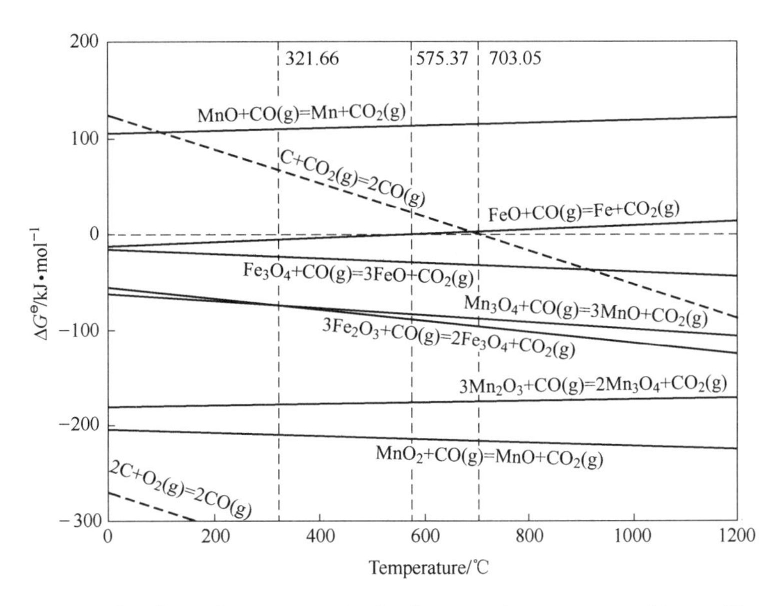

Fig. 4 The thermodynamics graph of reductions of manganese and iron oxide

Quantitative evaluations of the studied reactions from the perspective of thermodynamics were characterized. It can be seen from Fig. 4 that the reactions of MnO_2, Mn_2O_3, Mn_3O_4 and MnO (Eqs. (10) – (12)) can be achieved at room temperature. The reactions of Fe_2O_3, FeO can be performed at room temperature under standard conditions, and the temperature should be controlled at 575℃. According to the thermodynamic analysis of Fig. 4, Mn_xO_y can be reduced into MnO and Fe_xO_y is reduced into FeO at room temperature.

The microwave heating curves of the raw material, with a coal proportion of 10%, total material of 50g and microwave power of 200W, are characterized and the results are illustrated in Fig. 5. It can be seen from Fig. 5 that the mixed materials could be heated to 800℃ from room temperature in 6min under the low microwave power density(4W/g) with the highest heating rate at 372℃/min and the average heating rate at 165.2℃/min. It can be concluded that the mixed pyrolusite powder, which contains coal, has the good characteristic of microwave absorption.

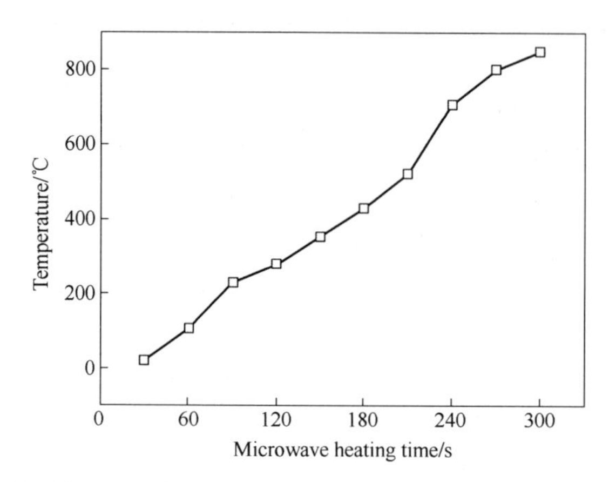

Fig. 5 Microwave heating curve of the raw material with 10% coal

Fig. 6 shows the relationship between the reduction ratio of Mn(η_{Mn}) and the holding time(T), when the reductant proportion of pyrolusite is 10%, and the holding time is 40min. It can be seen from Fig. 6 that the reduction ratio of pyrolusite increases gradually from 16.56% to 97.2% with increase in microwave heating temperature from 400℃ to 800℃. It can be concluded that with the temperature increasing, the $\Delta G^{\ominus}$ value of the reaction $MnO_2 \rightarrow Mn_2O_3 \rightarrow Mn_3O_4 \rightarrow MnO$ is obviously decreasing(Figs. 3 and 4), which indicates that the thermodynamic condition becomes better; moreover, while the temperature is high, the dynamic condition is also markedly improved.

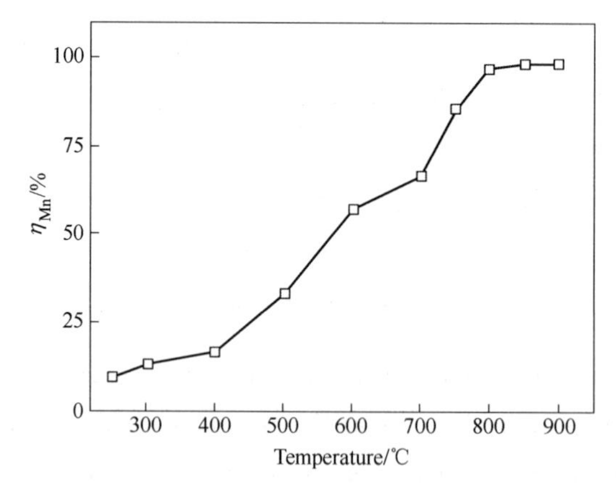

Fig. 6 Influence of reduction temperature on the reduction ratio of pyrolusite

The relationship between the reduction ratio of Mn(η_{Mn}) and the holding time(T) of the pyrolusite under microwave heating are obtained, and the results are illustrated in Fig. 7. It can be seen from Fig. 7 that the reduction ratio of Mn increases with the extension of the microwave holding time. After holding for 40min, the reduction ratio increases slowly because of the amount of un - reduced MnO_2 and the remaining reductant are reduced.

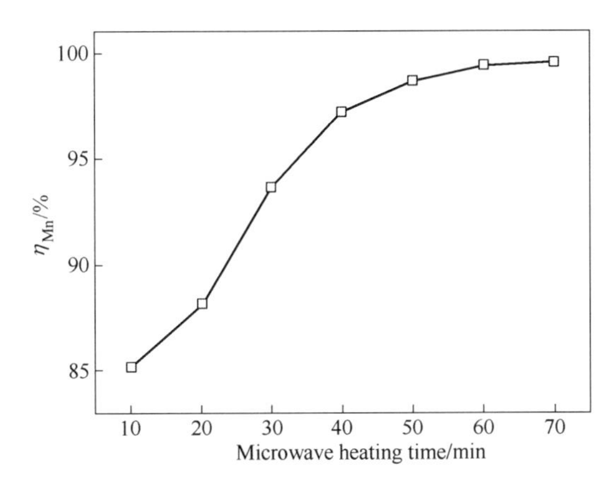

Fig. 7 Influence of holding time on the reduction ratio of pyrolusite

The crystal structures of the raw materials after microwave heating are characterized by XRD, and the results are illustrated in Fig. 8. It can be seen that MnO, Fe_3O_4, $MnSiO_4$ and SiO_2 are the major phase compositions in the microwave treated samples. In addition, a minor amount of $MnSiO_4$ is also present. The XRD results of the microwave reduced product show that the XRD patterns of the reference MnO and all peaksmatch the standard spectra of MnO(JCPDS card No. 89 – 4835). The strongest and second preferential orientation of(200) and(111) planes of the reduced MnO are observed at $2\theta = 40.577°$ and $\theta = 39.950°$, respectively. It can be seen from Fig. 8 that the total Mn of raw materials completely transforms to MnO, and Fe_2O_3 is reduced to Fe_3O_4. Moreover, it can be seen from Figs. 6 and 7 that Fe_2O_3 transforms to Fe_3O_4 almost completely at 800℃ (Eqs. (8) and(15)), and there was no Fe(Ⅱ) produced.

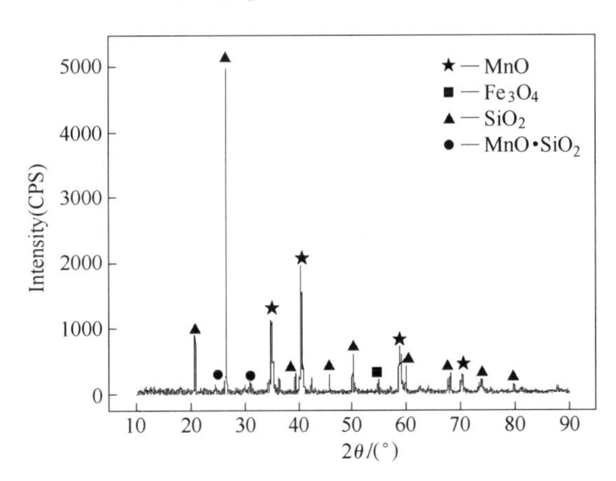

Fig. 8 XRD of the product prepared after microwave heating at 800℃ for 40min

The results also indicate that the low – grade pyrolusite was carbothermally reduced using microwave heating with lower temperature and shorter processing time. The major advantages of using microwave heating for industrial processing are rapid heat transfer, volumetric and selective heating, compactness of equipment, speed of switching on and off and a pollution – free environment.

The pyrolusite before and after microwave reduction at the microwave heating temperature of 800℃ and the holding time of 40min are characterized by SEM and EDAX techniques, and the result as shown in Figs. 9 and 10, respectively. Compared with the untreated raw materials (Fig. 9 (a)), from the SEM in Fig. 9(b), the results indicate that the size distribution of the microwave treated samples is wide (0.3 – 20μm), and the product seems to be hard agglomerates, which may have been formed by the sintering of small particles. EDAX analyses of microwave treated pyrolusite are carried out to estimate the elemental composition of microwave heating prepared MnO, and the results are shown in Fig. 10. It was observed from Fig. 10(a) and (b) that the microwave treated product consists of Mn and Fe, and minor amounts of Si, Au, Ca and Al.

(a) (b)

Fig. 9 SEM of the raw materials before and after microwave irradiation at 800℃ for 40min

(a) Raw materials; (b) Microwave treated samples

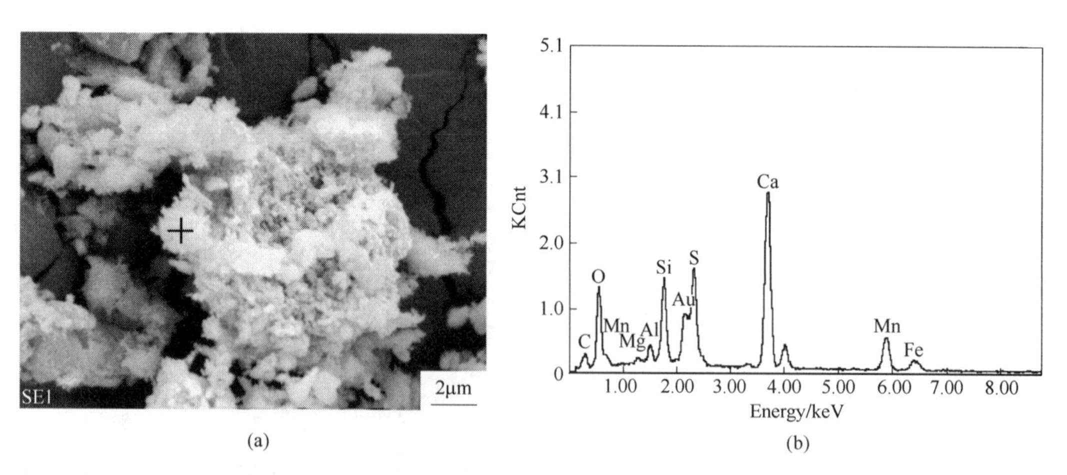

(a) (b)

Fig. 10 EDS of the product prepared after microwave heating at 800℃ for 40min

(a) SEM of EDAX analysis; (b) EDAX analysis results of the red area

4 Conclusions

The pyrolusite was microwave pre – reduced using coal at varying heating times. The optimum conditions for experimental parameters of microwave pre – reduced pyrolusite were obtained with the proportion of coal of 10% ,process temperature of 800℃ ,and holding time of 40min. Under these optimum conditions,the reduction ratio of MnO_2 to MnO in the pyrolusite was 97. 2% . Fe_2O_3 transforms to Fe_3O_4 completely and there was no Fe (Ⅱ) produced from the microwave heating process. The reduction product was formed by sintering of small particles. Compared with the traditional high temperature(1000 – 1350℃) method of pre – reduced pyrolusite,the microwave heating technique had the characteristics such as short time, low temperature, low consumption and high quality. Based on the above mentioned results,this method can be applied effectively and efficiently to the carbothermal reduction processes of low – grade pyrolusite.

Acknowledgements

Financial supports from the Key Projects in the National Science & Technology Pillar Program during the Twelfth Five – year Plan Period(No. 2015BAB17B00),the Specialized Research Fund for the Doctoral Program of Higher Education(No. 20125314120014),the Applied Foundation Fund of Yunnan Province of China(No. 2012FD015),the Yunnan Provincial Science,and Technology Innovation Talents scheme – Technological Leading Talent(No. 2013HA002) and the award for new academic graduate student of Yunnan province are acknowledged.

References

[1]Sahoo R N,Naik P K,Das S C. Leaching of manganese from low – grade manganese ore using oxalic acid as reductant in sulphuric acid solution[J]. Hydrometallurgy,2001,62(3):157 – 163.

[2]Ismail A A,Ali E A,Ibrahim I A,et al. A comparative study on acid leaching of low grade manganese ore using some industrial wastes as reductants[J]. The Canadian Journal of Chemical Engineering,2004,82(6):1296 – 1300.

[3]Zhang Y,You Z,Li G,et al. Manganese extraction by sulfur – based reduction roasting – acid leaching from low – grade manganese oxide ores[J]. Hydrometallurgy,2013,133:126 – 132.

[4]Staudhammer K P,Murr L E. Characterization of natural pyrolusite by electron microscopy[J]. Contributions to Mineralogy and Petrology,1974,45(3):251 –256.

[5]Zhang W,Cheng C Y. Manganese metallurgy review. Part Ⅰ:leaching of ores/secondary materials and recovery ofelectrolytic/chemical manganese dioxide[J]. Hydrometallurgy,2007,89(3):137 – 159.

[6]Das J N. The dielectric and piezoelectric behaviour of pyrolusite(polycrystalline ore of MnO_2)[J]. Zeitschrift für Physik,1959,155(4):465 –471.

[7] Xun W, Dong L Y. Selected – control hydrothermal synthesis of alpha – and beta – MnO_2 single crystal nanowires[J]. J. Am. Chem. Soc. ,2002,124(12):2880 –2881.

[8]Yamashita T,Vannice A. NO decomposition over Mn_2O_3 and Mn_3O_4[J]. J. Catal. ,1996,163(1):158 – 168.

[9]Park T S,Jeong S K,Hong S H,et al. Selective catalytic reduction of nitrogen oxides with NH_3 over natural manganese ore at low temperature[J]. Ind. Eng. Chem. Res. ,2001,40(21):4491 –4495.

[10]Armstrong A R,Bruce P G. Synthesis of layered $LiMnO_2$ as an electrode for rechargeable lithium batteries[J]. Nature,1996,381(6582):499 –500.

[11] Thackeray M M. Manganese oxides for lithium batteries[J]. Prog. Solid State Chem. ,1997,25(1):1 –71.

[12]Welham N J. Activation of the carbothermic reduction of manganese ore[J]. Int. J. Miner. Process,2002,67

(1):187 - 198.

[13] Ammundsen B, Paulsen J. Novel lithium - ion cathode materials based on layered manganese oxides[J]. Adv. Mater. ,2001,13(12 - 13):943 - 956.

[14] Das J N. Study of the semi - conducting properties of pyrolusite[J]. Zeitschrift für Physik,1958,151(3): 345 - 350.

[15] Ostrovski O, Olsen S E, Tangstad M, et al. Kinetic modelling of MnO reduction from manganese ore[J]. Can. Metall. Q. ,2002,41(3):309 - 318.

[16] Rankin W J, Wynnyckyj J R. Kinetics of reduction of MnO in powder mixtures with carbon[J]. Metall. Mater. Trans. B,1997,28(2):307 - 319.

[17] Akdogan G, Eric R H. Kinetics of the solid - state carbothermic reduction of wessel manganese ores[J]. Metall. Mater. Trans. B,1995,26(1):13 - 24.

[18] Rankin W J, Deventer J S. The kinetics of the reduction of manganous oxide by graphite[J]. J. South. Afr. Inst. Min. Metall. ,1980,80(7):239 - 247.

[19] Gao Y, Olivas - Martinez M, Sohn H Y, et al. Upgrading of low - grade manganese ore by selective reduction of iron oxide and magnetic separation[J]. Metall. Mater. Trans. B,2012,43(6):1465 - 1475.

[20] Cheng Z, Zhu G, Zhao Y. Study in reduction - roast leaching manganese from low - grade manganese dioxide ores using cornstalk as reductant[J]. Hydrometallurgy,2009,96(1):176 - 179.

[21] Kononov R, Ostrovski O, Ganguly S. Carbothermal reduction of manganese oxide in different gas atmospheres [J]. Metall. Mater. Trans. B,2008,39(5):662 - 668.

[22] Momade F W Y, Momade Z G. Reductive leaching of manganese oxide ore in aqueous methanol - sulphuric acid medium[J]. Hydrometallurgy,1999,51(1):103 - 113.

[23] Tian X, Wen X, Yang C, Liang Y. Reductive leaching of manganese from low - grade manganese dioxide ores using corncob as reductant insulfuric acid solution[J]. Hydrometallurgy,2010,100(3):157 - 160.

[24] Berg K L, Olsen S E. Kinetics of manganese ore reduction by carbon monoxide[J]. Metall. Mater. Trans. B, 2000,31(3):477 - 490.

[25] Hancock H. Use of coal and lignite to dissolve manganese dioxide in acidic solutions[J]. Chem. Ind. ,1986: 569 - 571.

[26] Chen J, Tian P F, Song X A, et al, Microstructure of solid phase reduction on manganese oxide ore fines containing coal by microwave heating[J]. J. Iron Steel Res. Int. ,2010,17(3):13 - 20.

[27] Moradkhani D, Malekzadeh M, Ahmadi E. Nanostructured MnO_2 synthesized via methane gas reduction of manganese ore and hydrothermal precipitation methods[J]. Trans. Nonferrous Met. Soc. China, 2013, 23(1): 134 - 139.

[28] Jana R K, Pandey B D. Ammoniacal leaching of roast reduced deep - sea manganese nodules[J]. Hydrometallurgy,1999,53(1):45 - 56.

[29] Sahoo R N, Naik P K, Das S C. Leaching of manganese from low - grade manganese ore using oxalic acid as reductant in sulphuric acid solution[J]. Hydrometallurgy,2001,62(3),157 - 163.

[30] Kholmogorov A G, Zhyzhaev A M, Kononov U S, et al, The production of manganese dioxide from manganese ores of some deposits of the Siberian region of Russia[J]. Hydrometallurgy,2000,56(1):1 - 11.

[31] Raghavan R, Upadhyay R N. Innovative hydrometallurgical processing technique for industrial zinc and manganese process residues[J]. Hydrometallurgy,1999,51(2):207 - 226.

[32] Bafghi M S, Zakeri A, Ghasemi Z, et al. Reductive dissolution of manganese ore in sulfuric acid in the presence of iron metal[J]. Hydrometallurgy,2008,90(2):207 - 212.

[33] Das S C, Sahoo P K, Rao P K. Extraction of manganese from low - grade manganese ores by $FeSO_4$ leaching

[J]. Hydrometallurgy,1982,8(1):35 – 47.

[34] Jiang T,Yang Y,Huang Z,et al. Leaching kinetics of pyrolusite from manganese – silver ores in the presence of hydrogen peroxide[J]. Hydrometallurgy,2004,72(1):129 – 138.

[35] Veglio F,Toro L. Fractional factorial experiments in the development of manganese dioxide leaching by sucrose in sulphuric acid solutions[J]. Hydrometallurgy,1994,36(2):215 – 230.

[36] Hariprasad D,Dash B,Ghosh M K,et al. Leaching of manganese ores using sawdust as a reductant[J]. Miner. Eng. ,2007,20(14):1293 – 1295.

[37] Trifoni M,Toro L. Reductive leaching of manganiferous ores by glucose and H_2SO_4:effect of alcohols[J]. Hydrometallurgy,2001,59(1):1 – 14.

[38] Ye Q X,Zhu H B,Peng J H,et al. Preparation of reduced iron powders from mill scale with microwave heating:optimization using response surface methodology[J]. Metall. Mater. Trans. B,2013,44(6):1478 – 1485.

[39] Vereš J,Jakabský Š,Lovás M. Zinc recovery from iron and steel making wastes by conventional and microwave assisted leaching[J]. Acta Montanistica Slovaca,2011,16(3):185 – 191.

[40] Vereš J,Lovás M,Jakabský Š,et al. Characterization of blast furnace sludge and removal of zinc by microwave assisted extraction[J]. Hydrometallurgy,2012,129:67 – 73.

Pilot – scale Production of Titanium – rich Material Using Ilmenite Concentrates as Raw Materials by Microwave Reduction

Wei Li, Jinhui Peng, Shenghui Guo, Libo Zhang, Hongying Xia, Bingguo Liu

Abstract: Pilot – scale production of titanium – rich material using ilmenite concentrate as raw materials by microwave reduction was studied on the basis of the laboratory scale experiments. Optimum experimental conditions are as follows: pellets of 20kg, reductive agent of coke 14%, additives(Na_2SO_4 + S + NaCl + Fe) 5%, reduction temperature of 1100℃ similar to 1150℃ and reduction time of 90 min. Under the optimum conditions, TiO_2 of metallic pellets was 53.38%, TiO_2 grade of primary titanium material was 72.01%, yield of TiO_2 was 67.5%, and the recovery rate of TiO_2 was 90.1%. The XRD analysis indicated that the metallic pellets contained TiO_2, FeO · $2TiO_2$, Fe, $MgTi_2O_5$ and SiO_3^{2-}. Titanium mainly exists in the form of Ti^{4+}, which is beneficial to the process of acid leaching.
Keywords: microwave reduction; ilmenite concentrates; titanium – rich material

1 Introduction

The mineral ilmenite($FeTiO_3$) is the main source of titanium dioxide which is widely used as a white pigment. The common method of treatment is thermal reduction of the ilmenite to form TiO_2 and elemental iron followed by a leach to remove the iron. The reduction of ilmenite concentrate plays an important role in the titanium industry. Numerous experiments have established that ilmenite concentrate usually needs high reductive temperature or needs additives to improve its reactivity when it is directly reduced[1,2]. Carbothermic reduction of ilmenite at temperatures below 1200℃ produces metallic iron and reduced form of oxides[3,4]. The reactivity of ilmenite can also be improved by using pre – oxidization process, increasing the rate of ilmenite reduction and the rate of leaching[5-8]. Pre – oxidization is now a broadly adopted practice in the processing of ilmenite ore for production of TiO_2 pigment and metallic titanium. Wang and Yuang[9] described the reductive degree and rate of Bama ilmenite concentrate by graphite at the temperatures from 850℃ to 1400℃. The reduction degree and reaction rate of the ilmenite increased with temperature rise. The higher the temperature was, the faster the reaction rate was.

The ilmenite deposit in Panzhihua region, Sichuan, China accounts for 35% of the titanium resource in the world, and for approximately of 92% in China[10]. So, it is very important to utilize the ilmenite resources efficiently for development of the titanium industry. However, due to the higher contents of CaO and MgO and complex mineralogy in ilmenite in Panzhihua region, it is very difficult to upgrade the ilmenite to titanium – rich slag, which limits the development and utilization of ilmenite deposit in Panzhihua region; it is urgent to develop new processing technologies[6,11-13].

In recent years there has been a growing interest of microwave heating in mineral treatment. Advantages in utilizing microwave technologies for processing materials include penetrating radiation, controlled electric field distribution and selective and volumetric heating[14]. Recently, a number of potential application of microwave processing have been investigated, including microwave assisted ore grinding, microwave assisted carbothermic reduction of metal oxides, microwave assisted drying and anhydration, microwave assisted mineral leaching, microwave assisted roasting and smelting of sulphide concentrate, microwave assisted pretreatment of refractory gold concentrate, microwave assisted spent carbon regeneration, coke making and activated carbon production, and microwave assisted waste management, etc[15-24].

For microwave processing of ilmenite, Itoh et al. described the microwave oxidation of rutile extraction process, in which rutile is extracted from a natural ilmenite ore by oxidation and magnetic separation followed by leaching with diluted acid[25]. Kelly and Rowson investigated microwave reduction of oxidized ilimenite concentrate[26]. Tong et al. evaluated the economic values of industrial applications of carbothermic reduction of metals oxide by microwave heating, showing that the cost is lowered about 15% - 50% compare to that of conventional method[27]. Cutmore et al. investigated dielectric properties of some minerals[28]. Microwave absorbing characteristics of ilmenite concentrate with different proportions of carbonaceous reduction agents were investigated by authors' group[13], which further confirms the feasibility of microwave reduction of ilmenite concentrate. All of these investigations present encouraging results.

However, to our best knowledge, most of the microwave processing is investigated by laboratory scale microwave equipments; similarly, there is litter information about pilot - scale production of titanium - rich material by microwave reduction.

The objective of present study is to investigate the feasibility of pilot - scale production of titanium - rich material using ilmenite concentrates as raw materials on the basis of laboratory scale experiments by pilot - scale microwave heating equipment, which has been developed and researched by authors' group.

2 Experimental

2.1 Materials and methods

The raw ilmenite concentrates was obtained from Panxi, Sichuan province, China. The composition of ilmenite was listed in Table 1. The SEM photographs of ilmenite concentrates before and after microwave reduction were shown in Fig. 1.

Table 1 Chemical compositions of ilmenite concentrate

Compositions	TFe	TiO_2	CaO	MgO	MnO	SiO_2	Al_2O_3	others
Content(mass fraction)/%	33.18	47.85	1.56	6.56	0.75	5.6	3.16	1.34

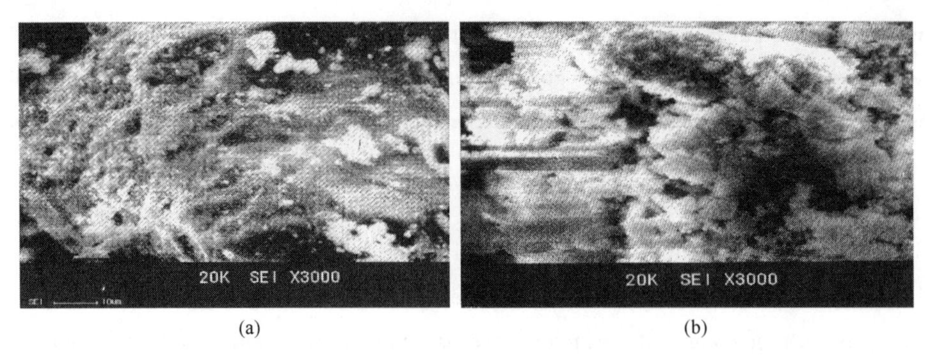

(a) (b)

Fig. 1 SEM photographs of the ilmenite concentrate before(a) and after reduction(b)

2.2 Features of pilot – scale microwave heating equipment

A microwave system typically consists of a generator to produce the microwaves, a waveguide to transport the microwaves and an applicator (usually a cavity) to manipulate microwaves for a specific purpose and a control system (tuning, temperature, power, etc.). Schematic diagram of the pilot – scale microwave heating equipment (Fig. 2) has been depicted in our previously published papers[19,20].

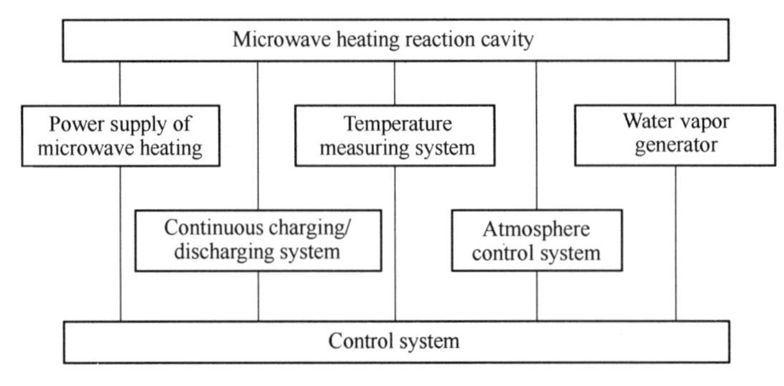

Fig. 2 Schematic diagram of the pilot – scale microwave heating equipment

3 Results and discussion

3.1 Characteristics of ilmenite concentrate

It can be seen from Table 1 that the contents of CaO and MgO were higher, and due to its compact structure of particles of ilmenite concentrate (Fig. 1 (a)), it is very difficult to upgrade the ilmenite to titanium – rich slag, because calcium and magnesium will form a layer of "barriers" to impede chemical reaction during the reduction process. Ilmenite concentrate used in present study is a kind of refractory mineral. Through microwave reduction, particle surface of reductive product is pores developed, and the structure is loose (Fig. 1 (b)), confirming the feasibility of microwave reduction of ilmenite concentrate.

3.2 Investigations on different carbonaceous reductive agents

In present study, three kinds of reductive agents (coconut shell based – activated carbon, coke and anthracite) have been used. Process conditions were as follows: Microwave power of 20kW, reduction time of 1.5h, additive (Na_2SO_4 + S + NaCl + Fe) 5%, binder 3%, reduction temperature of 1000℃ ±20℃. The ratio of coconut shell based – activated carbon, coke and anthracite to mineral was 20%, 14% and 15%, respectively, which was all used 5kg during the reduction process. The reduction results were listed in Table 2.

Table 2 Results of different carbonaceous reductive agents by microwave reduction

Reductive agents	TiO_2 grade of metalized pellets/%	Primary titanium – rich materials		
		TiO_2 grade/%	Yield/%	Recovery rate/%
Coconut shell based activated carbon	50.55	69.57	66.4	89.98
Coke	51.99	69.15	66.8	90.08
Anthracite	50.20	66.97	66.1	88.18

It can be seen from Table 2 that TiO_2 grade of metalized pellets obtained by using coke as reductive agent is better than that of coconut shell based – activated carbon, while the TiO_2 grade of primary titanium – rich materials obtained by coconut shell based – activated carbon and coke as reductive agents is very close, which is 69.57% and 69.15%, respectively. The TiO_2 grade of primary titanium – rich materials obtained by anthracite as reductive agent is the lowest. Recovery rate is also the highest by using coke as reductive agent; furthermore, from the perspective of cost evaluation, the distribution amount of coke to concentrates is lower about 6% than that of coconut shell based – activated carbon, which is the lowest among three carbonaceous reductive agents, which is beneficial to lowering the production cost of industrialization process. So, coke is chosen to be as the reductive agent in the present study.

3.3 Investigations on temperature of microwave reduction

Effects of temperature of microwave reduction on TiO_2 grades of metalized pellets and primary titanium – rich materials were investigated after coke had been chosen to be the reductive agent. Process conditions were as follows: microwave power 20kW, coke 14%, reduction time 90min, additive (Na_2SO_4 + S + NaCl + Fe) 5%, binder (sodium silicate) 3% and feed 5kg. Temperature ranges of investigation were: 950 – 1000℃, 1000 – 1100℃, 1100 – 1150℃ and 1150 – 1200℃. Results were listed in Table 3.

Table 3 Results of different temperatures by microwave reduction

Temperature range/℃	Weights of pellets/kg	TiO_2 grade of metalized pellets/%	Primary titanium – rich materials			Loss rate of external carbon/%
			Grade/%	Yield/%	Recovery/%	
950 – 1000	5	49.95	67.05	61.3	82.28	36.8
1000 – 1100	5	50.02	70.90	64.0	90.85	40.5
1100 – 1150	5	53.24	72.36	66.7	89.29	41.6
1150 – 1200	5	54.07	73.62	66.5	89.18	42.3

TiO_2 grades of metalized pellets and primary titanium - rich materials are lower at temperature range of 950 - 1000℃, which are increased at temperature range of 1000 - 1100℃, also substantial increased at temperature range of 1100 - 1150℃, while, there is no obvious increase at temperature range of 1150 - 1200℃. Therefore, 1100 - 1150℃ is chosen to be the reduction temperature range. The pilot - scale experimental results are in agreement with results of laboratory scale experiments. XRD analyses of products at different temperature ranges of 950 - 1050℃, 1050 - 1100℃, 1100 - 1150℃ and 1150 - 1200℃ are shown in Figs. 3 - 6, respectively.

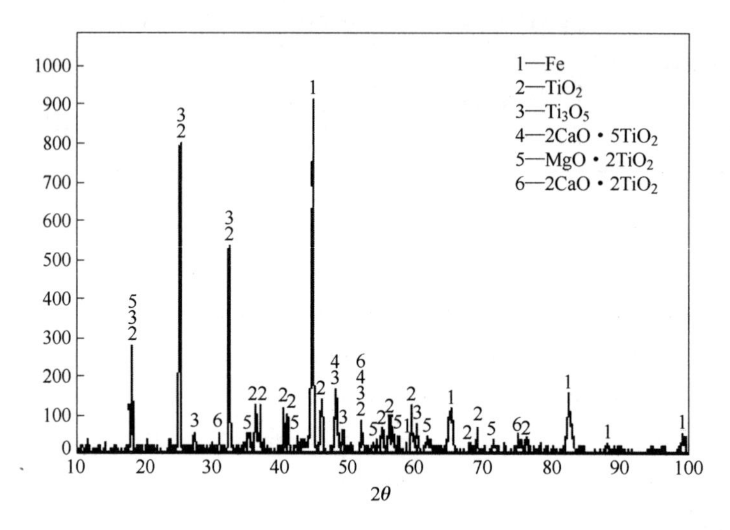

Fig. 3 X - ray diffraction pattern of reductive product by microwave reduction at 950 - 1050℃

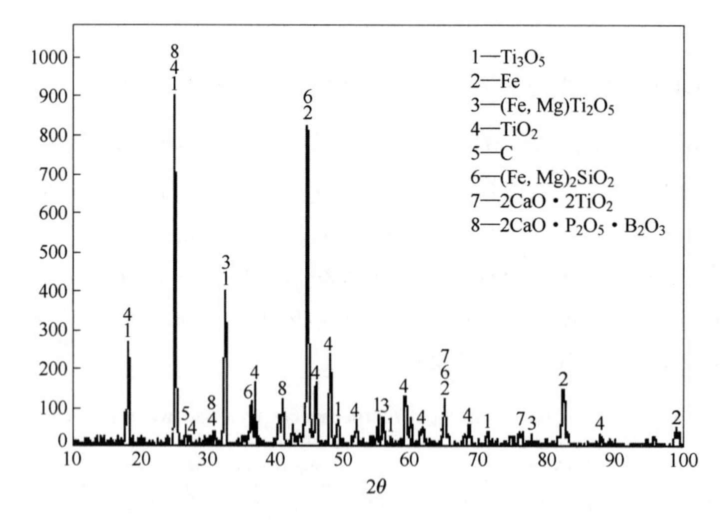

Fig. 4 X - ray diffraction pattern of reductive product by microwave reduction at 1050 - 1100℃

Experiments of different feed of 5kg, 10kg, 20kg, 30kg, 40kg and 60kg were investigated. Process conditions were as follows: microwave power 20kW, coke 14%, reduction time 90min, and reductive temperature range of 1150℃ ±20℃. Results were listed in Table 4.

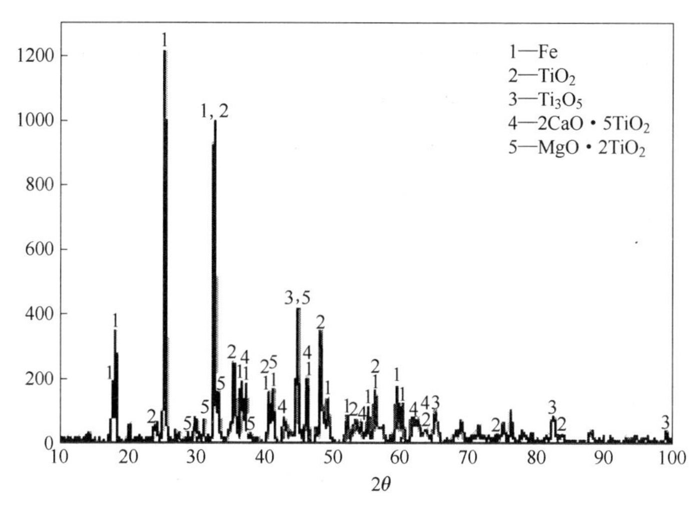

Fig. 5 X – ray diffraction pattern of reductive product by microwave reduction at 1100 – 1150℃ investigations on feed of materials

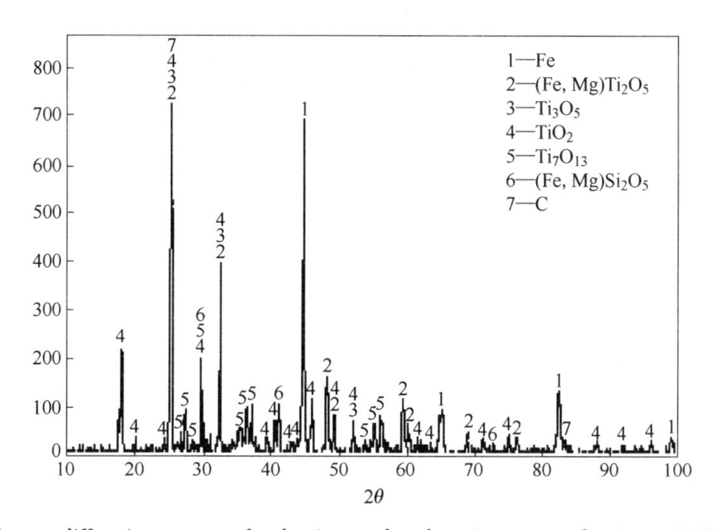

Fig. 6 X – ray diffraction pattern of reductive product by microwave reduction at 1150 – 1200℃

Table 4 Results of different weight of materials by microwave reduction

Weight/kg	TiO_2 grade of metalized pellets/%	Primary titanium – rich materials			Loss rate of external carbon/%
		Grade/%	Yield/%	Recovery rate/%	
5.0	53.92	72.47	66.8	90.30	41.0
10	53.61	71.28	67.2	89.70	33.5
20	50.10	70.19	63.3	88.68	31.0
30	51.16	69.50	66.4	90.2	34.8
40	50.86	69.20	67.5	91.8	34.0
60	49.91	67.39	68.5	92.5	42.0

It is shown that TiO_2 grades of metalized pellets of feeds of 5kg,10kg,20kg,30kg and 40kg are higher;all of them are more than 50%. TiO_2 grade of metalized pellets for 60kg feed is close to 50%, metallization rate is 89.2%, reaching the reductive effect basically. TiO_2 grades of primary titanium - rich materials feeds of 5kg,10kg,20kg,30kg and 40kg are also higher; all of them are more than 70%. TiO_2 grade of primary titanium - rich materials is lowered; however, the yield is the highest among all of them. Therefore, microwave reduction equipment set up can meet feed of 60kg.

3.4 Verification investigation

In order to verify that the reliability of the test parameters and rationality, the verification experiments were investigated under the conditions of reductive agent of coke, reductive temperature of 1100 - 1150℃, additives(Na_2SO_4 + S + NaCl + Fe) and reaction time of 90min. Results were listed in Table 5, TiO_2 grade of metalized pellets is 53.38%, and metallization rate of metalized pellets is 92.1%, which demonstrates the feasibility of pilot scale production of titanium - rich materials from ilmenite concentrate by microwave reduction.

Table 5 Results of validating experiment by microwave reduction

Optimum conditions	TiO_2 grade of metalized pellets/%	Metallization rate of metalized pellets/%	Primary titanium - rich materials			Loss rate of external carbon/%
			Grade/%	Yield/%	Recovery rate /%	
20kg of pellets; reductive agent of coke; Reductive temperature of 1150℃ + 20℃; Additives (Na_2SO_4 + S + NaCl + Fe); reaction time of 90min	53.38	92.1	72.01	67.5	91.0	50.8

4 Conclusions

Through pilot - scale experiments of production of titanium - rich material using ilmenite concentrate as raw materials by microwave reduction, the following conclusions can be drawn:

(1) Metallization rate of metalized pellets is 92.1%, TiO_2 grade of metalized pellets and primary titanium - rich materials is 53.38% and 72.01%, respectively, yield of primary titanium - rich materials is 67.5%, the recovery rate of TiO_2 is 90.1% under the process conditions: 20kg of feed, coke 14%, reductive temperature of 1100 - 1150℃, additives(Na_2SO_4 + S + NaCl + Fe) 5% and reaction time of 90min.

(2) XRD analyses show that main phase are TiO_2, $FeO \cdot 2TiO_2$, Fe, $MgTi_2O_5$ and salts silicate. Ti exists in the form of Ti^{4+}, which can not be solved by diluted acid, in line with ideas of getting high quality titanium - rich materials from slag by acid leaching during posterior process.

(3) Experimental results demonstrate the feasibility of pilot scale production of titanium - rich materials from ilmenite concentrate by microwave reduction.

(4) The semi – industrial experiment needs to be carried out, in order to study the stability and economic feasibility of the product index from present process, which is already underway.

Acknowledgements

The authors would like to express their gratitude for the financial support of the Major Program of National Natural Science Foundation of China (Grant No. 51090385).

References

[1] Kucukkaragoz C S, Eric R H. Solid state reduction of a natural ilmenite[J]. Minerals Engineering, 2006, 19 (3): 334 – 337.

[2] Gupta S K, Rajakumar V, Grieveson P. Kinetics of reduction of ilmenite with graphite at 1000 to 1100℃ [J]. Metallurgical Transactions B, 1987, 18(4): 713 – 718.

[3] Tawil S Z, Morsi I M, Francis A A. Kinetics of solid – state reduction of ilmenite ore[J]. Canadian Metallurgical Quarterly, 1993, 32(4): 281 – 288.

[4] Tawil S Z, Morsi I M, Yehia A, et al. Alkali reductive roasting of ilmenite ore[J]. Canadian Metallurgical Quarterly, 1996, 35(1): 31 – 37.

[5] Park E, Ostrovski O. Effects of preoxidation of titania – ferrous ore on the ore structure and reduction behavior [J]. ISIJ International, 2004, 44(1): 74 – 81.

[6] Roberts J M C. Mineralogical Magazine 1971, 125: 548.

[7] Sarker M K, Rashid A, Kurny A S W. Kinetics of leaching of oxidized and reduced ilmenite in dilute hydrochloric acid solutions[J]. International Journal of Mineral Processing, 2006, 80(2): 223 – 228.

[8] Zhang G, Ostrovski O. Effect of preoxidation and sintering on properties of ilmenite concentrates[J]. International Journal of Mineral Processing, 2002, 64(4): 201 – 218.

[9] Wang Y, Yuan Z. Reductive kinetics of the reaction between a natural ilmenite and carbon[J]. International Journal of Mineral Processing, 2006, 81(3): 133 – 140.

[10] Xiong D A. Physical Separation in Science and Engineering, 2004, 13: 119 – 126.

[11] Francis A A, El – Midany A A. An assessment of the carbothermic reduction of ilmenite ore by statistical design[J]. Journal of Materials Processing Technology, 2008, 199(1): 279 – 286.

[12] Mackey T S. Upgrading ilmenite into a high – grade synthetic rutile[J]. JOM, 1994, 46(4): 59 – 64.

[13] Guo S, Li W, Peng J, et al. Microwave – absorbing characteristics of mixtures of different carboniceous reducing agents and oxidized ilmenite[J]. International Journal of Mineral Processing, 2009, 93(3): 289 – 293.

[14] Ku H S, Siu F, Siores E, et al. Applications of fixed and variable frequency microwave (VFM) facilities in polymeric materials processing and joining[J]. Journal of Materials Processing Technology, 2001, 113(1): 184 – 188.

[15] Al – Harahsheh M, Kingman S W. Microwave – assisted leaching—a review[J]. Hydrometallurgy, 2004, 73 (3): 189 – 203.

[16] Haque K E. Microwave energy for mineral treatment processes—a brief review[J]. International Journal of Mineral Processing, 1999, 57(1): 1 – 24.

[17] Kingman S W. Recent developments in microwave processing of minerals[J]. International Materials Reviews, 2006, 51(1): 1 – 12.

[18] Lester E, Kingman S, Dodds C, et al. The potential for rapid coke making using microwave energy[J]. Fuel, 2006, 85(14): 2057 – 2063.

[19] Li W, Peng J, Zhang L, et al. Pilot – scale extraction of zinc from the spent catalyst of vinyl acetate synthesis by microwave irradiation[J]. Hydrometallurgy, 2008, 92(1): 79 – 85.

[20] Li W, Peng J, Zhang L, et al. Preparation of activated carbon from coconut shell chars in pilot – scale microwave heating equipment at 60kW[J]. Waste Management, 2009, 29(2): 756 – 760.

[21] Pickles C A. Microwaves in extractive metallurgy: Part 1—review of fundamentals[J]. Minerals Engineering, 2009, 22(13): 1102 – 1111.

[22] Pickles C A. Microwaves in extractive metallurgy: Part 2—a review of applications[J]. Minerals Engineering, 2009, 22(13): 1112 – 1118.

[23] Standisti N, Worner H. Microwave application in the reduction of metal oxides with carbon[J]. Journal of Microwave Power and Electromagnetic Energy, 1990, 25(3): 177 – 180.

[24] Xia D K, Pickles C A. Applications of microwave energy in extractive metallurgy, a review[J]. CIM Bulletin, 1997, 90(1011): 96 – 107.

[25] Itoh S, Suga T, Takizawa H, et al. Application of 28GHz microwave irradiation to oxidation of ilmenite ore for new rutile extraction process[J]. ISIJ International, 2007, 47(10): 1416 – 1421.

[26] Kelly R M, Rowson N A. Microwave reduction of oxidised ilmenite concentrates[J]. Minerals Engineering, 1995, 8(11): 1427 – 1438.

[27] Tong Z F, Bi S W, Yang Y H. Journal of Materials and Metallurgy, 2004, 3: 117 – 120(in Chinese).

[28] Cutmore N, Evans T, Crnokrak D, et al. Microwave technique for analysis of mineral sands[J]. Minerals Engineering, 2000, 13(7): 729 – 736.

Optimization of Preparation for Co_3O_4 by Calcination from Cobalt Oxalate Using Response Surface Methodology

Bingguo Liu, Jinhui Peng, Libo Zhang, Rundong Wan, Shenghui Guo, Liexing Zhou

Abstract: The conditions of technique to prepare Co_3O_4 by calcination from cobalt oxalate were optimized using response surface methodology. A quadratic equation model for decomposition rate was built and effects of main factors and their corresponding relationships were obtained. The statistical analysis of the results showed that in the range studied the decomposition rate of cobalt oxalate was significantly affected by the calcination temperature and calcination time. The optimized calcination conditions were as follows: the calcination temperature 680.86K, the calcination time 44.7min, and the mass of material 3.04g, respectively. Under these conditions the decomposition rate of cobalt oxalate was 99.06%. The validity of the model was confirmed experimentally and the results were satisfactory.

Keywords: response surface methodology; cobalt oxalate; calcination; Co_3O_4

1 Introduction

Cobalt oxide(Co_3O_4) is widely used in industries for the manufacture of lithium ion batteries[1-4], gas sensors[3], magnetic materials[5] and catalysts[6] as a result of their advanced electromagnetic properties and catalytic activity. Due to increasing use of the compound in various fields, various techniques have been developed for the preparation of Co_3O_4 powders, which include thermal decomposition method[7-9], sol-gel process[10], hydrothermal method[11,12], and solvothermal[13]. Among these methods, the conventional thermal decomposition of cobalt oxalate is being commercially applied at present, due to the process simplicity. However, this traditional approach "one variable at a time(OVAT)" is well accepted, in which the effect of each experimental factor is investigated by altering the level of one factor at a time while maintaining the level of the other factors constant. Furthermore, this technique is not only time and work demanding, but completely lacks in representing the effect of interaction between different factors. In order to solve these problems, it is necessary to find a multivariate statistic technique for optimization of preparation processes.

Response surface methodology(RSM) might be a useful method to optimize preparation processes. RSM is a collection of mathematical and statistical techniques useful for analyzing the effects of several independent variables[14]. The main advantage of RSM is the reduced number of experimental trials needed to evaluate multiple parameters and their interactions[15]. It can deal with multivariate experimental design strategy, statistical modeling and process optimization[16]. Several previous researchers have proved that RSM was a powerful statistical tool in process optimization, and it has just recently been applied to optimize the process parameters for biosorption of metals[17] or dyes[18] from synthetic solutions. However, as far as known to the authors, there have been very few studies to optimize the preparation of Co_3O_4 from cobalt oxalate by the RSM approach.

Thus, in the present work, we focus on the optimization of the calcaination of cobalt oxalate by electric oven. The calcination temperature, calcination time and mass of sample were determined as the main three influencing factors in the experiments. The decomposition rate was selected as the response affected by the variables. The central composite design (CCD, one part of RSM package) was selected to study simultaneously the effects of three influencing variables on the response. RSM was used to determine the optimal condition and an empirical model correlating the decomposition rate to the three variables was then developed.

Nomenclature

M	Final weight of cobalt oxalate
M_0	Initial weight of cobalt oxalate
W_t	Theory value of weight loss of cobalt oxalate

Greek symbol

β	Heating rates
β_i	Linear coefficient
β_{ii}	Quadratic coefficient
β_{ij}	Interaction coefficient
β_0	Constant
χ_a	Calcination temperature
χ_b	Calcination time
χ_c	Mass of sample
χ_{ij}	Independent variables
γ	Decomposition rate of cobalt oxalate

2 Experimental

2.1 Materials and thermal decomposition behavior

The cobalt oxalate used in the study was obtained from Shang Hai Sinopharm Chemical Reagent Co., Ltd., of which the particle size was less than 150μm. The thermal decomposition behavior ofcobalt oxalate was investigated at four heating rate $\beta = 5, 10, 15, 20$K/min under air atmospheres using thermogravimetric analysis. TG (thermogravimetry) and DTG (derivative thermogravimetry) curves for non – isothermal decomposition under air atmospheres at various heating rate are presented in Figs. 1 and 2, respectively. The results show that there are two weight loss stages (stages Ⅰ and Ⅱ) in the TG curves, corresponding to the two peaks in DTG curves. For the TG curves when $\beta = 10$K/min, stage Ⅰ begins at 445.7K and stops at 479.5K, accompanied by 19.35% weight loss, which is attributed to the dehydration of cobalt oxalate. Stage Ⅱ begins at 479.5K and stops at 630.0K, accompanied by 36.93% weight loss, which is attributed to the dehydration of cobalt oxalate and the decomposition of residues[19]. As can be seen, the weight loss is very significant for cobalt oxalate in rang for 550.0 – 650.0K, which indicated that the decomposition reaction runs quickly.

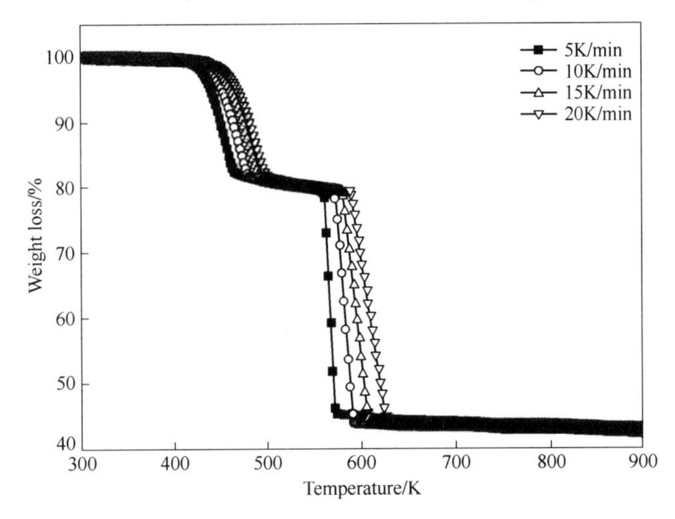

Fig. 1 TG curves of cobalt oxalate under air atmosphere at the heating rate of 5K/min, 10K/min, 15K/min and 20K/min

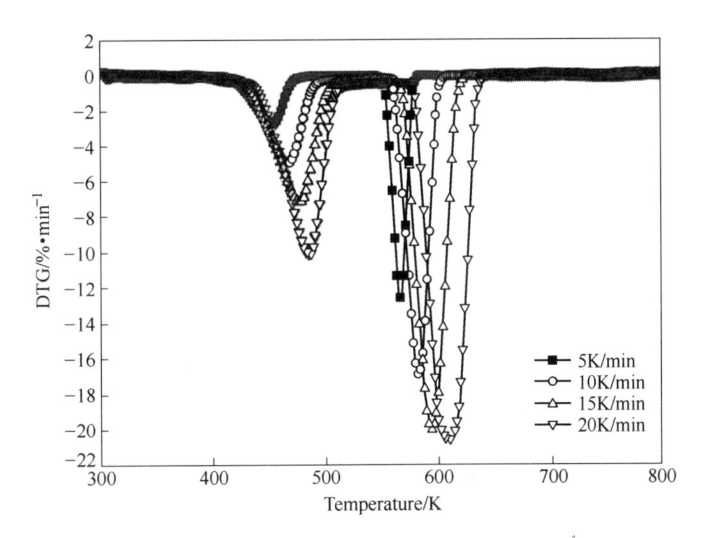

Fig. 2 DTG curves of cobalt oxalate under air atmosphere at the heating rate of 5K/min, 10K/min, 15K/min and 20K/min

2.2 Calcaination experiments

The calcination experiments were carried out at the different temperatures, calcination times and masses of sample. Initially, the tubular electric oven was preheated at 30K/min until the desired temperature was reached. Then the cobalt oxalate was weighed and placed inside the ceramic reactor which was located in the center of the conventional tubular electric oven. During the reaction, the temperature was monitored by a temperature controller system, namely the PID (proportional – integral – derivative) controller. Several cycles of experiment were repeated. For each cycle, a reaction was performed for a fixed duration, once the fixed duration is over, the experiment was stopped immediately. The product was moved out from the tubular reactor and put into the drier rapidly. They were naturally cooled to the room temperature. The final weight(M) of sample was weighted subsequently, from which the decomposition rate was calculated based on the following equation:

$$\gamma = \frac{M_0 - M}{M_0 \times W_t} \tag{1}$$

where M and M_0 are final weight and initial weight of cobalt oxalate, respectively; W_t means theory value of weight loss, which is the total weight loss of cobalt oxalate during the thermal decomposition based on the TG/DTG analyses; γ is decomposition rate of cobalt oxalate.

2.3 Designing experiment using response surface methodology

On the basis of initial decomposition results and analysis of TG/DTG, RSM was employed to optimize the calcination conditions in order to obtain a high decomposition rate, and CCD was employed to design the experiments. This method helps to optimize the effective parameters with a minimum number of experiments and also analyze the interaction between the parameters and results[20]. In this study, the effects of three independent variables, χ_a (calcination temperature), χ_b (calcination time), and χ_c (mass of sample), at two level were investigated using central composite design. The rank of values associated with the variables: calcination temperature was 608K and 683K, calcination time ranged between 30min and 50min, and mass of sample ranged between 2g and 5g.

Generally, a second - order polynomial model with main, quadratic and interaction terms can be developed to fit the experimental data obtained from the experimental runs conducted on the basis of CCD. The experimental data obtained from the designed experiment were analyzed by the response surface regression procedure using the following second - order polynomial equation:

$$\gamma = \beta_0 + \sum_{i=1}^{n} \beta_i \chi_i + \sum_{i=1}^{n} \beta_{ii} \chi_i^2 + \sum_{i<j}^{n} \beta_{ij} \chi_i \chi_j \tag{2}$$

where γ is the predicated response; β_0 is a constant; β_i is the i - th linear coefficient; β_{ii} is the i - th quadratic coefficient; β_{ij} is ij - th interaction coefficient; $\chi_i \chi_j$ are independent variables.

2.4 XRD analysis

The X - ray diffraction analysis of the final solid product under optimization conditions were carried out using D/max - 2200 Diffractometer(Japan) with Ni - filtered CuK_α radiation under air atmospheres. The identification of the completeness of the Co_3O_4 was made by comparing the diffraction peaks of each compound in the sample with the ones of the standard Co_3O_4. If the diffraction pattern of the final solid product satisfactorily matched with that of the standard Co_3O_4, it means that the decomposition of cobalt oxalate is complete.

3 Results and discussion

3.1 Data analysis and evaluation of the model by RSM

The experiments were conducted based on the design matrix under the defined conditions and the response obtained from the experimental runs is shown in Table 1.

Table 1 Experimental design matrix and results for the response surface of decomposition rate of cobalt oxalate

Run	Calcination variables			Decomposition, $\gamma/\%$
	Calcination temperature, χ_a/K	Calcination time, χ_b/min	Mass of sample, χ_c/g	
1	603	30	2	95.63
2	683	30	2	97.33
3	603	50	2	96.69
4	683	50	2	98.31
5	603	30	5	81.67
6	683	30	5	96.32
7	603	50	5	88.09
8	683	50	5	96.96
9	575.73	40	3	82.32
10	710.27	40	3	99.19
11	643	23.18	3	89.01
12	643	56.82	3	97.82
13	643	40	0.98	97.24
14	643	40	6.02	87.12
15	643	40	3.5	97.34
16	643	40	3.5	94.55
17	643	40	3.5	97.6
18	643	40	3.5	95.44
19	643	40	3.5	96.57
20	643	40	3.5	96.21

Table 1 shows the total number of 20 experiments as per CCD method. The experimental sequence was randomized in order to minimize the effects of the uncontrolled factors. Six experiments were repeated in order to estimate the experimental error. According to the sequential model, the sum of squares can be obtained, and the models were selected based on the highest order polynomial where the additional terms were significant and the models were not aliased[21]. The responses of decomposition rate were considered in studying the effect of process variables. The response and the independent variables were used to develop an empirical model after excluding the insignificant terms, which is presented by Eq. (3):

$$\gamma = 96.22 + 4.04\chi_a + 1.75\chi_b - 3.07\chi_c - 0.73\chi_a\chi_b + 2.52\chi_a\chi_c + 0.63\chi_b\chi_c - 1.53\chi_a^2 - 0.59\chi_b^2 - 1.03\chi_c^2 \quad (3)$$

The quality of the model developed was evaluated based on the correlation coefficient value[22]. The R^2 value for Eq. (3) was 0.9257, which indicated that 92.57% variability of the total variation in the decomposition rate was attributed to the experimental variables studied. The closer the R^2 value to unity, the better the model will be as it will give predicted vlues which are closer to the actual values for the response. The R^2 of 0.9257 for Eq. (3) was considered relatively high, indicating that there was a good agreement between the experimental decomposition rate of cabalt oxalate and the predicted one from this model.

Furthermore, anlysis of variance(ANOVA, also a part of RSM) was futher carried out to justify the adequacy of the model. The ANOVA for the quadratic model for decomposition rate is presented in Table 2. The model's adequacy was tested through the lack - of - fit F - test, in which the residual error was compared to the pure error(Table 2). According to the software analysis, "Lack of fit F - value" of 5.02 implies that the lack of fit was not significat realative to the pure error due to noise(Table 2). The "Model F - value" of 13.84 implies that the model was significant and there was only a 0.02% chance that a "Model F - value" this large could occur due to noise. Values of "Prob > F" less than 0.05 indicates that the model terms are significant[23], whereas the values greater than 0.1000 are not significant.

Table 2 Analysis of variance(ANOVA) for response surface quadratic model for decomposition rate

Source	Sum of squares	D_f	Mean square	F - value	Prob > F
Model	499.14	9	55.46	13.84	0.0002
Residual	40.08	10	6.69		
Lack of fit	33.43	5	1.33	5.02	0.0505
Pure error	6.65	5			
Cor total	539.22	19			

Note: $R^2 = 0.9257$, $R^2_{adj} = 0.8588$, adequate precision = 12.974 > 4.

The checking of model adequacy is an important part of the data analysis procedure, since it would give poor or misleading results if it is an inadequate fit[24]. Multivariable linear regression was used to calculate the coefficients of the second - order polynomial equation and the regression coefficients, whose significance was determined by the P - value, summarized in Table 3. In this case, χ_a, χ_b, χ_c, χ_a^2 and the interaction terms($\chi_a \cdot \chi_c$) were significant whereas the interaction terms($\chi_a \cdot \chi_b$, $\chi_b \cdot \chi_c$) were insignificant to the response.

Table 3 Regression coefficients of polynomial function for the response surface of decomposition rate of cobalt oxalate

Term	Regression coefficient	P - value	Term	Regression coefficient	P - value
Model	96.22	0.0002	$\chi_a\chi_c$	2.52	0.0051
χ_a	4.04	<0.0001	$\chi_b\chi_c$	0.63	0.3962
χ_b	1.75	0.0090	χ_a^2	-1.53	0.0158
χ_c	-3.07	0.0002	χ_b^2	-0.59	0.2892
$\chi_a\chi_b$	-0.73	0.3251	χ_c^2	-1.03	0.0801

3.2 Response surface analysis

Fig. 3 shows the predicted values versus the experimental values for decomposition rate of cobalt oxalate. Actual response values were measured for a particular run, and the predicted values were evaluated from the model and generated by using the approximating equtions. As can be seen, the predicted value obtained was quite close to the experimental values, indicating that the model developed was reasonable.

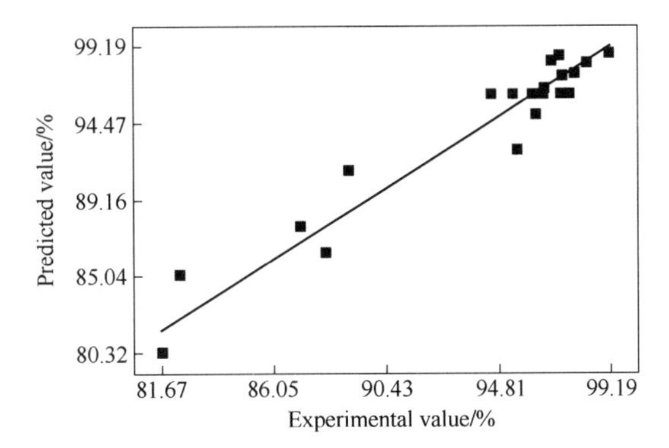

Fig. 3 Predicted vs. experimental decomposition rate of cobalt oxalate

The best way to visualize the influence of the independent variables on the response is to draw surface response plots of the model[25]. The three - dimensional response surfaces which were constructed to show the effects of the calcination of cobalt oxalate variables on decomposition rate using the fitted quadratic polynomial equation obtained from regression analysis were shown in Figs. 4 and 5.

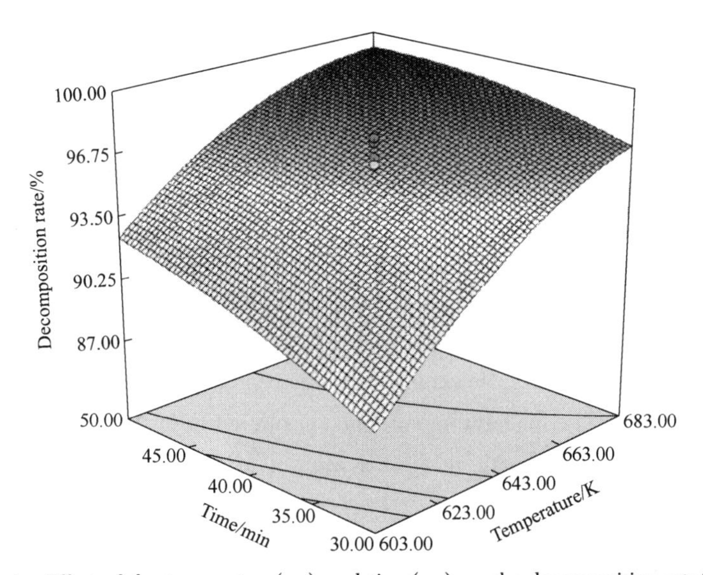

Fig. 4 Effect of the temperature(χ_1) and time(χ_2) on the decomposition rate(γ)

Fig. 4 shows the effect of calcination temperature and calcination time on decomposition rate at the fixed mass of sample of 3.5g. It was observed that the decomposition rate significantly in-

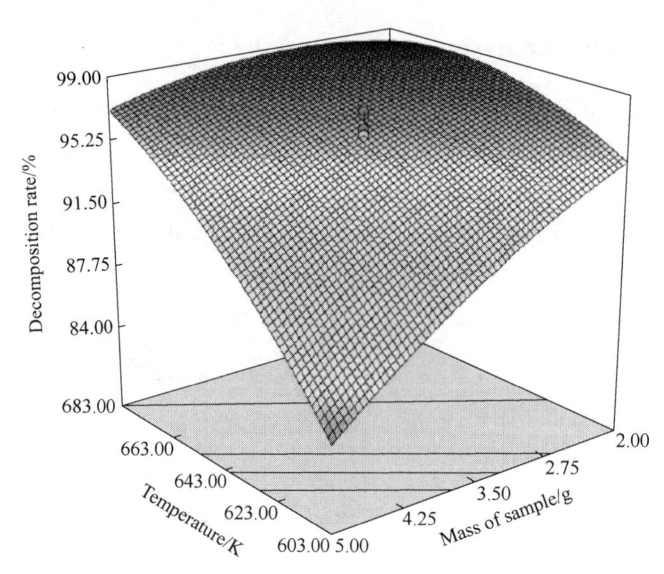

Fig. 5 Effect of the temperature(χ_1) and mass of sample(χ_3) on the decomposition rate(γ)

creased with increasing calcination temperature. Increasing the calcination temperature up to 683. 00K gave an enhanced effect on the decomposition rate, as the maximum predicated value of 98. 31% was achieved. The figure reveals that the effect of the calcination temperature on the decomposition rate was more significant than calcination time.

Fig. 5 shows the three - dimensional display of the response surface plot of the decomposition rate as function of the calcination temperature and mass of sample, with the calcination time set at 40 min. The decomposition rate is seen to increase with a decrease in mass of sample within the experimental range studied. The optimum value was selected based on the maximum decomposition rate. The depiction shows that the decomposition rate gave the maximum value 99. 19%.

3. 3 Optimal conditions and verification of the model

Thus, based on the above model, the optimal condition for decomposition rate was at 680. 86K, 44. 7min, and 3. 04g and the decomposition rate was 99. 06%. In order to confirm the optimized conditions, the accuracy of the model was validated with experiments under conditions of optimum. An experiment was carried out with parameters as suggested by the model. The conditions used in the confirmatory experiment were as follows: calcination temperature 681K, calcination time 45min, and mass of sample 3. 04g, the giving a decomposition rate of 99. 58%, which concurred with the model prediction. The model, therefore, be considered to fit the experimental data very well in these experimental conditions; with an error margin of only 0. 53%. Therefore, the model is acceptably valid.

3. 4 XRD analysis

Results of X - ray diffraction studies of the products on optimization conditions are shown in Fig. 6. The results show that cobalt oxide was the most solid product identified in which diffraction patterns satisfactorily matched with that of Co_3O_4 and few decomposition products or reaction intermediates were identified in the XRD studies[26,27]. Furthermore, it indicated that it is feasible to prepare Co_3O_4 by calcination from cobalt oxalate under optimum conditions.

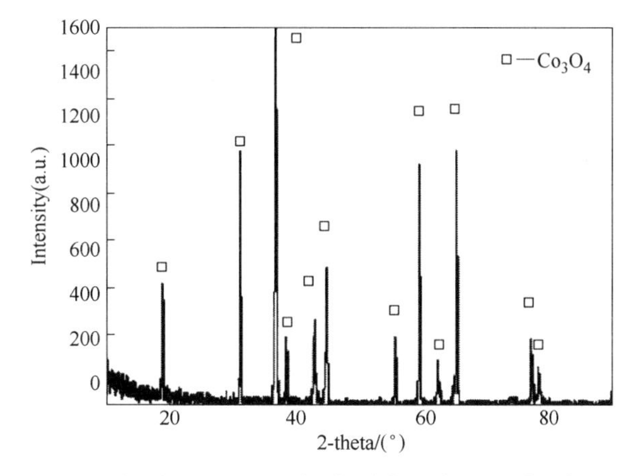

Fig. 6 The XRD pattern for the decomposition final solid products under the optimization conditions

4 Conclusions

This study showed that response surface methodology was a suitable approach to optimize conditions for achieving maximum decomposition rate of cobalt oxalate. The experimental and predicted values were very close, which reflected the correctness and applicability of RSM. In this case, the value of the determination coefficient (R^2 = 0. 9257) indicated that nearly 7. 43% of the variability of total variations werenot explainable by the model. The value of the adjusted determination coefficient was 0. 8588, showing a relatively high significance. Using RSM to optimize experiments, the optimal condition was found to be at 680. 86K, 44. 7min, and 3. 04g, respectively. Under these conditions, the predicted value of decomposition rate of 99. 06% was in good agreement with the actual experimental values (99. 58%). The Co_3O_4 prepared under the optimum conditions was characterized by XRD, from which the diffraction pattern satisfactorily matched with that of the standard Co_3O_4.

Acknowledgements

Financial support for this work from the National Natural Science Foundation Council of China (50734007), Technology Project in Yunnan Province (2007GA002) and Analysis and Testing Foundation of Kunming University of Science and Technology (2008 – 16) are gratefully acknowledged. Also authors thank Kunming Jin – Hui – Tong Wireless and Microwave Sensor Institute for the cooperation.

References

[1] Wang G X, Chen Y, Konstantinov K, et al. Investigation of cobalt oxides as anode materials for Li – ion batteries [J]. Journal of Power Sources, 2002, 109(1): 142 – 147.

[2]Yuan Z,Huang F,Feng C,et al. Synthesis and electrochemical performance of nanosized Co_3O_4[J]. Materials Chemistry and Physics,2003,79(1):1 -4.

[3]Li W Y,Xu L N,Chen J. Co_3O_4 nanomaterials in lithium - ion batteries and gas sensors[J]. Advanced Functional Materials,2005,15(5):851 -857.

[4]Xu X,Gao Y,Liu G,et al. Optimization of supercritical carbon dioxide extraction of sea buckthorn(Hippophaë thamnoides L.) oil using response surface methodology[J]. LWT - Food Science and Technology,2008,41(7):1223 -1231.

[5]Ichiyanagi Y,Kimishima Y,Yamada S. Magnetic study on Co_3O_4 Nanoparticles[J]. Journal of Magnetism and Magnetic Materials,2004,272:E1245 -E1246.

[6]Tang C W,Kuo C C,Kuo M C,et al. Influence of pretreatment conditions on low - temperature carbon monoxide oxidation over CeO_2/Co_3O_4 catalysts[J]. Applied Catalysis A:General,2006,309(1):37 -43.

[7]Ardizzone S,Spinolo G,Trasatti S. The point of zero charge of Co_3O_4 prepared by thermal decomposition of basic cobalt carbonate[J]. Electrochimica Acta,1995,40(16):2683 -2686.

[8]Liao C F,Liang Y,Chen H H. Preparation and characterization of Co_3O_4 by thermal decomposition from cobalt oxalate[J]. The Chinese Journal of Nonferrous Metals,2004,14(12):2131 -2136.

[9]Wang W W,Zhu Y J. Microwave - assisted synthesis of cobalt oxalate nanorods and their thermal conversion to Co_3O_4 rods[J]. Materials Research Bulletin,2005,40(11):1929 -1935.

[10]Gao J,Zhao Y,Yang W,et al. Sol - gel preparation and characterization of Co_3O_4 nanocrystals[J]. Journal of University of Science and Technology Beijing,10(1):54 -57.

[11]Cote L J,Teja A S,Wilkinson A P,et al. Continuous hydrothermal synthesis and crystallization of magnetic oxide nanoparticles[J]. Journal of Materials Research,2002,17(09):2410 -2416.

[12]Meskin P E,Baranchikov A E,Ivanov V K,et al. Synthesis of Nanodisperse Co_3O_4 powders under hydrothermal conditions with concurrent ultrasonic treatment[C]//Doklady Chemistry. MAIK Nauka/Interperiodica,2003,389(1):62 -64.

[13]Nethravathi C,Sen S,Ravishankar N,et al. Ferrimagnetic nanogranular Co_3O_4 through solvothermal decomposition of colloidally dispersed monolayers of α - cobalt hydroxide[J]. The Journal of Physical Chemistry B,2005,109(23):11468 -11472.

[14]Myer R H,Montgomery D C. Response Surface Methodology:Process and Product Optimization Using Designed Experiment[M]. New York:John Wiley and Sons,2002:343 -350.

[15]Chen M J,Chen K N,Lin C W. Optimization on response surface models for the optimal manufacturing conditions of dairy tofu[J]. Journal of Food Engineering,2005,68(4):471 -480.

[16]Fu J,Zhao Y,Wu Q. Optimising photoelectrocatalytic oxidation of fulvic acid using response surface methodology[J]. Journal of Hazardous Materials,2007,144(1):499 -505.

[17]Kiran B,Kaushik A,Kaushik C P. Response surface methodological approach for optimizing removal of Cr(VI) from aqueous solution using immobilized cyanobacterium[J]. Chemical Engineering Journal,2007,126(2):147 -153.

[18]Shrivastava S,Divecha J,Madamwar D. Response surface methodology for optimization of medium for decolorozation of textile dye Direct Black 22 by a novel bacterial consortium[J]. Bioresour Technol,2008,99:562 -569.

[19]Yang Y P,Huang K L,Liu R S,et al. Preparation of rod - like and polyhedron - like Co_3O_4 powders via hydrothermal treatment followed by decomposition[J]. Journal of Central South University:Science and Technology,2006,37(6):1103 -1106.

[20]Azargohar R,Dalai A K. Production of activated carbon from Luscar char:experimental and modeling studies

[J]. Microporous and Mesoporous Materials,2005,85(3):219 - 225.

[21] Tan I A W, Ahmad A L, Hameed B H. Optimization of preparation conditions for activated carbons from coconut husk using response surface methodology[J]. Chemical Engineering Journal,2008,137(3):462 - 470.

[22] Hill W J, Hunter W G. A review of response surface methodology: a literature survey[J]. Technometrics, 1966,8(4):571 - 590.

[23] Ahmad A L, Low S C. Optimization of membrane performance by thermal - mechanical stretching process using responses surface methodology[J]. Sep. Purif. Technol,2009,66:180 - 181.

[24] Körbahti B K, Rauf M A. Determination of optimum operating conditions of carmine decoloration by UV/H_2O_2 using response surface methodology[J]. Journal of Hazardous Materials,2009,161(1):281 - 286.

[25] Raymond H, Myers, A I. Response surface methodology[J]. Tecimometrics,1989,312:432 - 438.

[26] Garavaglia R, Mari C M, Trasatti S. Physicochemical characterization of Co_3O_4 preparaed by thermal decomposition Ⅱ: Response to solution pH[J], Surf. Technol. ,1983,23:41 - 47.

[27] Deshmukh P, Mankhand T R, Prasad P M. 1978, Decomposition characteristics of cobalt - cobaltic oxide[J]. Indian J. Technol. ,16:311 - 316.

Prediction Model of Ammonium Uranyl Carbonate Calcination by Microwave Heating Using Incremental Improved Back – Propagation Neural Network

Yingwei Li, Jinhui Peng, Bingguo Liu, Wei Li, Daifu Huang, Libo Zhang

Abstract: The incremental improved back – propagation(BP) neural network prediction model was put forward, which was very useful in overcoming the problems, such as long testing cycle, high testing quantity, difficulty of optimization for process parameters, many training data probably were offered by the way of increment batch and the limitation of the system memory could make the training data infeasible, which existed in the process of calcinations for ammonium uranyl carbonate(AUC) by microwave heating. The prediction model of the nonlinear system was built, which could effectively predict the experiment of microwave calcining of AUC. The predicted results indicated that the contents of U and U^{4+} were increased with increasing of microwave power and irradiation time, and decreased with increasing of the material average depth.

Keywords: microwave heating; AUC; BP neural network; prediction; increment

1 Introduction

Uranium is the material foundation for the development of nuclear industry; it has great strategic significance to protections of the national, the energy and the environmental securities. In the production process of uranium nuclear fuel, ammonium uranyl carbonate (AUC) and ammonium diuranate (ADU) are the intermediate products of uranium nuclear fuel, these are finally calcined to produce U_3O_8, as demonstrated[1-4].

Microwave calcination is one of the possible techniques to calcine AUC. The calcinations involve long testing cycle, large quantities of material for testing, and difficulties in optimizing a large number of process parameters, such as input power, calcination time, and average depth of AUC. Application of back – propagation (BP) neural network technique, with its ability for non – linear mapping, is an appropriate method to build the simulation model and predict the experimental process.

The traditional BP algorithm needs long convergent time and sometimes the convergent results cannot be obtained because of local minimum areas. The improved BP neural network based on Levenberg – Marquardt (L – M) optimization technique overcomes the limitations such as long training duration with a high num – ber of iterations, as demonstrated[5-7]. Aiming at the problems existing in the process of training the network, such as many training data probably were offered by the way of increment batch and the limitation of the system memory could make the training data infeasible when the sample scale was large, the incremental BP neural network is put forward, as demonstrated[8-13]. The incremental improved BP neural network non - linear system was build, which was used to predict the process results of microwave calcining AUC.

2 Incremental improved back - propagation neural network

The BP neural network is made up of an input layer, an output layer and a number of hidden layers in which neurons are connected to each other with modifiable weighted interconnections (Fig. 1). And its useful properties and capabilities are following:

Nonlinearity: The neuron is basically a nonlinear device. As a result, the BP neural network, made up of neurons, is nonlinear. This property of BP neural network enables better modeling capability of the nonlinear systems such as the microwave drying process.

Adaptability: The BP neural network has a capability to adapt its synaptic weights to changes in thesurrounding environment.

Input - output mapping: The BP neural network learns from the examples by constructing an input - output mapping for the problem. It tends to implicitly match the input vector to the output vector, and does not require a model to be stated.

Adaptability: The BP neural network has a capability to adapt its synaptic weights to changes in the surrounding environment.

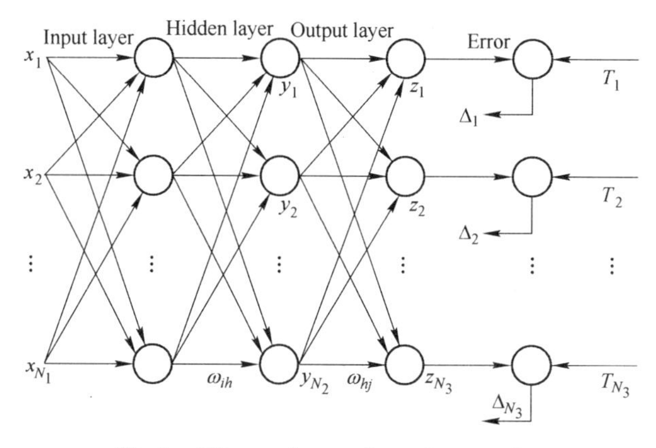

Fig. 1 BP neural network configureuration

In Fig. 1 x_i is the input; y_h is the output of one node in the hidden layers; z_j is the output of one node in the output layer; T_j is the target output; ω_{ih} is the weight between the ith node in the input layer and the hth node in the hidden layers; ω_{hj} is the weight between the hth node in the hidden layers and the jth node in the output layer; N_1 is the number of node in the input layer; N_2 is the number of node in the hidden layers; N_3 is the number of node in the output layer.

The BP neural network consists of two steps: the learning and the prediction. The first one is the forward phase where the activations are propagated from the input to the output layer. The second one is the backward phase where the error between the observed actual value and the desired nominal value in the output layer is propagated backward in order to modify the weights and bias values.

The performed functions are expressed as the following equations:

$$y_h = f\left(\sum_{i=1}^{N_1} \omega_{ih} x_i + \theta_h\right) \tag{1}$$

The function of one node in hidden layers:

$$Z_j = f\left(\sum_{h=1}^{N_2} \omega_{hj} y_h + \gamma_j\right) = f\left[\sum_{h=1}^{N_2} \omega_{hj} f\left(\sum_{i=1}^{N_1} \omega_{ih} x_i + \theta_h\right) + \gamma_j\right] \tag{2}$$

The function of one node in output layer:

$$E = \frac{1}{2}\sum_{j=1}^{N_3} (T_j - z_j)^2 \tag{3}$$

The sigmoid transfer function:

$$f(x) = \frac{1}{1 + e^{-x}} \tag{4}$$

where θ_h is the threshold value of the hth node; γ_j is the threshold value of the jth node.

The L - M algorithm, as demonstrated[5,14-18] is designed to approach second - order training speed without having to computethe Hessian matrix. The L - M algorithm uses this approximation to the Hessian matrix in the following Newton - like update:

$$x_{i+1} = x_i - (J^T J + \mu I)^{-1} J^T E \tag{5}$$

where J is the Jacobian matrix that contains first derivatives of the network errors with respect to the weights and biases; E is a vector of network errors. The J acobian matrix can be computed through a standard back - propagation technique that is much less complex than computing the Hessian matrix. μ is a scalar, when μ is zero, this is just Newton's method, using the approximate Hessian matrix. When μ is large, this becomes gradient descent with a small step size.

The incremental learning is implemented by adjusting the weights of the BP neural network, as demonstrated[8-11]. The knowledge effective extent is settled based on the prior knowledge; the weight vector can be changed in the effective extent when the accuracy of the learned knowledge is unchanged. The weight is adjusted through the fixed network structure when the new sample is provided; it can make the indication value approach to the target value, and then learn the new sample knowledge. So the network can not only learn the new sample knowledge but also held the original knowledge.

With the incremental learning, a scaling factor s which scales down all weight adjustments is introduced, so all the weights are within bounds. The learning rule is:

$$\Delta\omega_{ab}(k) = s(k)\eta\delta_b(k)O_a(k) \tag{6}$$

where $\Delta\omega_{ab}$ is the weight between the ath node and the bth node of all the network layers; $\eta(0 < \eta < 1)$ is a trial - independent learning rate; δ_b is the error gradient at the bth node; O_a is the activation level at the ath node; the parameter k is the kth iteration.

3 The prediction model of incremental improved BP neural network

3.1 The experimental design and results

In the experimental process, three input variables were used: microwave power x_1, irradiation time

x_2 and average depth x_3, two output variables: the contents of U $\sum$U and U^{4+} $\sum U^{4+}$. The experimental design and results were listed in Table 1.

Table 1 The experimental design and the results

Num	Power/W	Time/min	Average depth/mm	$\sum U^{4+}$/%	$\sum$ U/%
1	120	10	8.6	24.08	82.47
2	280	10	8.6	26.77	83.06
3	460	10	8.6	27.43	84.1
4	600	10	8.6	28.18	84.28
5	700	10	8.6	28.45	84.68
6	700	4	8.6	6.81	71.2
7	700	6	8.6	26.25	81.25
8	700	8	8.6	27.85	83.78
9	700	10	8.6	28.45	84.68
10	700	12	8.6	28.49	84.7
11	700	14	8.6	28.95	84.7
12	700	10	5.7	32.53	84.69
13	700	10	7.2	32.39	84.62
14	700	10	8.6	32.26	84.58
15	700	10	10	32.02	84.52
16	700	10	11.5	31.76	84.47
17	700	10	12.9	31.52	84.4
18	700	10	14.4	31.36	84.37
19	700	10	15.8	31.28	84.32
20	700	10	17.2	31.02	84.28

3.2 Data preprocessing and network training

The BP neural network based on L – M algorithm is described in MATLAB environment, using the corresponding toolbox. When the traditional BP network training reaches to a certain extent, the incremental improved BP neural network can avoid the error sum squares no longer be updated, the phenomenon of network paralysis and the network not be trained, be out of the local minimum when adjusting the network parameters, make the network be converged rapidly, and the training data cannot be provided for one – time, choose the representative samples to train the network in the case of occupying less memory source, as demonstrated[10,13].

The collected data units are inconsistent, the sample data must be pre – treated in order to train conveniently and reflect the interrelations of the various factors preferably. The data in Table 1 are used as the training samples to be pre – treated, when the network training is finished, the network is tested and the non – training sample data are used to predict the results, the error between the output results and the measured data is computed, if the error is within the regulated scope, the neural network model is available, it can be carried on simulation.

In the training process, the number of layers of the hidden layer is determined and the network weights and threshold values are obtained through computing the minimum value of the error function, the optimum number of units of the hidden layers is 14 after training the network.

4 The fitting and the analysis

The incremental improved BP neural network is used to verify and predict the input samples and the prediction samples, the predicted output data are compared with the actual output data shown in Tables 2 and 3, indicating the neural network are feasible.

Table 2 The predicted results with the data in Table 1

Num	Power/W	Time/min	Average depth/mm	$\sum U^{4+}$/%	$\sum U$/%
1	280	10	8.6	26.77	83.06
2	460	10	8.6	27.43	84.1
3	600	10	8.6	28.17	84.29
4	700	10	8.6	28.46	84.64
5	700	6	8.6	26.25	81.25
6	700	8	8.6	27.85	83.78
7	700	10	8.6	28.46	84.66
8	700	12	8.6	28.48	84.72
9	700	10	8.6	32.26	84.58
10	700	10	11.5	31.76	84.47
11	700	10	14.4	31.35	84.35
12	700	10	17.2	31.06	84.26

Table 3 The predicted results with no experiments

Num	Power/W	Time/min	Average depth/mm	$\sum U^{4+}$/%	$\sum U$/%
1	300	10	8.6	26.87	83.26
2	400	10	8.6	27.23	83.89
3	500	10	8.6	27.77	84.19
4	650	10	8.6	28.26	84.54
5	700	5	8.6	26.15	81.16
6	700	7	8.6	27.18	82.48
7	700	9	8.6	28.26	84.16
8	700	11	8.6	28.26	84.69
9	700	10	7.2	32.38	84.61
10	700	10	10	32.03	84.52
11	700	10	12.9	31.52	84.41

Fig. 2 shows the predicted impacts of the irradiation time, microwave power and material average depth on the contents of U and U^{4+}. The red "●" indicates the experimental results in the actual conditions. The black "□" indicates the experimental results in the following conditions: time: 6min, 8min, 10min and 12 min, average depth: 8.6mm, power: 700W. The green "○" indicates the experimental results in the following conditions: time: 10min, average depth: 8.6mm, 11.5mm, 14.4mm and 17.2mm, power: 700W. The blue "△" indicates the experimental results in the following conditions: time: 10min, average depth: 8.6mm, power: 280W, 460W, 600W and 700W.

Fig. 2(a) indicates the relationship among the irradiation time, the microwave power and the contents of U^{4+}, the purple "◇" indicates the experimental results in the predicted conditions: time: 5min, 7min, 9min and 11min, average depth: 8.6mm, power: 700W. Fig. 2(b) indicates the relationship among the irradiation time, the material average depth and the contents of U^{4+}, the purple "◇" indicates the experimental results in the predicted conditions: time: 10min, average depth: 7.2mm, 10mm, 12.9mm and 15.8mm, power: 700W. Fig. 2(c) indicates the relationship among the material average depth, the irradiation time and the contents of U^{4+}, the purple "◇" indicates the experimental results in the predicted conditions: time: 10min, average depth: 8.6mm, power: 300W, 400W, 500W and 650W.

Fig. 2(d) indicates the relationship among the irradiation time, the microwave power and the contents of U, the purple "◇" indicates the experimental results in the predicted conditions: time: 5min, 7min, 9min and 11min, average depth: 8.6mm, power: 700W. Fig. 2(e) indicates the relationship among the irradiation time, the material average depth and the contents of U, the purple "◇" indicates the experimental results in the predicted conditions: time: 10 min, average depth: 7.2mm, 10mm, 12.9mm and 15.8mm, power: 700W. Fig. 2(f) indicates the relationship among the material average depth, the irradiation time and the contents of U, the purple "◇" indicates the experimental results in the predicted conditions: time: 10min, average depth: 8.6mm, power: 300W, 400W, 500W and 650W.

Fig. 2(a) – (c) indicate the following results: when irradiation time and average depth were held constant, the percentage of U^{4+} was increased with the microwave power increasing. Increasing the microwave power from 280W to 700W gave an enhanced effect on the percentage of U^{4+}, as the percentages of U^{4+} 26.77% – 28.45% were achieved. Under constant irradiation time and the microwave power, the percentage of U^{4+} was decreased with the material average depth increasing. Increasing the material average depth from 10mm to 17.2mm gave no obvious effect on the percentage of U^{4+}, as the percentages of U^{4+} 32.02% – 31.02% were achieved. And under constant microwave power and the material average depth, the percentage of U^{4+} was increased with the irradiation time increasing. Increasing the irradiation time from 6min to 12min gave obvious effect on the percentage of U^{4+}, as the percentages of U^{4+} 26.25% – 28.49% were achieved. The microwave power and the irradiation time had the significant effect on the percentage of U^{4+}, the material average depth had no obvious effect on the percentage of U^{4+}.

Fig. 2(d) – (f) indicate the following results: under the condition of unchanged the irradiation time and the material average depth; the percentage of U was increased with the microwave power

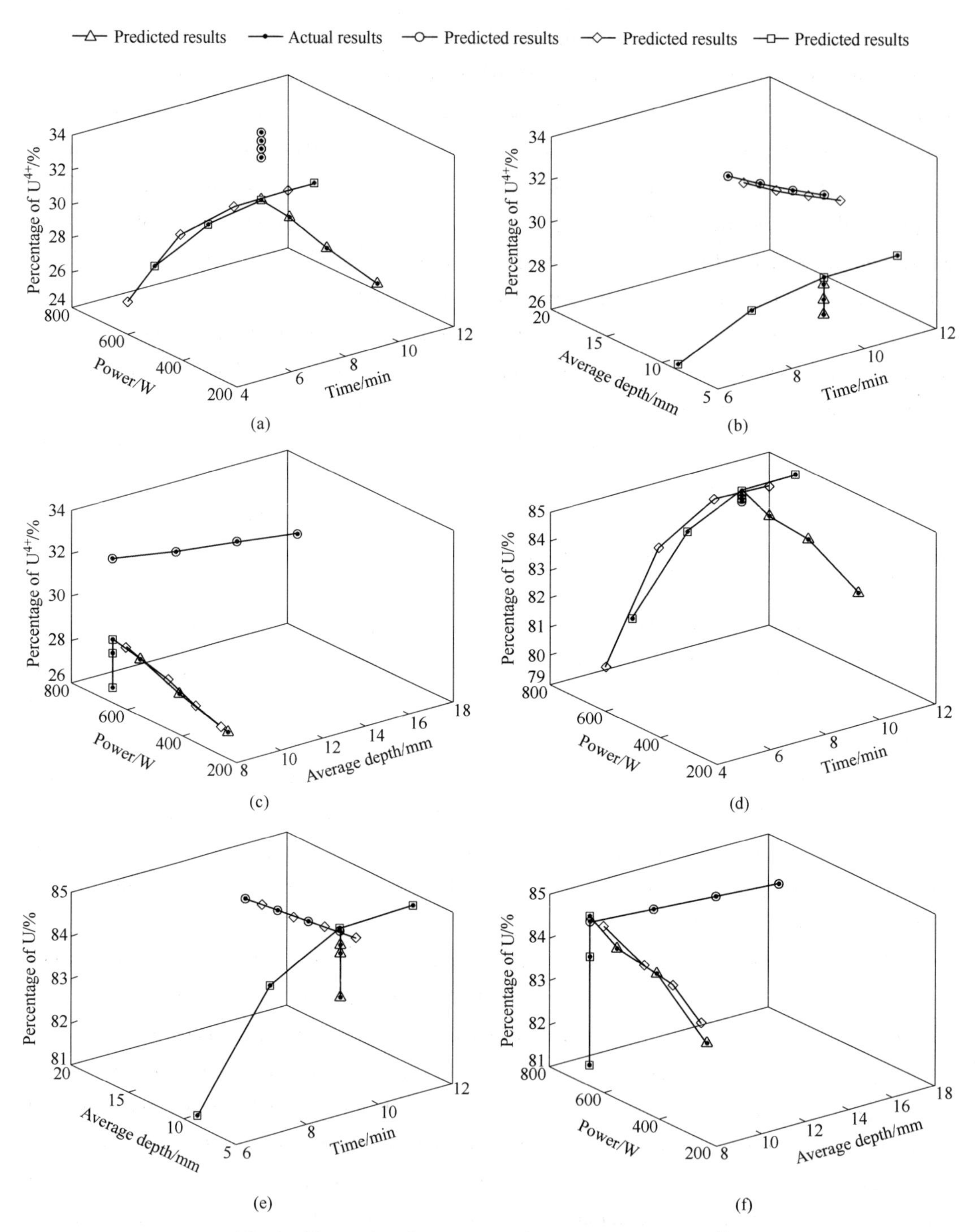

Fig. 2 The predicted impacts on the contents of U and U^{4+}

increasing. Increasing the microwave power from 280W to 700W gave an enhanced effect on the percentage of U, as the percentages of U 83.06% – 84.68% were achieved. Under the condition of unchanged the irradiation time and the microwave power, the percentage of U was decreased with the material average depth increasing, Increasing the material average depth from 10mm to 17.2mm gave no obvious effect on the percentage of U, as the percentages of U 84.52% –

84. 28% were achieved. Under the condition of unchanged the microwave power and the material average depth, the percentage of U was increased with the irradiation time increasing. Increasing the irradiation time from 6 min to 12 min gave obvious effect on the percentage of U, as the percentages of U 81. 25% - 84. 7% were achieved. The microwave power and the irradiation time had the significant effect on the percentage of U, the material average depth had no obvious effects on the percentage of U.

The prediction results with no experiments indicate that the prediction model of microwave calcining ADU using incremental improved BP neural network is available and accuracy to the prediction processing: When the irradiation time was 10min and the average depth was 8. 6mm, increasing the microwave power from 300W to 650W gave an enhanced effect on the percentage of U^{4+}, as the percentages of U^{4+} 26. 87% - 28. 26% were achieved in the scope of 26. 77% - 28. 46%. When the irradiation time was 10min and the microwave power was 700W, increasing the average depth from 7. 2mm to 12. 9mm gave no obverse effect on the percentage of U^{4+}, as the percentages of U^{4+} 32. 38% - 31. 52% were achieved in the scope of 32. 53% - 31. 36%. When the average depth was 8. 6mm and the microwave power was 700W, increasing the irradiation timefrom 5min to 11min gave obverse effect on the percentage of U^{4+}, as the percentages of U^{4+} 26. 15% - 28. 46% were achieved in the scope of 6. 81% - 28. 49%.

When the irradiation time was 10min and the average depth was 8. 6mm, increasing the microwave power from 300W to 650W gave an enhanced effect on the percentage of U, as the percentages of U 83. 26% - 84. 54% were achieved in the scope of 83. 06% - 84. 68%. When the irradiation time was 10min and the microwave power was 700W, increasing the average depth from 7. 2mm to 12. 9mm gave no obverse effect on the percentage of U, as the percentages of U 84. 61% - 84. 41% were achieved in the scope of 84. 69% - 84. 37%. When the average depth was 8. 6mm and the microwave power was 700W, increasing the irradiation time from 5min to 11min gave obverse effect on the percentage of U, as the percentages of U 81. 16% - 84. 69% were achieved in the scope of 71. 2% - 84. 7%.

The results indicate that (1) AUC can accept the microwave energy; (2) microwave heating can quickly decompose AUC; (3) under different experimental conditions, the percentages of U and U^{4+} in the production can meet the experimental demands. The BP neural network model is used in microwave calcining AUC to predict the experimental results, the predicted results and the measured value are fitted well, in different experimental conditions the trends of the predicted experimental results are matched to the trends of the measured value, indicating the feasibility of the BP neural network prediction model used in the experiment of microwave calcining AUC.

5 Conclusions

In the experiment of microwave calcining AUC, the contents of U and U^{4+} increased with increasing of microwave power and irradiation time, and decreased with increasing of the material average depth. And the changed microwave power and the irradiation time have the significant effect on the contents of U and U^{4+}, the material average depth has no obverse effect on the contents of U and

U^{4+}.

The BP neural network based on L – M algorithm can avoid the error sum squares no longer be updated when the traditional BP network training reaches to a certain extent, emerges the phenomenon of network paralysis, causes the network not be trained, it can be out of the local minimum when adjusting the network parameters, the network can be converged rapidly. The BP neural network based on incremental learning can effectively solve the problem such as the training data cannot be provided for one – time; choose the representative samples to train the network in the case of occupying less memory source.

The incremental improved BP neural network, having the faster convergence, the better prediction accuracy and the better fitting results, can make up the shortage of the traditional BP neural network. This demonstrates the feasibility of the BP neural network to predict the experiment of microwave calcining AUC.

Acknowledgements

The authors would like to express their gratitude for the financial support of the Key Project of Nature Science Foundations of China (No. 50734007), Technology Project in Yunnan Province (No. 2007GA002).

References

[1] Ayaz B, Bilge A N. The possible usage of ex – ADU uranium dioxide fuel pellets with low – temperature sintering[J]. J. Nucl. Mater. 2000:280, 45 – 50.

[2] Pei B F, Duan D Z. ADU technology and its effect on product characteristics[J]. Nucl. Power Eng. Technol, 1994, 7:41 – 50.

[3] Yang J H, et al. Microwave process for sintering of uranium dioxide[J]. J. Nucl. Mater, 2004, 325:210 – 216.

[4] Kim Y S. A thermodynamic evaluation of the U – O system from UO_2 to U_3O_8[J]. J. Nucl. Mater, 2000, 279: 173 – 180.

[5] Kermani B G, Schiffman S S. Performance of the Levenberg – Marquardt neural network training method in electronic nose applications[J]. Sens. Actuators B, 2005, 110:13 – 22.

[6] Lera G, Pinzolas M. Neighborhood based Levenberg – Marquardt algorithm for neural network training[J]. IEEE Trans. Neural Netw, 2002, 13:1200 – 1203.

[7] Su G L, Deng F P. On the improving back – propagation algorithms of the neural networks based on MATLAB language: a review[J]. Bull. Sci. Technol, 2003, 19:130 – 135.

[8] Ghosh J, Nag A C. Knowledge enhancement and reuse with radial basis function networks[J]. Neural Netw., 2002, 2:1322 – 1327.

[9] Zhang J, Morris A J. A sequential learning approach for single hidden layer neural networks[J]. Neural Netw., 1998, 11:65 – 80.

[10] Fu L M. Incremental back – propagation learning network[J]. IEEE Trans. Neural Netw., 1996, 7:757 – 761.

[11] Karayiannis N B, Mi G W. Growing radial basis neural networks: merging supervised and unsuper – vised learning with network growth techniques[J]. IEEE Trans. Neural Netw, 1997, 8:1492 – 1506.

[12] Parekh R, et al. Constructive neural network learning algorithms for pattern classification [J]. IEEE Trans. Neural Netw., 2000, 11:436 – 451.

[13] Li Y W, et al. Question classification based on incremental modified Bayes[C]//Proceedings of the 2008 2nd International Conference on Future Generation Communication and Networking, FGCN 2008: 149 - 152.

[14] Adeloye A J, Munari A D. Artificial neural network based generalized storage yield reliability models using the Levenberg - Marquardt algorithm[J]. J. Hydrol, 2006, 362: 215 - 230.

[15] Elif D U, Inan G. Multilayer perceptron neural networks to compute quasistatic parameters of asymmetric coplanar waveguides[J]. Neurocomputing, 2004, 62: 349 - 365.

[16] Bezerra E M, et al. Artificial neural network (ANN) prediction of kinetic parameters of (CRFC) composites [J]. Comput. Mater. Sci., 2008, 44: 656 - 663.

[17] Hossein M. Long - term prediction of chaotic time series with multi - step prediction horizons by a neural network with Levenberg - Marquardt learning algorithm[J]. Chaos Soliton Fract. 2009, 41: 1975 - 1979.

[18] Vijander S, et al. ANN - based estimator for distillation using Levenberg - Marquardt approach[J]. Eng. Appl. Artif. Intell. 2007, 20: 249 - 259.

Investigation on Phase Transformation of Titania Slag Using Microwave Irradiation

Guo Chen, Zengkai Song, Jin Chen, C. Srinivasakannan, Jinhui Peng

Abstract: The work addresses phase transformation of titania slag using microwave irradiation. Properties of samples before and after microwave irradiation including thermal stability, crystal structures, microstructure, surface chemical functional groups and molecular structures, have been investigated by TG/DTA, XRD, SEM, FT - IR and Raman, respectively. The results of TG/DTA showed that titania slag have two phase transformation, one at 578.0℃ and another at 850.0℃. It was confirmed that at roasting temperature in excess of 600℃, anatase starts to transform as rutile. The property changes can be attributed to microwave irradiation, which causes the crystal transformation.

Keywords: synthetic rutile; titania slag; microwave irradiation; phase transformation

1 Introduction

Titanium is the ninth most abundant chemical element and the fourth most abundant usable metallic element in the Earth's crust[1]. Rutile, anatase and brookite have higher titanium contents than other titaniferous minerals, which are in fact trimorphs[2,3]. These three distinct minerals are composed of different crystalline forms, but with similar chemical structure[4,5]. Titania pigment is manufactured either using the sulfate process, in which the ore is reacted with concentrated sulfuric acid, or the chloride process, in which the ore is chlorinated to form titanium tetrachloride, which are then reoxidized to form pigments[6]. Rutile TiO_2 is more popularly utilized due to its high refractive index, whiteness, brightness, thermal stability and chemical inertness, which finds wide range of applications, including the manufacture of white pigment[7,8]. About 60% of natural rutile is used for the manufacture of titanium metal, since they are light in weight, high in tensile strength and corrosion - resistant, widely used in the manufacture of aircraft, spacecraft and medical prostheses[9,10]. The increasing use of chloride process for producing titanium dioxide pigments has motivated the search for a more abundant and cheaper source of raw material than the presently used titania slag. Hence, development of more efficient and effective method to utilize titania slag is imperative.

Microwave irradiation technology is one of effective routes in realizing energy saving and green production[11]. The main difference between microwave heating and conventional heating systems is in the way the heat is generated[12]. Compared with conventional heating techniques, the main advantages of microwave processing is reduction in processing time and energy consumption, since microwave heating is both internal and volumetric[13]. Since microwave heating is energy efficient and facilitate rapid and uniform heating, it is favorable to the process economics. Microwave energy has been widely used in several fields of applications both at research as well as commercial scale industrial mineral processing[14,15]. Although microwave irradiation has been utilized to treat titania slag for preparation of synthetic rutile, relevant literature is very limited[16]. The objectives of present work are to: (1) preparation of synthetic rutile from high titania slag under microwave irradiation; (2) assess properties of samples before and after microwave irradiation, including thermal stability, crystal structures, microstructure, surface chemical functional groups and molecular structures, systematically.

2 Experiment

2.1 Materials

Titania slag was obtained from Kunming city, Yunnan province, China. The slag contains 72.33% TiO_2, 17.79% Ti_2O_3, and 5.26% FeO. The slag also contains 1.04% MnO, 2.75% Al_2O_3, 2.30% MgO, 2.57% SiO_2 and minor elements such as S, P and C. The titania slag was analyzed for element content in accordance with the National Standard of the People's Republic of China (GB/T).

2.2 Characterization

The thermogravimetry and corresponding differential thermal analysis were carried out simultaneously using a thermal gravimetric analyzer (NETZSCH STA 409, Germany). X – ray diffraction (XRD) patterns were recorded on a X – ray diffractometer (D/Max 2200, Rigaku, Japan) using CuK_α radiation (k = 0.15418nm) and a graphite monochromator for the diffracted beam. The Raman spectra of samples were performed at room temperature using a confocal microprobe Raman system (Renishaw Ramascope System 1000, UK) with an air – cooled charge – coupled device (CCD) detector. A 514nm argon laser was used for excitation. Backscattered Raman signals were collected through a microscope and holographic notch filters in the range of 100 – 1000cm^{-1} with a spectral resolution of 2cm^{-1}. Fourier transformation infrared spectra were collected using FT – IR spectrometer (8700, Nicolet, USA). The angle of incidence of the IR beam was 45° and 100 scans were collected at a resolution of 4cm^{-1} and averaged using the OMNIC spectroscopic software. A scanning electron microscope (XL30ESEM – TMP, Philips, Holland) was used to observe the microstructure morphology of the titania slag and synthetic rutile processed.

2.3 Instrumentation

The microwave irradiation experiments were carried out in a self – made microwave muffle furnace. An industrial microwave muffle furnace typically consists of a magnetron to produce the microwaves, a waveguide to transport the microwaves, a resonance cavity to manipulate microwaves for a specific purpose, and a control system to regulate the temperature and microwave power. The power supply of the microwave muffle furnace was two magnetrons at 2.45GHz frequency and 1.5kW power, which was cooled by water circulation. The inner dimensions of the multi – mode microwave resonance cavity were 260mm in height, 420mm in lengthen and 260mm in width. The temperature is measured using a Type K thermocouple, placed at closest proximity to the sample. Thermocouple provides feedback information to the control panel that controls the power to the magnetron, controlling the temperature of the sample during the microwave irradiation process in order to prevent the sample from overheating.

2.4 Procedure

100g of titania slag was loaded on a ceramics boat and placed inside a stainless steel tubular reactor which was heated to a temperature of 120℃ at a heating rate of 5℃/min. The samples were held at 120℃ for duration of 120min in order to completely remove the moisture from the samples. The dried samples were transferred to the microwave irradiation apparatus and heated to 950℃ and held at that temperature for duration of 60min.

3 Results and discussion

The TG and the DTA curves of the samples are shown in Fig. 1. According to TG/DTA curves, titania slag has three weight change stages. The first step is a dehydration step, characterized by an endothermic DTA peak in the temperature range 100 – 110℃. The weight loss accompanying this step amounts to about 0.08% in accordance with calculated weight loss of 0.09% attributed to the complete dehydration[17,18]. The $Fe_3Ti_3O_{10}$ decomposed in the second step, indicates weight gain of 0.87% due to the formation of anatase TiO_2 from $Fe_3Ti_3O_{10}$, accompanied by a very sharp exothermic DTA peak at 578.0℃. Anatase TiO_2 is thermally stable up to 600℃, which transformed in the third step with weight gain of 2.99% in agreement with the formation of rutile TiO_2, characterized by an exothermic DTA peak at 850℃. The final product corresponds to synthetic rutile. TG/DTA curves also indicate the corresponding temperatures to the theoretical weight gain, which formed the basis for selecting the minimum roasting temperature of 950℃ and 60min. On the basis of experimental results, the following mechanism of the weight gain of titania slag could be proposed:

$$2Ti_2O_3 + O_2 \longrightarrow 4TiO_2 \quad (1)$$

$$2Ti_3O_5 + O_2 \longrightarrow 6TiO_2 \quad (2)$$

The following mechanism of decomposition reaction of $Fe_3Ti_3O_{10}$ and the anatase rutile crystal transformation of TiO_2 could be proposed:

$$Fe_3Ti_3O_{10} \longrightarrow 3TiO_2(\text{Anatase}) + Fe_3O_4 \quad (3)$$

$$TiO_2(\text{Anatase}) \longrightarrow TiO_2(\text{Rutile}) \quad (4)$$

The crystal structures of titania slag are characterized by XRD, and the results are illustrated in Fig. 2. It is found from Fig. 2(a) that the main phase of titania slag is $Fe_3Ti_3O_{10}$ and the anatase TiO_2 match well with those of reference XRD data. However, XRD pattern shows that rutile TiO_2 also exist in the samples. These results indicate that titania slag consists of multi – crystalline phases. The strongest diffraction peaks of $Fe_3Ti_3O_{10}$ and anatase TiO_2 are observed around 25.45° and 25.28°, respectively. Fig. 2(b) shows the X – ray diffraction of pattern of titania slag after microwave irradiation at 950℃ for 60min[19,20]. It can be observed from Fig. 2(b) that the diffraction

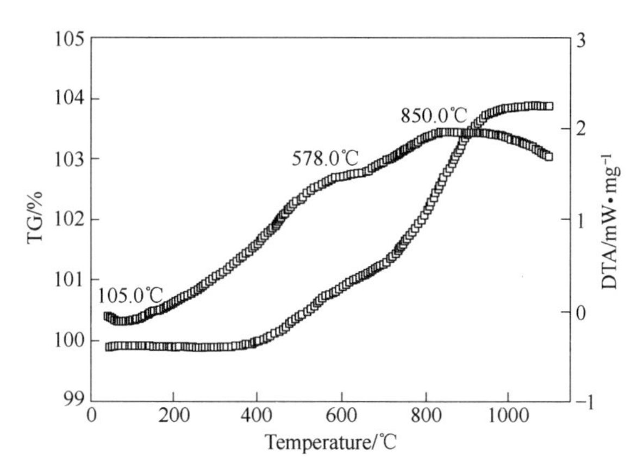

Fig. 1 TG/DTA curves of titania slag

peaks are smooth and clear, and the strong peaks of microwave treated samples appear at 27.44°, and hypostrong peaks appear at 36.09°, 41.25° and 54.33°, which are similar to the standard PDF card of rutile TiO_2. The narrow diffraction peaks also indication that the samples have a very good crystalline structure[21]. Based on the above results, beginning of the transformation of anatase TiO_2 into rutile can be confirmed at temperature in excess of 600℃[22,23]. Furthermore, the intensity of anatase phase decreases while the intensity of rutile phase increases with increasing microwave roasting temperature. At a microwave roasting temperature of 950℃, the anatase TiO_2 transformation into rutile TiO_2 is complete.

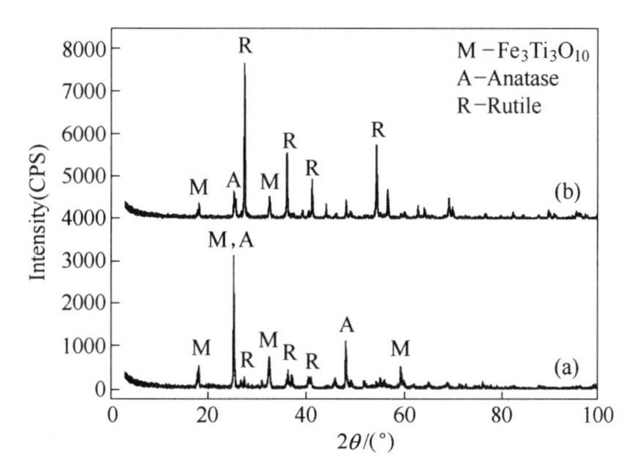

Fig. 2 XRD of titania slag before and after microwave irradiation at 950℃ for 60min

(a) Raw materials; (b) Microwave treated samples

The anatase structure belongs to the tetragonal space group I4/amd. The unit cell contains two TiO_2 units with Ti ions at (0,0,0) and (0,1/2,1/4), and O ions at (0,0,u), (0,0,$-u$), (0,1/2,$u+1/4$), and (0,1/2,$1/2-u$)[24]. Factor group analysis indicates there are six Raman active modes: $1A_{1g}+2B_{1g}+3E_g$. In contrast, the rutile structure belongs to the P42/mnm tetragonal space group. The unit cell is defined by the lattice vectors a and c and contains two TiO_2 munits with Ti ions at (0,0,0) and (1/2,1/2,1/2), and O ions at $\pm(u,u,0)$ and $\pm(1/2+u,1/2-u,1/2)$. Therefore, there are four Raman active modes: $A_{1g}+B_{1g}+B_{2g}+E_g$. The Raman spectrum of the samples before and after microwave irradiation is characterized and the results are shown in Fig. 3. From the Raman spectra in Fig. 3 (a), the frequencies of Raman bands identified as 155.2cm^{-1} are assigned to the symmetric stretching vibrations of Ti_3O_5. The bands at 195.8cm^{-1} are assigned to the wagging vibrations of O – Ti – O bonds of Ti_2O_3. The bands at 393.7cm^{-1} can be attributed to the symmetric bending vibrations of O – Ti – O bonds. The bands at 515.5cm^{-1} can be attributed to the asymmetric bending vibrations of O – Ti – O bonds, while the bands at 637.3cm^{-1} can be attributed to the symmetric stretching vibrations of O – Ti – O bonds, based on the factor group analysis. These bands agree well with those in previous reports for anatase powder and single crystals[24]. As shown in Fig. 3 (b), microwave treated samples have one weak and broad peak at 515.2cm^{-1}, two intense bands at 442.5cm^{-1} and 611.6cm^{-1}, and one weak and broad peak at 244.6cm^{-1}. The peak at 442.5cm^{-1} and 611.6cm^{-1} are due to the rutile TiO_2. The phase transition of titania slag from anatase TiO_2 to rutile TiO_2 occurs at temperatures above 950℃ and it is characterized by the disappearance of the very strong peak at 637.3cm^{-1} which is Raman active in the anatase TiO_2 form. The results of X – ray diffraction analysis are supported by the Raman spectra of synthetic rutile. The Raman spectroscopy results showed that microwave treated sample is a crystalline material, and the intensity of Raman spectrum of the rutile TiO_2 increased with increase in microwave irradiation temperature.

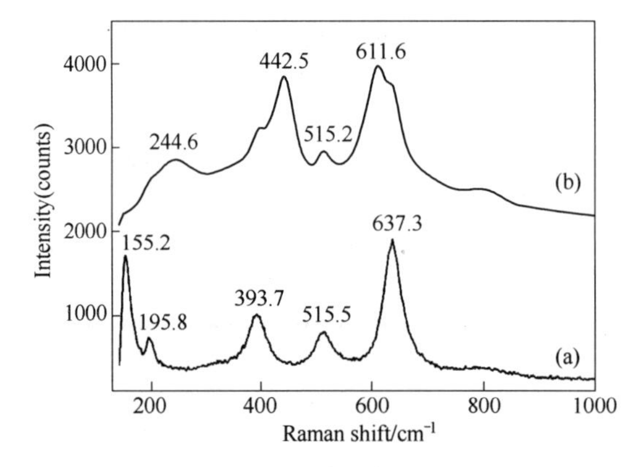

Fig. 3 Raman spectra of titania slag before and after microwave irradiation at 950℃ for 60min
(a) Raw materials; (b) Microwave treated samples

The surface chemical functional groups of the samples are characterized using FT – IR, and the spectra results are shown in Fig. 4. The titania slag sample in Fig. 4(a), shows absorption bands at 3426.9cm^{-1}, 1624.3cm^{-1}, 1089.5cm^{-1} and 493.3cm^{-1}. The absorption band at 3426.9cm^{-1} can be attributed to the O – H stretching vibrations. The bands at 1624.3 cm^{-1} is probably due to more strongly adsorbed H_2O molecules on the minerals surface. The absorption band at 1089.5cm^{-1} is assigned to the bending vibrations of O – H. The bands between 400cm^{-1} and 1000cm^{-1} could be assigned to the vibrations of metal ions. The peak at 493.3cm^{-1} can be attributed to the stretching vibrations of octahedral metal ion in the TiO_2 units. The FT – IR spectrum of the prepared synthetic rutile roasting at 950℃ for 60min is shown in Fig. 4(b). The most obvious change in the spectrum is the bands at 3426. 9cm^{-1}, 1624. 3cm^{-1} and 1089. 5cm^{-1} which are barely visible, while the band at 493. 3cm^{-1} peaks individually blueshift to 529. 1cm^{-1}. Apparently it can be attributed to the phase transformation from the anatase into the rutile of TiO_2 in that temperature range.

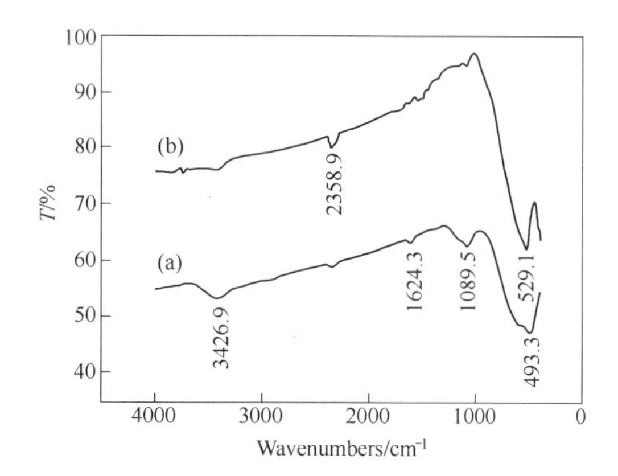

Fig. 4 FT – IR spectra of titania slag before and after microwave irradiation at 950℃ for 60min

(a) Raw materials; (b) Microwave treated samples

SEM of titania slag before and after microwave irradiation is characterized, and the results are shown in Fig. 5. From the SEM image in Fig. 5(a), the results indicate that the surface structures of primary particles have a tighter and smoother surface morphology, with more small pits and striations appearing on the titania slag surface. The microwave treated samples appears irregular with a complex acicular structure. The pores of samples surfacecould have been opened through microwave irradiation, which lead to increased surface area. Point and area analyses are conducted using EDAX attachment of SEM on the particle surfaces. EDAX analyses of microwave treated titania slag are carried out to estimate the elemental composition of synthetic rutile, and the results are shown in Fig. 6. The EDAX spectra of corresponding particles are obtained with SEM (Fig. 6(a)). The EDAX analysis results indicate that the sample consists of a certain amount of Fe and Ti, and other elements such as Si, Al, Mg and Mn.

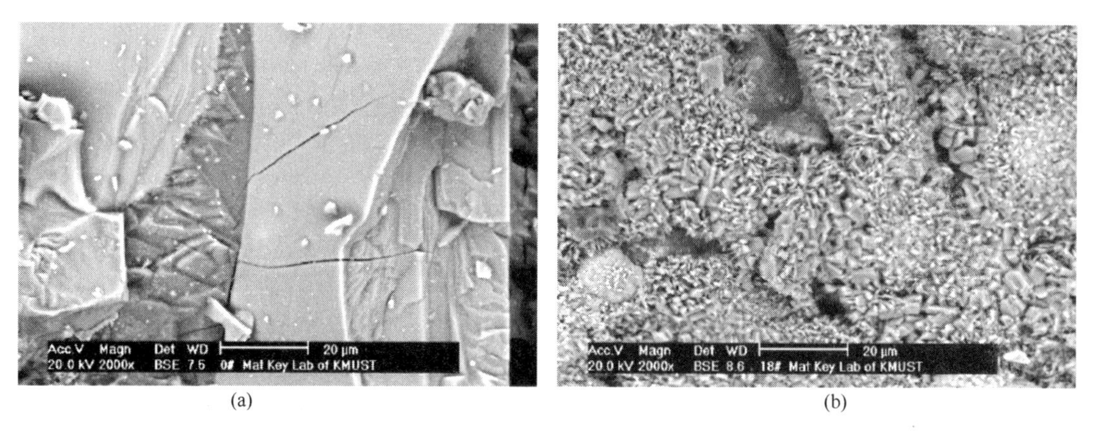

Fig. 5 SEM of titania slag before and after microwave irradiation at 950℃ for 60min

(a) Raw materials; (b) Microwave treated samples

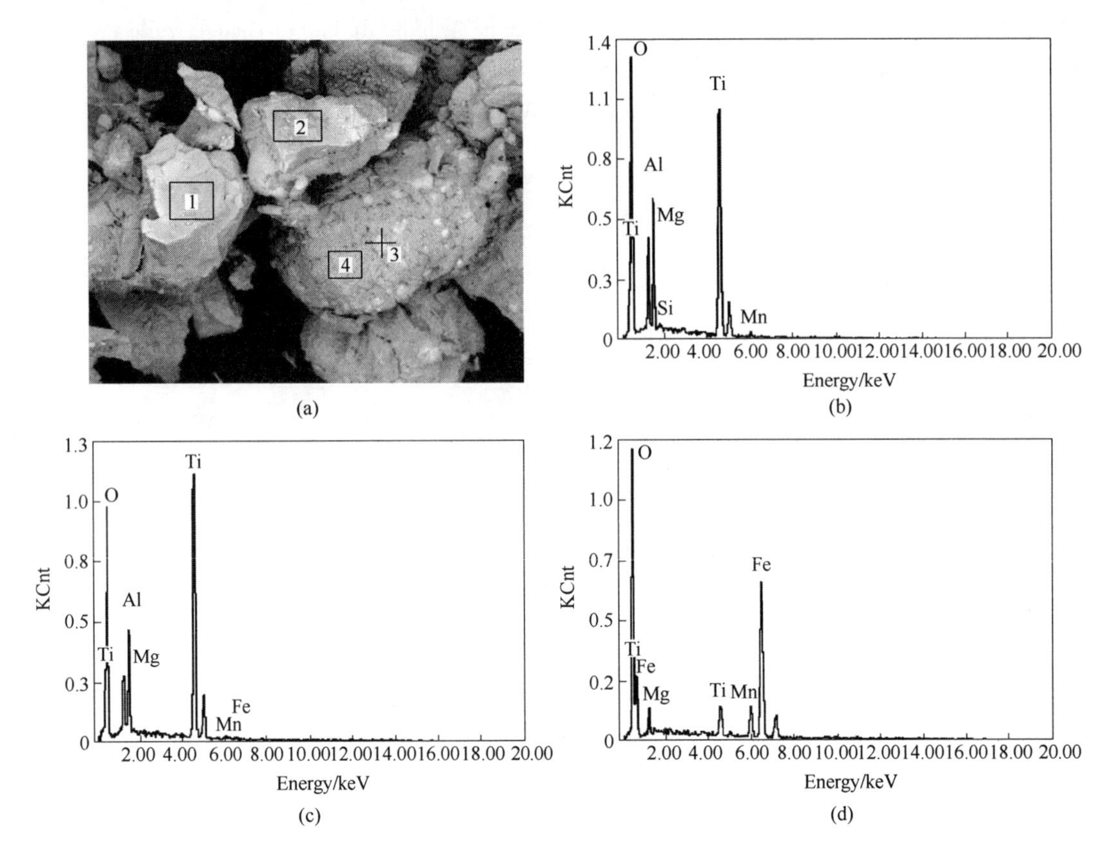

Fig. 6 EDAX spectra of synthetic rutile

(a) SEM; (b) District 1; (c) District 2; (d) Spot 3

4 Conclusions

In summary, paper offers a convenient method for the preparation of synthetic rutile under microwave irradiation. This technique is interesting not only for the shorter processing time and low temperature, but also for the possibility to control the microstructure and crystal structural properties of samples. Based on the XRD results, it was confirmed that, at microwave roasting temperature higher than 600℃, anatase TiO_2 starts to transform into rutile TiO_2. The FT – IR absorption peak at 493.3 cm^{-1} can be attributed to the stretching vibrations of octahedral metal ion in the TiO_2 units. The experimental results demonstrate that microwave irradiation technique can be effectively and efficiently applied to the preparation of synthetic rutile.

Acknowledgements

The authors acknowledge the financial supports from the National Scientific Foundation of China (No. 51090385), the International S&T Cooperation Program of China (No. 2012DFA70570), Specialized Research Fund for the Doctoral Program of Higher Education (No. 20125314120014) and the Applied Foundation Fund of Yunnan Province of China (No. 2012FD015).

References

[1]Chen G,Peng J H,Chen J. Miner. Metall. Process,2011,28:44 - 48.

[2]Chen Y,Marsh M,Williams J S,et al. Production of rutile from ilmenite by room temperature ball - milling - induced sulphurisation reaction[J]. Journal of Alloys and Compounds,1996,245(1):54 - 58.

[3]Samal S,Mohapatra B K,Mukherjee P S,et al. Integrated XRD EPMA and XRF study of ilmenite and Titania slag used in pigment production[J]. Journal of Alloys and Compounds,2009,474(1):484 - 489.

[4]Chen G,Chen J,Peng J,et al. Dissociation of Ti_2O_3 from titania slag under mechanical activation[J]. Journal of Alloys and Compounds,2011,509(24):L244 - L247.

[5]Mo W,Deng G Z. Titanium Metallurgy[M]. Beijing:Publishing Press of Metallurgical Industry,1998.

[6]Bessinger D,Geldenhuis J M A,Pistorius P C,et al. J. Non - Cryst. Solids,2001,282:132 - 142.

[7]Pistorius P C,Motlhamme T. Oxidation of high - titanium slags in the presence of water vapour[J]. Minerals Engineering,2006,19(3):232 - 236.

[8]Xue T,Wang L,Qi T,et al. Decomposition kinetics of titanium slag in sodium hydroxide system[J]. Hydrometallurgy,2009,95(1):22 - 27.

[9]Zhang Y,Qi T,Zhang Y. A novel preparation of titanium dioxide from titanium slag[J]. Hydrometallurgy,2009,96(1):52 - 56.

[10]Mazaheri M,Razavi Hesabi Z,Sadrnezhaad S K. Scr. Mater,2008,59:139 - 142.

[11]Vereš J,Lovás M,Jakabský Š,et al. Characterization of blast furnace sludge and removal of zinc by microwave assisted extraction[J]. Hydrometallurgy,2012,129:67 - 73.

[12]Yang K,Peng J,Srinivasakannan C,et al. Preparation of high surface area activated carbon from coconut shells using microwave heating[J]. Bioresource Technology,2010,101(15):6163 - 6169.

[13]Guo S,Li W,Peng J,et al. Microwave - absorbing characteristics of mixtures of different carbonaceous reducing agents and oxidized ilmenite[J]. International Journal of Mineral Processing,2009,93(3):289 - 293.

[14]Chen G,Chen J,Peng J H,Wang R D. Trans. Nonferrous Met. Soc. China, 2010,20:s198 - s204.

[15]Li W, Zhang L, Peng J, et al. Tobacco stems as a low cost adsorbent for the removal of Pb(Ⅱ) from wastewater:equilibrium and kinetic studies[J]. Industrial Crops and Products,2008,28(3):294 - 302.

[16]Li W,Peng J,Zhang L,et al. Preparation of activated carbon from coconut shell chars in pilot - scale microwave heating equipment at 60kW[J]. Waste Management,2009,29(2):756 - 760.

[17]Welham N J. Mechanically induced reduction of ilmenite($FeTiO_3$) and rutile(TiO_2) by magnesium[J]. Journal of Alloys and Compounds,1998,274(1):260 - 265.

[18]Setoudeh N,Saidi A,Welham N J. Carbothermic reduction of anatase and rutile[J]. Journal of Alloys and Compounds,2005,390(1):138 - 143.

[19]Chen G,Xiong K,Peng J,et al. Optimization of combined mechanical activation - roasting parameters of titania slag using response surface methodology[J]. Advanced Powder Technology,2010,21(3):331 - 335.

[20]Yu J,Chen Y. One - dimensional growth of TiO_2 nanorods from ilmenite sands[J]. Journal of Alloys and Compounds,2010,504:S364 - S367.

[21]Setoudeh N,Saidi A,Welham N J. Effect of elemental iron on the carbothermic reduction of the anatase and rutile forms of titanium dioxide[J]. Journal of Alloys and Compounds,2005,395(1):141 - 148.

[22]Delogu F. A mechanistic study of TiO_2 anatase - to - rutile phase transformation under mechanical processing conditions[J]. Journal of Alloys and Compounds,2009,468(1):22 - 27.

[23]Rezaee M,Mousavi Khoie S M. Mechanically induced polymorphic phase transformation in nanocrys - talline TiO_2 powder[J]. Journal of Alloys and Compounds,2010,507(2):484 - 488.

[24]Xiao P,Zheng S B,You J L. Spectrosc. Spectr. Anal,2007,27:936 - 939.

Leaching of Palladium and Rhodium from Spent Automobile Catalysts by Microwave Roasting

Shixing Wang, Anran Chen, Zebiao Zhang, Jinhui Peng

Abstract: Leaching of palladium and rhodium from spent automobile catalysts by microwave roasting was investigated. The results indicated that leaching efficiency was obviously improved, the roasting temperature was remarkably reduced and the leaching time was shortened by microwave roasting. Microwave destroys the coatings on platinum group metals and increases the effective contact area of the solid – liquid reaction. The leaching rate was accelerated at low microwave roasting temperatures. The leaching ratio of Pd and Rh was about 99.74% and 94.79%, respectively.

Keywords: leaching; palladium; rhodium; spent automobile catalysts; microwave

1 Introduction

Platinum group metalsare widely used in various industrial applications, such as electronic, catalytic converters, thermocouples, fuel cells, petroleum refining and numerous laboratory equipments, because of its unique physical and chemical properties. Every year, more than 60% of the platinum group metals were used in automobile catalytic converters[1,2].

There is an increasing attention towards their recovery from spent automobile catalytic converters. The automobile catalytic converters were also called the "mobile platinum group metals mine". The recycling method of platinum group metals has received much attention in recent years[3]. The recycling methods include pyrometallurgical, hydrometallurgical and volatilization methods[4,5]. During the typical recovery process, platinum group metals are extracted from catalytic converters by smelting them with collector metals such as nickel[6] or leaching them by strong acids [7]. These leaching methods are typically time consuming, high cost, high toxicity, and tedious process. Platinum group metals located in the surface may migrate into the inside and translate into the oxide, when the automobile catalytic converter was long – term used under high temperature. In order to overcome the difficulty and improve the leaching ratio, the coatings must be destroyed[8,9]. It is necessary to develop non – traditional method for the leaching of platinum group metals from the automobile catalytic converters.

In the metallurgy field, the microwave roasting has unique advantage. The major advantages of microwave metallurgy are volumetric heating, rapid heat transferring, selective heating, automatic controlling and environmentally friendly process. Moreover, it can cause molecular vibration for polar molecule at the molecular level, material ionization and interactive metathesis, phase transformation and the metallurgical process such as redox reaction. So microwave has wide industrial application prospect [10-12].

Microwave is widely used in the process such as material drying, pretreatment process of ores, metallurgical analysis and melting materials[13,14]. The extraction of platinum group metals by microwave has been reported. Jafarifar leached platinum and rhenium from spent catalyst using microwave – assisted heating at a power of 150W with aqua regia[15]. The leaching ratio of Pt and Re was 96.5% and 94.2%. As a rapidly developed green metallurgical method, microwave overcomes the disadvantages of pyrometallurgical and hydrometallurgical process for the leaching of platinum group metals from the automobile catalytic converters.

Herein, we leached platinum group metals from spent automobile catalytic converters by microwave roasting. The chemical compositions of the spent catalytic converter were characterized by X – ray fluorescence (XRF) and the contents of platinum group metals in the catalysts were analyzed by fire assaying.

2 Experimental

2.1 Materials and reagent

The spent catalyst was cracked and the main particle size is 2.2μm. The main chemical composition of particles is cordierite. Platinum group metals are coated on the surface of the cordierite. Sodium chlorate and sodium bisulfate monohydrate were obtained from Tianjin Guangfu Fine Chemical Research Institute, P. R. China and were used as leaching reagents. The mass ratio of sodium chlorate to sodium bisulfate monohydrate is 1∶5.

2.2 Characterization

The crystalline structure of the spent catalysts was characterized by X – ray powder diffraction (XRD) using a Philips PW1710 diffractometer. The morphology of the spent catalysts was examined using a JEOL JIB – 4500 scanning electron microscope (SEM). The chemical composition of the spent catalysts was confirmed by X – ray fluorescence (XRF) using a Rigaka ZSX100e diffractometer.

2.3 Experimental apparatus

The microwave roasting equipment used in the study is an industry microwave. Microwave power is 1.5kW. Microwave frequency is 2450MHz. The top temperature is 1200℃.

2.4 Procedure

The spent catalysts were cracked and milled usinga ball mill. A main particle size of 2.2μm was achieved. After the spent catalysts (20g) were uniformly blended with leaching reagentaccording to the ratio, the mixture was put into the ceramic crucible. Then, the ceramic crucible was coated by thermal insulation materials and put into microwave reactor. The microwave power is 300 – 330W. After roasted, the sinter was cooled and dispersed into water. The content of platinum group metals in the leaching residue was analyzed. The leaching solution was measured with ultraviolet spectrophotometer using a SHIMADZU UV2550.

3 Results and discussion

3.1 Characterization of the spent automobile catalysts

The crystalline phase of the spent catalysts was studied by X – ray powder diffraction(XRD) using a Philips PW1710 diffractometer. The angular 2θ working range was from 13° to 120° with a 0.020° step size. The automobile catalysts are composed of cordierite($2MgO \cdot 2Al_2O_3 \cdot 5SiO_2$) and alumina with some other oxides such as CeO_2, Cr_2O_3 etc. Crystallographic parameters of the different phases are summarized in Table 1 and Fig. 1.

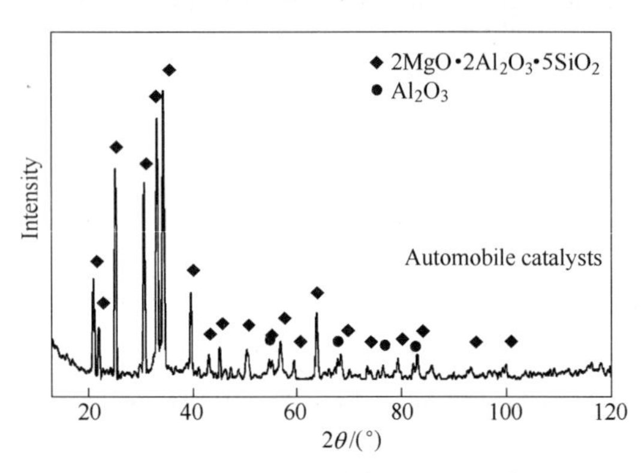

Fig. 1 XRD pattern of the spent catalysts

Table 1 Crystalline phase of the spent automobile catalysts

Phase	Space group
Al_2O_3	Orthorhombic
MgO	Cubic
SiO_2	Tetragonal
CeO_2	Cubic

The chemical composition of the spent catalysts was confirmed by X – ray fluorescence(XRF) using a Rigaku ZSX100e diffractometer. The content of platinum group metals in the spent catalysts was confirmed by fire assaying. The chemical composition of the spent catalysts was showed in Table 2.

Table 2 Chemical composition of the spent catalysts

Composition	Al/%	O/%	Si/%	Al/%	Mg/%	Ce/%	Zr/%	C/%	Pd/$g \cdot t^{-1}$	Rh/$g \cdot t^{-1}$
Content	30.97	28.26	16.66	4.71	7.60	2.82	2.20	1.43	1700	240

3.2 Effect of the holding time on the leaching ratio

Fig. 2(a) showed the effect of the holding time for the leaching ratio when the proportion of the leachingreagent and the spent catalysts was 8:1, and the reaction temperature was 573K under microwave roasting. The experiment results show that the leaching ratio of Pd and Rh was 93.97% and 92.91% after holding for 60min. The leaching ratio of rhodium was effectively improved when the holding time was 60min. Moreover, the leaching ratio of Pd was 93.97% after holding for 30min. Because of the active chemical properties, palladium reacts preferentially with the leaching reagent. The lower leaching ratio of rhodium is due to its chemical inertness. The leaching ratio of rhodium gradually decreased after holding for 60min. Rhodium reacted with the leaching reagent and generated the rhodium sulfate. Rhodium sulfate selectively absorbed microwave to cause local overheating and decomposed into rhodium oxide after holding time 60min under microwave roasting. So, the optimum holding time is 60min. Fig. 2(b) showed the effect of the holding time for the leaching ratio when the proportion of the leachingreagent and the spent catalysts was 8∶1, and the reaction temperature was 573K under conventional roasting in muffle furnace. The maximum leaching ratio of Pd and Rh was obtainedafter roasted 120min by conventional roasting. The leaching ratio of Pd and Rh was higher under microwave roasting than that under conventional roasting. Fig. 2 showed that microwave roasting obviously shortened the leaching time and improved the leaching efficiency. The issue of hydrometallurgical processes is that platinum group metals are hardly dissolved in ordinary acids because of their chemical inertness and several hours are required for the complete dissolution of platinum group metals[16-19]. Compared with hydrometallurgical processes, the leaching time was shortened by microwave roasting.

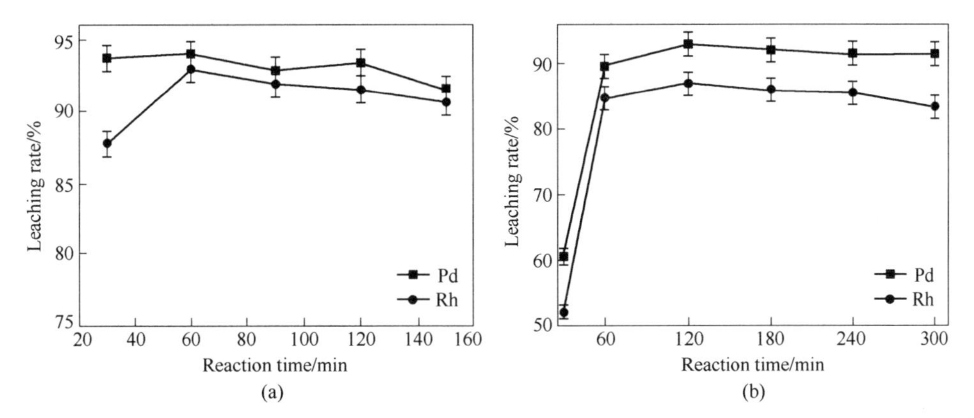

Fig. 2 Effect of the holding time on the leaching ratio of platinum group metals

(a) Microwave roasting; (b) Conventional roasting

3.3 Effect of the roasting temperature on the leaching ratio

Fig. 3 showed the effect of the roasting temperature on the leaching ratio when the proportion of the leaching reagent and the spent catalysts was 8∶1, and the holding time was 60min. With increasing of the roasting temperature, the leaching ratio of rhodium and palladium increased. The leaching ratio of palladium was 99.36% and the leaching ratio of rhodium was 93.32% when the sample was heated up to 773K and kept for 60min by microwave roasting. Hydrometallurgical or pyrometallur-

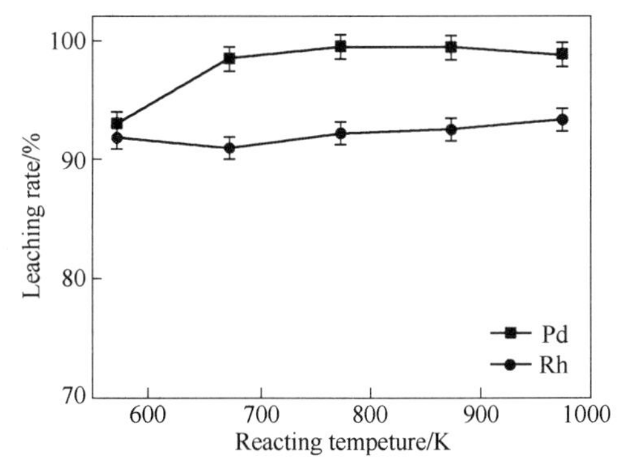

Fig. 3 Effect of the roasting temperature on the leaching ratio of platinum group metals

gical processes were widely used in platinum group metals leaching[20,21]. The potential of pyrometallurgical processes is high leaching ratio as well as coping with the impurities found in catalysts[22,23]. However, operating temperatures of pyrometallurgical processes are in the range from 1773K to 1923K. Our results showed that optimum roasting temperature was 773K by microwave and was remarkably lower than that of pyrometallurgical processes.

3.4 Effect of the proportion of the leaching reagent and the spent catalysts on the leaching ratio

Fig. 4 showed the effect of the proportion of the leachingreagent and the spent catalysts on the leaching ratio when the reaction temperature was 773K and the holding time was 60min. Based on the experiment, the effect of the leaching reagent on the leaching ratio of palladium was inconspicuous, and the leaching ratio of rhodium increased with increasing the proportion of the leaching reagent and the spent catalysts. Fig. 4 showed the leaching ratio of Pd and Rh was 99.74% and 94.79% when the proportion of the leaching reagent and the spent catalysts was 12:1.

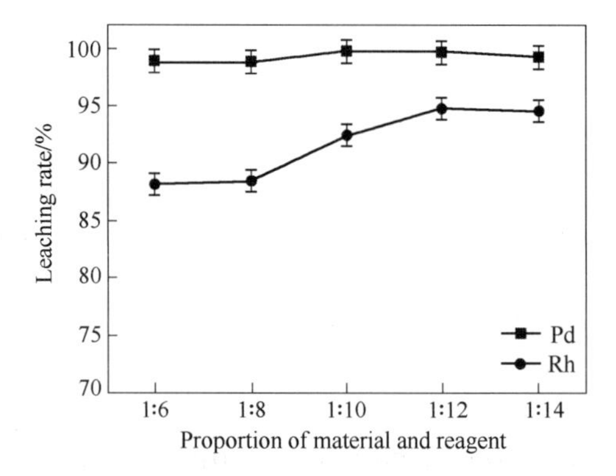

Fig. 4 Effect of the proportion of the leaching reagent and the spent catalysts on the leaching ratio of platinum group metals

3.5 The leaching mechanism of platinum group metals from spent catalysts by microwave roasting

In generally, platinum group metals located in the washcoat surface may migrate into the inside and translate into the oxide under high temperature condition. Therefore, it is necessary to destroy the coatings on platinum group metals for enhancing the leaching ratio. Fig. 5 showed the XRD pattern of the sample after microwave roasting. The intensity of the main phases ($2MgO \cdot 2Al_2O_3 \cdot 5SiO_2$) was remarkably lowered and some new phase was found after microwave roasting. Because the structure of the spent automobile catalysts had been damaged during the leaching process by microwave roasting. The morphology of the spent automobile catalysts and the leaching residue was presented in Fig. 6. The spent automobile catalysts were nearly spherical particle. A lot of cracks on the surface of the leaching residue after microwave roasting indicated that the molecular vibration was caused by microwave, compared with the spent automobile catalysts and the leaching residue after conventional roasting. The cracks increase the contact area between the leaching reagent and the spent catalysts. This is an efficient way to reduce the reagent consumption and increase the leaching ratio.

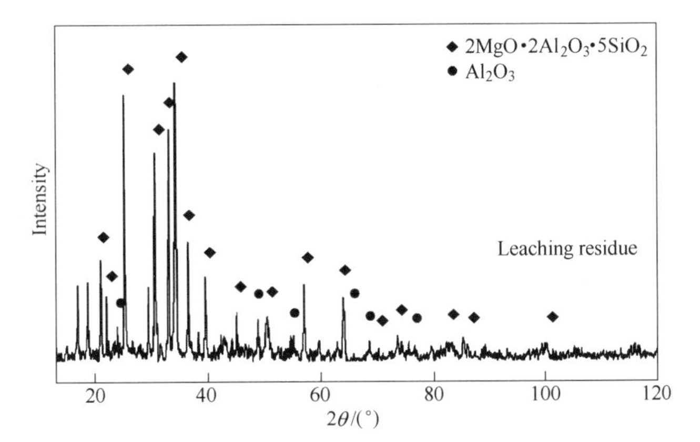

Fig. 5 XRD pattern of the leaching residue

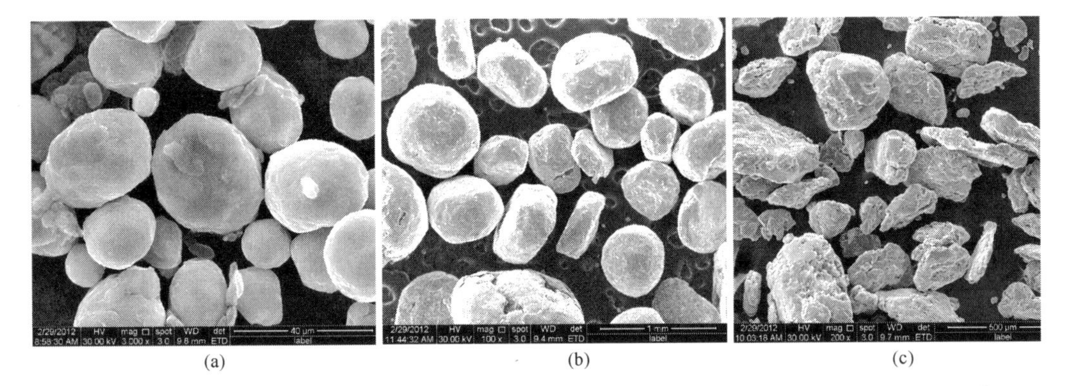

Fig. 6 SEM images of the spent automobile catalysts(a), the leaching residue after conventional roasting(b) and the leaching residue after microwave roasting(c)

The temperature - rise curve of the spent automobile catalysts and the reagents was shown in Fig. 7. We found the heated rate of the spent automobile catalysts was lower than that of the leac-

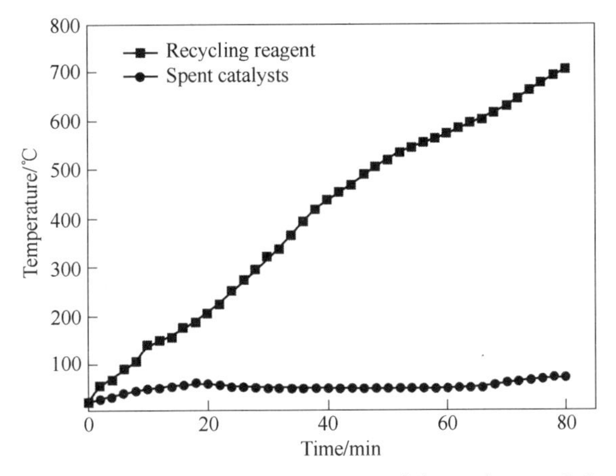

Fig. 7 The temperature - rise curve of the spent automobile catalysts and the leaching reagent

hing reagents. The reason is that the spent automobile catalysts do not absorb 2450MHz microwave energy appreciably. The selective dielectric heating of microwave enhanced the heat flow. It was the key factor to reduce the roasting temperature and time.

4 Conclusions

Microwave was used to leach Pd and Rh from the spent automobile catalysts. The leaching ratio of Pd and Rh was 99.74% and 94.79%, respectively, when the reaction temperature was 773K, the holding time was 60min and the proportion of the recycling reagent and the spent catalysts was 12 : 1. The results showed that the roasting temperature was remarkably reduced and the leaching time was shortened by microwave roasting. XRD and SEM illustrated that microwave roasting could cause the particle breakage. Microwave roasting may be an effective way to increases the leaching ratio and to reduce roasting temperature and time. We anticipate that microwave energy has widely potential applications in the recovery of precious metals from spent automobile catalysts in the near future.

References

[1] Bencs L, Ravindra K, Van Grieken R. Methods for the determination of platinum group elements originating from the abrasion of automotive catalytic converters[J]. Spectrochimica Acta Part B: Atomic Spectroscopy, 2003, 58 (10): 1723 - 1755.

[2] Kašpar J, Fornasiero P, Hickey N. Automotive catalytic converters: current status and some perspectives[J]. Catalysis Today, 2003, 77(4): 419 - 449.

[3] Bernardis F L, Grant R A, Sherrington D C. A review of methods of separation of the platinum group metals through their chloro - complexes[J]. Reactive and Functional Polymers, 2005, 65(3): 205 - 217.

[4] Mishra R K. A review of platinum group metals recovery from automobile catalytic converters[J]. Precious Metals 1993, 1993: 449 - 474.

[5] Kayanuma Y, Okabe T H, Maeda M. Metal vapor treatment for enhancing the dissolution of platinum group metals from automotive catalyst scraps[J]. Metallurgical and Materials Transactions B, 2004, 35(5): 817 - 824.

[6] Han S, Wu X, Wang H, et al. Research process on platinum group metals recovery from spent automobile catalyst[J]. Min. Metall, 2010, 19: 80 - 83.

[7] Jimenez D A D, Pinedo R, Ruiz D L I, et al. Recovery by hydrometallurgical extraction of the platinum - group metals from car catalytic converters[J]. Minerals Engineering, 2011, 24(6): 505 - 513.

[8] Benson M, Bennett C R, Harry J E, et al. The recovery mechanism of platinum group metals from catalytic converters in spent automotive exhaust systems[J]. Resources, Conservation and Recycling, 2000, 31(1): 1 - 7.

[9] Moldovan M, Gomez M M, Palacios M A. Determination of platinum, rhodium and palladium in car exhaust fumes[J]. Journal of Analytical Atomic Spectrometry, 1999, 14(8): 1163 - 1169.

[10] Bolinski L, Distin P A. Platinum group metals recovery from recycled autocatalyst by aqueous processing[J]. Publications of the Australasian Institute of Mining and Metallurgy, 1992, 9: 92.

[11] Menéndez J A, Arenillas A, Fidalgo B, et al. Microwave heating processes involving carbon materials[J]. Fuel Processing Technology, 2010, 91(1): 1 - 8.

[12] Nai S M L, Kuma J V M, Alam M E, et al. Using microwave - assisted powder metallurgy route and nano - size reinforcements to develop high - strength solder composites[J]. Journal of Materials Engineering and Perform-

ance,2010,19(3):335 - 341.

[13] Djingova R,Heidenreich H,Kovacheva P,et al. On the determination of platinum group elements in en - vironmental materials by inductively coupled plasma mass spectrometry and microwave digestion[J]. Analytica Chimica Acta,2003,489(2):245 - 251.

[14] Lekse J W,Stagger T J,Aitken J A. Microwave metallurgy:synthesis of intermetallic compounds via microwave irradiation[J]. Chemistry of Materials,2007,19(15):3601 - 3603.

[15] Jafarifar D,Daryanavard M R,Sheibani S. Ultra fast microwave - assisted leaching for recovery of platinum from spent catalyst[J]. Hydrometallurgy,2005,78(3):166 - 171.

[16] Chen J,Huang K A. New technique for extraction of platinum group metals by pressure cyanidation[J]. Hydrometallurgy,2006,82(3):164 - 171.

[17] Cunningham C E. Pre - recovery - system and recovery - system losses of platinum group metals contained in automotive catalytic converters[J]. Conservation & Recycling,1985,8(3):343 - 357.

[18] Jimenez de Aberasturi D,Pinedo R,Ruiz D L I,et al. Recovery by hydrometallurgical extraction of the platinum - group metals from car catalytic converters[J]. Minerals Engineering,2011,24(6):505 - 513.

[19] Baghalha M,Khosravian G H,Mortaheb H R. Kinetics of platinum extraction from spent reforming catalysts in aqua - regia solutions[J]. Hydrometallurgy,2009,95(3):247 - 253.

[20] Mishra R K,Reddy R G. Pyrometallurgical processing and recovery of precious metals from autocatalysts using plasma arc smelting[J]. Precious Metals,1986:230.

[21] Gibbon A,Harry J E,Hodge D. The plasma process for the recovery of the platinum metals from automobile catalysts[J]. Proceedings of International Symposium on Plasma Chemistry 8,Tokyo,1987,D11:03.

[22] Burnham R F,Harry J E,Gibbon A. Plasma arc furnaces[P]. European Patent,1983,0096493A2.

[23] Saville J. Process for the extraction of platinum group metals[P]. US Patent,1987,4685963.

Leaching Zinc from Spent Catalyst: Process Optimization Using Response Surface Methodology

Zhengyong Zhang, Jinhui Peng, C. Srinivasakannan, Zebiao Zhang, Libo Zhang, Y. Fernández, J. A. Menéndez

Abstract: The spent catalyst from vinyl acetate synthesis contains large quantity of zinc. The present study attempts to leach zinc using a mixture of ammonia, ammonium carbonate and water solution, after microwave treatment. The effect of important parameters such as leaching time, liquid/solid ratio and the ammonia concentration was investigated and the process conditions were optimized using surface response methodology (RSM) based on central composite design (CCD). The optimum condition for leaching of zinc from spent catalyst was identified to be a leaching time of 2.50h, a liquid/solid ratio of 6 and ammonia concentration 5.37mol/L. A maximum of 97% of zinc was recovered under the optimum experimental conditions. The proposed model equation using RSM has shown good agreement with the experimental data, with a correlation coefficient (R^2) of 0.95. The samples were characterized before and after leaching using X-ray diffraction (XRD), nitrogen adsorption and scanning electron microscope (SEM).

Keywords: spent catalyst; ammonia leaching; response surface methodology; optimization; zinc

1 Introduction

Increasing concerns of environment and ore shortage has led to considerable efforts being made to recover valuable metals from various waste materials[1-3]. Fresh catalyst used in vinyl acetate synthesis production contains up to 35wt% zinc acetate in the activated carbon[4]. The catalyst losses its effectiveness progressively due to poisoning of the active sites, rendering ineffective for further utilization. The spent catalyst still contains considerable amount of zinc occurring in the form of acetate, oxide or metallic form or as organic compound which were by-products of vinyl acetate synthesis[5]. In China, around 9000t of the spent activated carbon is generated from vinyl acetate synthesis per year as solid waste. These spent catalysts do not have commercial value and are usually dumped as landfills. The hazardous nature of the spent catalyst is a potential threat to the environment, as various chemicals can possibly leach to surrounding areas from the dump site.

Few of the possible ways for regeneration of the spent catalyst have been attempted in the recent years[5], which includes removing organic substances by organic solvents or with carbon dioxide in the supercritical phase, leaching of zinc with a solution of hydrochloric acid with solutions of HCl and HNO_3 assisted by microwave energy. The maximum% of zinc removal using the above methods was reported to be less than 94.7%. As more than 5% of the zinc is still left with the activated carbon it significantly influences the absorption capacity of regenerated activated carbon. Our earlier publication[4] reported the enhanced recovery of zinc from the spent catalyst, using a two step being pretreatment by microwave irradiation, followed by leaching by a mixture of ammonia, ammonium bicarbonate and water. The % leaching was reported to improve on account of organic substances removal, achieving an effective separation of activated carbon from zinc. The present study attempts to identify extraction conditions that could possibly maximize the zinc removal by optimizing the process conditions, by designing the experiments using response surface methodology (RSM).

RSM is one of the relevant multivariate techniques which can deal with multivariant experimental design strategy, statistical modeling and process optimization[6-11]. It is used to examine the relationship between one or more response variables and a set of quantitative experimental variables or factors. This method is often employed after the vital controllable factors are identified and to find the factor settings that optimize the response. Designs of this type are usually chosen when a curvature in the response surface is suspected. The process optimization of the microwave induced zinc recovery from the spent catalyst of vinyl acetate process has not been reported in literature. Hence the present work intends to assess the effects of variables such as ammonia concentration, leaching time and liquid/solid ratio to identify the optimum conditions using a central composite design (CCD). The characteristics of sample before and after leaching are assessed using the advanced analytical instruments such as X-ray diffraction (XRD), nitrogen adsorption and scanning electron microscope (SEM).

2 Materials and methods

2.1 Materials and apparatus

The spent catalyst of vinyl acetate synthesis was obtained from a chemical plant (Yunnan Yunwei Group Co., Ltd.) in Yunnan province, China. The composition of spent catalyst is presented in Table 1. Prior to leaching, the samples were treated with microwave and heated to a temperature of 900℃ for 30min using a self-made microwave oven. The zinc content of pretreated sample decreased from 8.76% to 5.55% due to lose of zinc during the pretreatment process. SEM (Philips XL30ESEM-TMP) analysis of the sample was utilized to assess the microstructure of sample before and after leaching. XRD (Rigaku diffractometer [D/max 2500] using CuK_{α} radiation) analysis was performed to assess the changes in composition of samples, while (Brunauer-Emmett-Teller) BET analysis was used to assess the surface area and pore structure.

Table 1 Compositions of spent catalyst (mass fraction, %)

C	Zn	Fe	Si	Ca	Mg
81.71	8.76	0.5	0.15	0.015	0.007

2.2 Experimental methods

The leaching solution was prepared by dissolving ammonium carbonate in an ammonia solution. A desired liquid to solid mass ratio was prepared by mixing distilled water to the ammonium carbonate solution. A total of 20 experiments were conducted with 20g of sample in a batch mode using a glass reactor of volume 250mL. The contents of reactor were well stirred using a digital and thermostatic magnetic stirrer at an agitation speed of 800r/min, so as to keep the contents of the reaction well stirred and suspended. The contents of reactor were filtered, upon completion of the experiment and the filtrate was analyzed for the zinc content using EDTA titration[12]. The amount of zinc leached was estimated using Eq. (1).

$$\% \text{ leaching} = m_1/m_0 \times 100 \quad (1)$$

where m_0 and m_1 correspond to zinc content of sample before and after leaching.

2.3 Experimental design

RSM helps to optimize the process, influenced by number of operating parameters with a minimum number of experiments as well as to analyze the interaction between the parameters. Ammonia concentration(χ_1), leaching time(χ_2) and liquid/solid ratio(χ_3) were chosen as the independent variables with their levels and ranges shown in Table 2. Table 3 shows the actual values of the independent variables at which the experiments were conducted to estimate the response variable the % leaching(Y). The chosen independent variables used in process optimization were coded according to Eq. (2),

$$\chi_i = \frac{\chi_i - \chi_0}{\Delta\chi} \quad (2)$$

where χ_i was the dimensionless coded value of the independent variable; χ_0 was the value of χ_i at the center point; $\Delta\chi$ was the step change value. The % leaching of zinc was the response variable of the experimental conditions in the design of experiments. The leaching time was varied from 0.32h to 3.68h, the liquid/solid ratio was varied from 1.98 to 7.02, the ammonia concentration was varied from 0.64mol/L to 7.36mol/L. A total of 20 experiments consisting of 8 factorial points, 6 axial points and 6 replicates at the central points were performed[13]. Experimental results obtained from the CCD model were described in the form as given in Eq. (3):

$$Y = \beta_0 + \sum_{i=1}^{n} \beta_i \chi_i + \sum_{i=1}^{n} \beta_{ii} \chi_i^2 + \sum_{i<j}^{n} \beta_{ij} \chi_i \chi_j \quad (3)$$

where β_0 was the value for the fixed response at the central point of the experiment; β_i, β_{ii} and β_{ij} were the linear, quadratic and cross product coefficients, respectively. The analyses of variance (ANOVA) and response surfaces were performed using the Design - Expert software (version 7.1.5) from Stat - Ease Inc., USA. The optimized leaching conditions, for regeneration of the spent catalyst were estimated using the software's numerical and graphical optimization tools.

Table 2 Independent variables and their levels used for central composite rotatable design

Variables	Symbol	Range and levels				
		-1.682	-1	0	+1	+1.682
Ammonia concentration/mol · L^{-1}	χ_1	0.64	2.00	4.00	6.00	7.36
Leaching time/h	χ_2	0.32	1.00	2.00	3.00	3.68
Liquild/solid ratio	χ_3	1.98	3.00	4.50	6.00	7.02

3 Results and discussion

3.1 Response analysis and interpretation

The results of experiments were shown in Table 3 and the % leaching of zinc was found to range from 55% to 99% in response to the variation in the experimental conditions. According to the sequential model sum of squares were selected based on the highest order polynomials, where the additional terms were significant[13].

Table 3 Experimental design matrix and results

Run	Leaching variables			% Leaching, Y/%
	Ammonia concentration/mol · L^{-1}	Leaching time	Liquid/solid ratio	
1	2	1.00	3.00	69.39
2	6	1.00	3.00	90.10
3	2	3.00	3.00	82.40
4	6	3.00	3.00	95.83
5	2	1.00	6.00	76.63
6	6	1.00	6.00	98.98
7	2	3.00	6.00	88.65
8	6	3.00	6.00	97.16
9	0.64	2.00	4.50	54.93
10	7.36	2.00	4.50	92.86
11	4	0.32	4.50	82.15
12	4	3.68	4.50	93.72
13	4	2.00	1.98	90.15
14	4	2.00	7.02	93.67
15	4	2.00	4.50	91.61
16	4	2.00	4.50	91.62
17	4	2.00	4.50	92.08
18	4	2.00	4.50	91.92
19	4	2.00	4.50	91.29
20	4	2.00	4.50	92.15

The ANOVA of quadratic model is presented in Table 4 which proves the validity of the model. The ammonia concentration has the greatest effect on % leaching with the highest F - value of 143.16 whereas liquid/solid ratio and leaching time were found to be less significant. The model F - value of 26.19 implies the significance of the model. The validation of model was an important part of the data analysis procedure, since an inadequate model could lead to misleading results[14]. For the fixed model, adequate precision can be ensured with a signal to noise ratio greater than 4. A adequate precision ratio of 18.25 indicates the ability of model to precisely navigate through the design space. Not all the effects of parameters on % leaching were significant, while values of Prob > F less than 0.05 indicate that the model terms were significant. In this case, χ_1, χ_2, χ_3 and the interaction terms χ_1^2, $\chi_1\chi_2$ were significant model terms whereas the others were insignificant. In order to enhance the effect of significant parameters, the insignificant parameters were eliminated. The final equation in terms of coded factors was shown in Eq. (4) as,

$$Y = 91.72 + 9.43\chi_1 + 3.54\chi_2 + 2.17\chi_3 - 2.64\chi_1\chi_2 - 5.64\chi_1^2 \quad (4)$$

Table 4 Analysis of variance (ANOVA) for response surface quadratic model for leaching rate

Source	Sum of squares	Degree of freedom	Mean square	F - Value	p - value prob > F
Model	1999.47	9	222.16	26.19	< 0.0001
χ_1	1214.46	1	1214.46	143.16	< 0.0001
χ_2	171.59	1	171.59	20.23	0.0011
χ_3	64.23	1	64.23	7.57	0.0204
$\chi_1\chi_2$	55.75	1	55.75	6.57	0.0282
$\chi_1\chi_3$	1.35	1	1.35	0.16	0.6988
$\chi_2\chi_3$	9.11	1	9.11	1.07	0.3244
χ_1^2	457.89	1	457.89	53.95	< 0.0001
χ_2^2	6.52	1	6.52	0.77	0.4011
χ_3^2	7.73	1	7.73	0.91	0.3622
Residual	84.83	10	84.83		

Note: $R^2 = 0.95$; $R^2_{adj} = 0.92$; adequate precision = 18.25 (> 4).

The R^2 value for Eq. (4) was found to be 0.95 close to unity, indicating the good agreement between the experimental and the predicted % leaching, which do not show any significant non - linear pattern (S - shaped curve) indicating non - normality in the error term. Fig. 1 shows an approximate linearity confirming normality of the data[14].

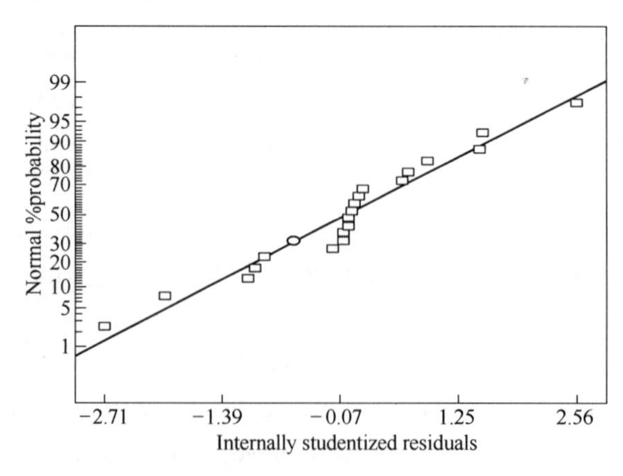

Fig. 1 Normal % probability versus internally studentized residuals

3.2 Process optimization

The experimental and predicted % leaching is shown in Fig. 2. The figure shows a close proximity of the model prediction with the experimental data signifying the validity of the regression model. Fig. 3 shows three - dimensional plot of the ammonia concentration and leaching time on % leaching. It shows that the % leaching increases significantly with increase in the ammonia concentration in the leaching solution. An increase in ammonia concentration from 0.64mol/L to 7.36mol/L increased the % leaching remarkably from 54.93% to 92.86%. The preheating of spent catalyst using microwave heating to a temperature of 900℃ facilitate reopening of the pores in the activated carbon, through which the mixture of ammonia diffuses through to form the zinc ammonium carbonate complex. The chemical leaching reaction can be defined as follows:

$$ZnO + (NH_4)_2CO_3 + (i-1)NH_3 \rightleftharpoons Zn(NH_3)_iCO_3 + H_2O \tag{5}$$

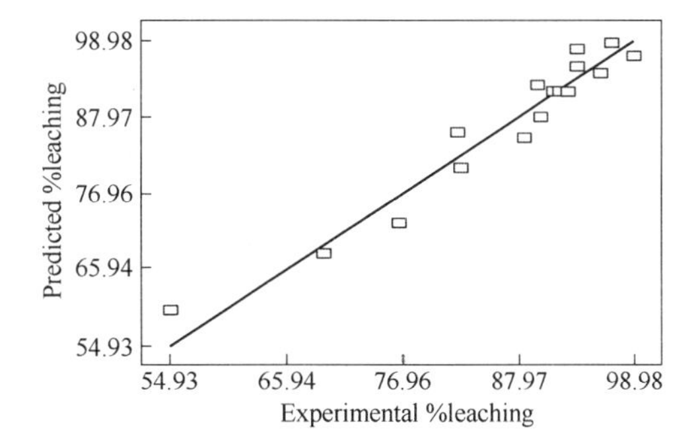

Fig. 2 Comparison of model prediction with the experimental data

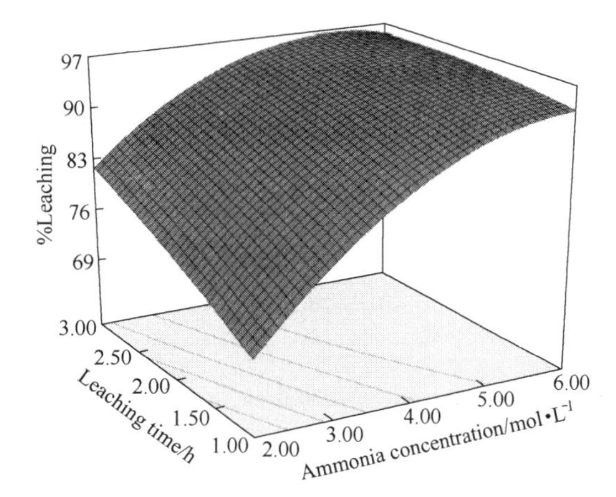

Fig. 3 Effect of ammonia concentration and leaching time on % leaching (liquid/solid ratio: 4.5)

In the above reaction as the ammonia concentration increases in solution, the complex $Zn(NH_3)_iCO_3$ becomes more stable due to higher value of i in the complex. The leaching agent diffuses through the pores of the activated carbon (support material) and reacts with the zinc oxide adsorbed on the activated carbon to form a complex $Zn(NH_3)_iCO_3$, which diffuses back to the bulk of the solution. An increase in the concentration of ammonia on one hand would increase the rate of reaction and on the other hand reduce the mass transfer resistance aiding diffusion of ammonia to the site of zinc oxide, facilitating higher % leaching. The increase in % leaching with increase in leaching time could be attributed to the increase in extent of reaction, as the reaction rate was mass transfer controlled[15]. A significant increase in the % leaching was observed with increase in the leaching time at low ammonia concentration, while at higher ammonia concentration it was less significant. Contrary to lower ammonia concentration, at higher ammonia concentrations the rate of reaction as well as the mass transfer diffusion rates would be significantly higher, which appropriately explains the reduced dependence of % leaching with increase in leaching time at higher ammonia concentrations.

Fig. 4 shows the combined effect of liquid/solid ratio and leaching time on the % leaching. It can be observed that an increase in both liquid/solid ratio as well as the leaching time increases the % leaching. It was found experimentally that a minimum liquid to solid ratio of 1.5 was necessary, as part of the solution was adsorbed by the spent catalyst, resulting in a viscous slurry with poor mass transfer between the phases. Hence a minimum liquid/solid ratio of 1.98, reasonably higher than the minimum of 1.5, has been chosen in the present study. At low liquid to solids ratios high viscosity and segregation of solids as lumps affected uniform mixing, resulting in poor mass transfer rate between the solution and the spent catalyst, yielding low % Zn leaching. The increase in liquid to solid ratio reduced the viscosity of the solution which improved the mass transfer rates, while simultaneously increasing the stoichiometric ratio of reactants, aiding formation of more stable zinc ammonium carbonate complex, resulting in a higher % leaching. However, increasing liquid/solid ratio from 1.98 to 7.02 in ammonia leaching did not significantly change zinc removal, as compared to the % zinc removal with increase in ammonia concentration (Fig. 3). It is economical to utilize a lower liquid/solid ratio as higher liquid/ solid ratio results in a dilute solution[16]. So, it is desirable to use a lower liquid to solid ratio as the concentration of Zn in the solution can be maximized, which would ease the operation of Zn recovery from the leached solution. Fig. 4 also shows an increase in % leaching with increase in the leaching time for different liquid solid ratios. However the effect of leaching time on the % recovery was varied, with lower liquid/solid ratio, shows a significant effect while higher liquid/solid ratio shows only a marginal effect. The marginal effect at higher liquid/solid ratio can be attributed to the better mass transfer conditions which in signifies the effect of time.

With the objective being to maximize the % leaching at the lowest ammonia concentration, leaching time and liquid/solid ratio, the optimum leaching conditions were identified using the Design-Expert software. It reports the optimum conditions to be an ammonia concentration of 5.37mol/L, leaching time of 2.5h and liquid/solid ratio of 6, which resultant % zinc leaching of 97.06% (Table 5).

Table 5 Optimum leaching conditions with model validation

Ammonia concentration /mol · L^{-1}	Leaching time /h	Liquid/solid ratio	% Leaching	
			Predicted	Experimental
5.37	2.5	6	98.27	97.06

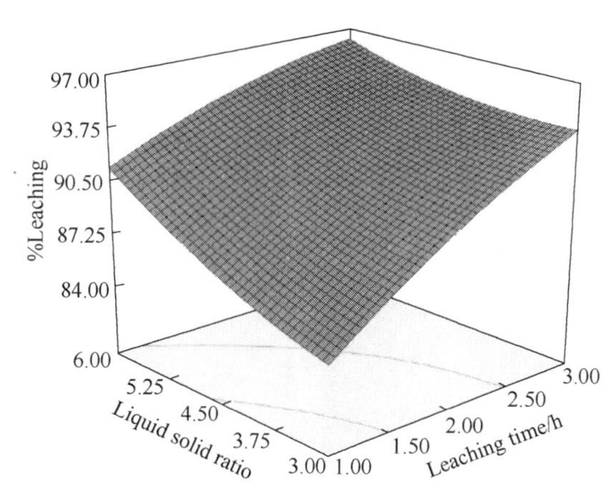

Fig. 4 Effect of liquild/solid ratio and leaching time on % leaching(ammonia concentration 4mol/L)

3.3 Characterization of sample before and after being leaching

The spent catalyst samples before and after leached were characterized by XRD technique and results were shown in Fig. 5. It was observed from Fig. 5 that the spent catalyst after microwave pretreatment was mainly composed of zinc oxide and activated carbon, while the diffraction peak of zinc oxide disappeared in the samples after the leaching as the zinc was removed by leaching with ammonia solution. The key reactions occurring during the microwave pretreatment could be summarized as follows[17]

$$Zn(CH_3COO)_2 \cdot 2H_2O \longrightarrow Zn(CH_3COO)_2 + 2H_2O\uparrow \quad (6)$$

$$4Zn(CH_3COO)_2 + 2H_2O \longrightarrow Zn_4O(CH_3COO)_6 + 2H_2O\uparrow \quad (7)$$

$$Zn_4O(CH_3COO)_6 + 3H_2O \longrightarrow 4ZnO + 3CH_3COCH_3\uparrow + 3CO_2\uparrow \quad (8)$$

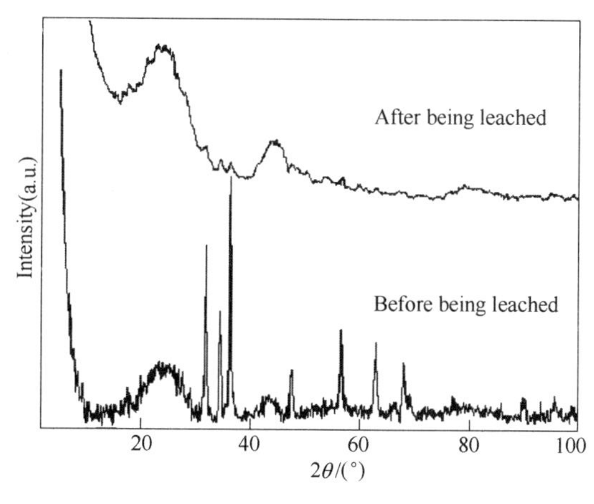

Fig. 5 XRD pattern of sample before and after leaching

The SEM analysis of sample before and after microwave pretreatment was shown in Fig. 6. Fig. 6 (a) shows large existence of ZnO material on the surface and pores of the activated carbon visible in the form of white color. Fig. 6(b) corresponds to the SEM analysis of the sample after extraction, which shows that most of the pores were open with far less ZnO material in the activated carbon evidenced by reduction in the white color.

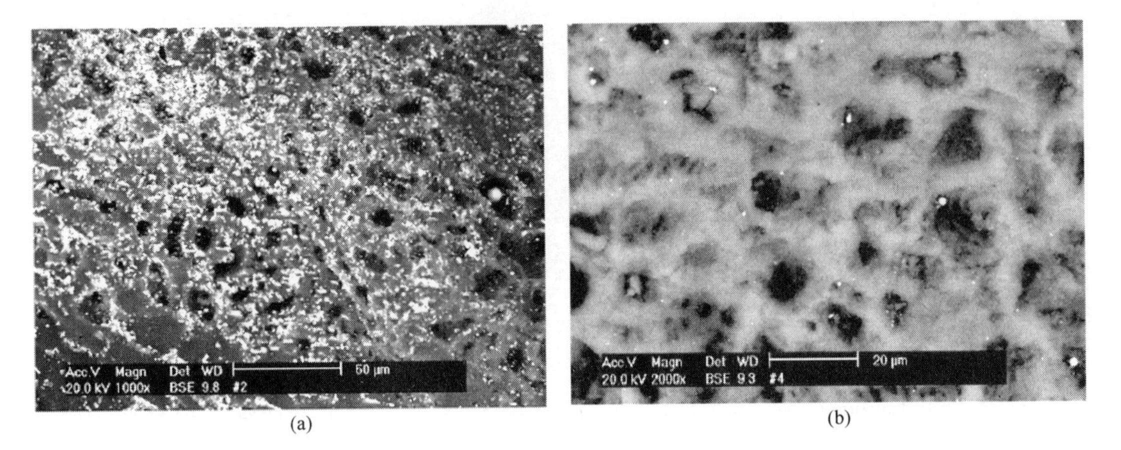

(a) (b)

Fig. 6 SEM image of sample before and after ammonia leaching

(a) Before ammonia leaching; (b) After ammonia leaching

As a further evidence to the opening of pores and removal of zinc from the spent catalyst, nitrogen adsorption – desorption isotherms at 77K[18,19] of the sample before and after leaching were performed using the surface area analyzer, (Autosorp – 1), and the results were shown in Fig. 7. It can be seen that the N_2 adsorption volume of sample increased after leaching, as compared to samples before leaching. Further the leached sample offers a high adsorption capacity at relatively low pressures, which indicates the presence of a well – developed microporous structure, which could be attributed to the removal of entrapped zinc from the microp – ores of the spent catalyst. From the nitrogen adsorption isotherms, the specific surface area (SBET) and total pore volume of sample after being leached were $922m^2/g$ and $0.63cm^2/g$, while those of sample before being leached were $824m^2/g$ and $0.58cm^2/g$, respectively.

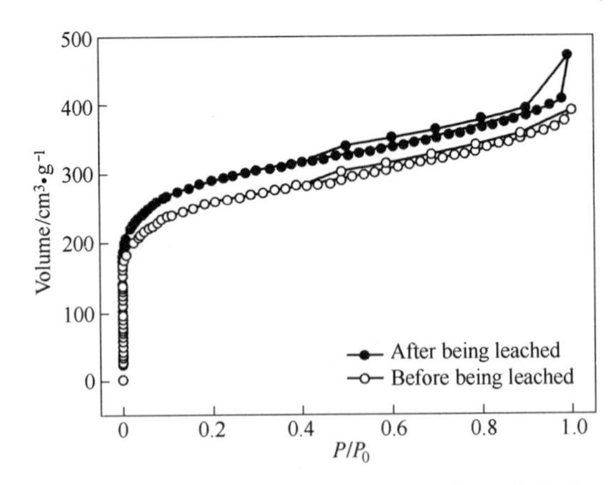

Fig. 7 Nitrogen adsorption – desorption isotherms at 77K of sample before and after leaching

4 Conclusions

The present study was aimed to explore the effects of ammonia concentration, leaching time and liquid/solid ratio on the % leaching of zinc from spent vinyl acetate catalyst and to optimize the process conditions using RSM. The proposed quadratic model agrees well with the experimental data, with correlation coefficient(R^2) of 0.95. The ammonia concentration was found to have significant effect on the % leaching while other parameters showed little effect. Based on "Design - Expert" software, the optimum conditions were identified to be an ammonia concentration of 5.37mol/L, a leaching time of 2.5h, a liquid/solid ratio of 6, with the % zinc leaching of 97%. Further the XRD, nitrogen adsorption and SEM analysis of samples before and after leaching indicated that most zinc was removed from the spent catalyst, and quite a number of pores were opened.

Acknowledgements

The authors would like to express their gratitude to the China International Science and Technology Cooperation Program(No. 2008DFA91500), the International Collaboration Project of Yunnan Provincial Science and Technology Department(No. 2006GH01) and the Science and Technology Development program of Yunnan Environmental Protection Bureau for financial support. Y. Fernández is also grateful to CSIC of Spain and the European Social Fund(ESF) for financial support under the PhD grant I3P - BDP - 2006.

References

[1] Kar B B, Datta P, Misra V N. Spent catalyst: secondary source for molybdenum recovery[J]. Hydrometallurgy, 2004, 72(1): 87 - 92.

[2] Huang Y, Tanaka M. Analysis of continuous solvent extraction of nickel from spent electroless nickel plating baths by a mixer - settler[J]. Journal of Hazardous Materials, 2009, 164(2): 1228 - 1235.

[3] Barakat M A, Mahmoud M H H, Mahrous Y S. Recovery and separation of palladium from spent catalyst[J]. Applied Catalysis A: General, 2006, 301(2): 182 - 186.

[4] Li W, Peng J, Zhang L, et al. Pilot - scale extraction of zinc from the spent catalyst of vinyl acetate synthesis by microwave irradiation[J]. Hydrometallurgy, 2008, 92(1): 79 - 85.

[5] Dabek L. Sorption of zinc ions from aqueous solutions on regenerated activated carbons[J]. Journal of Hazardous Materials, 2003, 101(2): 191 - 201.

[6] Bezerra M A, Santelli R E, Oliveira E P, et al. Response surface methodology(RSM) as a tool for optimization in analytical chemistry[J]. Talanta, 2008, 76(5): 965 - 977.

[7] Secula M S, Suditu G D, Poulios I, et al. Response surface optimization of the photocatalytic decolorization of a simulated dyestuff effluent[J]. Chemical Engineering Journal, 2008, 141(1): 18 - 26.

[8] Gönen F, Aksu Z. Use of response surface methodology(RSM) in the evaluation of growth and copper(Ⅱ) bioaccumulation properties of Candida utilis in molasses medium[J]. Journal of Hazardous Materials, 2008, 154(1): 731 - 738.

[9] Fu J, Zhao Y, Wu Q. Optimising photoelectrocatalytic oxidation of fulvic acid using response surface methodology[J]. Journal of Hazardous Materials, 2007, 144(1): 499 - 505.

[10] Hameed B H, Tan I A W, Ahmad A L. Preparation of oil palm empty fruit bunch - based activated carbon for removal of 2,4,6 - trichlorophenol: Optimization using response surface methodology[J]. Journal of Hazardous Materials, 2009, 164(2): 1316 - 1324.

[11]Baçaoui A,Yaacoubi A,Dahbi A,et al. Optimization of conditions for the preparation of activated carbons from olive - waste cakes[J]. Carbon,2001,39(3):425 - 432.

[12]Peng B,Gao H M,Chai L Y,et al. Leaching and recycling of zinc from liquid waste sediments[J]. Transactions of Nonferrous Metals Society of China,2008,18(5):1269 - 1274.

[13]Tan I A W,Ahmad A L,Hameed B H. Optimization of preparation conditions for activated carbons from coconut husk using response surface methodology[J]. Chemical Engineering Journal,2008,137(3):462 - 470.

[14]Körbahti B K,Rauf M A. Determination of optimum operating conditions of carmine decoloration by UV/H_2O_2 using response surface methodology[J]. Journal of Hazardous Materials,2009,161(1):281 - 286.

[15]Dutra A J B,Paiva P R P,Tavares L M. Alkaline leaching of zinc from electric arc furnace steel dust[J]. Minerals Engineering,2006,19(5):478 - 485.

[16]Chen A,Jia X,Long S,et al. Alkaline leaching Zn and its concomitant metals from refractory hemimorphite zinc oxide ore[J]. Hydrometallurgy,2009,97(3):228 - 232.

[17]Paraguay D F,Estrada L W,Acosta N D R,et al. Growth,structure and optical characterization of high quality ZnO thin films obtained by spray pyrolysis[J]. Thin Solid Films,1999,350(1):192 - 202.

[18]Prakash Kumar B G,Shivakamy K,Miranda L R,et al. Preparation of steam activated carbon from rubberwood sawdust(Hevea brasiliensis) and its adsorption kinetics[J]. Journal of Hazardous Materials,2006,136(3):922 - 929.

[19]Berenguer R,Marco - Lozar J P,Quijada C,et al. Effect of electrochemical treatments on the surface chemistry of activated carbon[J]. Carbon,2009,47(4):1018 - 1027.

Green Evaluation of Microwave - assisted Leaching Process of High Titanium Slag on Life Cycle Assessment

Guo Chen, Jin Chen, Jinhui Peng, Rundong Wan

Abstract: A greenness evaluation index and a system of microwave - assisted leaching method were established. The effects of the life cycle assessment variables such as the resource consumption, environment impact, cost, time and quality were investigated systematically, and the concept of green degree was applied in the production of synthetic rutile. An analytic hierarchy process was utilized to assess matrix of greenness evaluation. The Gauss - Seidel iterative matrix method was employed to solve the assessment matrix, obtaining the weights and membership functions of all evaluation indexes. A fuzzy decision - making method was applied to build the greenness evaluation model, and then the scores of green degree in microwave - assisted leaching process were obtained. The greenness evaluation model was applied to the life cycle assessment of microwave - assisted leaching process. The results show that microwave - assisted leaching process has advantages over the conventional ones, in respect of energy - consumption, processing time and environmental protection.

Keywords: life cycle assessment; greenness evaluation; microwave - assisted leaching; high titanium slag

1 Introduction

Titanium dioxide pigment accounts for more than 92% consumption of titanium minerals[1]. Rutile is the major raw material for the production of TiO_2 pigment and its main applications include manufacture of paint, paper and plastics. However, the supply of high grade natural rutile is limited[2-4]. Thus, producing synthetic rutile from abundantly available high titanium slag becomes a major alternative. How to produce high grade synthetic rutile with low energy consuming and less environment polluting becomes more and more emergent[5,6]. A new process, which eliminates the environmental pollution from production source and maintains titanium resources sustainable development, prepares synthetic rutile from high titanium slag by microwave - assisted leaching, which removes partial iron content and produces high grade TiO_2[7-9].

Compared with conventional heating techniques, the main advantage of using microwave heating is that microwave heating is both internal and volumetric heating. Since there is a high efficiency to convert electricity energy to electromagnetic energy and no thermal conductivity mechanism involved, the heating is very rapid, uniform and highly energy efficient, resulting in energy savings and shortening the processing time[10-13]. In addition, additional advantages include greater control of the microwave heating process, no direct contact between the heating source and heated materials and reduced equipment size and waste[14,15].

The Life Cycle Assessment (LCA) is one of the most widely used and internationally accepted methods for the evaluation of the environmental impacts of products and systems[16]. An LCA calculates the environmental burden of products extraction from raw materials through manufacturing, transporting, utilizing and disposing, and identifies the particular stage of the life cycle which causes the maximum environmental damage. Recently, many statistical and mathematical methods have been developed for LCA process[17,18]. The Analytic Hierarchy Process (AHP) is a general problem - solving method that is useful in making complex decisions such as multi - criteria decisions, based on variables that do not have exact numerical consequences. The AHP approach determines the weights qualitatively by constructing multi - level decision structures and forms pair wise comparison matrices[19].

The focus of this research was to compare the preparation conditions of synthetic rutile from high titanium slag of high grade synthetic rutile content by microwave - assisted leaching and conventional leaching. The effects of the fuzzy comprehensive assessment variables, in which five decision - making aspect factors: resource consumption, environmental impact, cost, time and quality were considered simultaneously.

2 Experimental

2.1 Materials

High titanium slag was prepared from ilmenite ore by carbothermal reduction in an electric arc furnace. The chemical composition of the high titanium slag was presented in Table 1. The slag contains 64.33% TiO_2, 25.79% Ti_2O_3, and 5.26% FeO. This particular slag was considered as a high grade slag since it contains a relatively great amount of titanium. The slag also contains 1.04% MnO, 2.75% Al_2O_3, 2.30% MgO, 2.57% SiO_2 and minor elements such as S, P and C.

Table 1 Chemical composition of high titanium slag (%)

TiO_2	Ti_2O_3	FeO	Al_2O_3	SiO_2	MnO	MgO	S	P	C
64.33	25.79	5.26	2.75	2.57	1.04	2.30	0.049	0.014	0.049

2.2 Instruments

The high titanium slag was treated using the self - made microwave - assisted leaching equipment. Our microwave system consists of a magnetron, a power controller, a matched load, a wave guide, and a multi - mode cavity. Schematic diagram of the microwave heating equipment was shown in Fig. 1. The microwave power supply for the microwave heating equipment consists of 2 magnetrons, which are cooled by water circulation, of 2.45GHz frequency and 1.5kW power; A polytetrafluoroethene column container, 300mm (inner diameter) × 200mm (in length), was positioned at the center of the microwave stainless steel oven; an attached infrared pyrometer (Marathon Series, Raytek, USA), which was used to monitor the temperature of the sample, has the circular crosswire focusing on the sample cross – section.

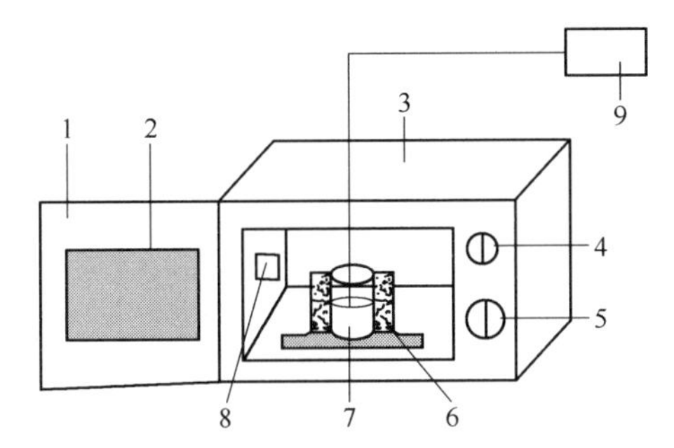

Fig. 1 Schematic diagram of the self – made microwave – assisted leaching equipment
1—oven door;2—observation door;3—microwave multi – mode cavity;4—time;5—power controller;
6—fireproof materials;7—raw materials;8—ventilation hole;9—temperature measurement system

2.3 Procedure

Prior to the use, high titanium slag was crushed and sieved to obtain particles of size less than 0.2mm. Subsequently, high titanium slag was loaded on a ceramics boat which was placed inside a stainless steel tubular reactor, whose internal diameter is 38mm. The samples were heated to 120℃ at a heating rate of 5℃/min in the drying oven and held at this temperature for 2h. After drying, the samples were cooled to room temperature. The total mass of sample taken was 100g was placed in the muffle furnace and heated to 950℃ for 60min, and then it was naturally cooled in the furnace to room temperature. After reaction, the sample, about 80g, was then subject to microwave – assisted acid leaching. The microwave – assisted conditions were as follows: hydrochloric acid concentration 20% , reaction temperature 80℃ , solid – liquid ratio of 1∶10 and reaction time 60min, simultaneously microwave – assisted leaching conditions were compared with the conventional acid leaching conditions. Thus, synthetic rutile was prepared from high titanium slag by microwave – assisted leaching process.

3 Analytic hierarchy process

3.1 Assessment of hierarchical structure

Like many other methods, AHP allows decision makers to create a model for a complex problem with a hierarchical structure[20,21]. In this study, the top level of the hierarchy structure on life cycle assessment is the overall goal of the leaching process of high titanium slag. The following levels describe the tangible criterion and sub criterion that contribute to the goal. The bottom level is formed by the alternatives to make evaluations in terms of criterion. Hierarchical structure on life cycle assessment of leaching process of high titanium slag was shown in Fig. 2.

3.2 Pairwise comparison

As soon as the AHP logic hierarchical structure is formulated, one should first yield the judgment

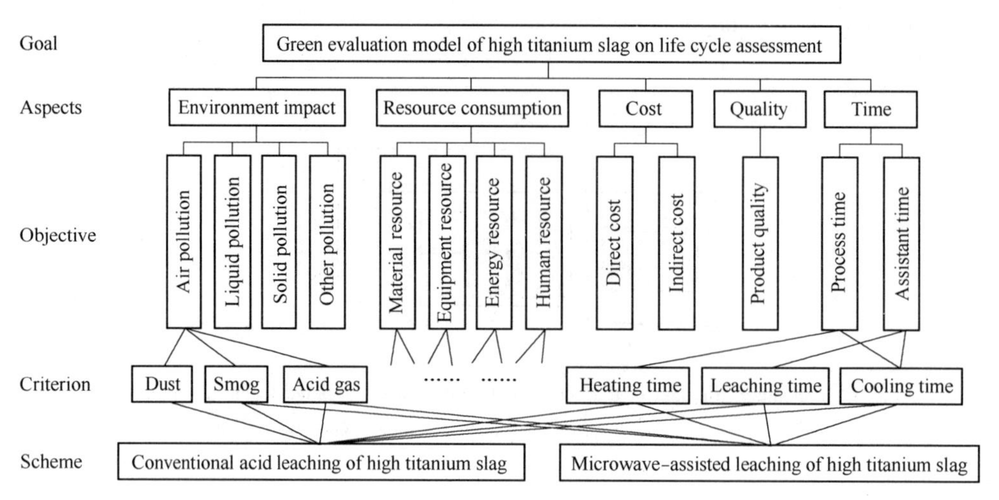

Fig. 2 Hierarchical structure on life cycle assessment of leaching process of high titanium slag

matrices based on pairwise comparison of all elements in each hierarchy level with respect to the higher hierarchy level according to certain criteria of comparison within set scales[22]. The Saaty's scales of pairwise comparisons are given in Table 2[23].

Table 2 Scale of pairwise comparisons

Importance	Definition	Explanation
1	Of equal value	Two activities contribute equally to the objective
3	Slightly more value	One is slightly in favor over another
5	Essential or strong value	One is strongly in favor over another
7	Very strong value	Dominance of one element proved in practice
9	Extreme value	The highest order dominance of one element over another
2,4,6,8	Intermediate values	When compromise is needed between the two adjacent judgments

For each criterion and weight of the hierarchy structure all the relevant elements are compared in pairwise comparison matrices as follows:

$$\boldsymbol{P}=\begin{bmatrix} 1 & \dfrac{w_1}{w_2} & \cdots & \dfrac{w_1}{w_n} \\ \dfrac{w_2}{w_1} & 1 & \cdots & \dfrac{w_2}{w_n} \\ \vdots & \vdots & \ddots & \vdots \\ \dfrac{w_n}{w_1} & \dfrac{w_n}{w_2} & \cdots & 1 \end{bmatrix} \tag{1}$$

where $\boldsymbol{P}$ is comparison pairwise matrix; w_1 is the relative importance of criterion 1; w_2 is the relative importance of criterion 2; w_n is the relative importance of criterion n.

An eigenvalue method and the Gauss – Seidel iterative matrix are used to calculate the relative weights of elements in each pairwise comparison matrix. The biggest eigenvalue is obtained from following equation:

$$\det(\boldsymbol{P} - \lambda_{max}\boldsymbol{I}) = 0 \tag{2}$$

where λ_{max} is the biggest eigenvalue of matrix $\boldsymbol{A}$; $\boldsymbol{I}$ is unit matrix. Then the relative weights($\boldsymbol{A}$) of matrix($\boldsymbol{P}$) is obtained from following equation:

$$(\boldsymbol{P} - \lambda_{max}\boldsymbol{I}) \times \boldsymbol{A} = 0 \tag{3}$$

The biggest eigenvalue was obtained by solving the eigenvalue Eq. (2) and showed in Eq. (4). The relative weights was obtained by solving and normalizing the eigenvalue Eq. (3)

$$\lambda_{max} = \frac{1}{n}\sum_{i=1}^{n}\left\{\frac{\sum_{j=1}^{n} a_{ij} \times w_j}{w_i}\right\} \tag{4}$$

3.3 Definition of assessment matrices

Assessment aspects matrix is expressed as:

$$\boldsymbol{U} = \{U_1, U_2, \cdots, U_i\} \tag{5}$$

where i is the number of assessment aspects.

Assessment objective matrix is expressed as:

$$\boldsymbol{U}_i = \{U_{i1}, U_{i2}, \cdots, U_{ij}\} \tag{6}$$

where j is the number of assessment aspects of $\boldsymbol{U}_i$.

Assessment criterion matrix is expressed as:

$$\boldsymbol{U}_{ij} = \{U_{ij1}, U_{ij2}, \cdots, U_{ijk}\} \tag{7}$$

where k is the number of assessment criterion of $\boldsymbol{U}_{ij}$.

3.4 Definition of weight matrices

The weight of assessment aspects is expressed as:

$$\boldsymbol{A} = \{a_1, a_2, \cdots, a_i\} \tag{8}$$

where a_i is the i th weight of assessment aspects and the boundary conditions can be given as follows:

$$\begin{cases} 0 < a_i \leqslant 1 \\ \sum_{i=1}^{i} a_i = 1 \end{cases} \tag{9}$$

The weights of assessment objective are expressed as:

$$\boldsymbol{A}_i = \{a_{i1}, a_{i2}, \cdots, a_{ij}\} \tag{10}$$

where a_{ij} is the j th weight of assessment objective of a_i and the boundary conditions can be given as follows:

$$\begin{cases} 0 < a_{ij} \leqslant 1 \\ \sum_{j=1}^{j} a_{ij} = 1 \end{cases} \tag{11}$$

The weights of assessment criterion are expressed as:

$$A_{ij} = \{a_{ij1}, a_{ij2}, \cdots, a_{ijk}\} \tag{12}$$

where a_{ijk} is the k th weight of assessment criterion of a_{ij} and the boundary conditions can be given as follows:

$$\begin{cases} 0 < a_{ijk} \leqslant 1 \\ \sum_{k=1}^{k} a_{ijk} = 1 \end{cases} \tag{13}$$

3.5 Definition of assessment matrices

The fourth level of fuzzy assessment matrices are expressed as

$$R_{ij} = \begin{bmatrix} r_{i11} & r_{i12} & \cdots & r_{i1n} \\ r_{i21} & r_{i22} & \cdots & r_{i2n} \\ \vdots & \vdots & \ddots & \vdots \\ r_{im1} & r_{im2} & \cdots & r_{imn} \end{bmatrix} \tag{14}$$

where r_{imn} is the value of the fuzzy subset membership function between m th assessment criterion and n th assessment criterion.

The third level of fuzzy assessment matrices are expressed as

$$\boldsymbol{R}_{ij} = \boldsymbol{A}_{ij} \circ \boldsymbol{R}_{ijk} \tag{15}$$

where $\boldsymbol{R}_{ijk}$ is the fourth level fuzzy assessment matrices; $\boldsymbol{A}_{ij}$ is the weight matrices of assessment criterion; $\circ$ gives the fuzzy relation comprehensive algorithms.

The second level of fuzzy assessment matrices are expressed as

$$\boldsymbol{R}_{i} = \boldsymbol{A}_{i} \circ \boldsymbol{R}_{ij} \tag{16}$$

where $\boldsymbol{R}_{ij}$ is the third level fuzzy assessment matrices; $\boldsymbol{A}_{i}$ is the weight matrices of assessment objective.

The first level of fuzzy assessment matrices are expressed as

$$\boldsymbol{R} = \boldsymbol{A} \circ \boldsymbol{R}_{i} \tag{17}$$

where $\boldsymbol{R}_{i}$ is the second level fuzzy assessment matrices; $\boldsymbol{A}$ is the weight matrices of assessment aspects.

3.6 Definition of green grade matrix

In order to obtain a final result, the following expression is used to represent the green grade matrix set of each assessment aspects:

$$\boldsymbol{V} = \{V_1, V_2, \cdots, V_i\} \tag{18}$$

where V_i is the th assessment grade of assessment aspects.

To obtain sensible comparisons, the expert judging and multi – fuzzy assessment theory were applied in the life cycle assessment process. Accordingly, the fuzzy assessment set $\boldsymbol{V}$ can be expressed by the green grade of the assessment aspects as follows:

$$\boldsymbol{V} = \{1.0, 0.8, 0.7, 0.6, 0.5\} \tag{19}$$

3.7 Green evaluation results

The green evaluation results of high titanium slag under microwave - assisted leaching and conventional acid leaching are obtained from following equation, which can be derive from Eq. (17) and Eq. (19):

$$Y = \boldsymbol{V}\boldsymbol{R}^{\mathrm{T}} \times 100 \tag{20}$$

4 Results and discussion

In order to demonstrate the universal application of the greenness evaluation index and assessment system of leaching process of high titanium slag for life cycle assessment, microwave - assisted leaching process and conventional acid leaching process are compared with each other under same evaluation system. The weight values are used and the assessment matrices of both microwave - assisted leaching and conventional acid leaching are shown in Tables 3 and 4, respectively. The comparison of these two leaching process can be considered as a decision making model since it guides decision makers to choose best alternative for the experiment.

Table 3 The life cycle assessment of conventional acid leaching of high titanium slag

Assessment aspects	Weight	Assessment objective	Weight	Assessment criterion	Weight	Assessment matrices				
Environment impact	0.25	Air pollution	0.30	Dust	0.30	0.2	0.3	0.2	0.2	0.1
				Smog	0.30	0.1	0.3	0.2	0.2	0.2
				Acid gas	0.40	0.2	0.3	0.3	0.1	0.1
		Liquid pollution	0.25	Leaching liquid	0.45	0.2	0.3	0.1	0.3	0.1
				Washing liquid	0.25	0.2	0.3	0.2	0.2	0.1
				Absorbing liquid	0.30	0.3	0.2	0.3	0.1	0.1
		Solid pollution	0.25	Waste slag	0.40	0.1	0.4	0.2	0.2	0.1
				Additive	0.35	0.2	0.2	0.2	0.3	0.1
				Adhesive	0.25	0.1	0.3	0.4	0.2	0
		Other pollution	0.20	Noise	0.25	0.4	0.1	0.1	0.3	0.1
				Thermal radiation	0.30	0.5	0.2	0.1	0.1	0.1
				Other	0.45	0.2	0.3	0.1	0.3	0.1
Resource consumption	0.25	Material resource	0.35	Materials consume	0.35	0.1	0.3	0.2	0.4	0
				Materials utilization	0.40	0.1	0.3	0.2	0.2	0.2
				Materials recovery rate	0.25	0.2	0.5	0.2	0.1	0
		equipment resource	0.20	Equipment utilization	0.70	0.3	0.3	0.2	0.1	0.1
				Automation	0.30	0.4	0.2	0.2	0.1	0.1
		Energy resource	0.30	Energy utilization	0.55	0.2	0.2	0.4	0.1	0.1
				Renewable resource	0.35	0.1	0.4	0.3	0.1	0.1
				Surplus energy	0.10	0.2	0.2	0.4	0.1	0.1
		Human resource	0.15	Professional	0.55	0.1	0.3	0.4	0.1	0.1
				Administrator	0.25	0.2	0.4	0.1	0.2	0.1
				Knowledge	0.20	0.1	0.5	0.3	0.1	0

Continues Table 3

Assessment aspects	Weight	Assessment objective	Weight	Assessment criterion	Weight	Assessment matrices				
Cost	0. 2	Direct cost	0. 60	Design cost	0. 45	0. 2	0. 3	0. 3	0. 1	0. 1
				Materials cost	0. 30	0. 5	0. 2	0. 1	0. 1	0. 1
				Processing cost	0. 25	0. 2	0. 4	0. 1	0. 2	0. 1
		Indirect cost	0. 40	Maintenance fee	0. 55	0. 2	0. 3	0. 2	0. 2	0. 1
				Governance fee	0. 30	0. 3	0. 1	0. 4	0. 1	0. 1
				Waste disposal fee	0. 15	0. 3	0. 2	0. 2	0. 2	0. 1
Quality	0. 15	Product quality	1. 0	Product content	0. 40	0. 1	0. 3	0. 4	0. 1	0. 1
				Microstructure	0. 40	0. 2	0. 5	0. 1	0. 2	0
				Surface area	0. 20	0. 3	0. 1	0. 1	0. 3	0. 2
Time	0. 15	Process time	0. 65	Heating time	0. 35	0. 3	0. 1	0. 3	0. 2	0. 1
				Cooling time	0. 65	0. 1	0. 2	0. 4	0. 1	0. 2
		Assistant time	0. 35	Leaching time	0. 35	0. 3	0. 1	0. 3	0. 1	0. 2
				Cooling time	0. 65	0. 3	0. 4	0. 1	0. 2	0

Table 4 The life cycle assessment of microwave – assisted leaching of high titanium slag

Assessment aspects	Weight	Assessment objective	Weight	Assessment criterion	Weight	Assessment matrices				
Environment impact	0. 25	Air pollution	0. 30	Dust	0. 30	0. 4	0. 3	0. 2	0. 1	0
				Smog	0. 30	0. 5	0. 3	0. 1	0. 1	0
				Acid gas	0. 40	0. 3	0. 3	0. 2	0. 1	0. 1
		Liquid pollution	0. 25	Leaching liquid	0. 45	0. 4	0. 2	0. 2	0. 1	0. 1
				Washing liquid	0. 25	0. 5	0. 2	0. 2	0. 1	0
				Absorbing liquid	0. 30	0. 3	0. 3	0. 2	0. 2	0
		Solid pollution	0. 25	Waste slag	0. 40	0. 2	0. 4	0. 2	0. 1	0. 1
				Additive	0. 35	0. 4	0. 1	0. 3	0. 1	0. 1
				Adhesive	0. 25	0. 4	0. 2	0. 1	0. 2	0. 1
		Other pollution	0. 20	Noise	0. 25	0. 3	0. 3	0. 1	0. 2	0. 1
				Thermal radiation	0. 30	0. 4	0. 3	0. 2	0. 1	0
				Other	0. 45	0. 3	0. 1	0. 4	0. 1	0. 1
Resource consumption	0. 25	Material resource	0. 35	Materials consume	0. 35	0. 5	0. 1	0. 2	0. 1	0. 1
				Materials utilization	0. 40	0. 4	0. 1	0. 3	0. 1	0. 1
				Materials recovery rate	0. 25	0. 5	0. 2	0. 1	0. 1	0. 1
		Equipment resource	0. 20	Equipment utilization	0. 70	0. 3	0. 2	0. 3	0. 1	0. 1
				Automation	0. 30	0. 4	0. 1	0. 3	0. 1	0. 1
		Energy resource	0. 30	Energy utilization	0. 55	0. 4	0. 1	0. 2	0. 3	0
				Renewable resource	0. 35	0. 3	0. 4	0. 1	0. 1	0. 1
				Surplus energy	0. 10	0. 3	0. 3	0. 3	0. 1	0
		Human resource	0. 15	Professional	0. 55	0. 3	0. 2	0. 3	0. 1	0. 1
				Administrator	0. 25	0. 2	0. 5	0. 1	0. 1	0. 1
				Knowledge	0. 20	0. 3	0. 4	0. 2	0. 1	0

Continues Table 4

Assessment aspects	Weight	Assessment objective	Weight	Assessment criterion	Weight	Assessment matrices				
Cost	0.2	Direct cost	0.60	Design cost	0.45	0.6	0.1	0.1	0.1	0.1
				Materials cost	0.30	0.5	0.2	0.1	0.1	0.1
				Processing cost	0.25	0.4	0.2	0.3	0.1	0
		Indirect cost	0.40	Maintenance fee	0.55	0.2	0.4	0.2	0.1	0.1
				Governance fee	0.30	0.4	0.1	0.3	0.2	0
				Waste disposal fee	0.15	0.5	0.1	0.2	0.1	0.1
Quality	0.15	Product quality	1.0	Product content	0.40	0.3	0.2	0.3	0.1	0.1
				Microstructure	0.40	0.3	0.3	0.3	0.1	0
				Surface area	0.20	0.3	0.1	0.4	0.1	0.1
Time	0.15	Process time	0.65	Heating time	0.35	0.4	0.3	0.1	0.1	0.1
				Cooling time	0.65	0.3	0.3	0.2	0.1	0.1
		Assistant time	0.35	Leaching time	0.35	0.4	0.1	0.3	0.1	0.1
				Cooling time	0.65	0.3	0.3	0.1	0.2	0.1

From the above theoretical work, the detailed calculation processes can be carrying out according to fuzzy comprehensive assessment method and analytic hierarchy process. The fuzzy comprehensive assessment results are shown in Table 5. The results show that microwave – assisted leaching process have advantages over the conventional one, with respect to energy – consumption, processing time and environmental protection, since microwave – assisted leaching is very rapid, uniform and highly energy efficient. Thus, microwave – assisted leaching process is better than conventional acid leaching process, which fulfill with the aim of green manufacture.

Table 5 Fuzzy comprehensive assessment results for microwave – assisted leaching and conventional acid leaching

Scheme	Green degree	Scheme	Green degree
Microwave – assisted leaching	86.51	Conventional acid leaching	81.28

5 Conclusions

(1) The effects of the life cycle assessment variables such as the resource consumption, environment impact, cost, time and quality were investigated systematically.

(2) The green degree of microwave – assisted leaching is 86.51, a relatively high value indicating that microwave – assisted leaching is better than conventional acid leaching, especially in terms of environment impact and resource consumption, which is consistent with the aim of life cycle assessment.

(3) Microwave – assisted leaching process of high titanium slag has a potential to provide a new method to prepare synthetic rutile with high efficiency and low energy consumption.

References

[1] Wang J, Ma T, Zhang Z, et al. Investigation on the transition crystal of ordinary rutile TiO_2 powder by microwave irradiation in hydrogen peroxide solution and its sonocatalytic activity[J]. Ultrasonics Sonochemistry, 2007, 14(5): 575 - 582.

[2] Bessinger D, Geldenhuis J M A., Pistorius P C, et al. The decrepitation of solidified high titania slags[J]. Journal of Non - Crystalline Solids, 2001, 282(1): 132 - 142.

[3] Dondi M, Cruciani G, Balboni E, et al. Titania slag as a ceramic pigment[J]. Dyes and Pigments, 2008, 77(3): 608 - 613.

[4] Wang M, Li L, Zhang L, et al. Effect of oxidization on enrichment behavior of TiO_2 in titanium - bearing slag[J]. Rare Metals, 2006, 25(2): 106 - 110.

[5] Li C, Liang B, Guo L. Dissolution of mechanically activated Panzhihua ilmenites in dilute solutions of sulphuric acid[J]. Hydrometallurgy, 2007, 89(1/2): 1 - 10.

[6] Samal S, Rao K K, Mukherjee P S, et al. Statistical modelling studies on leachability of titania - rich slag obtained from plasma melt separation of metallized ilmenite[J]. Chemical Engineering Research and Design, 2008, 86(2): 187 - 191.

[7] Pistorius P C, Motlhame T. Oxidation of high - titanium slags in the presence of water vapour[J]. Minerals Engineering, 2006, 19(3): 232 - 236.

[8] Cristallo G, Roncari E, Rinaldo A, et al. Study of anatase - rutile transition phase in monolithic catalyst V_2O_5/TiO_2 and $V_2O_5 - WO_3/TiO_2$[J]. Applied Catalysis A: General, 2001, 209(1/2): 249 - 256.

[9] Mo W, Deng G, Luo F. Ti Metallurgy[M]. Beijing: Metallurgical Industry Press, 1998: 161 - 165.

[10] Li W, Peng J H, Zhang L B, et al. Pilot - scale extraction of zinc from the spent catalyst of vinyl acetate synthesis by microwave irradiation[J]. Hydrometallurgy, 2008, 92(1/2): 79 - 85.

[11] Chang Y C, Dong S K. Microwave induced reactions of sulfur dioxide and nitrogen oxides in char and anthracite bed[J]. Carbon, 2001, 39(8): 1159 - 1166.

[12] Cutmore N, Evans T D, Crnokrak A, Stoddard S. Microwave technique for analysis of mineral sands[J]. Minerals Engineering, 2000, 13(7): 729 - 736.

[13] Lester E D, Sam K M, Chris D, John P. The potential for rapid coke making using microwave energy[J]. Fuel, 2006, 85(14/15): 2057 - 2063.

[14] Huang M, Peng J H, Yang J J, et al. Microwave cavity perturbation technique for measuring the moisture content of sulphide minerals concentrates[J]. Minerals Engineering, 2007, 20(1): 92 - 94.

[15] Li W, Zhang L B, Peng J H, et al. Effects of microwave irradiation on the basic properties of wood - ceramics made from carbonized tobacco stems impregnated with phenolic resin[J]. Industrial Crops and Products, 2008, 28(2): 143 - 154.

[16] Renou S, Thomas J S, Aoustin E, Pons M N. Influence of impact assessment methods in wastewater treatment LCA[J]. Journal of Cleaner Production, 2008, 16(10): 1098 - 1105.

[17] Khoo H H. Life cycle impact assessment of various waste conversion technologies[J]. Waste Management, 2009, 29(6): 1892 - 1900.

[18] Wittmaier M, Langer S, Sawilla B. Life cycle impact assessment of various waste conversion technologies[J]. Waste Management, 2009, 29(5): 1732 - 1738.

[19] Saaty T L. Applications of analytical hierarchies[J]. Mathematics and Computers in Simulation, 1979, 21(1): 1 - 20.

[20] Emshoff J R, Saaty T L. Applications of the analytic hierarchy process to long range planning processes[J]. European Journal of Operational Research, 1982, 10(2): 131 - 143.

[21] Xu Y, Zhang Y A. Online credit evaluation method based on AHP and SPA[J]. Communications in Nonlinear Science and Numerical Simulation, 2009, 14(7): 3031 - 3036.

[22] Millet D, Bistagnino L, Lanzavecchia C, Camous R, Poldma T. Does the potential of the use of LCA match the design team needs[J]. Journal of Cleaner Production, 2007, 15(4): 335 - 346.

[23] Saaty T L. How to make a decision: the analytic hierarchy process[J]. European Journal of Operational Research, 1990, 48(1): 9 - 26.

Removing Chlorine of CuCl Residue from Zinc Hydrometallurgy by Microwave Roasting

Shuaidan Lu, Yi Xia, Changyuan Huang, Guoqin Wu,
Jinhui Peng, Shaohua Ju, Libo Zhang

Abstract: Most Zn hydrometallurgy factories adopt Cu_2SO_4 as a dechlorination reagent from zinc solution nowadays, thus much CuCl residue is produced. The existing process of treating this residue was washed with water or sodium carbonate solution, which would cause a lot of troubles to water treatment and waste discharge. A method of microwave roasting is adopted in this work for dechlorination of the CuCl residue. A 1.5kW microwave roasting equipment with dust collection and tail gas adsorption systems was set up and applied during the experiment. By investigating temperature, heat preservation time, moisture of raw material and grain size of samples on the dechlorination effect, the optimal experimental condition was obtained: the samples with 2% moisture and −100 mesh grain size, microwave roasting at 400℃ for 2h, the Cl content turned from 14.27% to 1.35% and the dechlorination rate are as high as 90%, while those with conventional heating are only 60% − 80%. The phase changes of the roasting process are also investigated with X − ray diffraction.

Keywords: microwave roasting; zinc hydrometallurgy; dechlorination; CuCl residue

1 Introduction

Cl element can be accumulated gradually in the solution of Zinc hydrometallurgical system. When Cl content in electrolysis solution comes up to 100mg/L, the stability of electrodeposits process would be severely affected[1]. It would not only accelerate the consumption of cathode and anode plate, elevate the power consumption, but also cause serious corrosion of device, increase production costs and lower the quality of electrolytic zinc[2]. Thus, Cl^- contained in the solution should be removed before electrodeposits.

In present, there are two methods of dechlorination which are commonly used in zinc hydrometallurgy in China. One is roasting the raw material with high Cl content[3,4], the other is removing Cl from zinc sulfate solution by adding Cu^+ to form CuCl[5]. Now most of zinc hydrometallurgical smelters select the latter method for its high removal rate and low operation cost. However, the CuCl residue produced during this process is still a hazardous substance hard to treat. Most smelters washed the residue with distilled water and alkaline solution such as Na_2CO_3[3]. While washing the zinc dross simply with distilled water at 95℃ for a short period of time such as half an hour and at a suitable solid − liquid ratio between 1∶2 − 1∶10, about 80% chlorine found in the dross can be removed[6]. CuCl containing in the residue turned to NaCl solution subsequently, which would take a lot of troubles to water treatment and waste discharge.

As a new green metallurgical method, microwave metallurgy has been developed as noticeable advanced discipline. The worldwide researchers have done many research in fields of the heat transfer of material assisted by microwave, microwave pre - treatment[7,8], drying[9,10], roasting[11,12], strengthening of the processes such as reducing and leaching[13-16], and the microwave roasting has become one of the most hot research fields[17,18]. Microwave roasting shows incomparable advantages that normal heating cannot present: selective heating, rapid heating rate, high heating efficiency[19,20]. Secondly, to chemical reaction, microwave roasting shows catalysis ability, reduces the reacting time and also improves the energy saves. Furthermore, no gas will be produced by microwave so that microwave treatment is an effective path to achieve clean production in metallurgical industry[21-23]. Besides, material treated under microwave can instantly gain or loss the heat resource, for that it can be easily automatically controlled. Yixin Hua's work shows that the copper extractions from the copper sulfide concentrate roasted with microwave heating are as high as 90% - 96.6%, while those with conventional heating are only 71% - 77.4%[24]. R. K. Amankwah utilized microwaves in the roasting of a carbonaceous sulphidic gold concentrate, while the gold extraction values after cyanidation were over 96% and these were similar to those obtained by conventional roasting, the microwave roasting has the main advantages: both the total carbon removal rates and the heating rates were higher and the specific energy consumptions were lower than in conventional roasting[25]. Chang Yongfeng utilized microwaves in the roasting of the laterite mixed with active carbon and the temperature can reach approximate 1000℃ in 6.5min, while the nickel recovery can reach about 90% and iron recovery is less than 30%[26]. However, studies of the dechlorination of the CuCl residue by microwave roasting haven't been found yet.

In this study, a method of microwave roasting will be adopted for dechlorination of the CuCl residue. A 1.5kW microwave roasting equipment with dust collection and tail gas adsorption system will be set up. By investigating temperature, heat preservation time, moisture of raw material, and grain size of samples on dechlorination effect, the best experimental condition can be obtained.

2 Material and methods

2.1 Experimental material

The CuCl residue cake with high water content of 18.10% was supplied by a zinc hydrometallurgy smelter in Yunnan province. The chemical elemental analysis of the microwave dried sample is showed in Table 1. The Cl content of dechlorination CuCl residue cake is showed as 14.27%.

Table 1 Element chemical analysis of dechlorination CuCl residue cake

Composition	Zn	Cu	Cl	S
Content/%	9.55	54.68	14.27	4.74

The samples were dried with treating time 10min in 3kW microwave drying device and investigated with X - ray diffraction.

As is it showed in Fig. 1, after drying for 10min, the main phase of the samples are Cu_2O, $CuCl$, $Cu_2Cl(OH)_3$ and $CaSO_4(H_2O)_2$. By microwave heating (1.28kW), the temperature rising results of sample (100g) in experimental process is showed in Fig. 2.

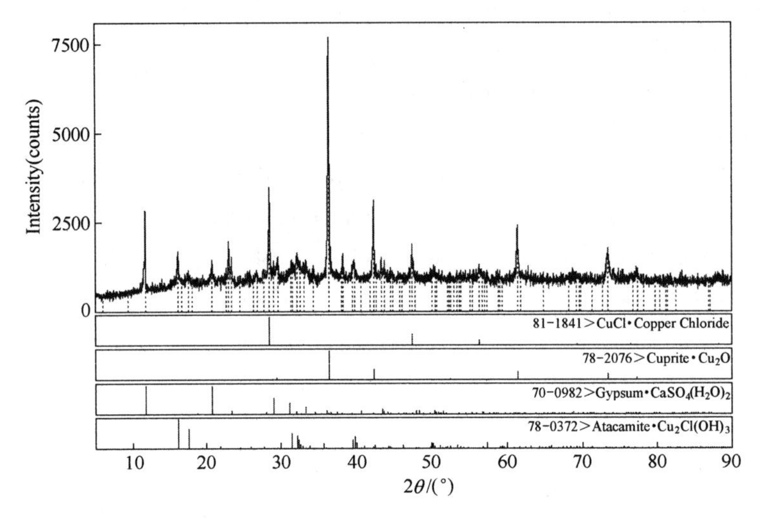

Fig. 1 X - ray diffraction pattern of microwave dried CuCl residue

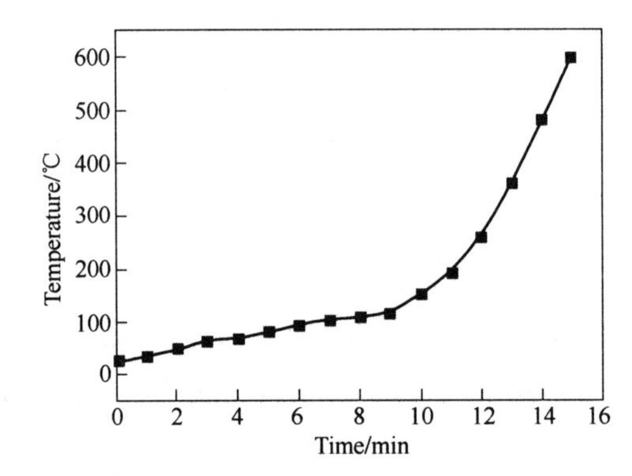

Fig. 2 The heating curve of CuCl residue under microwave irradiation

It is showed in Fig. 2 that after 15min of microwave heating, the temperature of the CuCl residue was heated up to 600℃. So that we can see that the CuCl residue has the hyperactive response to the microwaves.

2.2 Experimental device

A microwave reactor system (1.5kW) developed by the Key Laboratory of Unconventional Metallurgy was used in this research. The formation of device is showed in Fig. 3.

2.3 Analysis method

The method of Cl^- selective electrode was use for its high accuracy, simple procedure and anti -

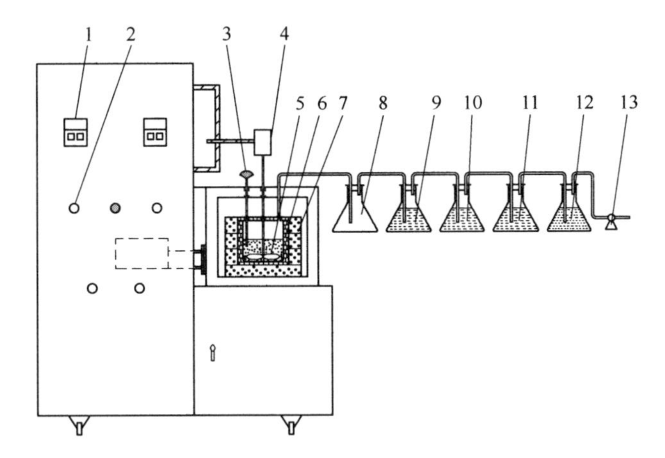

Fig. 3 The experimental microwave high - temperature reactor system

1—instrument display; 2—controller; 3—thermocouple; 4—stirrer; 5—experimental material; 6—crucible; 7—heat preservation material; 8—dust collecting bottle; 9 - 11—gas collecting system; 12—buffer bottle; 13—micro pump

interference. The specific procedure is as following: firstly, the grinded samples were weighted and dissolved in diluted nitric acid; then 1mL dissolved solution was sampled and controlling the pH at about 5.5; adding 5mL of sodium citrate solution as the strong ion agent, and finally measure the potential value by using Cl^- selective electrode and mercury - mercurous sulfate as reference electrode. Then the content of Cl in solid material can be calculated.

Atomic absorption spectrometry (PE AA700, PerkinElmer) was used for detecting the content of Cu and Zn in samples. XRD (XRD - 7000, Shimadzu Corporation) was selected to study the phase of samples.

2.4 Experimental method

100g dried and grinded sample was put in mullite crucible with heat preservation material around and all of them were carried in microwave reactor. Then the power of microwave system, stirring system and off - gas absorption system were turned on. The temperatures which were preset in 300℃, 350℃, 400℃, 450℃ and 500℃ changed with time and was recorded. The samples were sampled after roasting for 2h. In the roasting process, the samples were stirred fully. After microwave treatment, material loss was recorded and the Cl content and Cu content of samples were tested.

In order to study the dechlorination effectivity by microwave roasting, different moisture (6%, 5%, 2%, 0%), grain sizes (- 200, - 100, - 60, - 20mesh) of samples under microwave treatment were studied.

2.5 Experimental process

In this work, a microwave roasting process was designed, which was demonstrated as Fig. 4.

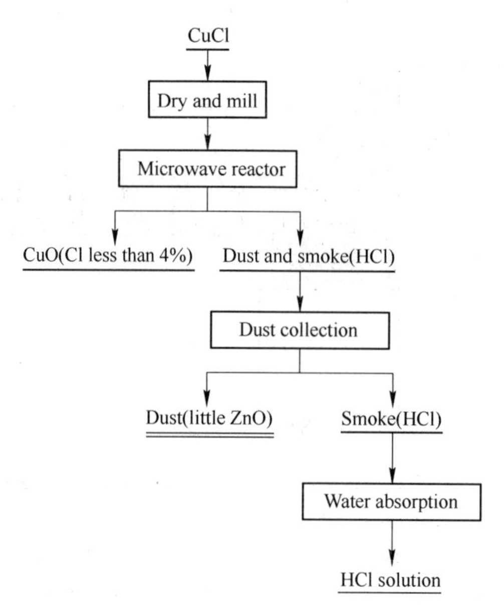

Fig. 4 The experimental process of dechlorination

3 Results and discussion

3.1 Dechlorine effect of temperature

The process of removing chlorine of CuCl residue by microwave roasting can be considered a oxidation reaction. The effect of temperature which had great influences on the dechlorination efficiency was investigated under the experimental condition with microwave power: 1.5kW, roasting time: 2h, stirring rate: 10r/min and grain size: -100mesh. The experimental result is shown in Fig. 5.

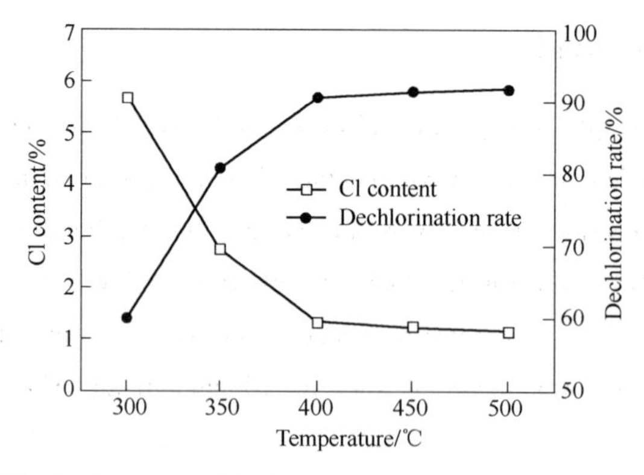

Fig. 5 Variation of dechlorination rate of Cu residue and Cl content at different temperature roasting for 2h

It can be seen that the Cl content and dechlorination rate at different temperature after roasting for 2h. With the temperature getting higher, dechlorination rate increased strongly. When the temperature reach to 400℃, dechlorination rate achieved to a high value as more than 90%. In contrast, the dechlorination rate at 300℃ was only 60%.

3.2 Dechlorine effect of roasting time

In this group experiments, 100g of CuCl residue was employed for each run. The dechlorination rate is plotted against roasting time for different roasting temperatures (300℃, 350℃ and 400℃) in Fig. 6.

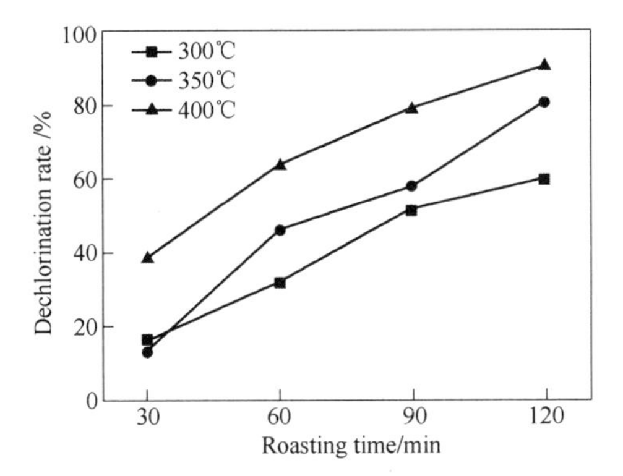

Fig. 6 Variation of dechlorination rate with different roasting time

It is showed in Fig. 6 that the dechlorination rate with different heat preservation temperature. The longer roasting time was the higher dechlorination rate would be. With the roasting time changing from 30min to 120min, at 400℃, the dechlorination increased more than 50 %.

3.3 Dechlorine effect of moisture content of CuCl residue

In order to investigate moisture content of CuCl residue on the dechlorination effect, the roasting experiments were conducted with different moisture content (0%, 2%, 5% and 6%) of CuCl residue. The microwave power is 1.5kW and the roasting time is 2h. The relationship between dechlorination rate and moisture content of samples is showed in Fig. 7.

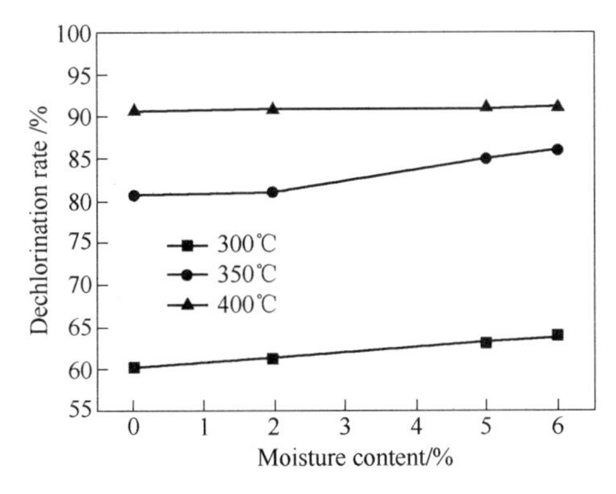

Fig. 7 Variation of dechlorination rate with different moisture content

According to the figure, moisture content can present a certain effect to the dechlorination rate. With the moisture content of samples increasing, dechlorination rate slightly improved. For that, certain amount water in sample could be helpful to dechlorination. However, the increased moisture could reduce the efficiency of microwave heating.

3.4 Dechlorination effect of grain size

The effect of moisture content on the microwave roasting of CuCl residue was investigated though a group of experiments which were under the conditions with moisture content of CuCl residue: 0%, microwave power: 1.5kW, roasting time: 2h and stirring rate: 10r/min. The experimental result is shown in Fig. 8.

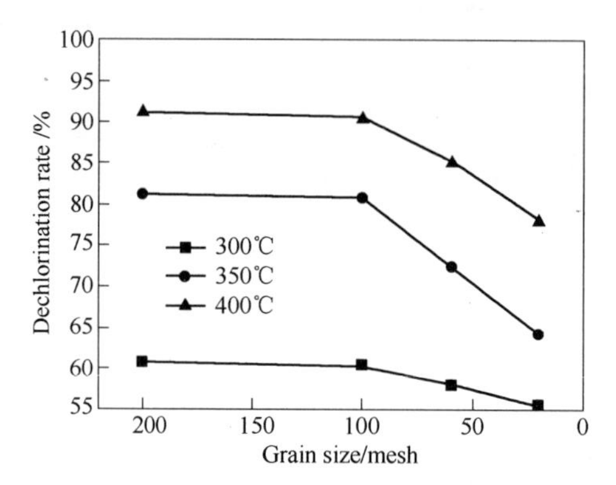

Fig. 8 The variation of dechlorination rate with different grain size

Fig. 8 shows the trend that samples with different grain size can also affect the dechlorination rate. With the sample grain size getting smaller, the dechlorination rate gets higher. When the sample grain size reached −100 mesh, the dechlorination has been at a high level, and when the sample grain size was bigger than 100 mesh, no obvious improvement of Cl removal was observed.

3.5 The phase changes and dechlorination rate of the roasting process in optimal conditions

By investigating temperature, heat preservation time, moisture of raw material and grain size of samples on the dechlorination effect, the optimal experimental condition was obtained: the samples with 2% moisture and −100 mesh grain size, microwave roasting at 400℃ for 2h.

The samples after microwave roasting were investigated with X-ray diffraction. As is it showed in Fig. 9, after microwave roasting, the main phase of the samples is CuO.

The contrast of Fig. 1 and Fig. 9 shows that the peak of chloride is getting weaker and the main phase transforms into CuO after roasting at 400℃ for 2h. It infers that CuCl in CuCl residue has been removed by transforming into CuO.

4 Conclusions

(1) Temperature, heat preservation time, initial moisture content of material and material grain size in the dechlorination procedure was investigated. Dechlorination rate was higher with the temperature getting higher; over 350℃ with 2h heat preservation, dechlorination rate can achieve over 75%; grain size in roasting process showed significant effects on dechlorination rate. Because water was removed in heating process, the initial moisture content of material rarely showed effect on de-

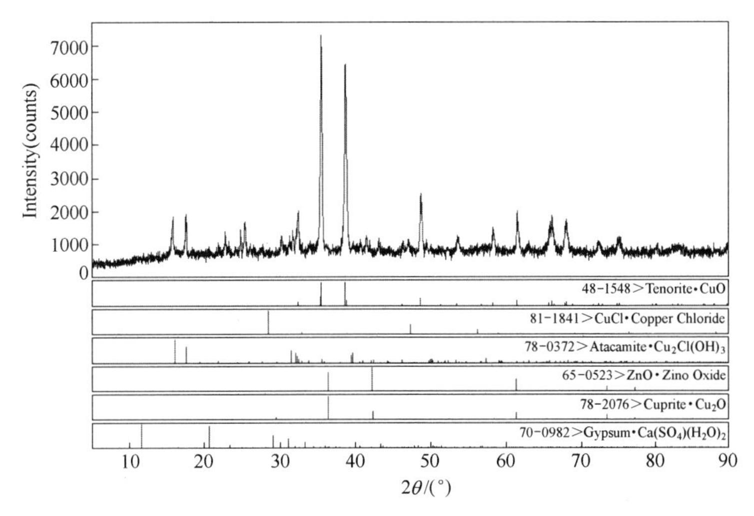

Fig. 9 XRD pattern of CuCl residue after roasting at 400℃ for 2h at 1.5kW

chlorination rate.

(2) Under microwave treatment, the most suitable experimental procedure was when the moisture content, grain size of samples were 2% and −100mesh, samples were heated at a 400℃ and fully stirred for 2h. Cl content decreased from 14.27% to less than 1.35%, and the dechlorination rate achieved over 90%.

(3) The method removing Cl of CuCl residue from zinc hydrometallurgy by microwave roasting shows incomparable advantages that the Cl removal rates and the heating rates are higher and the specific energy consumptions are lower than in conventional roasting and has wide industrial prospect.

Acknowledgements

The authors gratefully acknowledge the National Natural Science Foundation of China (No. 51104073) for funding this work. The project was also funded as a 2012 transformation item of sci-tech achievements by Ministry of Industry and Information Technology of People's Republic of China.

References

[1] Mathewson C H. Zinc: the Science and Technology of the Metal, its Alloys and Compounds[M]. New York: Reinhold Pub. Corp, 1959: 176 − 178.

[2] Jha M K, Kumar V, Singh R J. Review of hydrometallurgical recovery of zinc from industrial wastes[J]. Resources, Conservation and Recycling, 2001, 33(1): 1 − 22.

[3] Sahin F C, Derin B, Yucel O. Chloride removal from zinc ash[J]. Scandinavian Journal of Metallurgy, 2000, 29(5): 224 − 230.

[4] Yasushi Y, Koichiro S. Treatment of waste containing heavy metal and chlorine and device therefor: JP 10156313A[P]. 19980616.

[5] Fernand J, Joseph B. Process for the elimination of chloride from zinc sulphate solutions: US4005174

[P]. 19770125.

[6] Güresin N, Topkaya Y A. Dechlorination of a zinc dross[J]. Hydrometallurgy, 1998, 49(1-2): 179-187.

[7] Jorjani E, Chapi H G, Khorami M T. Ultra clean coal production by microwave irradiation pretreatment and sequential leaching with HF followed by HNO_3[J]. Fuel Processing Technology, 2011, 92(10): 1898-1904.

[8] Swart A J, Mendonidis P. Evaluating the effect of radio-frequency pre-treatment on granite rock samples for comminution purposes[J]. International Journal of Mineral Processing, 2013, 120: 1-7.

[9] Huang M, Peng J H, Yang J J, Wang J Q. Microwave cavity perturbation technique for measuring the moisture content of sulphide minerals concentrates[J]. Minerals Engineering, 2007, 20(1): 92-94.

[10] Li Y, Lei Y, Zhang L B, Peng J H, Li C L. Microwave drying characteristics and kinetics of ilmenite[J]. Transactions of Nonferrous Metals Society of China, 2011, 21(1): 202-207.

[11] Hiromi Y, Yuka T, Yuki H, Mizushina Y. Microwaveroasting effects on the oxidative stability of oils and molecular species of triacylglycerols in the kernels of pumpkin (Cucurbita spp.) seeds[J]. Journal of Food Composition and Analysis, 2006, 19(4): 330-339.

[12] Uysal N, Sumnu G, Sahin S. Optimization of microwave-infrared roasting of hazelnut[J]. Journal of Food Engineering, 2009, 90(2): 255-261.

[13] Xia H Y, Peng J H, Niu H, Huang M Y, Zhang Z Y, Zhang Z B, Huang M. Non-isothermal microwave leaching kinetics and absorption characteristics of primary titanium-rich materials[J]. Transactions of Nonferrous Metals Society of China, 2010, 20(4): 721-726.

[14] Isabel S S P, Helena M V M S. Selective leaching of molybdenum from spent hydrodesulphurisation catalysts using ultrasound and microwave methods[J]. Hydrometallurgy, 2012, 129-130: 19-25.

[15] Guo S H, Li W, Peng J H, Niu H, Huang M Y, Zhang L B, Zhang S M, Huang M. Microwave-absorbing characteristics of mixtures of different carbonaceous reducing agents and oxidized ilmenite[J]. International Journal of Mineral Processing, 2009, 93(3-4): 289-293.

[16] Zhai X J, Wu Q, Fu Y, Ma L Z, Fan C L, Li N J. Leaching of nickel laterite ore assisted by microwave technique[J]. Transactions of Nonferrous Metals Society of China, 2010, 20: 77-81.

[17] Zhang N B, Bai C G, Ma M Y, Li Z Y. Preparation of $BaAl_2O_4$ by microwave sintering[J]. Transactions of Nonferrous Metals Society of China, 2010, 20(10): 2020-2025.

[18] Huang X H, Huang X Y, Mao H K, Yin Z X. The study on microwave magnetic roasting plus magnetic seperation and acid pickling to enrich Nb of low-grade niobium minerals[J]. Applied Mechanics and Materials, 2012, 182-183: 17-22.

[19] Hidaka H, Saitou A, Honjou H, Hosoda K, Moriya M, Serpone N. Microwave-assisted dechlorination of polychlorobenzenes by hypophosphite anions in aqueous alkaline media in the presence of Pd-loaded active carbon[J]. Journal of Hazardous Materials, 2007, 148(1-2): 22-28.

[20] Liu X T, Zhao W, Sun K, Zhang G X, Zhao Y. Dechlorination of PCBs in the simulative transformer oil by microwave-hydrothermal reaction with zero-valent iron involve[J]. Chemosphere, 2011, 82(5): 773-777.

[21] Jou C J G, Hsieh S C, Lee C L, Lin C, Huang H W. Combining zero-valent iron nanoparticles with microwave energy to treat chlorobenzene engineers[J]. Journal of the Taiwan Institute of Chemical, 2010, 41(2): 216-220.

[22] Takashima H, Karches M, Kanno Y. Catalytic decomposition of trichloroethylene over Pt-/Ni-catalyst under microwave heating[J]. Applied Surface Science, 2008, 254(7): 2023-2030.

[23] Ito M, Ushida K, Nakao N, Kikuchi N, Nozaki R, Asai K, Washio M. Dechlorination of poly(vinyl chloride) by microwave irradiation I: a simple examination using a commercial microwave oven[J]. Polymer Degradation and Stability, 2006, 91(8): 1694-1700.

[24] Hua Y X, Cai C J, Cui Y. Microwave - enhanced roasting of copper sulfide concentrate in the presence of $CaCO_3$[J]. Separation and Purification Technology, 2006, 50(1): 22 - 29.

[25] Amankwah R K, Pickles C A. Microwave roasting of a carbonaceous sulphidic gold concentrate[J]. Minerals Engineering, 2009, 22(13): 1095 - 1101.

[26] Chang Y F, Zhai X J, Fu Y, Ma L Z, Li B C, Zhang T A. Phase transformation in reductive roasting of laterite ore with microwave heating[J]. Transactions of Nonferrous Metals Society of China, 2008, 18(4): 969 - 973.

Dechlorination of Zinc Dross by Microwave Roasting

Yaqian Wei, Jinhui Peng, Libo Zhang, Shaohua Ju, Yi Xia, Qin Zheng, Yajian Wang

Abstract: Traditional treatments of zinc dross have many disadvantages, such as complicated recovering process and serious environmental pollution. In this work, a new process of chlorine removal from zinc dross by microwave was proposed for solving problem of recycling the zinc dross. With better ability of absorbing the microwave than zinc oxide, the main material in zinc dross, chlorides, can be heated and evaporated rapidly during microwave roasting. Various parameters including roasting temperature, duration time and stirring speed were optimized. The microstructure of roasted materials was characterized by X - ray diffraction (XRD) and scanning electron microscope (SEM). The content of the chloride was analyzed by the method of chlorine ion selective electrode. The experiments indicate that the best duration time is 60min with a stirring speed of 15r/min during the microwave roasting process. The dechlorination rate reaches peak value of 88% at 700℃. The chlorine is removed as HCl gas when water vapor is used as activating agent, which means that it can be recovered into hydrochloride acid.

Keywords: zinc dross; microwave roasting; dechlorination; resources recovery

1 Introduction

During casting process of cathode zinc, some NH_4Cl would be added into the furnace, for releasing the ZnO film covered around the zinc liquid drop and increasing the direct recovery rate during zinc casting. However, some scum, about 3% - 4% of the zinc ingots, will be still observed during the procedure. The zinc, zinc oxide, zinc chloride and zinc oxy - chloride are the major components of the scum. Many hydrometallurgical zinc plants utilize the method of grinding and screening zinc dross to recover zinc metal. Then, the residue mainly containing ZnO, Zn, $ZnCl_2$ and Zn_2OCl_2 is called as zinc dross[1].

How to recover valuable metal from this kind of zinc dross is still a big challenge due to its high content of Cl^- (0.5% - 2%). Cl element will be dissolved into the solution when leached by sulfuric, which leads some serious problems to electrowinning of cathode zinc, like corrosion of electrode materials and decreasing the quality of zinc cathode[2,3].

At present, the main process of dechlorinate of zinc dross are as follows:

(1) Washing zinc dross with water or soda alkali solution.

The principle of these two methods mainly lies in that zinc chloride can be easily dissolved into these solutions. The dechlorination rate of these methods could be as high as 95%. However, these processes would consume a lot of water, and produce lots of refractory waste water containing Cl^-[4].

(2) Roasting zinc dross with kiln or multiple hearth furnace.

Chloride can be volatilized easily at high temperature, or oxidized by O_2/H_2O to release Cl_2/HCl. The dechlorination rate of these methods is approximately 60% - 70%. But the roasting temperature is over 800℃, it consumes much coal/nature gas, and produces massive waste gas which hard to be treated.

Thus, finding a clean way to remove chlorine of zinc dross efficiently with less energy consumption is still one of the key problems in zinc hydrometallurgy.

Microwave heating process is a green energy supply method which can selectively transfer the required energy to the reaction molecules or atoms by the means of dielectric loss of the material themselves. Jones, et al. [5] concluded that microwave heating offers a number of advantages over conventional heating such as: non – contact heating, rapid and selective heating[6], volumetric heating, quick start – up and stopping[7], heating starts from interior of the material[8,9], and higher level of safety and automation[10,11]. Thus, the researchers have done many researches in fields of drying, roasting, oxidation, leaching strengthening of the processes by microwave energy[12-14].

In this research, a new process of chlorine removal from zinc dross by microwave was proposed for solving the problem of recycling the zinc dross. The samples were analyzed by XRD and SEM, and the dielectric characters of ZnO and zinc dross were detected too. Then, the effect of the temperature, duration time and stirring speed on the dechlorination effect of zinc dross was investigated. Finally, in order to find out the reaction mechanism of dechlorination process, the tail gases from roasting process in water vapor atmosphere or in air atmosphere were both analyzed by gas chromatography.

2 Experimental

2.1 Material

The zinc dross used in this investigation was provided by a zinc hydrometallurgical plant in Yunnan province, China. The chemical composition is given in Table 1, and the particle size of zinc dross is < 150μm. The mass fraction of zinc powder is about 5%. Other ingredients are mainly ZnO. The Cl content of the zinc dross is shown as 1.11%. The X – ray diffraction (XRD) pattern of the zinc dross as received is shown in Fig. 1, which demonstrates that ZnO, Zn and Zn_2OCl_2 are the major components in the zinc dross.

Table 1 Chemical Composition of zinc dross

Composition	Cl	Zn	Others
Content (mass fraction)/%	1.11	85.00	13.89

The dielectric constants of zinc dross and zinc oxide were measured by Dielectric kit System of Püschner Microwave Power Systems. The dielectric property (ε', ε'' and $\tan\delta$) of materials is an important physical indicator in the field of microwave chemistry[15,16]. It can largely describe the behavior characteristics of materials in microwave heating process. The result is shown in Table 2.

Table 2 Dielectric constants of zinc dross and zinc oxide

Item	Zinc dross	Zinc oxide	Item	Zinc dross	Zinc oxide
Dielectric constant ε'	2.137	1.222	Loss factor ε''	0.064	0.001

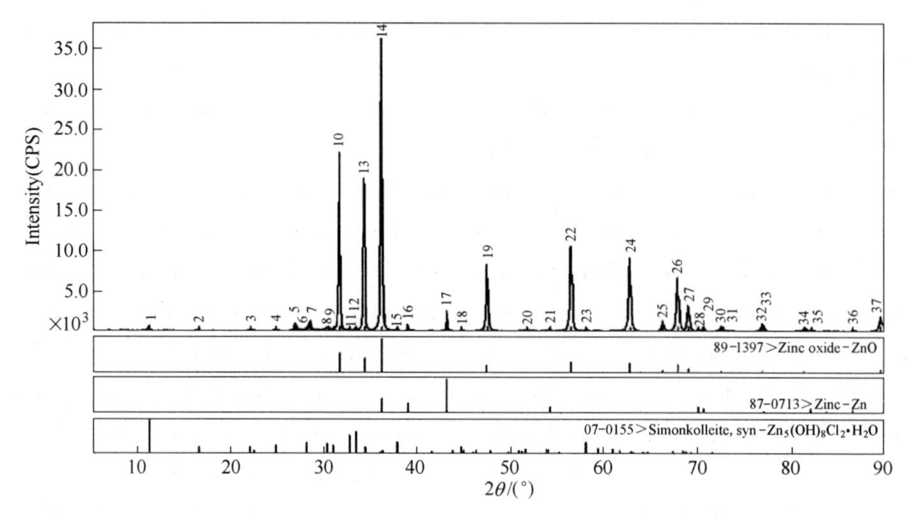

Fig. 1 X - ray diffraction pattern of zinc dross

Table 2 shows that the dielectric constant and loss factor of zinc dross both are much higher than that of zinc oxide. This means that the ability of absorbing microwave of zinc dross is higher than that of zinc oxide[17,18].

2.2 Microwave roasting device

A microwave reactor system(3kW,2450MHz) developed by the Key Laboratory of Unconventional Metallurgy of Ministry of Education, China was employed in this research. The formation of device is shown in Fig. 2.

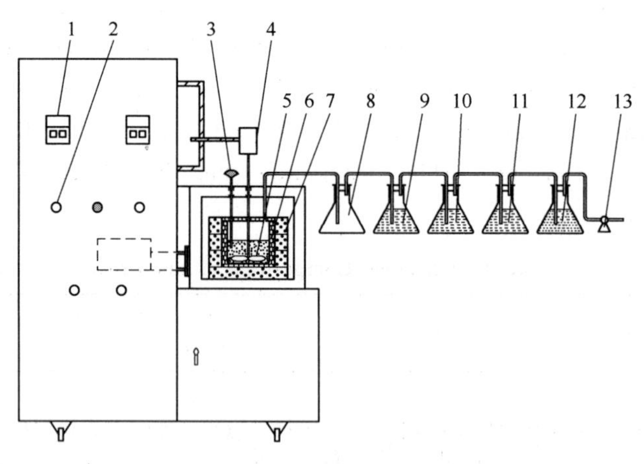

Fig. 2 Formation of microwave high - temperature reactor system

1—instrument display; 2—controller; 3—thermocouple; 4—stirrer; 5—experimental material; 6—crucible; 7—heat preservation material; 8—dust collecting bottle; 9 - 11—gas collecting system; 12—buffer bottle; 13—micro pump

2.3 Procedures

100g zinc dross was placed in clay crucible with heat preservation material around and then subjected to microwave irradiation in aforesaid microwave reactor. The stirring system and smoke absorption system were turned on sequentially, the material was stirred by a 4IK25GN – C type stirrer, and the stirring speed was held at 5 – 25r/min.

During microwave roasting, the smoke was sampled four times (2 times per hour) with a 100mL glass syringe from the first conical flask of the gas absorption system. Then the gas sample was injected into a tinsel gas collecting bag. After microwave treating, the Cl content of samples was analyzed, and the dechlorination rate was calculated.

The whole flow sheet of microwave roasting for declorination is shown in Fig. 3.

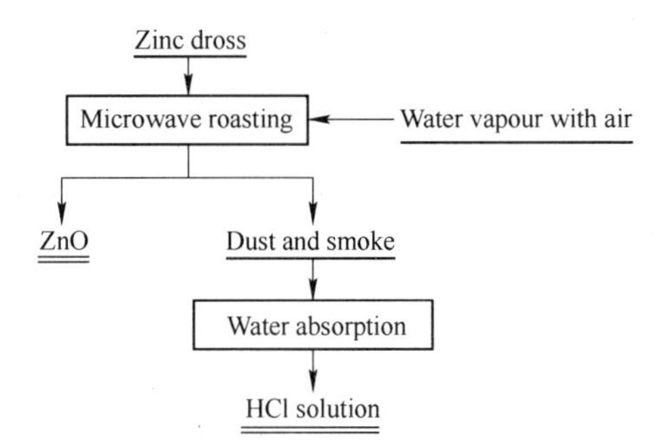

Fig. 3 Flow sheet for microwave treatment and recovering HCl

2.4 Analysis method

The method of analyzing the chloride ion in material is very complicated and difficult. There are many dissolved methods, such as water – boil method, nitric acid dissolved method, alkali dissolved method, mixed acid (HNO_3 and HF) dissolve method[19-21]. The detection methods include ion selective electrode and chlorine – mercury nitric volumetric method. Based on the above experimental comparison, the Cl^- selective electrode was selected as an appropriate test method for its high accuracy, simple procedure and anti – interference. The procedure is as follows: firstly, the samples were weighted and dissolved in diluted nitric acid; then 1mL dissolved solution was sampled and the pH was controlled in 5.5; 5mL of sodium citrate solution was added as the strong ion agent, and finally the potential value was measured by using Cl^- selective electrode and mercury – mercurous sulfate as reference electrode. Then the content of Cl in solid material can be calculated[22].

The morphological changes during microwave roasting process were detected by scanning electron microscopy (SEM, JEOL Ltd., JSM – 6360LV). The dust and smoke generated from roasting process were analyzed by gas chromatography (GC – 2010, Shimadzu Corporation). The gas products can be determined.

3 Results and discussion

3.1 Microwave heating characteristics of zinc dross

In the experiment, 100g zinc dross was placed into the microwave reactor for heating under power of 1.5kW. The heating rate is shown in Fig. 4.

It is seen that heating rate of the zinc dross under microwave irradiation is very fast, because zinc

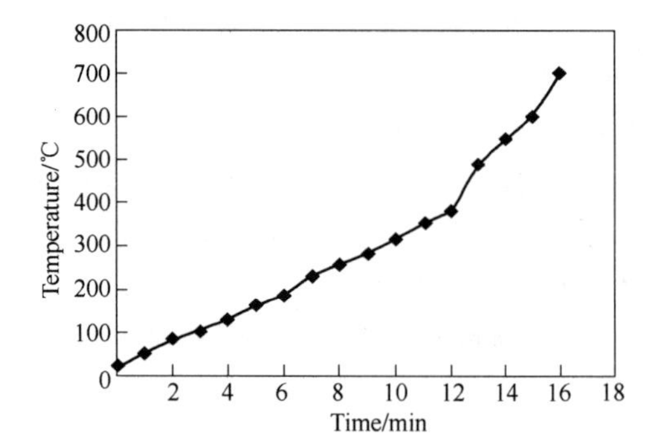

Fig. 4 Heating rate of zinc dross

dross is a hyperactive material in microwave. It has good characteristic of absorbing microwave by a bigger dielectric loss factor(ε''), which indicates the ability of transforming microwave energy into internal energy (heat energy)[23]. The process of heating the material to 700℃ only takes 16min.

3.2 Effect of temperature on roasting

In each run, 100g of zinc dross was placed in a clay crucible and then submitted to microwave irradiation for a definite time at various temperatures. The effect of temperature on the dechlorination efficiency was investigated under microwave power of 1.5kW, and stirring rate of 15r/min. The experimental result is shown in Fig. 5.

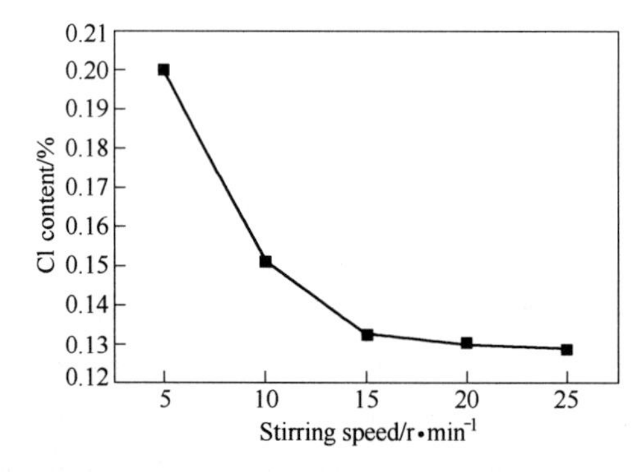

Fig. 5 Relation of temperature and dechlorination rate under different duration time

The oxidative roasting of zinc dross is a complex process. Depending on temperature and atmosphere, Cl^- may be reacted to different products such as Cl_2 and HCl. The gas chromatography pattern of gas samples obtained under atmosphere of water vapor with air is shown in Fig. 6.

It is obvious that the dechloridation rate of zinc dross significantly depends on the roasting temperature of zinc dross. The process of removing chlorine of zinc dross by microwave roasting can be considered as an oxidation reaction. And the permittivity and the dielectric loss factor increase with temperature[24]. So, an increase of temperature could enhance dechloridation reaction obviously. Therefore, the experiment at 700℃ after 80min will result in about 0.13% of the chloride content of zinc dross, while the dechloridation rate reaches to 88.06%. However, samples could be sintered if roasting temperature is too high.

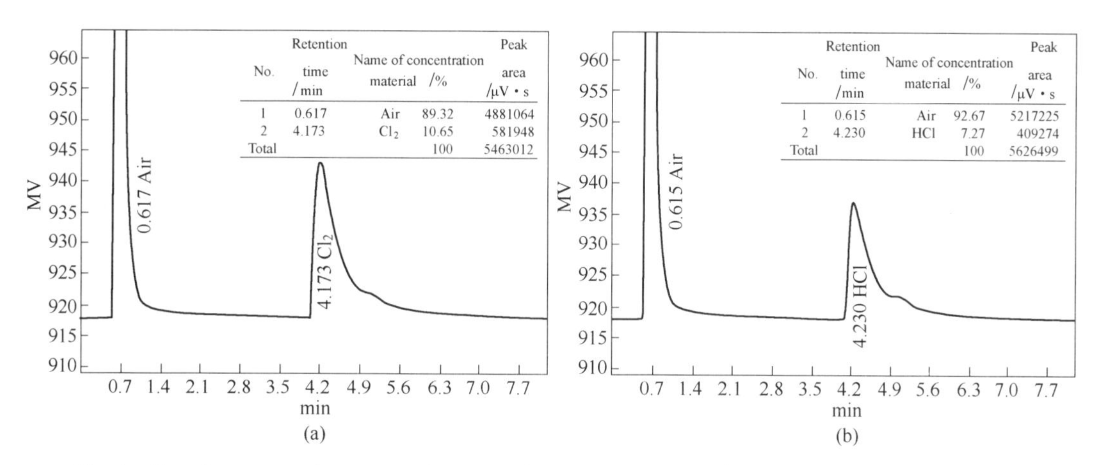

Fig. 6 Gas chromatography pattern of gas sample obtained under air(a) and water vapor with air(b)

From the results of gas chromatography shown in Fig. 6, the tail gas obtained for microwave roasting of zinc dross is mainly $Cl_2(g)$ in air atmosphere. The reaction equation is

$$2ZnCl_2 + O_2 \xlongequal{} 2ZnO + 2Cl_2 \quad (1)$$

While in air atmosphere with water vapor, the tail gas obtained from microwave roasting of zinc dross is mainly HCl(g). The reaction equation is

$$ZnCl_2 + H_2O \xlongequal{} ZnO + 2HCl \quad (2)$$

The HCl gas was generated during the microwave roasting in air atmosphere with water vapor. It's easy to be absorbed into water to generate hydrochloride. Thus, the process not only achieves the goal of dechlorination, but also feasible to recover the removed chlorine into hydrochloride.

The microstructure of zinc dross before and after dechlorination was investigated by SEM and the results are presented in Fig. 7.

Fig. 7 SEM micrographs of zinc dross before(a) and after(b) microwave treatment

As shown in Fig. 7, there are few pore structures on the surface of the original sample, and the morphology is denser. While a lot of pores structure appears obviously on the surface of the sample after microwave roasting. The reason could be attributed to the characteristic of microwave heating. The interior temperature of the sample is higher than the temperature the surface, which lead to chloridion volatilized from interior to superficies while $ZnCl_2$ phase transforms into ZnO phase.

3.3 Effect of duration time on roasting

In this group experiment, 100g zinc dross was employed for each run under the experimental condition with microwave power of 1.5kW for different duration time of 0min, 20min, 40min, 60min and 80min, respectively. When the sample was heated to the target temperature, the timer started. The dechlorination rate is plotted against duration time for various roasting temperatures in Fig. 8.

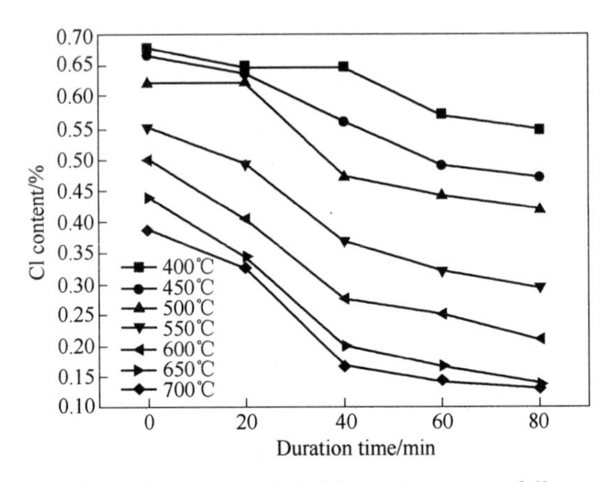

Fig. 8 Relation of duration time and dechlorination rate at different temperature

It can be seen from Fig. 8 that the dechlorination rate increases sharply in early 40min, but gradually slows down after 60min. At 700℃, when the duration time extend from 0min to 60min, the chloride content of zinc dross reduce from 0.39% to 0.14%. The appropriate duration time ensures the sufficient contact of zinc dross and the atmosphere, and makes it almost to the status of dynamical equilibrium when the reaction temperature and the stirring rate were constant.

Subsequently, it is evident that prolonging the duration time had no noticeable effect on the dechlorination. Conversely, it could lead to more volatilization of dust and low rate of zinc recovery. So, the optimal duration time for dechlorination of zinc dross is 60min in this work.

3.4 Effect of stirring rate on roasting

In order to investigate stirring rate on the dechlorination effect of zinc dross, the roasting experiments were conducted with different stirring rate (5r/min, 10r/min, 15r/min, 20r/min and 25r/min) under the microwave power of 1.5kW and the roasting time of 60min.

Fig. 9 indicates that the stirring rate can present a certain effect on the dechlorination rate. With the stirring rate increasing, dechlorination rate slightly improved. The chloride content of zinc dross reduced from 0.20% to 0.13% when the stirring speed increased from 5r/min to 15 r/min.

Due to the viscosity of zinc dross material at high temperatures, high speed stirring can make the stirring paddle provide certain shear force and mechanical activation enhancement to the zinc dross. The dust could be floated up in these experiments when the stirring speed is over 15r/min,

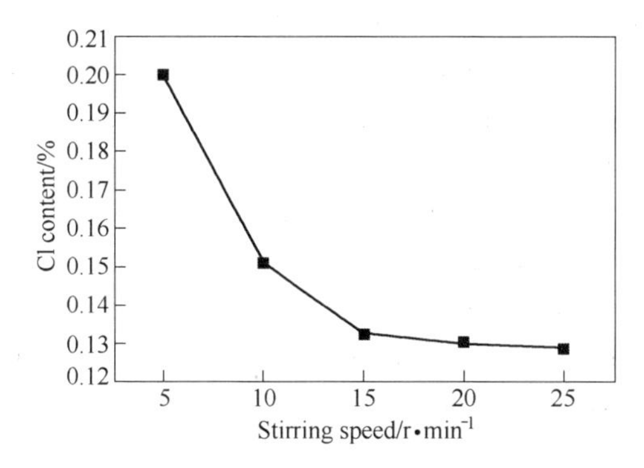

Fig. 9 Effect of stirring rate on the dechlorination of zinc dross at 700℃

which would increase the dust containing the smoking and decrease the zinc recovery. The material and air atmosphere should be contacted fully to obtain a high level of dechlorination efficiency. However, faster stirring speed over 15r/min cannot obviously improve the dechlorination rate. Synthesizes all situations, the optimal stirring speed is 15r/min.

4 Conclusions

(1) According to the dielectric property of zinc dross and zinc oxide, the material is heated rapidly under microwave radiation. The temperature of samples can reach 700℃ in 16min.

(2) The dechlorination is influenced by roasting temperature, duration time and stirring speed. Under optimal condition of roasting temperature of 700℃, duration time of 60min and stirring speed of 15r/min, the dechlorination rate is as high as 88%.

(3) The dechlorination effect of zinc dross from zinc hydrometallurgy by microwave roasting shows advantages such as high Cl removal rate, rapid heating and the chlorine can be recovered into hydrochloride, thus consuming lower energy, which extends a wide industrial prospect.

References

[1] Lu S, Xia Y, Huang C, Wu G, Peng J H, Ju S H, Zhang L B. Removing chlorine of CuCl residue from zinc hydrometallurgy by microwave roasting[J]. Journal of Central South University, 2014, 21: 1290 - 1295.

[2] Paik D J, Hong M H, Huh Y, Park J H, Chae H K, Park S H, Choun S Y. Metastable phases of dross particles formed in a molten zinc bath and prediction of soluble aluminum during galvannealing processes[J]. Metallurgical and Materials Transactions A, 2012, 43(6): 1934 - 1943.

[3] Vodyanitskii Y N, Plekhanova I O, Prokopovich E V, Savichev A T. Soil contamination with emissions of non - ferrous metallurgical plants[J]. Eurasian Soil Science, 2011, 44(2): 217 - 226.

[4] Güresin N, Topkaya Y A. Dechlorination of a zinc dross[J]. Hydrometallurgy, 1998, 49(1): 179 - 187.

[5] Jones D A, Lelyveld T P, Mavrofidis S D, Kingman S W, Miles N J. Microwave heating applications in environmental engineering—a review[J]. Resources, Conservation and Recycling, 2002, 34(2): 75 - 90.

[6] Hidaka H, Saitou A, Honjou H, Hosoda K, Moriya M, Serpone N. Microwave - assisted dechlorination of polychlorobenzenes by hypophosphite anions in aqueous alkaline media in the presence of Pd - loaded active carbon [J]. Journal of Hazardous Materials, 2007, 148(1): 22 - 28.

[7] Li Y, Lei Y, Zhang L B, Peng J H, Li C L. Microwave drying characteristics and kinetics of ilmenite[J]. Trans-

actions of Nonferrous Metals Society of China,2011,21(1):202 - 207.

[8]Xia H Y,Peng J H,Niu H,Huang M Y,Zhang Z Y,Zhang Z B,Huang M. Non - isothermal microwave leaching kinetics and absorption characteristics of primary titanium - rich materials[J]. Transactions of Nonferrous Metals Society of China,2010,20(4):721 - 726.

[9]Guo S H,Li W,Peng J H,Niu H,Huang M Y,Zhang L B,Zhang S M,Huang M. Microwave - absorbing characteristics of mixtures of different carbonaceous reducing agents and oxidized ilmenite[J]. International Journal of Mineral Processing,2009,93(3):289 - 293.

[10]Jou C J G,Hsieh S C,Lee C L,Linb C,Huang H W. Combining zero - valent iron nanoparticles with microwave energy to treat chlorobenzene[J]. Journal of the Taiwan Institute of Chemical Engineers,2010,41(2): 216 - 220.

[11]Foo K Y,Hameed B H. Adsorption characteristics of industrial solid waste derived activated carbon prepared by microwave heating for methylene blue[J]. Fuel Processing Technology,2012,99:103 - 109.

[12]Chang Y F,Zhai X J,Fu Y,Ma L Z,Li B C,Zhang T A. Phase transformation in reductive roasting of laterite ore with microwave heating[J]. Transactions of Nonferrous Metals Society of China,2008,18(4):969 - 973.

[13]Hua Y X,Cai C J,Cui Y. Microwave - enhanced roasting of copper sulfide concentrate in the presence of $CaCO_3$[J]. Separation and Purification Technology,2006,50(1):22 - 29.

[14]Liu X T,Zhao W,Sun K,Zhang G X,Zhao Y. Dechlorination of PCBs in the simulative transformer oil by microwave - hydrothermal reaction with zero - valent iron involved[J]. Chemosphere,2011,82(5):773 - 777.

[15]Li W,Peng J H,Zhang L B,Zhang Z B,Li L,Zhang S M,Guo S H. Pilot - scale extraction of zinc from the spent catalyst of vinyl acetate synthesis by microwave irradiation[J]. Hydrometallurgy,2008,92(1):79 - 85.

[16]Pickles C A. Microwaves in extractive metallurgy: part 1—review of fundamentals[J]. Minerals Engineering, 2009,22(13):1102 - 1111.

[17]Ju S H,Peng J H,Zhang L B. A method of microwave roasting process remove chlorine of CuCl residue from zinc hydrometallurgy: China:201210095568. 7[P]. 2012 - 08 - 01.

[18]Jha M K,Kumar V,Singh R J. Review of hydrometallurgical recovery of zinc from industrial wastes[J]. Resources,Conservation and Recycling,2001,33(1):1 - 22.

[19]Şahin F Ç,Derin B,Yücel O. Chloride removal from zinc ash[J]. Scandinavian Journal of Metallurgy,2000,29(5):224 - 230.

[20]Tsakiridis P E,Oustadakis P,Katsiapi A,Agatzini - Leonardou S. Hydrometallurgical process for zinc recovery from electric arc furnace dust(EAFD). part Ⅱ: downstream processing and zinc recovery by electrowinning [J]. Journal of Hazardous Materials,2010,179(1):8 - 14.

[21]Hossan M R,Dutta P. Analytical solution for temperature distribution in microwave heating of rectangular objects[C]. ASME,2011.

[22]Ren X L,Wei Q F,Hu S R,Wei S J. The recovery of zinc from hot galvanizing slag in an anion - exchange membrane electrolysis reactor[J]. Journal of Hazardous Materials,2010,181(1):908 - 915.

[23]Pickles C A. Microwave heating behaviour of nickeliferous limonitic laterite ores[J]. Minerals Engineering, 2004,17(6):775 - 784.

[24]Haque K E. Microwave energy for mineral treatment processes—a brief review[J]. International Journal of Mineral Processing,1999,57(1):1 - 24.

Removal of Fluorides and Chlorides from Zinc Oxide Fumes by Microwave Sulfating Roasting

Zhiqiang Li, Libo Zhang, Guo Chen, Jinhui Peng, Liexing Zhou, Shaohua Yin, Chenhui Liu

Abstract: Dechlorination and defluorination from zinc oxide dust by microwave sulfating roasting was investigated in this study. According to proposed reactions in the process, detailed experiments were systematically conducted to study the effect of roasting temperature, holding time, air and steam flow rates on the efficiency of the removal of F and Cl. The results show that 92.3% of F and 90.5% of Cl in the fume could be purified when the condition of the roasting temperature of 650℃, holding time at 60min, air flow of 300L/h and steam flow of 8mL/min was optimized. Our investigation indicates that microwave sulfating roasting could be a promising new way for the dechlorination and defluorination from zinc oxide dust.

Keywords: zinc oxide fumes; microwave sulfating roasting; removal of F and Cl

1 Introduction

Zinc is an important base non - ferrous metal required for various applications in metallurgical, chemical and textile industries[1]. It is mainly recovered from primary supplied concentrates[2,3]. However, the zinc concentrate resources decrease rapidly with the increasing consumption of zinc based products[4]. Before the exhaustion of sphalerite and marmatite, the secondary zinc oxide fume resources are applied to obtain the Zn, lead and other value metals based on the environmental and economic benefits[1,5-7].

In practice, the zinc oxide fumes are placed into hydrometallurgical leaching process after pretreatment. Lots of impurities, especially F^- and Cl^-, cause problems like anode corrosion, cathode corrosion and sticking of zinc[8,9], which are introduced into the electrowinning system through the leaching process. F and Cl ion concentrations in electrolytes have to be removed firstly to meet electrolysis requirements ($F < 80mg/L$, $Cl < 100mg/L$) in the zinc electrolysis process[10,11]. There are two conventional processes for the removal of F and Cl, namely pyrometallurgical roasting and caustic washing. The removal efficiency of high fluoride and chloride concentrations in multiple hearth furnaces, rotary kilns, and other roasting approaches is low, while removal via caustic washing results in generation large amounts of waste water, which demand subsequent waste water treatment system[12]. This necessitates development of better processes for dechlorination and defluorination from zinc oxide fume.

Microwave metallurgy is successfully being adopted for microwave drying, microwave - assisted grinding, microwave - assisted reduction, and microwave strengthening leaching, based on the uniqueness of microwave heating, which fully demonstrates microwave metallurgy as a highly efficient, clean, and green metallurgy technology[13-16]. The microwave sulfating roasting was applied to remove F and Cl from zinc oxide fumes, combined with the thermodynamic analysis of volatilization reaction, thermal hydrolysis reaction and sulfate reaction occurred to halides. Systematic experiments of defluorination and dechlorination by microwave sulfating roasting were conducted to investigate the effect of influencing parameters such as roasting temperature, holding time, air flow and steam flow, on the removal efficiency.

2 Materials and methods

2.1 Experimental materials

The zinc oxide fume was received from a fuming furnace smelting process, in Yunnan province in China. The main chemical composition of the zinc oxide dust is listed in Table 1.

Table 1 Chemical composition of zinc oxide dust sample

Composition	Zn	Pb	Ge	Cd	Fe	Sb	S	As	F	Cl	SiO_2	CaO
Content (mass fraction)/%	53.17	22.38	0.048	0.21	0.38	0.23	3.84	1.04	0.0874	0.0783	0.65	0.096

It can be observed that zinc and lead are the main components of the dust, which also presents small amounts of fluorine and chlorine which is harm to zinc electrolysis process. The dust is of extremely high recovery value, but the fluorine and chlorine have to be first removed.

Additionally, the content of S reached 3.84%, which should be handled carefully in the roasting process.

2.2 Experimental set-up and method

A schematic diagram of the microwave roasting system is shown in Fig. 1.

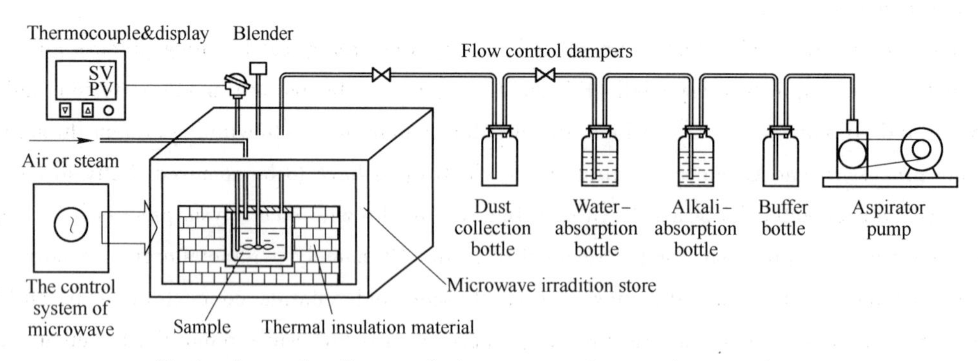

Fig. 1 Connection diagram of microwave roasting experiment equipment

The experimental facilities are made up of microwave reactor, air into system, stirring system and off-gas absorption system. A 3kW (2450MHz) microwave reactor is employed, which could realize continuously adjustable and automatic temperature control. The air into system consists of a Rota meter and a mini air compressor pump, controlling the flow rate of air or water vapor. The stirring speed of the stirring speed ranges from 0r/min to 160r/min to meet the required zinc oxide dust -

stirring intensity. The off - gas absorption system is applied to collect the flue dust and absorb the hazardous impurities, which is made up of a dust collection bottle, one water absorption bottles, an alkali absorption bottle and micro suction pump.

Firstly, 300g of samples was dried, ground and placed into the mullite crucible. The mullite crucible was the diameter of 90mm and the height of 120mm, owning a strong transparent wave performance and a good thermal shock resistance. The crucible was surrounded with heat preservation materials. Then, all of these were then transferred into the microwave reactor. Secondly, the starting of the experiment was marked by activating the microwave, stirring, and off - gas absorption systems. Thirdly, the material was microwave - roasted while air and water vapor passed into the reactor after the material was heated to the set roasting temperature. After a set time roasting, the samples were taken out and cooled in the air atmosphere. Then, the effect of different single factor on F and Cl removal efficiency was investigated.

2.3 Experimental mechanism

In the conventional roasting process, the increase in temperature converts halides into gas phase which are volatile, enabling their removal from the solid matrix. Halide in zinc oxide dust under high - temperature volatilization reaction is calculated using Eq. (1).

Combined with the XRD analysis of Fig. 2, element Zn mainly existed in the ZnO, ZnF_2 and $ZnCl_2$ phase, while Pb mainly existed in the PbO, PbF_2, $PbCl_2$ and PbS phases. At high temperature, PbS reacts with O_2 in air and produces SO_2, which provides the necessary reactant for sulfating reaction. SEM images of zinc oxide fume show in the Fig. 3. There are two different microstructures (white fluorescence structure and gray tiny particles) existed in the zinc oxide fume. To further explore the component differences in the two microstructures, an EDS line scan across the two different submicron - sized micro - region in the zinc oxide fume was shown in Fig. 4. From the EDS line, Zn was aggregated in the gray tiny particles while Pb, F, Cl and S were enriched in the flocculent structure. The aggregation of halides and PbS provides an positive condition for the microwave sulfating roasting.

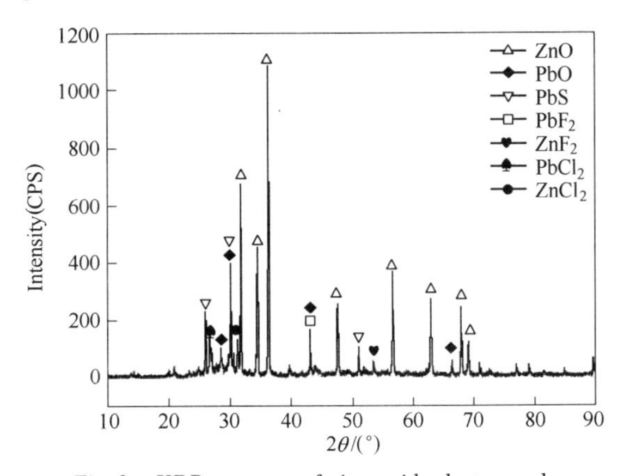

Fig. 2 XRD patterns of zinc oxide dust sample

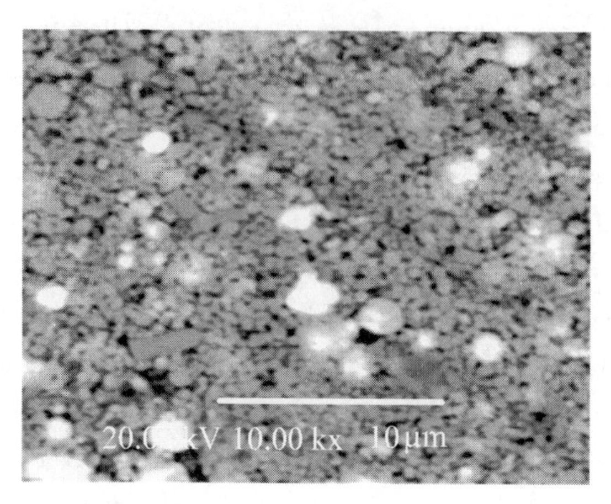

Fig. 3 SEM images of zinc oxide dust sample

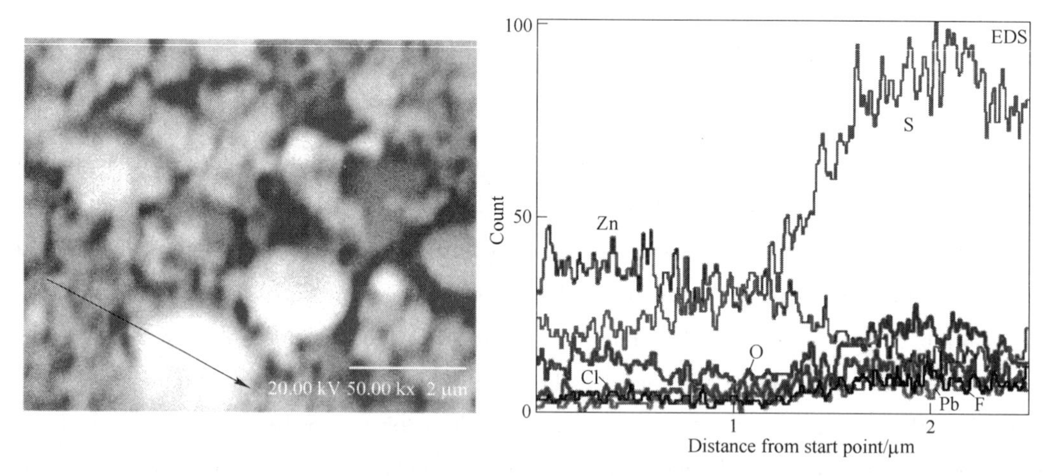

Fig. 4 EDS line scanning results of zinc oxide dust sample (red line)

From above analysis, air and steam were added to microwave roasting process to strengthen the halide thermal hydrolysis reaction (2) and sulfate reaction (3), in order to further reduce the reaction temperature, shorten the reaction time and improve the removal efficiency of F and Cl. The reactions were showed as follows:

$$MeX_2 \Longrightarrow MeX_2(g) \quad (1)$$

$$MeX_2 + H_2O \Longrightarrow MeO + 2HX(g) \quad (2)$$

$$MeX_2 + H_2O + 1/2O_2 + SO_2 \Longrightarrow MeSO_4 + 2HX(g) \quad (3)$$

where Me represents Pb^{2+} and Zn^{2+} and X denotes F^- and Cl^-.

The Gibbs free energies (kJ/mol) and equilibrium constant the of the Pb and Zn halides thermal hydrolysis reaction and the sulfate reaction are shown in Tables 2 and 3 respectively. From Tables 2 and 3, Gibbs free energy of the sulfate reaction is at its minimum and equilibrium constant is at its maximum, which means that sulfate reaction is most easily conducted in thermodynamics. The reaction becomes difficult while the Gibbs free energy increases gradually with the increase in temperature. Thus, the temperature of sulfate roasting does not have to be very high.

Table 2 Gibbs free energies of Pb and Zn halides chemical reactions (kJ/mol)

Matter	Volatile reaction				Thermal hydrolysis reaction				Sulfate reaction			
	600℃	700℃	800℃	900℃	600℃	700℃	800℃	900℃	600℃	700℃	800℃	900℃
$ZnCl_2$	3.76	0.91	-1.86	-4.54	27.44	18.65	18.27	21.38	-62.71	-43.79	-19.23	-9.31
$PbCl_2$	9.33	6.53	3.86	1.32	104.66	97.69	90.95	83.16	-67.90	-48.22	-27.36	-14.72
ZnF_2	26.47	22.61	18.8	15.12	0.33	-12.81	-25.77	-47.33	-89.82	-75.25	-63.27	-49.40
PbF_2	19.52	16.11	12.79	9.79	63.38	54.91	47.10	38.87	-109.28	-91.00	-71.20	-59.01

Table 3 Equilibrium constant of Pb and Zn halides chemical reactions

Matter	Volatile reaction			Thermal hydrolysis reaction			Sulfate reaction		
	600℃	700℃	800℃	600℃	700℃	800℃	600℃	700℃	800℃
$ZnCl_2$	0.121	0.699	2.719	0.023	0.10	0.13	5651.63	224.45	8.63
$PbCl_2$	5.06×10^{-3}	3.67×10^{-2}	0.172	5.46×10^{-7}	5.52×10^{-6}	3.74×10^{-5}	1.16×10^{4}	387.78	21.46
ZnF_2	2.36×10^{-7}	8.35×10^{-6}	1.48×10^{-4}	0.960	4.870	17.980	2.37×10^{5}	1.10×10^{4}	1203.2
PbF_2	1.30×10^{-5}	2.41×10^{-4}	2.48×10^{-3}	1.61×10^{-4}	1.13×10^{-3}	5.09×10^{-3}	3.46×10^{6}	7.68×10^{4}	2.93×10^{3}

The heating curves of 300g zinc oxide fume under microwave power of 1200W and 1800W are plotted in Fig. 5. The zinc oxide fume in microwave field reaches 800℃ within 8min, which provides a well thermodynamic and dynamic condition for removal F and Cl by microwave sulfating roasting.

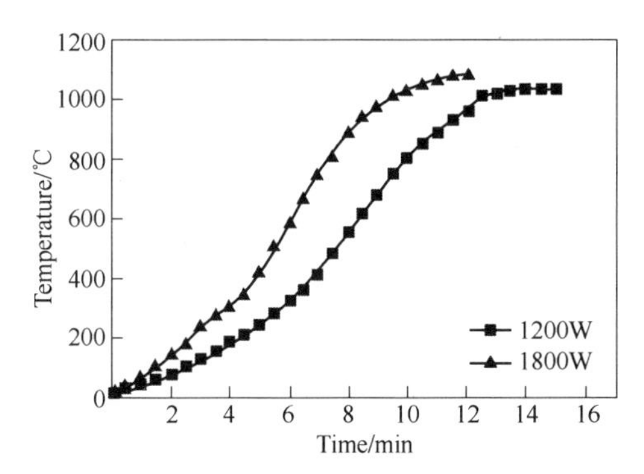

Fig. 5 Heating behavior of zinc oxide fumes by different microwave power

2.4 Calculation of F and Cl removal efficiency in zinc oxide dust

After the roasted samples were cooled to room temperature, fluorine ion selective electrode and silver chloride turbidimetric method[17] was used for determination of the fluorine and chlorine content. The raw materials were measured at 166.3mg/L fluorine and 145.4mg/L chlorine.

The fluorine and chlorine removal efficiency were mathematically expressed as,

$$\eta = \frac{M - M'}{M} \times 100\% \tag{4}$$

where M represents initial fluorine and chlorine content in zinc oxide dust; M' denotes F and Cl content in zinc oxide dust after roasting; η represents F and Cl removal efficiency.

3 Results and discussion

3.1 Single factor experiments

3.1.1 Effect of air and steam flow

Raw material containing 0.087% F and 0.078% Cl was used in the experiments on removal F and Cl via microwave roasting. Roasting temperature and time were constant at 650℃ and 60min, respectively. The samples were fully stirred at a speed of 120r/min. The results under different air flow rates(200L/h to 400L/h) are shown in Fig. 6. F and Cl removal efficiency increased gradually from 74.73% and 68.50% at 200L/h to 76.55% and 70.00% at 300L/h; the increase from 76.55% and 70.00% at 300L/h to 79.63% and 73.50% at 400L/h was more significant. However, the air flow rate should not be excessively high because the increases in air volume cause strong system heat dissipation, and the required temperature must be maintained. The optimum air flow velocity was 300L/h, as observed in the following experiment. The purpose of this experiment is to reduce energy consumption.

From Fig. 6, the microwave roasting in air atmosphere keeps a low removal efficiency of F and Cl. The former mechanism analysis shows that the microwave roasting in air atmosphere is mainly based on the volatile reaction, thanks to very little water in the air. According to the analysis of Gibbs free energy and equilibrium constant, the sulfate reaction is much easier to conduct than the volatile reaction. The water steam will be carried by the air to strengthen the thermal hydrolysis reaction and sulfate reaction, for the purpose of improving the removal efficiency.

Kept other roasting conditions in the same and let the water steam be carried by the 300L/h air flow. F and Cl removal efficiency at different steam flow rates(2mL/min to 12mL/min) are plotted in Fig. 7. Water vapor is observed to have a significant positive effect on the F and Cl removal rate. F and Cl removal rate increased respectively from 80.00% and 73.80% at 2mL/min to 92.3% and 90.07% at 8mL/min. When steam flow passes 8mL/min, the removal rate increase slowly. Besides, controlling the steam flow rate will reduce energy consumption.

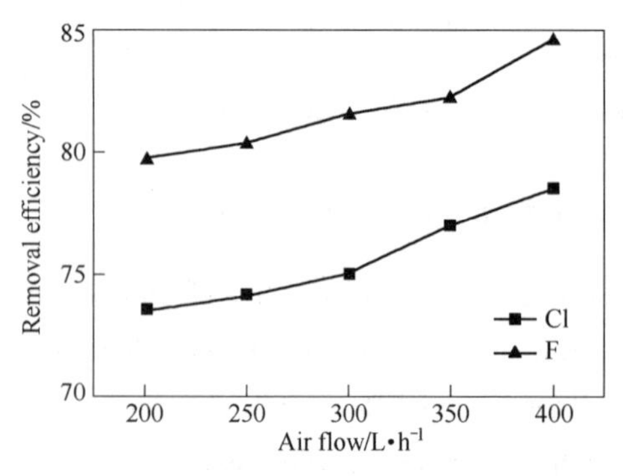

Fig. 6 The effect of removal F and Cl efficiency by air flow

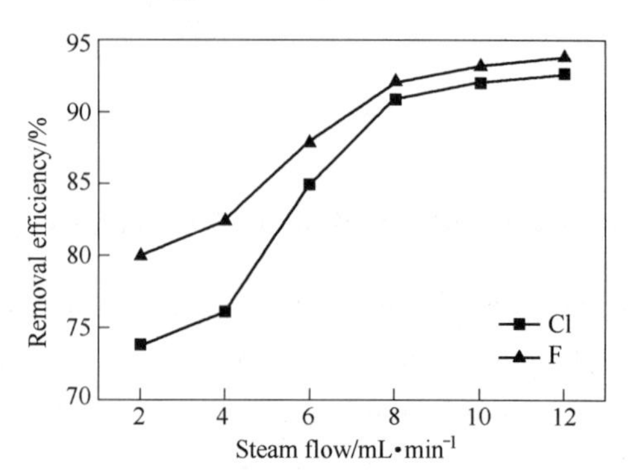

Fig. 7 The effect of removal F and Cl efficiency by steam flow

3.1.2 Effect of roasting temperature

300g of zinc oxide dust was fully stirred quickly for 1h at 120r/min under different temperatures. In the meantime, the water steam of 8mL/min was blown into the microwave reactor which was carried by an air flow of 300L/h. The results plotted in Fig. 8 indicate that the roasting temperature owns a significant effect on F and Cl removal efficiency, which increases rapidly from 65.75% and 57.5% at 500℃ to 92.27% and 90.50% at 650℃. The former mechanism analysis shows that the most favorable temperature of sulfate reaction is the temperature range from 500℃ to 650℃. Within the temperature range of 500℃ to 650℃, the free Gibbs energy change of sulfate reaction is less than 0, and the reaction can be carried out smoothly. When the temperature gets over 650℃, the removal efficiency increases very slowly. The equilibrium constant of sulfate reaction increasing with the temperature increasing (Table 3) indicates that the sulfate reaction is exothermic. It proves that the sulfate reaction does not need a very high temperature, which meets the former mechanism analysis. The microwave sulfating roasting can realize a very high F and Cl removal efficiency in a relatively low temperature.

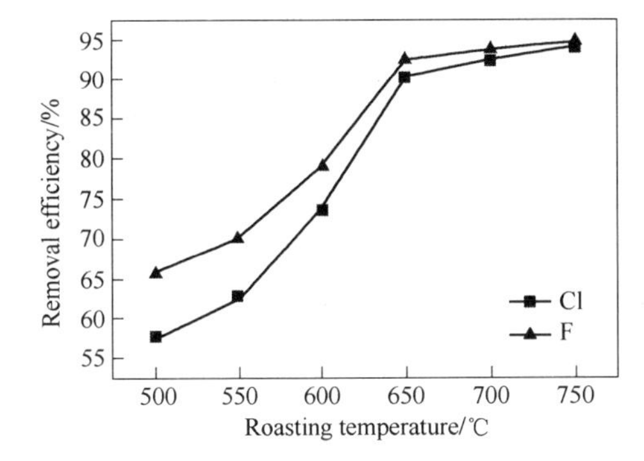

Fig. 8 The effect of removal F and Cl efficiency by roasting temperature

3.1.3 Effect of holding time

The removal of F and Cl from zinc oxide dust was examined with respect to time when an 8mL/min water steam was carried by air flow of 300 L/h, temperature was 650℃, and the samples were fully stirred at a speed of 120r/min. The results in Fig. 9 indicate that the removal efficiency of F and Cl significantly increases from 58.75% and 50.00% to 92.30% and 90.50% when the holding time

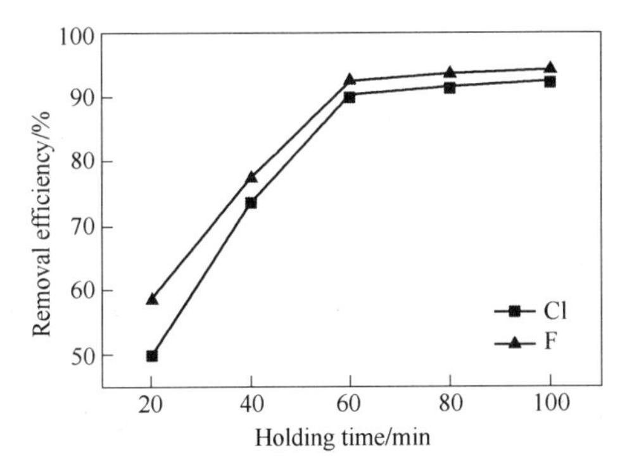

Fig. 9 The effect of removal F and Cl efficiency by holding time

ranges from 20min to 60min. The reactions are carried out more thoroughly with time. The rate increases slowly when holding time passes over 60min, which indicates that most F and Cl in the zinc oxide dust have been removed.

3.2 Parallel experiments

The results of single factor experiments indicate that F and Cl removal efficiency could reach a very high level, when a 8mL/min water steam was carried by air flow of 300L/h, the roasting temperature is 650℃, holding time is of 60min and stirring speed is at 120r/min. The results of three parallel experiments are plotted in Table 4. When the steam and air flow were respective 8mL/min and 300L/h, the roasting temperature is 650℃, holding time is of 60min and stirring speed is at 120r/min, F and Cl removal rate reach 92.3% and 90.5%. Besides, 12.8mg/L fluorine content and 13.8mg/L chlorine content in the roasted samples could satisfy the electrolysis requirements. XRD patterns of the roasted sample comparing with the raw sample were shown in Fig. 10. The peaks of halides in the raw sample were disappeared in the roasted sample, indicating the efficiency of the microwave sulfating roasting.

Table 4 Results of parallel experiments

Roasting temperature/℃	Holding time /min	Air flow /L·h^{-1}	Steam flow /mL·min^{-1}	Fluorine		Chlorine	
				Removal rate of F/%	F^- content /mg·L^{-1}	Removal rate of Cl/%	Cl^- content /mg·L^{-1}
650	60	300	8	92.3	12.8	90.5	13.8

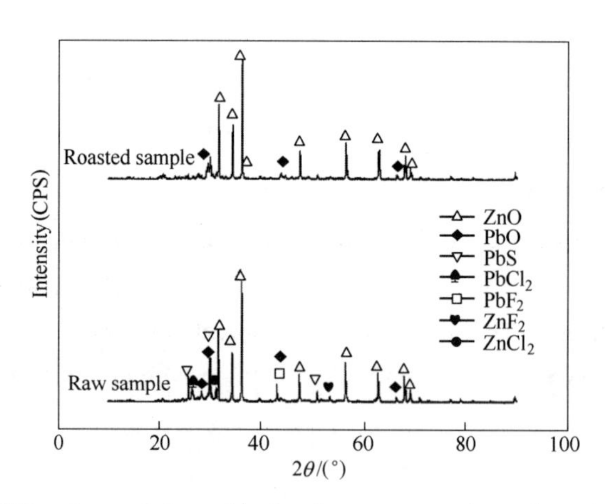

Fig. 10 XRD patterns of zinc oxide dust between roasted sample and raw sample

4 Conclusions

A new and effective process of defluorination and dechlorination by microwave sulfating roasting was investigated in details. The following conclusions are drawn:

(1) The results obtained in the present laboratory study show that zinc oxide dust received from fuming furnace could be treated effectively via microwave sulfating roasting, which carves out a new

and effective way for the defluorination and dechlorination.

(2) A nearly 92.3% defluorination efficiency and 90.5% dechlorination efficiency of zinc oxide dust could be achieved under the roasting temperature of 650℃, holding time of 60min, air flow of 300L/h and steam flow of 8mL/min, which would satisfy the requirements of the wet smelting electrolysis process.

(3) Compared with the microwave roasting in air atmosphere, the microwave sulfating roasting achieves better removal efficiency in a lower reaction temperature and a shorter holding time.

Acknowledgements

The authors are grateful for the financial support from the National Natural Science Foundation (No. 51104073), the National Technology Research and Development Program of China (No. 2013AA064003), and the Yunnan Province Young Academic Technology Leader Reserve Talents (2012HB008).

References

[1] Jha M K, Kumar V, Singh R J. Review of hydrometallurgical recovery of zinc from industrial wastes[J]. Resources, Conservation and Recycling, 2001, 33(1): 1 - 22.

[2] Pecina T, Franco T, Castillo P, Orrantia E. Leaching of a zinc concentrate in H_2SO_4 solutions containing H_2O_2 and complexing agents[J]. Minerals Engineering, 2008, 21(1): 23 - 30.

[3] Moradi S, Monhemius A J. Mixed sulphide - oxide lead and zinc ores: problems and solutions[J]. Minerals Engineering, 2011, 24(10): 1062 - 1076.

[4] Vahidi E, Rashchi F, Moradkhani D. Recovery of zinc from an industrial zinc leach residue by solvent extraction using D2EHPA[J]. Minerals Engineering, 2009, 22(2): 204 - 206.

[5] Dutra A J B, Paiva P R P, Tavares L M. Alkaline leaching of zinc from electric arc furnace steel dust[J]. Minerals Engineering, 2006, 19(5): 478 - 485.

[6] Peng N, Peng B, Chai L Y, et al. Recovery of iron from zinc calcines by reduction roasting and magnetic separation[J]. Minerals Engineering, 2012, 35: 57 - 60.

[7] Alfantazi A M, Moskalyk R R. Processing of indium: a review [J]. Minerals Engineering, 2003, 16(8): 687 - 694.

[8] Lashgari M, Hosseini F. Lead - Silver Anode Degradation during zinc electrorecovery process: chloride effect and localized damage[J]. Journal of Chemistry, 2013: 1 - 5.

[9] Güresin N, Topkaya Y A. Dechlorination of a zinc dross[J]. Hydrometallurgy, 1998, 49(1): 179 - 187.

[10] Şahin F Ç, Derin B, Yücel O. Chloride removal from zinc ash[J]. Scandinavian Journal of Metallurgy, 2000, 29(5): 224 - 230.

[11] Lan Y Z, Zhao Q R, Smith R W. Recovery of zinc from high fluorine bearing zinc oxide ore[J]. Mineral Processing and Extractive Metallurgy, 2006, 115(2): 117 - 119.

[12] Chen W S, Shen Y H, Tsai M S, et al. Removal of chloride from electric arc furnace dust[J]. Journal of Hazardous Materials, 2011, 190(1): 639 - 644.

[13] Pickles C A. Microwaves in extractive metallurgy: part 1—review of fundamentals[J]. Minerals Engineering, 2009, 22(13): 1102 - 1111.

[14] Pickles C A. Microwaves in extractive metallurgy: part 2—a review of applications[J]. Minerals Engineering,

2009,22(13):1112 - 1118.

[15] Nanthakumar B, Pickles C A, Kelebek S. Microwave pretreatment of a double refractory gold ore[J]. Minerals Engineering, 2007, 20(11):1109 - 1119.

[16] Haque K E. Microwave energy for mineral treatment processes—a brief review[J]. International Journal of Mineral Processing, 1999, 57(1):1 - 24.

[17] Zenki M, Iwadou Y. Repetitive determination of chloride using the circulation of the reagent solution in closed flow - through system[J]. Talanta, 2002, 58(6):1055 - 1061.

An External Cloak with Arbitrary Cross Section Based on Complementary Medium and Coordinate Transformation

Chengfu Yang, Jingjing Yang, Ming Huang, Zhe Xiao, Jinhui Peng

Abstract: Electromagnetic cloak is a device which makes an object "invisible" for electromagnetic irradiation in a certain frequency range. Material parameters for the complementary medium – assisted external cylindrical cloak with arbitrary cross section are derived based on combining the concepts of complementary media and transformation optics. It can make the object with arbitrary shape outside the cloaking domain invisible, as long as an "antiobject" is embedded in the complementary media layer. Moreover, we find that the shape, size and the position of the "antiobject" is dependent on the contour of the cloak and the coordinate transformation. The external cloaking effect has been verified by full – wave simulation.

Keywords: electromagnetic cloak; arbitrary cross section; complementary media; transformation optics

1 Introduction

Control of electromagnetic wave with metamaterials is of great topical interest, and is fuelled by rapid progress in electromagnetic cloaks[1-6]. Recent proposals for electromagnetic cloaking techniques include plasmonic cloaking due to scattering cancellation[7,8], transformation based cloaking[1,2], active cloaking[9], broadband exterior cloaking[10], transmission – line based cloaking[11], cloaking due to anomalous resonance[12,13], and so on. The scattering cancellation technique can be achieved for example by cancelling radiation from the induced dipole moments of the scatter by introducing another object, in which dipole moments of the opposite direction are induced. Transformation based cloaking techniques[14] rely on the transformation of coordinates, e. g., a point in the electromagnetic space is transformed into a special volume in the physical space, thus leading to the creation of the volume where electromagnetic fields do not exist, but are instead guided around this volume. The active cloaking uses sensors and active sources near the surface of the region, and could operate over broad bandwidths. Broadband exterior cloaking is based on three or more active devices. The devices, while not radiating significantly, create a "quiet zone" between the devices where the wave amplitude is small. Objects placed within this region are virtually invisible. Transmission – line technique[14] is based on the use of volumetric structures composed of two – dimensional or three – dimensional transmission – line networks. In these structures, the electromagnetic fields propagate inside transmission lines, thus leaving the volume between these lines effectively cloaked. Cloaking by anomalous resonance enables dielectric bodies of finite size to be perfectly cloaked by certain cylindrical arrangements of materials of positive and negative permittivities known as superlenses, but apparently not larger objects[15].

More recently, Lai, et al.[16] proposed a new recipe for an invisibility cloak, which is based on complementary media, composed of a dielectric core and an "antiobject" embedded inside a negative index shell. It can cloak an object with a prespecified shape and size within a certain distance outside the shell. In the foregoing investigations, however, the numerical simulations and parameter designs are devoted to circularly cylindrical invisibility cloak, which are cloaks with rotational symmetry. Toward the practical and flexible realizations of the electromagnetic cloaks, we present the general material parameters for the cylindrical complementary medium - assisted cloak with arbitrary cross sections, and validate them by numerical simulation. We show that the material parameters developed in this paper can be also specialized to the complementary medium - assisted cloak with regular shapes, such as circular, elliptical and square, which represents an important progress towards the realization of the cloak with arbitrary cross sections, based on complementary medium. Meanwhile, we compared the performance of the external cloaks based on linear and non - linear transformation, and some interesting phenomena are found.

2 Theoretical model

Combining the concepts of complementary media and transformation optics, material parameters for the 2D cloak with arbitrary geometries are derived. The schematic diagram of the space transformation is shown in Fig. 1, where three cylinders enclosed by contours $aR(\theta)$, $bR(\theta)$ and $cR(\theta)$ divide the space into three regions, i. e., the core material layer ($r'' < aR(\theta)$), the complementary layer ($aR(\theta) < r' < bR(\theta)$) and the outer air layer ($bR(\theta) < r' < cR(\theta)$). Here, $R(\theta)$ is an arbitrary continuous function with period 2π[17]. According to the coordinate transformation method, the permittivity $\varepsilon^{i'j'}$ and permeability $\mu^{i'j'}$ tensors of the transformation media can be written as[18,19]

$$\varepsilon^{i'j'} = \Lambda_i^{i'}\Lambda_j^{j'} |\det(\Lambda_i^{i'})|^{-1}\varepsilon^{ij}, \mu^{i'j'} = \Lambda_i^{i'}\Lambda_j^{j'} |\det(\Lambda_i^{i'})|^{-1}\mu^{ij} \quad (1)$$

where $\Lambda_i^{i'}$ is the Jacobian transformation matrix. It is just the derivative of the transformed coordinates with respect to the original coordinates. $|\det(\Lambda_i^{i'})|$ is the determinant of the matrix. ε^{ij} and μ^{ij} are the permittivity and permeability of the original space, respectively.

The complementary media is obtained by the coordinate transformation of folding the air layer into the complementary layer with the linear coordinate transformation of

$$r' = k_1 r + k_2 R(\theta), \theta' = \theta, z' = z \quad (2)$$

where $k_1 = (c - b)/(a - b)$, $k_2 = (a - c)b/(a - b)$.

And then, the Jacobian transformation matrix and its determinant can be obtained as

$$\Lambda_i^{i'} = [a_1, a_2, 0; b_1, b_2, 0; 0, 0, 1] \quad (3)$$

$$\det(\Lambda_i^{i'}) = a_1 b_2 - a_2 b_1 \quad (4)$$

where

$$a_1 = k_1 + k_2 R(\theta) y^2/r^3 - k_2 R'(\theta) xy/r^3, a_2 = -k_2 R(\theta) xy/r^3 + k_2 R'(\theta) x^2/r^3$$

$$b_1 = -k_2 R(\theta) xy/r^3 - k_2 R'(\theta) y^2/r^3, b_2 = k_1 + k_2 R(\theta) x^2/r^3 - k_2 R'(\theta) xy/r^3$$

$$R'(\theta) = \frac{d[R(\theta)]}{d\theta}$$

Substituting Eq. (3) and Eq. (4) into Eq. (1), we can obtained the permittivity and permeability tensors for the complementary layer as

$$\varepsilon^{i'j'} = \mu^{i'j'} \begin{bmatrix} (a_1^2 + a_2^2)/(a_1 b_2 - a_2 b_1) & (a_1 b_1 - a_2 b_2)/(a_1 b_2 - a_2 b_1) & 0 \\ (a_1 b_1 + a_2 b_2)/(a_1 b_2 - a_2 b_1) & (b_1^2 + b_2^2)/(a_1 b_2 - a_2 b_1) & 0 \\ 0 & 0 & 1/(a_1 b_2 - a_2 b_1) \end{bmatrix} \tag{5}$$

The core material is obtained by the coordinate transformation of compressing a large circle of air with contour $cR(\theta)$ into a small circle with contour $aR(\theta)$, which is formed by the coordinate transformation of

$$r'' = ar/c, \theta'' = \theta, z'' = z$$

And then, we can obtain the Jacobian transformation matrix and its determinant as

$$\Lambda_i^{i'} = [a/c, 0, 0; b_1, b_2, 0; 0, 0, 1] \tag{6}$$

$$\det(\Lambda_i^{i'}) = a^2/c^2 \tag{7}$$

Substituting Eq. (6) and Eq. (7) into Eq. (1), we can obtained the permittivity and permeability tensors for the core material as

$$\varepsilon^{i'j'} = \mu^{i'j'} = [1, 0, 0; 0, 1, 0; 0, 0, 1] \tag{8}$$

Suppose that an object of permittivity ε_0 and permeability μ_0 is located in the outer air layer. In order to make it invisible, we need to add an "antiobject" with parameters $\varepsilon'_0 = \varepsilon_0 \varepsilon^{i'j'}$ and $\mu'_0 = \mu_0 \mu^{i'j'}$ which optically cancel the object of ε_0 and μ_0, as shown in Fig. 1(b). It should be noted that the "antiobject" is mapped into the complementary layer according to Eq. (2), therefore, its image varies with the contour equation $R(\theta)$. The cloak is composed of the modified complementary layer embedded with the "antiobject" and the core material. Eqs. (5) and (8) give the general expressions of material parameters for the complementary medium - assisted cloak with arbitrary geometries. For special cases such as circular, elliptical and square cloaks, the contour equation $R(\theta)$ can be simplified by the procedure illustrated in[17] to obtain the corresponding material parameters. It means that the material parameters derived in this paper can be specialized to the formally designed complementary medium - assisted cloaks. It will be confirmed by full - wave simulation based on finite element software COMSOL Multiphysics in the next section.

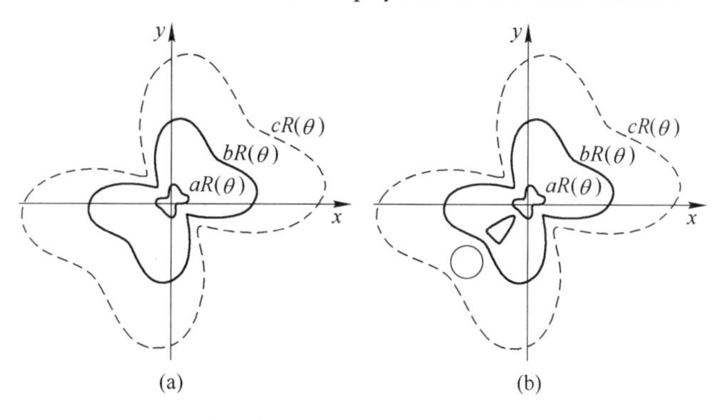

Fig. 1 The system composed of air layer($bR(\theta) < r' < cR(\theta)$), the complementary media layer ($aR(\theta) < r' < bR(\theta)$) and the core material layer($r'' < aR(\theta)$) that is optically equal to a large circle of air($r < cR(\theta)$)(a) and a scheme to cloak an object of ε_0, μ_0 by placing the "antiobject" of ε'_0, μ'_0 in the complementary media layer(b)

3 Simulation results and discussion

First we demonstrate the scheme shown in Fig. 1(a), i. e., the air layer($bR(\theta) < r < cR(\theta)$) and the complementary media layer($aR(\theta) < r' < bR(\theta)$) that is optically equal to a large circle of air ($r < cR(\theta)$). Geometry parameters used in the simulation is chosen $R(\theta) = \cos(4\theta) + \sin(2\theta) + 3$, $a = 0.1$, $b = 0.5$ and $c = 0.9$. We consider the case of cylindrical wave irradiation, of which the wavelength is $\lambda = 1$ unit. Fig. 2 shows the electric field distribution in the vicinity of the transformation region composed of a core material and a complementary medial layer. The line source with a current of 1A/m is located at(−3, −3). The absence of scattered waves clearly verifies the invisibility of the whole system.

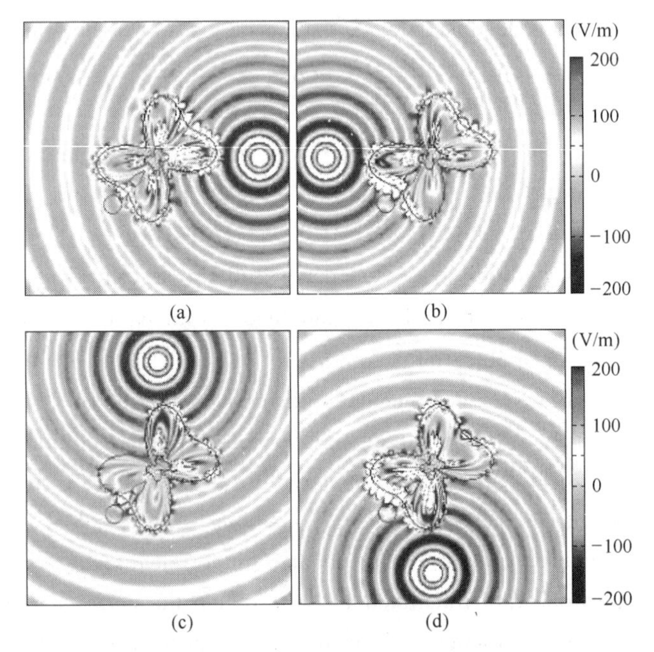

Fig. 2 The electric field (E_z) distributions in the vicinity of the core material($r'' < aR(\theta)$) and the complementary layer($aR(\theta) < r' < bR(\theta)$) under cylindrical wave irradiation

Next we demonstrate the scheme shown in Fig. 1(b), i. e., the cloaking of an object by placing its "antiobject" in the complementary layer. The dielectric object with radius $r = 0.3\lambda$, parameters $\varepsilon_0 = 2$, $\mu_0 = 1$ is centered at(−1.5, −1.5), as shown in Fig. 3(a), which also shows its scattering pattern under cylindrical wave irradiation. In order to make it invisible, we include an "antiobject" with parameters $\varepsilon_0' = 2\varepsilon^{i'j'}$ and $\mu_0' = 2\mu^{i'j'}$ into the complementary layer. The image of the "antiobject" is obtained according to the linear transformation of Eq. (2). The calculated electric field shown in Fig. 3(b) clearly demonstrates the "external" cloaking effect. It is worth noting that the object to be cloaked is placed outside the cloaking shell, and the cloaking effect comes from its "antiobject" embedded in the complementary media. In the space closely adjacent to the cloaked object, the cloaking effect doesn't exist, and the pressure fields are very strong due to the surface mode resonance induced by the multiple scattering of acoustic wave between object and the cloak device. We emphasize that there is no shape or size constraint on the object to be cloaked, as long as it fits into the region bounded by $r = bR(\theta)$ and $r = cR(\theta)$. In Fig. 3(c) – (f), we show the cloaking scheme of the two curved shell. In this case, the image of the "antiobject" is also a curved shell according Eq. (2). Fig. 3(c) is the scattering pattern of the dielectric shell of $\varepsilon_0 = 2, \mu_0 = 1$, which is fitted into the region bounded between $0.52R(\theta) < r < 0.58R(\theta)$. In Fig. 3(d) the dielectric shell is hidden by the cloak with an "antiobject" located between the contours of $r' = 0.42R(\theta)$ and $r' = 0.48R(\theta)$ in the complementary layer. The cylindrical wave pattern in Fig. 3(d) manifests the clocking effects. Next we consider the circular shell with parameters $\varepsilon_0 = -1$, $\mu_0 = 1$. The scattering pattern for such a shell shown in Fig. 3(e) is similar to that of metal shell. In such a case, the "antiobject" of the shell with parameters $\varepsilon_0' = -\varepsilon^{i'j'}, \mu_0' = \mu^{i'j'}$ is located in the complementary layer between the contours of $r' = 0.42R(\theta)$ and $r' = 0.48R(\theta)$. The electric field distribution in the vicinity of the cloak is shown in Fig. 3(f). Again, the cylindrical wave pattern manifests the cloaking effect.

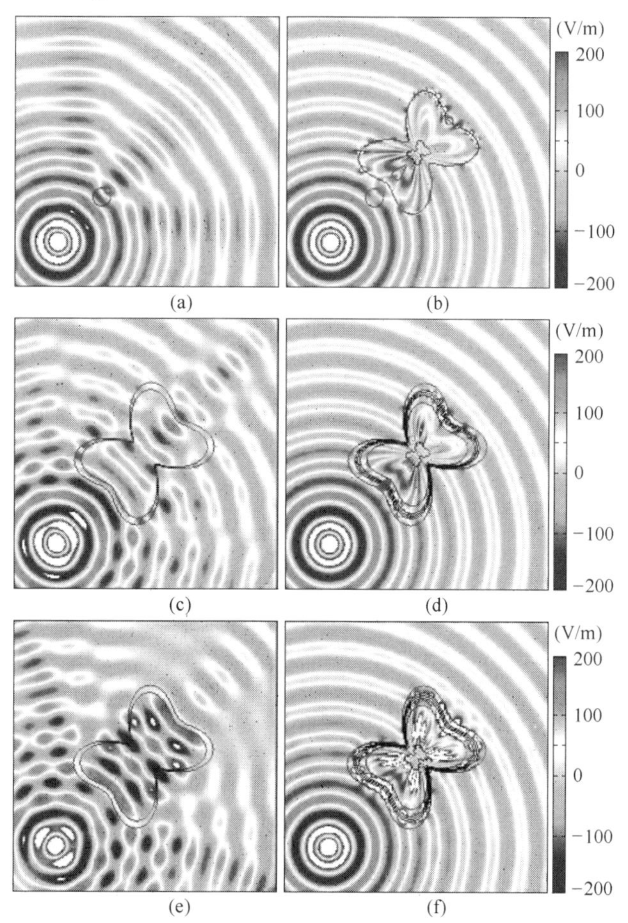

Fig. 3 Electric field distributions under cylindrical wave irradiation

(a) The circular dielectric object with $r = 0.3$ is centered at $(-1.5, -1.5)$;
(b) The object in (a) is hidden by the cloak with arbitrary shape; (c) The circular dielectric shell of $\varepsilon_0 = 2, \mu_0 = 1$;
(d) The shell in (c) is hidden by the cloak with embedded "antiobject" shell of $\varepsilon_0' = 2\varepsilon^{i'j'}, \mu_0' = \mu^{i'j'}$;
(e) The shell with parameters of $\varepsilon_0 = -1, \mu_0 = 1$;
(f) The shell in (e) is hidden by the cloak with embedded "antiobject" shell of $\varepsilon_0' = -\varepsilon^{i'j'}, \mu_0' = \mu^{i'j'}$

To investigate the interaction of the cloak with electromagnetic wave from different orientations, the current line source is located at four different positions in the computational domain, and the electric field distributions are simulated, as shown in Fig. 4. It can be clearly seen that the cylindrical waves are restored to the original wave fronts when passing through the cloak, and the circular dielectric object is perfectly hidden independent on the orientation of the incident electromagnetic wave.

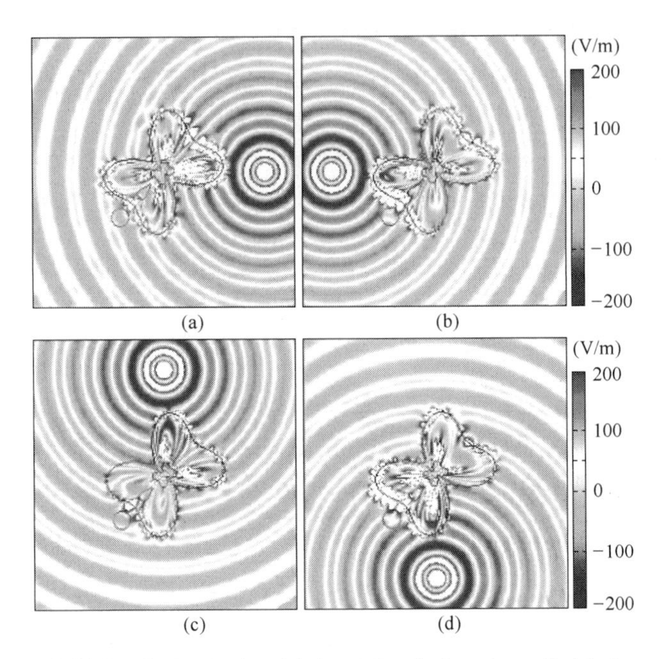

Fig. 4 Electric field distribution in the vicinity of the cloak under cylindrical wave irradiation
The line source is located at(3,0),(−3,0),(0,3) and(0,−3) for(a),(b),(c) and(d)

Fig. 5 shows the electric field distribution in the computation domain under TE wave irradiation. The incident TE wave with $\lambda = 1$ unit is from left to right. In Fig. 5(a), the absence of scattered waves clearly verifies the invisibility of the system composed of core material and the complementary layer. Fig. 5(b) shows the cloaking of the dielectric circular object. Fig. 5(c) and(d) shows the cloaking of the shell with $\varepsilon_0 = 2$ and $\varepsilon_0 = -1$, respectively. Although the incident TE waves are distorted in the transformation region, they restore their original wave fronts when passing through, and the cloaking effect is independent on the type of the exciting source.

Since artificial metamaterials are always lossy in real applications, it does make sense to investigate the effects of loss on the performance of the cloak. Electric field distributions of the cloaks with electric and magnetic – loss tangents($\tan\delta$) of 10^{-4}, 0.01, and 0.1 are displayed in panels (a), (b) and (c) of Fig. 6. In the case of $\tan\delta = 10^{-4}$ and $\tan\delta = 0.01$, the performance of the cloak is basically undisturbed, as shown in Fig. 6(a) and(b). In such cases, the effects of loss can be ignored. But when the loss tangent of the metamaterials is 0.1 or more than that, it deteriorates the performance of the cloak mainly in the transformation region and the forward – scattering region of the near field, as shown in Fig. 6(c).

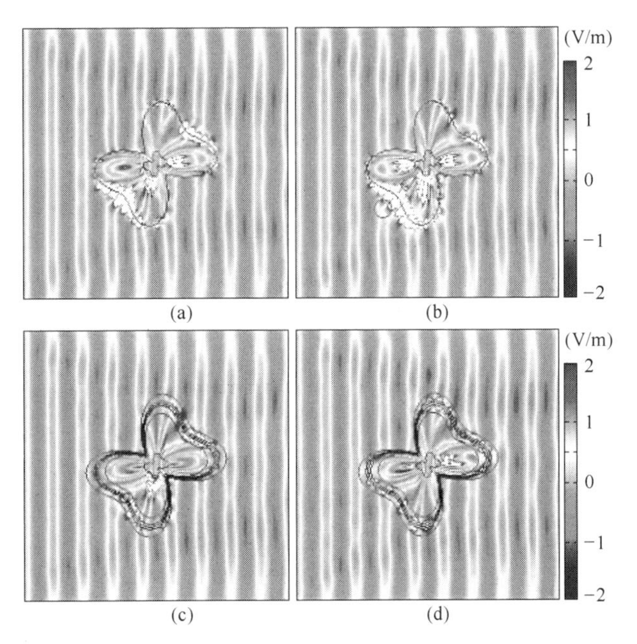

Fig. 5 Electric field distribution in the computation domain under TE wave irradiation

(a) The electric field distribution in the vicinity of the system composed of core material and the complementary layer; (b) The cloaking of the dielectric circular dielectric object; (c) The cloaking of the shell of $\varepsilon_0 = 2, \mu_0 = 1$; (d) The cloaking of the shell of $\varepsilon_0 = -1, \mu_0 = 1$

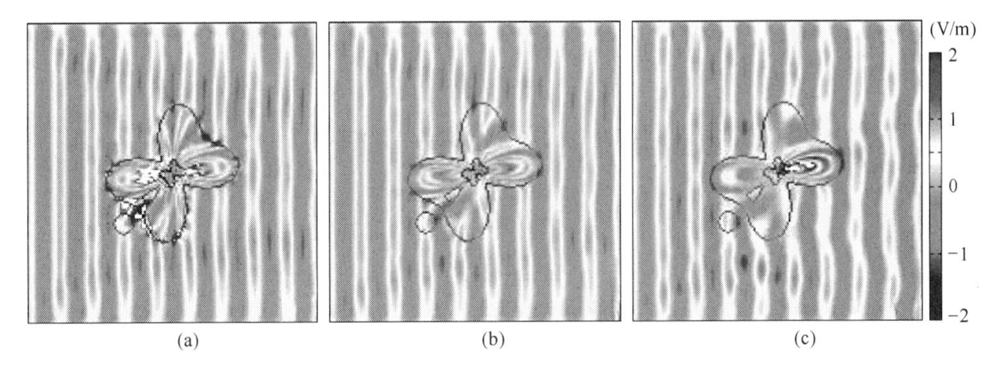

Fig. 6 Electric field distribution in the computation domain of the cloak with loss tangent of 10^{-4} (a), 0.01 (b) and 0.1 (c)

According to the procedure illustrated in Ref. [17], material parameters for the complementary medium - assisted cloaks with circular, elliptical and square cross sections can be obtained from Eqs. (5) and (8). The electric field distributions in the computation domain under cylindrical wave irradiation are simulated as shown in Fig. 7, which clearly demonstrate the generality of the material parameters developed in this paper for designing the complementary medium - assisted cloaks with arbitrary geometries. Besides, under TE plane wave irradiation, the cloaking effect can also be observed, as shown in Fig. 8. The simulation results for circular cloak are in good agreement with Ref. [16], which further confirms the effectiveness and the generality of the material parameters we developed.

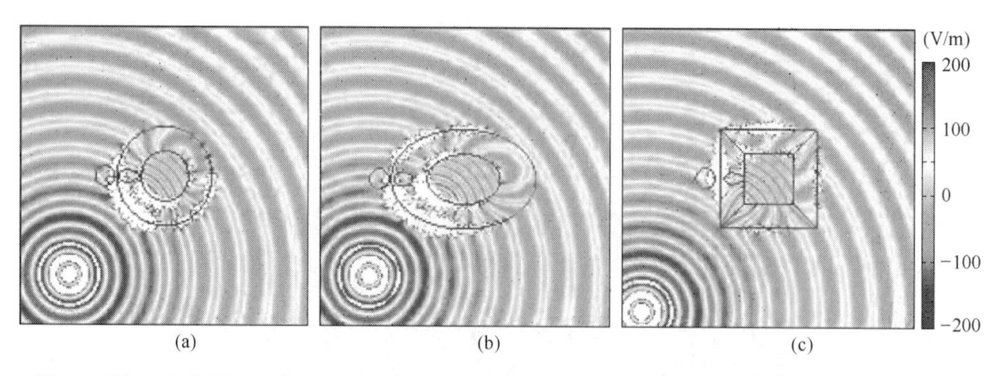

Fig. 7 Electric field distributions for the complementary medium – assisted cloak with circular(a), elliptical(b) and square(c) cross sections under cylindrical wave irradiation

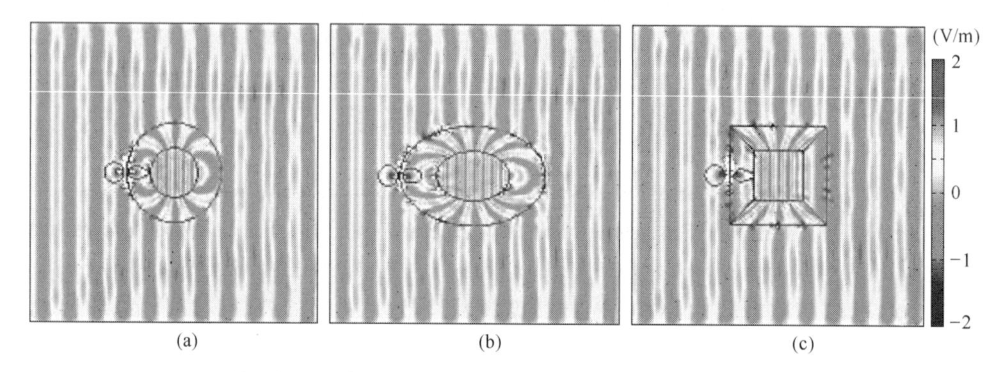

Fig. 8 Similar to Fig. 7, but for TE plane wave irradiation

Similarly, the external cloak based on nonlinear transformation is studied. Here, we consider two kinds of nonlinear transformation as shown below.

$$r' = b^2/r, \theta' = \theta, z' = z \tag{9}$$

$$r' = b^{m+1}(1 + a/r)^m/(a + b)^m, \theta' = \theta, z' = z \tag{10}$$

The material parameters can be obtained according to the procedure illustrated in section 2, and such results are not included herein for brevity. Next, we will discuss the characteristic of the external cloak based on linear and nonlinear transformation. Taking the case of a circular cloak as a special example, the simulation results are shown in Fig. 9, where panels (a) and (b) show the simulation results under linear transformation of Eq. (2), (c) and (d) show the cases under nonlinear transformation of Eq. (9), while (e) and (f) show the cases under nonlinear transformation of Eq. (10). From Fig. 9(a), (c) and (e), we can observe that to cloak the same dielectric object with radius $r = 0.2\lambda$, centered at $(-1.25, 0)$, the shape of the "antiobject" is quite different for linear and nonlinear transformations. Therefore, the image of the "antiobject" mapped into the complementary media layer is not only dependent on the contour of the cloak but also dependent on the coordination transformation, which is an interesting feature of the complementary medium – assisted external cloak. Meanwhile, from Fig. 9(b), (d) and (f), we can find that for the external cloaks in the same size and with the same "antiobject" of $r = 0.2\lambda$ centered at $(-0.75, 0)$, an object with much larger geometry size can be cloaked based on nonlinear transformation of Eq.

(10), as shown in Fig. 9(f). It is worth noting that in Eq. (10), we just choose $m = 4$ in our simulation; increasing the value of will enlarge the outer air layer in the transformation region, as a consequence, a much larger object can be cloaked for the given cloak size. It means that the nonlinear transformation has some advantages over the linear transformation for miniaturizing the size of the external cloak.

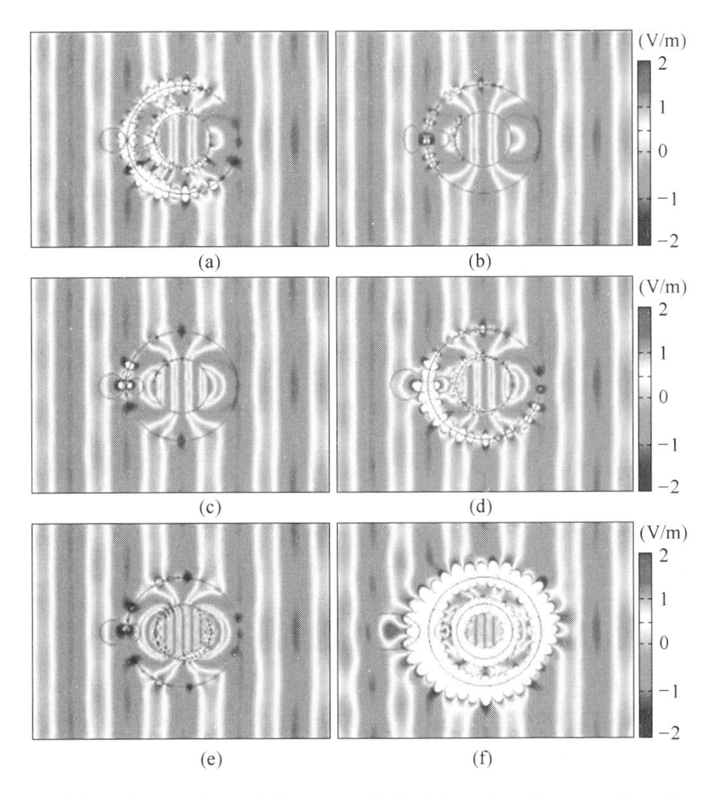

Fig. 9 Comparison of the characteristic of the external cloak based on linear and nonlinear transformation
(a) and (b) are corresponding to the linear transformation of Eq. (2);
(c) and (d) are corresponding to the nonlinear transformation of Eq. (9);
(e) and (f) are corresponding to the nonlinear transformation of Eq. (10)
In panels (a), (c) and (e), the dielectric object with radius $r = 0.2\lambda$, is centered at $(-1.25, 0)$;
in (b), (d) and (f), the "antiobject" with radius $r = 0.15\lambda$, is centered at $(-0.75, 0)$

4 Conclusion

Material parameters for the complementary medium – assisted cloak with arbitrary geometries are derived. The cloak is composed of the core material and the modified complementary layer embedded with the "antiobject", and it can make the object outside its domain invisible. We have investigated the influence of electric and magnetic – loss of the metamaterials on the performance of the device. Results show that for loss tangent less than 0.01, the performance of the cloak is basically undisturbed; increasing the loss tangent will disturb the forward – scattering region of the near field. This work has greatly improved the designing flexibility of the complementary medium – assisted cloak, since material parameters for the cloak with arbitrary geometries can be easily obtained for the given contour equations. We show that the material parameters can be also specialized to the 2D cloaks with regular shapes, such as circular, elliptical and square, which represents an important progress towards the realization of arbitrary shaped complementary medium – assisted cloak. Moreover, we find that the shape and size of the "antiobject" is dependent on the contour of the cloak and the coordination transformation. Interestingly, the object with much larger size can be hidden by the cloak based on non – linear transformation, which shows some advantages in open up an avenue for miniaturization in future cloak design.

Acknowledgements

This work was supported by the National Natural Science Foundation of China (Grant No. 60861002), Training Program of Yunnan Province for Middle – aged and Young Leaders of Disciplines in Science and Technology (Grant No. 2008PY031), the Research Foundation from Ministry of Education of China (Grant No. 208133), the Natural Science Foundation of Yunnan Province (Grant No. 2007F005M), Research Foundation of Education Bureau of Yunnan Province (Grant No. 07Z10875), and the National Basic Research Program of China (973 Program) (Grant No. 2007CB613606).

References

[1] Pendry J B, Schurig D, Smith D R. Controlling electromagnetic fields [J]. Science, 2006, 312 (5781): 1780 – 1782.

[2] Leonhardt U. Optical conformal mapping[J]. Science, 2006, 312(5781): 1777 – 1780.

[3] Schurig D, Mock J J, Justice B J, et al. Metamaterial electromagnetic cloak at microwave frequencies[J]. Science, 2006, 314(5801): 977 – 980.

[4] Leonhardt U, Tyc T. Broadband invisibility by non – Euclidean cloaking[J]. Science, 2009, 323 (5910): 110 – 112.

[5] Liu R, Ji C, Mock J J, et al. Broadband ground – plane cloak[J]. Science, 2009, 323(5912): 366 – 369.

[6] Valentine J, Li J, Zentgraf T, et al. An optical cloak made of dielectrics[J]. Nature Materials, 2009, 8(7): 568 – 571.

[7] Alù A, Engheta N. Achieving transparency with plasmonic and metamaterial coatings[J]. Physical Review E, 2005, 72(1): 016623.

[8] Alù A, Engheta N. Theory and potentials of multi – layered plasmonic covers for multi – frequency cloaking[J]. New Journal of Physics, 2008, 10(11): 115036.

[9] Miller D A B. On perfect cloaking[J]. Opt. Express, 2006, 14(25): 12457 – 12466.

[10] Vasquez F G, Milton G W, Onofrei D. Broadband exterior cloaking[J]. Optics Express, 2009, 17(17): 14800 – 14805.

[11] Alitalo P, Luukkonen O, Jylhä L, et al. Transmission – line networks cloaking objects from electromagnetic fields[J]. IEEE Transactions on Antennas and Propagation, 2008, 56(2): 416 – 424.

[12] Nicorovici N A, Milton G W, McPhedran R C, et al. Quasistatic cloaking of two – dimensional polarizable discrete systems by anomalous resonance[J]. Optics Express, 2007, 15(10): 6314 – 6323.

[13] Milton G W, Nicorovici N A P, McPhedran R C, et al. Solutions in folded geometries, and associated cloaking due to anomalous resonance[J]. New Journal of Physics, 2008, 10(11): 115021.

[14] Alitalo P, Tretyakov S. Electromagnetic cloaking with metamaterials [J]. Materials Today, 2009, 12 (3): 22 - 29.

[15] Bruno O P, Lintner S. Superlens - cloaking of small dielectric bodies in the quasistatic regime [J]. Journal of Applied Physics, 2007, 102(12): 124502.

[16] Lai Y, Chen H Y, Zhang Z Q, et al. Complementary media invisibility cloak that cloaks objects at a distance outside the cloaking shell [J]. Physical Review Letters, 2009, 102(9): 093901.

[17] Li C, Li F. Two - dimensional electromagnetic cloaks with arbitrary geometries [J]. Optics Express, 2008, 16 (17): 13414 - 13420.

[18] Schurig D, Pendry J B, Smith D R. Calculation of material properties and ray tracing in transformation media [J]. Optics Express, 2006, 14(21): 9794 - 9804.

[19] Yang J J, Huang M, Yang C F, et al. Metamaterial electromagnetic concentrators with arbitrary geometries [J]. Optics Express, 2009, 17(22): 19656 - 19661.

An Efficient 2 – D FDTD Method for Analysis of Parallel – Plate Dielectric Resonators

Bin Yao, Qinhong Zheng, Jinhui Peng, Runeng Zhong, Shenghui Li, Tai Xiang

Abstract: An efficient two – dimensional (2 – D) finite – difference time – domain (FDTD) method is presented for the analysis of parallel – plate dielectric resonator. Three – dimensional (3 – D) electromagnetic problems, with the description of z – dependence by k_z, can be solved by compact 2 – D FDTD method. Moreover, the perfect matched layer (PML), corresponding to the proposed FDTD algorithm, is also presented for the simulation of open system. Three representative examples—cylindrical, ring, and asymmetric parallel – plate resonators—are analyzed to verify the proposed compact 2 – D FDTD method.

Keywords: compact two – dimensional finite – difference time – domain (2 – D FDTD) method; parallel – plate dielectric resonator; perfect matched layer (PML); resonant frequency

1 Introduction

The parallel – plate dielectric resonator is a critical component of integral circuits. A number of approaches—finite element method (FEM), finite – difference time – domain (FDTD) method[1], multipole theory (MT) method[2], and so on—are efficient for the analysis of electromagnetic problems. Among them, FDTD method is the most versatile one. Recently, some researchers applied FDTD method coupled with some techniques for the analysis of a resonator. They are digital filtering, modern spectrum estimation technique, Prony analysis, Padé approximation, and Baker algorithm[3–7]. Benefitting from these techniques, the analysis time of the resonator is greatly reduced. However, these approaches are based on a three – dimensional (3 – D) mesh that needs numerous computational resources. References [8 – 10] proposed a compact two – dimensional (2 – D) FDTD algorithm for the analysis of axially symmetric geometry, or so – called body of revolution. This algorithm has been widely used. However, this algorithm is invalid for asymmetric geometry. On the other hand, Jung, et al. [11] presented a compact one – dimensional (1 – D) FDTD method for the analysis of a parallel – plate waveguide loaded with photonic crystals, and Moghaddam, et al. [12] proposed a compact 2 – D FDTD method coupled with Liao's multitransmitting boundary conditions for the analysis of subsurface interface radar. These methods can significantly reduce the computational costs.

In this letter, we propose a compact 2 – D FDTD algorithm coupled with corresponding perfect matched layer (PML) for the analysis of asymmetric parallel – plate dielectric resonators. First of all, with the description of z – dependence by k_z, the 2 – D electromagnetic equations in PML are derived. Then, as a demonstration, the derivation of the FDTD formulas is presented. Eventually, to validate the efficiency and accuracy of the proposed 2 – D FDTD method, three parallel – plate dielectric resonators are analyzed.

2 Theory

A parallel - plate dielectric resonator is produced by placing two relatively undefined metallic planes at the top and bottom of a dielectric resonator. Fig. 1 shows the general geometry of a asymmetric dielectric resonator, which is uniform in z - direction. The electromagnetic field components in this resonator can be expressed as

$$\boldsymbol{E}(x,y,z,t) = \boldsymbol{e}(x,y,t)\exp(-jk_z z) \tag{1}$$

$$\boldsymbol{H}(x,y,z,t) = \boldsymbol{h}(x,y,t)\exp(-jk_z z) \tag{2}$$

where $j = \sqrt{-1}$ and is the wavenumber in z - direction.

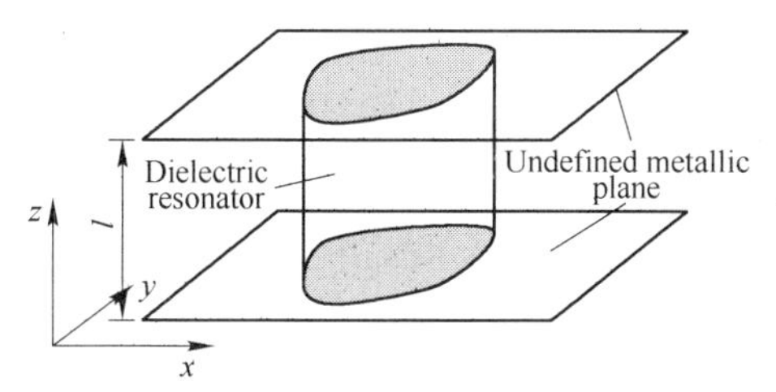

Fig. 1 General geometry of asymmetric parallel - plate dielectric resonator

Assuming that the metallic planes are perfect conductors and taking the consideration of z - directional uniformity, the z - dependence of the field is that appropriate to standing waves

$$A\sin k_z z + B\cos k_z z \tag{3}$$

If the planes are at $z = 0$ and $z = 1$, the boundary conditions can be satisfied at each surface only if

$$k_z = \frac{p\pi}{l} \quad (p = 0,1,2,\cdots) \tag{4}$$

where is the height of resonator in z - direction. Then, the field components can be written as

$$\boldsymbol{E}(x,y,z,t) = \boldsymbol{e}(x,y,t)\exp(-j\frac{p\pi}{l}z) \tag{5}$$

$$\boldsymbol{H}(x,y,z,t) = \boldsymbol{h}(x,y,t)\exp(-j\frac{p\pi}{l}z) \tag{6}$$

Moreover, the field components can also be expressed as[13]

$$E_x(x,y,z), E_y(x,y,z), H_z(x,y,z) = [E_x(x,y), E_y(x,y), H_z(x,y)]j\exp\left(-j\frac{p\pi}{l}z\right) \tag{7}$$

$$H_x(x,y,z), H_y(x,y,z), E_z(x,y,z) = [H_x(x,y), H_y(x,y), E_z(x,y)]j\exp\left(-j\frac{p\pi}{l}z\right) \tag{8}$$

Finally, instead of $\partial/\partial z$ by $-j(p\pi/l)$ and considering the additional factor j in Eq. (7), the curl equations of Maxwell's equations in the PML can be written as

$$\begin{bmatrix} \dfrac{\partial H_z}{\partial y} + \dfrac{p\pi}{l}H_y \\ -\dfrac{p\pi}{l}H_x - \dfrac{\partial H_z}{\partial x} \\ \dfrac{\partial H_y}{\partial x} - \dfrac{\partial H_x}{\partial y} \end{bmatrix} = j\omega\varepsilon_0\varepsilon_r^* \begin{bmatrix} \dfrac{s_y}{s_x} & 0 & 0 \\ 0 & \dfrac{s_x}{s_y} & 0 \\ 0 & 0 & s_x s_y \end{bmatrix} \cdot \begin{bmatrix} E_x \\ E_y \\ E_z \end{bmatrix} \tag{9}$$

$$\begin{bmatrix} \frac{\partial E_z}{\partial y} + \frac{p\pi}{l}E_y \\ -\frac{p\pi}{l}E_x - \frac{\partial E_z}{\partial x} \\ \frac{\partial E_y}{\partial x} - \frac{\partial E_x}{\partial y} \end{bmatrix} = j\omega\varepsilon_0\mu_r^* \begin{bmatrix} \frac{s_y}{s_x} & 0 & 0 \\ 0 & \frac{s_x}{s_y} & 0 \\ 0 & 0 & s_x s_y \end{bmatrix} \cdot \begin{bmatrix} H_x \\ H_y \\ H_z \end{bmatrix} \tag{10}$$

where $\varepsilon_r^* = \varepsilon_r + \sigma/(j\omega\varepsilon_0)$ and $\mu_r^* = \mu_r + \sigma_m/(j\omega\varepsilon_0)$ are the complex relative permittivity and permeability. $s_x = k_x + \sigma_x/(j\omega\varepsilon_0)$ and $s_y = k_y + \sigma_y/(j\omega\varepsilon_0)$ are the assumed tensor coefficients of the PML, which has been implemented according to Gedney[14]. The stability of the PML is similar to the discussion of Berenger[15]. Subsequently, the corresponding timedomain difference equations can be derived. For brevity, we only present the derivation of E_x.

Specifically, let

$$E'_x - \frac{1}{s_x}E_x \tag{11}$$

$$D_x = \varepsilon_0\varepsilon_r^* E'_x \tag{12}$$

Then, the first equation of (9) is rewritten as

$$\frac{\partial H_z}{\partial y} + \frac{p\pi}{l}H_y = j\omega s_y D_x \tag{13}$$

Finally, the time - domain difference equations of is

$$D_x^{n+1}(i+1/2,j) = \left(\frac{2\varepsilon_0 k_y - \sigma_y \Delta t}{2\varepsilon_0 k_y + \sigma_y \Delta t}\right) \cdot D_x^n(i+1/2,j) + \left(\frac{2\varepsilon_0 \Delta t}{2\varepsilon_0 k_y + \sigma_y \Delta t}\right) \cdot \left[\frac{H_z^{n+1/2}(i+1/2,j+1/2) - H_z^{n+1/2}(i+1/2,j-1/2)}{\Delta y} + \frac{p\pi}{l}H_y^{n+1/2}(i+1/2,j)\right] \tag{14}$$

$$E_x'^{n+1}(i+1/2,j) = \left(\frac{2\varepsilon_0\varepsilon_r - \sigma\Delta t}{2\varepsilon_0\varepsilon_r + \sigma\Delta t}\right) \cdot E_x'^{n+1}(i+1/2,j) + \frac{2}{2\varepsilon_0\varepsilon_r + \sigma\Delta t} \cdot [D_x^{n+1}(i+1/2,j) - D_x^n(i+1/2,j)] \tag{15}$$

$$E_x'^{n+1}(i+1/2,j) = E_x^n(i+1/2,j) + \left(k_x + \frac{\sigma_x\Delta t}{2\varepsilon_0}\right) \cdot E_x'^{n+1}(i+1/2,j) - \left(k_x - \frac{\sigma_x\Delta t}{2\varepsilon_0}\right) E_x^n(i+1/2,j) \tag{16}$$

where p is an input parameter that is corresponding to different mode series, such as $TE_{\delta 0}$, $TE_{\delta 1}$, $TE_{\delta 2}$, etc. Other corresponding time - domain difference equations of the remaining equations of (9) and (10) are similar to (14) - (16), which, for brevity, are not given here.

3 Numerical results

To test the algorithm, the first example is a parallel - plate cylindrical dielectric resonator. The geometry is shown in Fig. 2. In the computation, the mesh, time - step, and time iteration are chosen as $\Delta l = 0.25$mm, $\Delta t = \Delta l/(2c)$, and 2^{17} (= 131072), where c is the light speed in vacuum. To mitigate the undesirable late - time reflection, the differential Gaussian pulse is used as the excitation. The computed resonant frequencies, which are derived by using fast Fourier technique[10], are compared to the analytical results in Table 1. From the comparison, one can find a good agreement is achieved.

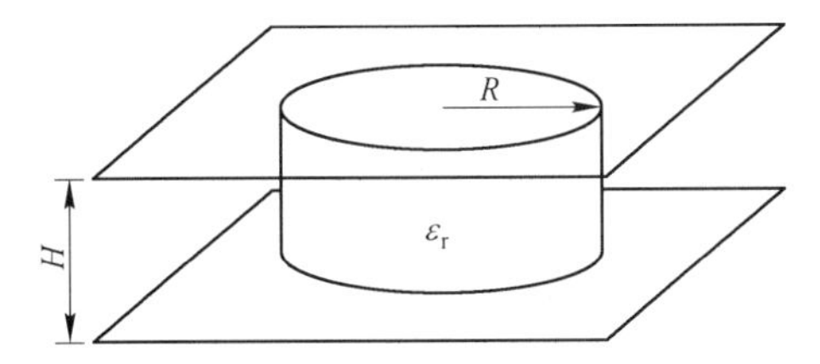

Fig. 2 Parallel – plate cylindrical dielectric resonator with radius $R = 5.25$mm, thickness $H = 4.62$mm, and relative permittivity $\varepsilon_r = 38$

Table 1 Comparison of resonant frequencies of parallel – plate cylindrical dielectric resonator computed by analytical method and proposed compact 2 – D FDTD method

Modes	Analytical results[10] /GHz	Proposed 2 – D FDTD		
		Resonant frequencies/GHz	p	Error/%
HEM211	7.4995	7.502	1	0.03
HEM121	8.3177	8.325	1	0.09
HEM311	9.0250	8.985	1	0.44
HEM221	9.7139	9.716	1	0.02
HEM212	11.8310	11.839	2	0.07
HEM312	12.8107	12.808	2	0.02
HEM331	13.3215	13.338	1	0.12

The second example is the analysis of a parallel – plate ring dielectric resonator as shown in Fig. 3. The mesh, time – step, and time iteration chosen in the program are the same as the example one. The calculated resonant frequencies are also compared to analytical results in Table 2. From the comparison, one can find the maximum error is no more than 0.90%.

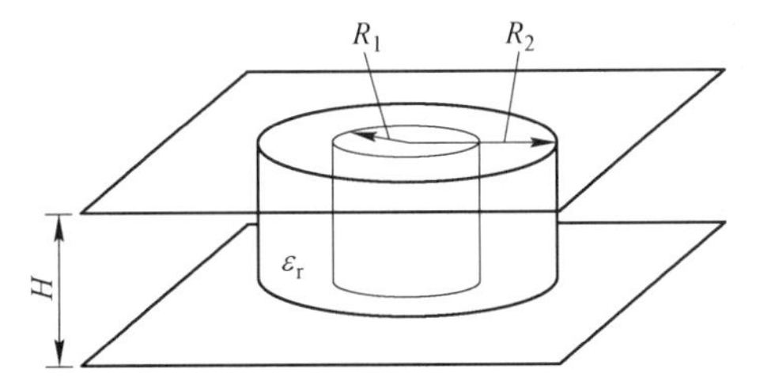

Fig. 3 Parallel – plate ring dielectric resonator with radius $R_1 = 2.5$mm, $R_2 = 5.25$mm, thickness $H = 4.62$mm, and relative permittivity $\varepsilon_r = 38$

Table 2 Comparison of resonant frequencies of parallel – plate ring dielectric resonator computed by analytical method and proposed compact 2 – D FDTD method

Modes	Analytical results[16] /GHz	Proposed 2 – D FDTD		
		Resonant frequencies/GHz	p	Error/%
HEM111	7.452	7.447	1	0.07
HEM211	8.144	8.143	1	0.01
HEM311	9.321	9.405	1	0.90

Continues Table 2

Modes	Analytical results[16] /GHz	Proposed 2 – D FDTD		
		Resonant frequencies/GHz	p	Error/%
HEM121	10. 525	10. 555	1	0. 29
HEM411	10. 739	10. 741	1	0. 02

Furthermore, to show the versatility of the proposed compact 2 – D FDTD algorithm, a fabricated asymmetrical resonator as shown in Fig. 4, which is constructed by a parallel – plate cylindrical dielectric resonator with a square air post, is analyzed by 3 – D FDTD method and proposed 2 – D FDTD method. The mesh, time – step, and time iteration chosen in the program are also the same as the example one. Also, the program is running in the same personal computer. The computed lowest seven resonant frequencies and elapsed CPU time are presented in Table 3. From the comparison, we can find the proposed 2 – D FDTD method achieves the same accuracy in less than 3. 7% CPU time and 6. 3% memory cost of 3 – D FDTD method.

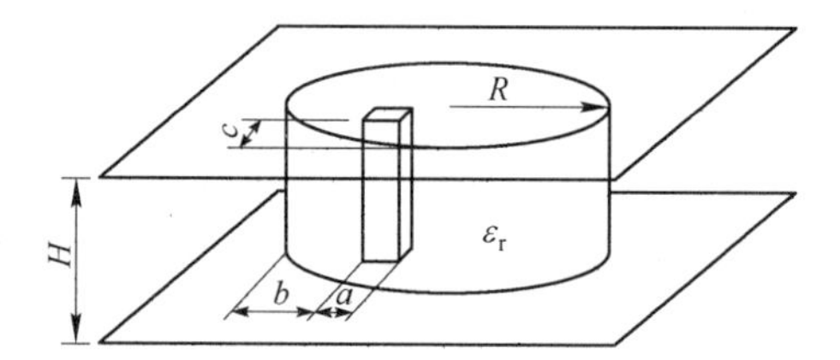

Fig. 4 Parallel – plate cylindrical dielectric resonator with a square air post
$R = 5.25$mm, $H = 4.0$mm, $a = 1.0$mm, $b = 2.0$mm, $c = 3.25$mm, and relative permittivity $\varepsilon_r = 38$

Table 3 Comparison of resonant frequencies of parallel – plate cylindrical dielectric resonator with a square air post computed by 3 – D FDTD method and proposed compact 2 – D FDTD method

3 – D FDTD method, the elapsed CPU time and memory requirement are 517min and 12. 11MB/GHz	Proposed 2 – D FDTD method, the elapsed CPU time and memory requirement are 18. 8min and 0. 76MB		3 – D FDTD method, the elapsed CPU time and memory requirement are 517min and 12. 11MB/GHz	Proposed 2 – D FDTD method, the elapsed CPU time and memory requirement are 18. 8min and 0. 76MB	
	Resonant frequencies/GHz	p		Resonant frequencies/GHz	p
6. 972	6. 990	1	8. 977	9. 021	1
7. 758	7. 764	1	9. 021	9. 058	1
8. 187	8. 197	1	9. 570	9. 585	1
8. 344	8. 271	1			

4 Conclusion

The proposed 2 – D FDTD method, with the description of z – dependence by k_z, is applied to the analysis of parallel – plate dielectric resonators. Three numerical examples—cylindrical, ring, and asymmetric parallel – plate dielectric resonators—are analyzed. From these examples one can find that the proposed 2 – D FDTD method has advantages of CPU time and memory saving over 3 – D

FDTD method for the calculation of the low - order resonant frequencies. Therefore, the proposed method is an efficient method for the analysis of parallel - plate dielectric resonators.

References

[1] Yee K S. Numerical solution of initial boundary value problems involving Maxwell's equations in isotropic media [J]. Antennas and Propagation, IEEE Transactions on, 1966, 14(3): 302 - 307.

[2] Zheng Q H, Xie F Y, Lin W G. Solution of three - dimensional Helmholtz equation by multipole theory method [J]. Journal of Electromagnetic Waves and Applications, 1999, 13(3): 339 - 357.

[3] Bi Z, Shen Y, Wu K, et al. Fast finite - difference time - domain analysis of resonators using digital filtering and spectrum estimation techniques [J]. Microwave Theory and Techniques, IEEE Transactions on, 1992, 40(8): 1611 - 1619.

[4] Pereda J A, Vielva L A, Vegas A, et al. Computation of resonant frequencies and quality factor of open dielectric resonators by a combination of the finite - difference time - domain (FDTD) and Prony's methods [J]. IEEE Microwave and Guided Wave Letters, 1992, 2(11): 431 - 433.

[5] Dey S, Mittra R. Efficient computation of resonant frequencies and quality factors of cavities via a combination of the finite - difference time - domain technique and the Padé approximation [J]. Microwave and Guided Wave Letters, IEEE, 1998, 8(12): 415 - 417.

[6] Guo W H, Li W J, Huang Y Z. Computation of resonant frequencies and quality factors of cavities by FDTD technique and Padé approximation [J]. IEEE Microwave and Wireless Components Letters, 2001, 11(5): 223 - 225.

[7] Zhang Y J, Zheng W H, Xing M X, et al. Application of fast Pade approximation in simulating photonic crystal nanocavities by FDTD technology [J]. Optics Communications, 2008, 281(10): 2774 - 2778.

[8] Navarro A, Nunez M J, Martin E. Study of TE 0 and TM 0 modes in dielectric resonators by a finite difference time - domain method coupled with the discrete Fourier transform [J]. Microwave Theory and Techniques, IEEE Transactions on, 1991, 39(1): 14 - 17.

[9] Navarro A, Nunez M J. FDTD method coupled with FFT: a generalization to open cylindrical devices [J]. Microwave Theory and Techniques, IEEE Transactions on, 1994, 42(5): 870 - 874.

[10] Shi S Y, Yang L Q, Prather D W. Numerical study of axisymmetric dielectric resonators [J]. Microwave Theory and Techniques, IEEE Transactions on, 2001, 49(9): 1614 - 1619.

[11] Jung K Y, Ju S, Teixeira F L. Application of the modal CFS - PML - FDTD to the analysis of magnetic photonic crystal waveguides [J]. IEEE Microwave and Wireless Components Letters, 2011, 21(4): 179 - 181.

[12] Moghaddam M, Yannakakis E, Chew W C, et al. Modeling of the subsurface interface radar [J]. Journal of Electromagnetic Waves and Applications, 1991, 5(1): 17 - 39.

[13] Xiao S, Vahldieck R. An efficient 2 - D FDTD algorithm using real variables [J]. IEEE Microwave and Wireless Components Letter, 1993, 3(5): 127 - 129.

[14] Gedney S D. An anisotropic perfectly matched layer - absorbing medium for the truncation of FDTD lattices [J]. Antennas and Propagation, IEEE Transactions on, 1996, 44(12): 1630 - 1639.

[15] Berenger J P. Perfectly matched layer for the FDTD solution of wave - structure interaction problems [J]. Antennas and Propagation, IEEE Transactions on, 1996, 44(1): 110 - 117.

[16] Kaneda N, Houshmand B, Itoh T. FDTD analysis of dielectric resonators with curved surfaces [J]. Microwave Theory and Techniques, IEEE Transactions on, 1997, 45(9): 1645 - 1649.

Efficient Analys is of Ridged Cavity by Modal FDTD Method

Bin Yao, Qinhong Zheng, Jinhui Peng, Runeng Zhong, Wansong Xu, Tai Xiang

Abstract: A modal finite - difference time - domain (FDTD) method is extended for the analysis of ridged cavities, which are uniform in the z - direction. Assuming that the end surfaces of cavity are the perfect conductor, thus, the fields along the z - axis can be described by k_z. Therefore, three - dimensional (3 - D) problems can be simulated by the use of a two - dimensional model. Besides, to achieve a faster computation, the field components are expressed by two pairs of equations——sine and cosine. To validate the utility and efficiency of proposed method, we analyzed two ridged cavities. Numerical results show that less than one - tenth memory and CPU requirements are needed by the modal FDTD as compared with conventional 3 - D FDTD method.

Keywords: modal finite - difference time - domain (FDTD) method; ridged cavity; resonant frequency

1 Introduction

A number of approaches, finite element method, finite - difference time - domain (FDTD) method, multipole theory method[1], and so on, are efficient for the analysis of electromagnetic problems. Among them, the FDTD method has been widely used since it was first introduced by Yee[2]. Some progresses have been made for a more efficient analysis of microwave cavity. The improvement is that some algorithms have been combined with FDTD method. These algorithms are Prony analysis, digital filtering, modern spectrum estimation technique, Padé approximation, and Baker algorithm[3-8]. Benefiting from these techniques, the computational time is greatly reduced. Nevertheless, there are two deficiencies among these methods. First, these methods are based on a three - dimensional (3 - D) model, which requires numerous computational resources. Second, these methods lack clear physical meaning and suitable mathematical demonstration. Particularly, the computational accuracy very much depends on the samples of signals. In some cases, the accurate results are even difficult to obtain. Recently, some compact two - dimensional (2 - D) FDTD methods have been widely used for full - wave analysis of wave - guided structures[9-12]. These methods utilize a 2 - D model to solve 3 - D problems. It takes the advantages of CPU time and memory saving over 3 - D methods. On the other hand, Jung, et al. [13] presented a modal one - dimensional (1 - D) FDTD method for the analysis of parallel - plate waveguide loaded with photonic crystals. It reduces both memory and CPU costs by one order.

In this paper, we proposed a modal 2 - D FDTD method for the analysis of ridged cavity, which is uniform in the z - direction. Assuming that the end surfaces of cavity are perfect conductors, the fields along the z - axis can be described by k_z. Therefore, 3 - D cavities can be simulated by the use of a 2 - D model. In addition, to avoid complex variables and therefore to imp rove the efficiency of proposed modal FDTD algorithm, the field equations are expressed in two pairs of equations—sine and cosine. Two ridged cavities are computed to validate the accuracy and efficiency of proposed modal FDTD method.

2 Theory

Although an electromagnetic cavity can be of any shape whatsoever, an import ant class of cavities is produced by placing end surfaces on a length of ridged waveguide. Fig. 1 shows a cavity, which has an arbitrary cross - section and is uniform in the z - direction. The electromagnetic field components in this cavity can be expressed as:

$$\boldsymbol{E}(x,y,z,t)=\boldsymbol{e}(x,y,t)\exp(-jk_z z) \tag{1}$$

$$\boldsymbol{H}(x,y,z,t)=\boldsymbol{h}(x,y,t)\exp(-jk_z z) \tag{2}$$

where k_z is the wave number in the z - direction.

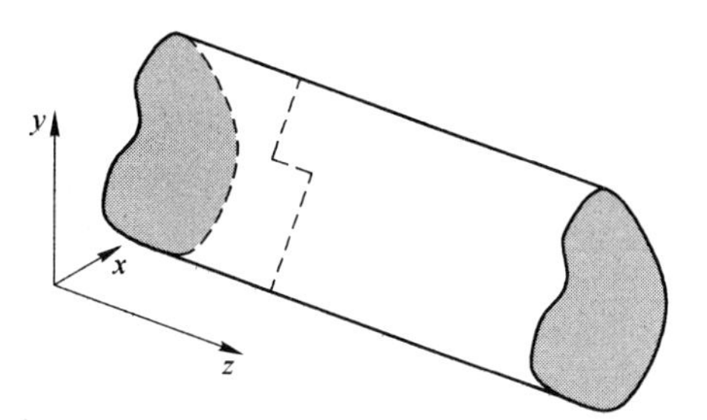

Fig. 1 Cavity with arbitrary cross - section

Assuming that the end surfaces of this cavity are perfect conductors and taking into consideration the z - directional uniformity, the z dependence of the field is appropriate to standing waves

$$A\sin k_z z + B\cos k_z z \tag{3}$$

If the plane boundary surfaces are at $z=0$ and $z=1$, the boundary conditions can be satisfied at each surface only if

$$k_z=\frac{p\pi}{l}\quad (p=0,1,2,\cdots) \tag{4}$$

where l is the length of cavity in the z - direction. Then, the field components can be written:

$$\boldsymbol{E}(x,y,z,t)=\boldsymbol{e}(x,y,t)\exp\left(-j\frac{p\pi}{l}z\right) \tag{5}$$

$$\boldsymbol{H}(x,y,z,t)=\boldsymbol{h}(x,y,t)\exp\left(-j\frac{p\pi}{l}z\right) \tag{6}$$

Besides, the field components can also be expressed as[14]

$$E_x(x,y,z),E_y(x,y,z),H_z(x,y,z) = [E_x(x,y),E_y(x,y),H_z(x,y)]j\exp\left(-j\frac{p\pi}{l}z\right) \tag{7}$$

$$H_x(x,y,z),H_y(x,y,z),E_z(x,y,z) = [H_x(x,y),H_y(x,y),E_z(x,y)]j\exp\left(-j\frac{p\pi}{l}z\right) \tag{8}$$

Finally, instead of $\frac{\partial}{\partial z}$ by $\frac{p\pi}{l}\cos\left(\frac{p\pi}{l}\right)$ or $\frac{p\pi}{l}\sin\left(\frac{p\pi}{l}\right)$, the curl equations of Maxwell's equations can be written:

$$\begin{cases} \dfrac{\partial H_z}{\partial y} - \dfrac{p\pi}{l}H_y = \varepsilon\dfrac{\partial E_x}{\partial t} + \sigma E_x \\ \dfrac{p\pi}{l}H_x - \dfrac{\partial H_z}{\partial x} = \varepsilon\dfrac{\partial E_y}{\partial t} + \sigma E_y \\ \dfrac{\partial H_y}{\partial x} - \dfrac{\partial H_x}{\partial y} = \varepsilon\dfrac{\partial E_z}{\partial t} + \sigma E_z \\ \dfrac{\partial E_z}{\partial y} + \dfrac{p\pi}{l}E_y = -\mu\dfrac{\partial H_x}{\partial t} - \sigma_m H_x \\ -\dfrac{p\pi}{l}E_x - \dfrac{\partial E_z}{\partial x} = -\mu\dfrac{\partial H_y}{\partial t} - \sigma_m H_y \\ \dfrac{\partial E_y}{\partial x} - \dfrac{\partial E_x}{\partial y} = -\mu\dfrac{\partial H_z}{\partial t} - \sigma_m H_z \end{cases} \tag{9}$$

Where σ is the electric conductivity; ε is the permittivity; σ_m is the magnetic conductivity; μ is the permeability. Subsequently, the time - domain difference equations of E_x and E_z can be derived as

$$E_x^{n+1}\left(i+\frac{1}{2},j\right) = \frac{1-\dfrac{\sigma(i+1/2,j)\Delta t}{2\varepsilon(i+1/2,j)}}{1+\dfrac{\sigma(i+1/2,j)\Delta t}{2\varepsilon(i+1/2,j)}} \cdot E_x^n\left(i+\frac{1}{2},j\right) + \frac{\dfrac{\Delta t}{\varepsilon(i+1/2,j)}}{1+\dfrac{\sigma(i+1/2,j)\Delta t}{2\varepsilon(i+1/2,j)}} \cdot \left[\frac{H_z^{n+1/2}\left(i+\frac{1}{2},j+\frac{1}{2}\right) - H_z^{n+1/2}\left(i+\frac{1}{2},j-\frac{1}{2}\right)}{\Delta y} - \frac{p\pi}{l}H_y^{n+1/2}\left(i+\frac{1}{2},j\right)\right] \tag{10}$$

$$E_z^{n+1}(i,j) = \frac{1-\dfrac{\sigma(i,j)\Delta t}{2\varepsilon(i,j)}}{1+\dfrac{\sigma(i,j)\Delta t}{2\varepsilon(i,j)}} \cdot E_z^n(i,j) + \frac{\dfrac{\Delta t}{\varepsilon(i,j)}}{1+\dfrac{\sigma(i,j)\Delta t}{2\varepsilon(i,j)}} \cdot \left[\frac{H_y^{n+1/2}\left(i+\frac{1}{2},j\right) - H_y^{n+1/2}\left(i-\frac{1}{2},j\right)}{\Delta x} - \frac{H_x^{n+1/2}\left(i,j+\frac{1}{2}\right) - H_x^{n+1/2}\left(i,j-\frac{1}{2}\right)}{\Delta y}\right] \tag{11}$$

where p is an input parameter, which corresponds to different mode series, such as $TE_{\delta 0}$, $TE_{\delta 1}$, and $TE_{\delta 2}$. Corresponding time - domain difference equations of the remaining equations of (9) are similar to those of (10) and (11), which, for brevity, are not given here.

3 Numeriacal results

To validate the proposed modal 2 - D FDTD method, we analyzed a typical air - loaded double - ridge cavity[15], which is made of perfect conductors. The geometry of the cavity is shown in Fig. 2. In the computation, the mesh, time step, and time iterations are chosen as $\Delta l = 0.05$cm, $\Delta t = \Delta l/(2c)$, and 2^{18} ($=262144$), where c is the light speed in vacuum. To avoid missing the mode, three random distributed differential Gaussian pulses are set as the excitation. After the program was run, one can derive a series time – domain response. Then, the resonant frequency can be calculated after applying fast Fourier transformation on the time domain response. The computed resonant frequencies are compared with the results of [15] and [16] in Table 1 and Fig. 3. From the comparison, one can see that a good agreement is achieved. For the convenience of knowing the characteristics of the ridged cavity, the dominant resonant frequencies versus d, w and h, as shown in Tables 2 and 3, are also calculated by the use of the proposed modal FDTD method and Ansoft's high – frequency structure simulator (HFSS).

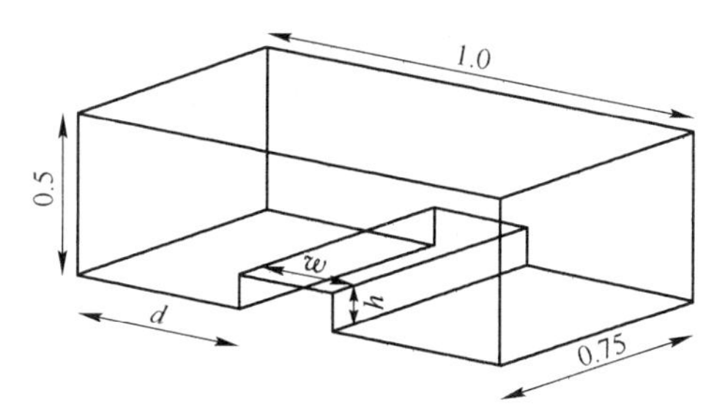

Fig. 2 Geometry of double – ridge cavity (All dimensions are in centimeter)

Table 1 Resonant frequencies (in gigahertz) computed by the use of a modal method when d = 0.4cm, w = 0.2cm, and h = 0.1

Mode	Ref. [15]	Ref. [16]	Proposed method		Mode	Ref. [15]	Ref. [16]	Proposed method	
			p	Resonant frequency				p	Resonant frequency
1	23.795	23.852	1	24.267	5	37.980	37.975	1	38.160
2	34.793	35.088	1	35.512	6	41.062	41.272	2	42.355
3	37.374	37.369	1	37.538	7	43.338	42.541	0	42.977
4	37.517	37.894	0	37.602	8	43.677	43.434	0	43.430

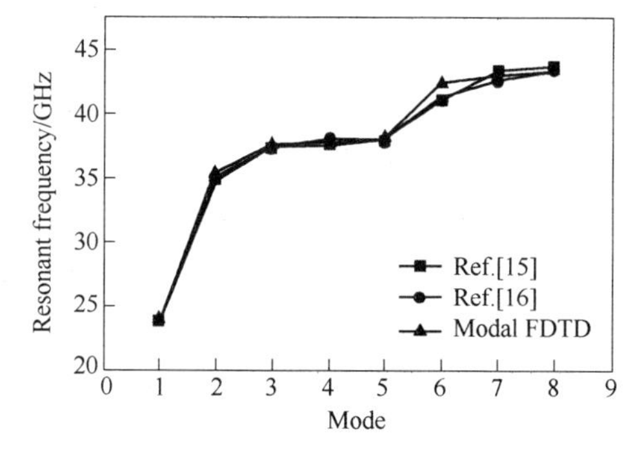

Fig. 3 The comparison of resonant frequency computed by the use of literatures and modal FDTD method

Table 2 The dominant resonant frequencies (in gigahertz) versus w and d computed by the use of modal FDTD method and HFSS when h = 0.1

w /cm	Method	d/cm								
		0.05	0.10	0.15	0.20	0.25	0.30	0.35	0.40	0.45
0.1	FDTD	25.091	24.990	24.867	24.742	24.624	24.530	24.460	24.417	24.409
	HFSS	25.0805	24.9865	24.8670	24.7485	24.6298	24.5359	24.4633	24.4181	24.4110

Continues Table 2

w /cm	Method	d/cm								
		0.05	0.10	0.15	0.20	0.25	0.30	0.35	0.40	0.45
0.2	FDTD	25.080	24.904	24.725	24.565	24.437	24.342	24.286	24.267	—
	HFSS	25.0755	24.9047	24.7324	24.5705	24.4497	24.3392	24.2972	24.2734	—
0.3	FDTD	24.949	24.734	24.542	24.386	24.277	24.212	24.190	—	—
	HFSS	24.9515	24.7341	24.5496	24.3908	24.2719	24.2136	24.1883	—	—
0.4	FDTD	24.757	24.547	24.378	24.258	24.185	24.162	—	—	—
	HFSS	24.7589	24.5477	24.3777	24.2495	24.1854	24.1651	—	—	—
0.5	FDTD	24.581	24.409	24.285	24.208	24.185	—	—	—	—
	HFSS	24.5775	24.4045	24.2829	24.2064	24.1741	—	—	—	—
0.6	FDTD	24.473	24.354	24.281	24.254	—	—	—	—	—
	HFSS	24.4683	24.3488	24.2750	24.2538	—	—	—	—	—
0.7	FDTD	24.462	24.397	24.375	—	—	—	—	—	—
	HFSS	24.4555	24.3917	24.3685	—	—	—	—	—	—
0.8	FDTD	24.560	24.545	—	—	—	—	—	—	—
	HFSS	24.5513	24.5353	—	—	—	—	—	—	—

Table 3 The dominant resonant frequencies (in gigahertz) versus w and h computed by the use of modal FDTD method when $d = 0.5\ \text{cm} - w/2$

w /cm	Method	h/cm							
		0.05	0.10	0.15	0.20	0.25	0.30	0.35	0.40
0.1	FDTD	24.771	24.403	23.938	23.423	22.904	22.401	21.915	21.431
	HFSS	24.7676	24.4197	23.9648	23.4417	22.9152	22.3913	21.9244	21.4276
0.2	FDTD	24.695	24.268	23.767	23.241	22.721	22.223	21.742	21.267
	HFSS	24.6953	24.2809	23.7739	23.2466	22.7239	22.2316	21.7521	21.2496
0.3	FDTD	24.647	24.190	23.675	23.147	22.630	22.136	21.660	21.184
	HFSS	24.6414	24.1859	23.6788	23.1530	22.6361	22.1394	21.6604	21.1770
0.4	FDTD	24.624	24.162	23.650	23.128	22.614	22.118	21.639	21.162
	HFSS	24.6161	24.1565	23.6487	23.1269	22.6155	22.1157	21.6359	21.1428
0.5	FDTD	24.629	24.185	23.691	23.176	22.667	22.166	21.678	21.185
	HFSS	24.6214	24.1702	23.6885	23.1781	22.6608	22.1629	21.6706	21.1769
0.6	FDTD	24.661	24.254	23.792	23.298	22.790	22.287	21.781	21.267
	HFSS	24.6525	24.2471	23.7936	23.2943	22.7892	22.2820	21.7736	21.2497
0.7	FDTD	24.720	24.375	23.967	23.508	23.014	22.502	21.976	21.422
	HFSS	24.7093	24.3655	23.9593	23.4992	23.0138	22.5019	21.9725	21.4136
0.8	FDTD	24.798	24.544	24.224	23.833	23.375	22.868	22.319	21.719
	HFSS	24.7860	24.5366	24.2161	23.8334	23.3770	22.8664	22.3182	21.7150

Furthermore, to show the efficiency of proposed modal FDTD method, another air – loaded ridged cavity, which is made of perfect conductors, is considered. The geometry of the cavity is illustrated in Fig. 4. The first 10 lowest resonant frequencies of this cavity, which have length $l = 10$cm and 20cm, are calculated by the use of HFSS, conventional 3 – D FDTD method (a homemade classical code), and proposed modal 2 – D FDTD method, respectively. In the FDTD simulation, the mesh, time step, and time iterations are chosen as $\Delta l = 0.5$cm, $\Delta t = \Delta l/(2c)$, and 2^{18}. The numerical results are compared in Table 4 and Fig. 5. From the comparison, one can find three highlights of the proposed modal FDTD method. The first one is that the results calculated by the use of the proposed modal FDTD method agree very well with those calculated by the use of HFSS and conventional 3 – D FDTD method.

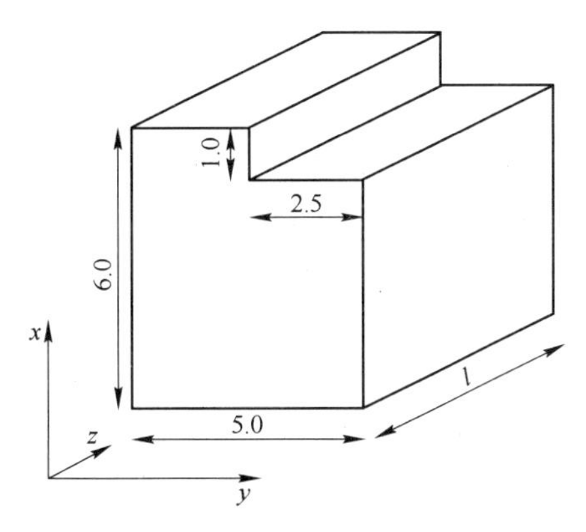

Fig. 4 Geometry of single – ridge cavity (All dimensions are in centimeter)

Table 4 Comparison of resonant frequencies (in gigahertz) computed by the use of HFSS, conventional 3 – D FDTD method, and modal FDTD method

Cavity length $l = 10$cm				Cavity length $l = 20$cm			
HFSS	Conventional 3 – D FDTD. the elapsed CPU time is 439s	Proposed 2 – D FDTD. the total elapsed CPU time is 96s for $p = 0, 1, 2, 3$		HFSS	Conventional 3 – D FDTD. the elapsed CPU time is 863s	Proposed 2 – D FDTD. the total elapsed CPU time is 96s for $p = 0, 1, 2, 3$	
		p	Resonant frequency			p	Resonant frequency
2.9718	2.9602	1	2.9606	2.6731	2.6596	1	2.6598
3.3784	3.3677	1	3.3683	2.9714	2.9602	2	2.9606
3.9458	3.9334	0	3.9425	3.1190	3.1068	1	3.1070
4.1568	4.1595	0	4.1595	3.3783	3.3677	2	3.3684
4.2402	4.2195	1	4.2199	3.4115	3.4006	3	3.4043
4.2607	4.2497	2	4.2584	3.7713	3.7614	3	3.7645
4.4192	4.4231	1	4.4235	3.9455	3.9333	0	3.9425
4.9717	4.9537	2	4.9615	4.0365	4.0136	1	4.0136
5.1258	5.1289	2	5.1362	4.1570	4.1595	0	4.1596
5.1765	5.1492	3	5.1856	4.2241	4.2268	1	4.2268

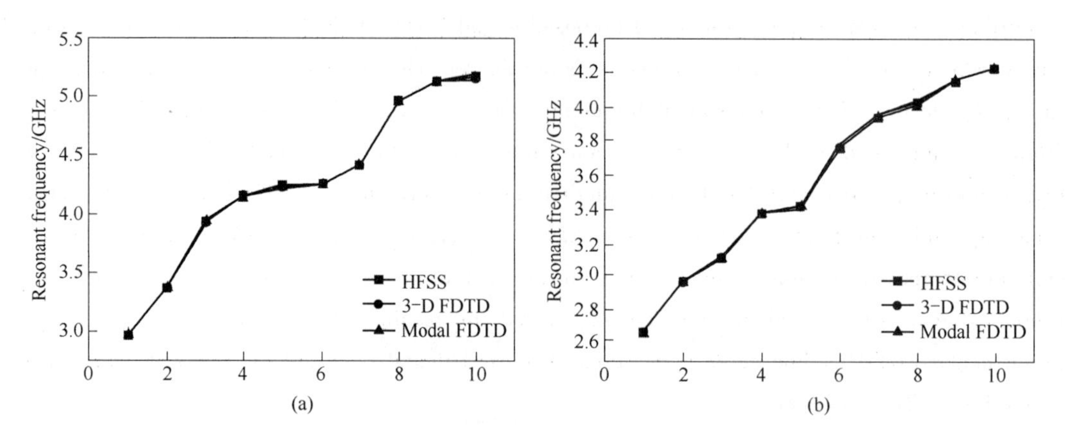

Fig. 5 The resonant frequency of the 10 lowest mode computed by the use of HFSS, 3 – D, and modal FDTD method
(a) When cavity length $l = 10$cm; (b) When cavity length $l = 20$cm

Second, the modal FDTD method achieves the same accuracy in only about 22% and 11% CPU time cost of a 3 – D FDTD method. The last one is that the modal FDTD method shows more CPU time saving when the cavity has a bigger length in the z – direction because no discretization is necessary.

Lastly, for the investigation of the convergence of proposed modal FDTD method, a hollow rectangular cavity with dimensions 3cm × 4cm × 5cm, which is made of perfect conductors, is computed by the use of 3 – D FDTD and modal FDTD method. The mesh and time step are chosen as $\Delta l = 0.25$cm and $\Delta t = \Delta l/(2c)$. The calculated errors can be derived by the comparison of calculated resonant frequency with analytical solutions[17]. Fig. 6 presents the calculated errors of the two lowest modes versus time iterations. It shows that 3 – D FDTD and modal FDTD can reach almost the same precision in the same time iterations.

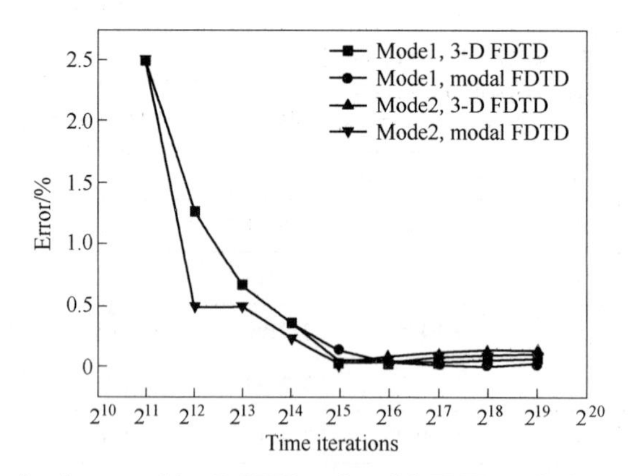

Fig. 6 The calculated errors of 3 – D FDTD and modal FDTD method versus time iterations

4 Conclusion

The proposed modal FDTD method, with the description along the z – axis by wave number k_z, could transform a 3 – D problem to a 2 – D model. Therefore, compared with conventional 3 – D

FDTD method, less CPU time and memory resource are required by modal FDTD. To validate the proposed method, we computed two ridged cavities. The numerical results show that little CPU time is needed by proposed modal FDTD method as compared with 3 – D FDTD method. In addition, the influence of different dimensions to the dominant resonant frequency is investigated.

Although the proposed modal FDTD method is superior to conventional 3 – D FDTD method on computing ridged cavity, it will be invalid if the z – direction of cavity is not uniform or the two end surfaces in the z – direction are not perfect conductors. In addition, the proposed method can be used in more round – shaped cavities. However, errors will occur when dealing with this kind of problem because the staircase approximation will be exerted to denote the round boundary in the $x-y$ plane.

Acknowledgements

This work was supported by the National Natural Science Foundation of China under Grant No. 51090385 and Yunnan Provincial Education Natural Science Foundation of China under Grant No. 2011Y308.

References

[1] Zheng Q H, Xie F Y, Lin W G. Solution of three – dimensional Helmholtz equation by multipole theory method [J]. Journal of Electromagnetic Waves and Applications, 1999, 13(3): 339 – 357.

[2] Yee K S. Numerical solution of initial boundary value problems involving Maxwell's equations in isotropic media [J]. IEEE Trans. Antennas Propag, 1966, 14(3): 302 – 307.

[3] Pereda J A, Vielva L A, Vegas A, et al. Computation of resonant frequencies and quality factors of open dielectric resonators by a combination of the finite – difference time – domain (FDTD) and Prony's methods [J]. IEEE Microwave and Guided Wave Letters, 1992, 2(11): 431 – 433.

[4] Bi Z, Shen Y, Wu K, et al. Fast finite – difference time – domain analysis of resonators using digital filtering and spectrum estimation techniques [J]. Microwave Theory and Techniques, IEEE Transactions on, 1992, 40(8): 1611 – 1619.

[5] Dey S, Mittra R. Efficient computation of resonant frequencies and quality factors of cavities via a combination of the finite – difference time – domain technique and the Padé approximation [J]. Microwave and Guided Wave Letters, IEEE, 1998, 8(12): 415 – 417.

[6] Guo W, Huang Y, Wang Q. Resonant frequencies and quality factors for optical equilateral triangle resonators calculated by FDTD technique and the Padé approximation [J]. Photonics Technology Letters, IEEE, 2000, 12(7): 813 – 815.

[7] Guo W, Li W, Huang Y. Computation of resonant frequencies and quality factors of cavities by FDTD technique and Padé approximation [J]. IEEE Microwave and Wireless Components Letters, 2001, 11(5): 223 – 225.

[8] Yang Y, Huang Y, Guo W, et al. Enhancement of quality factor for TE whispering – gallery modes in microcylinder resonators [J]. Optics Express, 2010, 18(12): 13057 – 13062.

[9] Xiao S, Vahldieck R, Jin H. Full – wave analysis of guided wave structures using a novel 2 – D FDTD [J]. Microwave and Guided Wave Letters, IEEE, 1992, 2(5): 165 – 167.

[10] Xu F, Wu K. A compact 2 – D finite – difference time – domain method for general lossy guiding structures [J]. Antennas and Propagation, IEEE Transactions on, 2008, 56(2): 501 – 506.

[11] Lu Q, Guo W, Bryne D, et al. Analysis of leaky modes in deep - ridge waveguides using the compact 2D FDTD method[J]. Electronics Letters, 2009, 45(13): 700 - 701.

[12] Lu Q, Guo W, Byrne D C, et al. Compact 2 - D FDTD method combined with Padé approximation transform for leaky mode analysis[J]. Journal of Lightwave Technology, 2010, 28(11): 1638 - 1645.

[13] Jung K, Ju S, Teixeira F. Application of the modal CFS - PML - FDTD to the analysis of magnetic photonic crystal waveguides[J]. Microwave and Wireless Components Letters, IEEE, 2011, 21(4): 179 - 181.

[14] Gwarek W, Morawski T, Mroczkowski C. Application of the FD - TD method to the analysis of circuits described by the two - dimensional vector wave equation[J]. Microwave Theory and Techniques, IEEE Transactions on, 1993, 41: 311 - 317.

[15] Venkatarayalu N V, Lee J F. Removal of spurious dc modes in edge element solutions for modeling three - dimensional resonators [J]. Microwave Theory and Techniques, IEEE Transactions on, 2006, 54 (7): 3019 - 3025.

[16] Jin J. The Finite Element Method in Electromagnetics[M]. John Wiley & Sons, 2014.

[17] Jackson J D, Jackson J D. Classical Electrodynamics[M]. New York: Wiley, 1962.

Parallel Algorithm for the Effective Electromagnetic Properties of Heterogeneous Materials on 3D RC Network Model

Jun Sun, Jinhui Peng, Ming Huang, Zhe Xiao, Jingjing Yang

Abstract: In order to study the effective electromagnetic properties of heterogeneous materials, computation of very large 3D RC (resistor – capacitor) network model is indispensable. The huge computations and long time – consuming exist in previous sequential algorithm, which hinder the widely application of the model. A novel parallel algorithm for the model is presented in this paper. Simulation results show that the parallel algorithm greatly reduces the computation time. It's found that the larger the size of the model is, the higher the efficiency of the parallel algorithm is. It is helpful to develop materials with required dielectric properties at radio and microwave frequencies.

Keywords: heterogeneous materials; 3D RC network; parallel algorithm

1 Introduction

Heterogeneous materials which contain both conductive and dielectric (insulating) phases are applied in a wide variety of fields, including metallurgical engineering and materials physics. How to develop heterogeneous materials with a required function has become a hotspot of modern materials science[1-6]. For example, in the microwave – assisted redox reactions of ilmenite, the carbonaceous reductant has dual functions, i. e., it acts as reactant, and improves the microwave absorbing properties of the mixture by increasing the ratio of the conductive phase[7]. The dielectric constant of heterogeneous materials will increase 1000 times by dispersing the conductive particles with diameter 10μm in the dielectric material (the thickness of the boundary layer is 10nm)[8].

Studying the effective electromagnetic properties of heterogeneous materials is of significant theoretical and technological interest. These properties can be reproduced by simulations of the electrical characteristics of large random networks with simple resistors and capacitors.

In 1999, Almond, et al. [9] showed that effective electromagnetic properties of heterogeneous materials can be modelled by a 2D RC network model based on the software SPICE. Later, an efficient algorithm for calculating the effective electromagnetic properties of the 2D RC network was proposed by Bouamrane, et al. [10], and the dielectric properties of the 2D RC network containing 32768 components were computed on computer PⅢ 500MHz. It has been found that the new algorithm only need 106. 5s. Meanwhile, it spent 2547s by using SPICE software.

Inspired by these works, our lab[11,12] has presented a 3D RC networks model, and show that it is more valuable compared with the 2D model. However, due to the complexity of the 3D model, the simulation is quite time consuming. For example, 50h are required for computing the dielectric properties of random networks containing 22440 components using one CPU of the Dawning TC4000A cluster supercomputer. This restricts the promotion and application of the model in the fields of microwave metallurgical engineering and material science. A parallel algorithm on the model is proposed in this paper, and implemented on eight CPUs of the supercomputer. We show that the computation time for the same model is only 42min. Furthermore, with the increase of the network size, the algorithm shows higher efficiency. This work has opened up an avenue for online designing heterogeneous material with required properties.

2 3D RC network model and parallel algorithm

2.1 3D RC network model

The frequency dependent effective conductivity and permittivity of heterogeneous materials are modelled using electrical networks consisting of randomly positioned resistors and capacitors. For understanding the modelling process, a $3 \times 3 \times 3$RC network model with $R = 1\text{k}\Omega$ and $C = 1\text{nF}$ is presented in the Fig. 1. The gray parts represent two parallel plates, the left one (node 0) connects to the ground, and the right one (the last node 28) connects to the constant current source of 1 ampere. Assuming that the point(1,1,1) is the origin and is marked as node 1, then any point(x,y,z) (excluding the parallel plates) is marked as node $M_p(x,y,z)$ in the network, where $M_p(x,y,z) = X_n[(z-1)Y_n + (y-1)] + x$. Here, X_n, Y_n and Z_n denote the size of the network along the axis of X, Y and Z, respectively. The total number of nodes of the network is N_p, where $N_p = X_nY_nZ_n + 1$. For example, in the $3 \times 3 \times 3$RC network, the total number of the network nodes is 28 and the point(2,2,2) is marked as node 14.

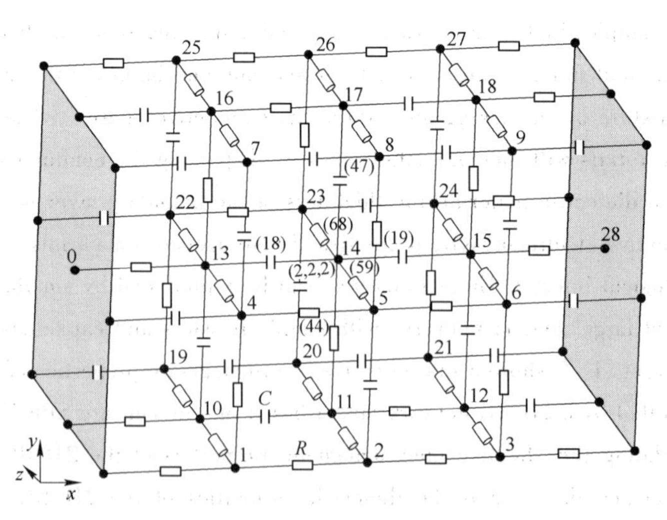

Fig. 1 The layout of 3D RC network model

Each node connects with other nodes is up to six in the 3D RC network model more than that of only four nodes in 2D RC network model[11]. Therefore, the existing 2D algorithm cannot be applied to the 3D RC network model. In our parallel algorithm on the 3D RC network model, marking edge is introduced, as shown in Fig. 2. Every node (except the nodes on the surface) connects with the other six nodes by six edges. Each edge is randomly placed a resistor or a capacitor. The six edges is marked as $P_r(x,y,z)$, $P_l(x,y,z)$, $P_u(x,y,z)$, $P_d(x,y,z)$, $P_b(x,y,z)$ and $P_a(x,y,z)$, respectively. The edges between the left plate and points $(1,y,z)$ are marked as $P_1(1,y,z)$ and the edges between the right plate and points (X_n,y,z) are marked as $P_2(X_n,y,z)$. Here

$$P_1(1,y,z) = Y_n(z-1) + y$$
$$P_2(X_n,y,z) = Y_nZ_n + (X_n-1)Y_nZ_n + Y_n(z-1) + y$$
$$P_r(x,y,z) = Y_nZ_n + (X_n-1)Y_n(z-1) + (X_n-1)(y-1) + x$$
$$P_l(x,y,z) = Y_nZ_n + (X_n-1)Y_n(z-1) + (X_n-1)(y-1) + x - 1$$
$$P_u(x,y,z) = 2Y_nZ_n + (X_n-1)Y_nZ_n + X_n(Y_n-1)(z-1) + X_n(y-1) + x$$
$$P_d(x,y,z) = 2Y_nZ_n + (X_n-1)Y_nZ_n + X_n(Y_n-1)(z-1) + X_n(y-2) + x$$
$$P_b(x,y,z) = 2Y_nZ_n + (X_n-1)Y_nZ_n + X_n(Y_n-1)Z_n + X_nY_n(z-1) + X_n(y-1) + x$$
$$P_a(x,y,z) = 2Y_nZ_n + (X_n-1)Y_nZ_n + X_n(Y_n-1)Z_n + X_nY_n(z-2) + X_n(y-1) + x$$

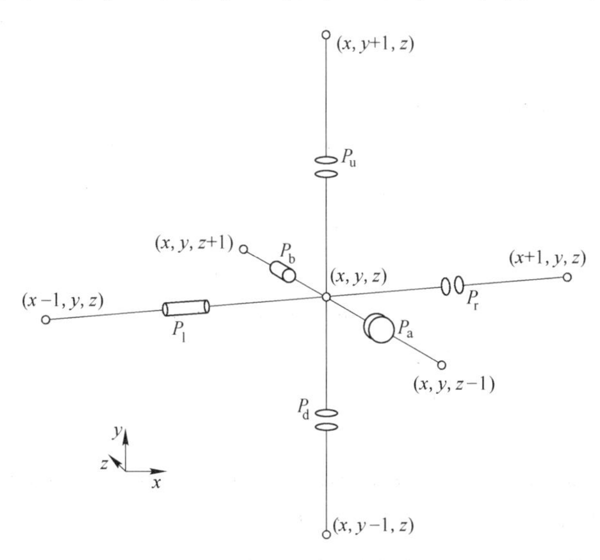

Fig. 2 Marking edge

2.2 Parallel algorithm

For implementing the parallel algorithm, several parameters are defined as follows. M is the number of frequency points in the frequency band of the simulation. Each frequency point is an element of array F, which is denoted as $F = [f_0, f_1, f_2, \cdots, f_{M-1}]^T$. N is the number of available CPUs in the parallel environment. *MyRank* is the current process ID; E is the admittance of the components on the edge, where $E = [e_1, e_2, e_3, \cdots, e_{Bn}]^T$. B_n is the total number of components in the model, where $B_n = X_n[Y_n(Z_n-1) + (Y_n-1)Z_n] + (X_n-1)Y_nZ_n + 2Y_nZ_n$. V is node voltage of the network, where $V = [v_1, v_2, v_3, \cdots, v_{Np}]$. I is node current, where $I = [i_1, i_2, i_3, \cdots, i_{Np-1}, i_{Np}] = [0,0,0,\cdots,0,1]^T$. $A_{Np\times Np}$ is coefficient matrix of equations for the network nodes, and its elements are $a_{s,t}(s, t = 1,2,3,\cdots,N_p)$.

The parallel algorithm is as follow:

Step 1: Initialize $i=0, k=0, A=0$.

Step 2: The process 0 generates B_n random numbers which are uniformly distributed and broadcasted to the other processes.

Step 3: If i is less than $[M/N]$, compute $k = MyRank + Ni, i = i+1$, otherwise go to step 8.

Step 4: If k is less than M, compute $G_R = 1/R, G_C = 1/R_p + 1/(2\pi f_k jC)$ (R_p is a large resistance that connected with capacitance C in parallel, and R_p is equal to $1G\Omega$). According to the ratio of the resistors in the network and the random numbers generated by the process 0, the components are placed on the edges of the network. Otherwise, go to Step 7.

Step 5: Based on the network model, A is constructed with nodes mark and edges mark. For all nodes (except the right plate): (1) If the node (x, y, z) connects with the left plate, then $a_{Mp(x,y,z),Mp(x,y,z)} = a_{Mp(x,y,z),Mp(x,y,z)} + e_{P_1(1,y,z)}$. Otherwise, $a_{Mp(x,y,z),Mp(x,y,z)} = a_{Mp(x,y,z),Mp(x,y,z)} + e_{Pl(x,y,z)}$, $a_{Mp(x,y,z),Mp(x-1,y,z)} = -e_{Pl(x,y,z)}$. (2) If the node (x,y,z) connects with the right plate, then $a_{Mp(x,y,z),Mp(x,y,z)} = a_{Mp(x,y,z),Mp(x,y,z)} + e_{P2(X_n,y,z)}$, $a_{Mp(x,y,z),Np} = -e_{P_2(Xn,y,z)}$. Otherwise, $a_{Mp(x,y,z),Mp(x,y,z)} = a_{Mp(x,y,z),Mp(x,y,z)} + e_{Pr(x,y,z)}$, $a_{Mp(x,y,z),Mp(x+1,y,z)} = -e_{Pr(x,y,z)}$. (3) If the node (x,y,z) is not on the bottom side of the 3D RC network (that is $(y-1)!=0$), then $a_{Mp(x,y,z),Mp(x,y,z)} = a_{Mp(x,y,z),Mp(x,y,z)} + e_{Pd(x,y,z)}$, $a_{Mp(x,y,z),Mp(x,y-1,z)} = -e_{Pd(x,y,z)}$. (4) If the node (x,y,z) is not on the top side of the 3D RC network ($(y+1) <= Y_n$), then $a_{Mp(x,y,z),Mp(x,y,z)} = a_{Mp(x,y,z),Mp(x,y,z)} + e_{Pu(x,y,z)}$, $a_{Mp(x,y,z),Mp(x,y+1,z)} = -e_{Pu(x,y,z)}$. (5) If the node (x,y,z) is not on the back side of the 3D RC network (that is $(z-1)!=0$), then $a_{Mp(x,y,z),Mp(x,y,z)} = a_{Mp(x,y,z),Mp(x,y,z)} + e_{Pa(x,y,z)}$, $a_{Mp(x,y,z),Mp(x,y,z-1)} = -e_{Pa(x,y,z)}$. (6) If the node (x, y, z) is not on the front side of the 3D RC network (that is $(z+1) <= Z_n$), then $a_{Mp(x,y,z),Mp(x,y,z)} = a_{Mp(x,y,z),Mp(x,y,z)} + e_{Pb(x,y,z)}$, $a_{Mp(x,y,z),Mp(x,y,z+1)} = -e_{Pb(x,y,z)}$. For the right plate (that is the last node) $a_{Np,l} = a_{l,Np}$ $(l = 1, 2, 3, \cdots, N_{p-1})$, $a_{Np,Np} = -a_{1,Np} - a_{2,Np} - a_{3,Np} - \cdots - a_{Np-1,Np}$.

Step 6: Solve the equation $AV = I$ using Gaussian elimination method without back substitution. Get the voltage of the right plate and send it to the process 0.

Step 7: Process synchronization, go to step 3.

Step 8: The process 0 saves data, exit the program.

3 Results and discussions

The algorithm is implemented in C and MPI, and simulation results were obtained by using Dawning TC4000A cluster supercomputer. Comparisons of computation time between different number of network components and different algorithms are made, and listed in Table 1. T_s represents the computation time based on sequential algorithm in reference[11]. T_p is the execution time of the parallel algorithm with p processors. It is obvious that the time of the parallel algorithm implemented on 1 CPU is about one tenth of sequential time. It means that the novel algorithm greatly improves the computational efficiency even if just one CPU is used. Table 2 shows the speedup S_p and efficiency E_p of parallel algorithms with different network components. Where, $S_p = T_l/T_p, E_p = S_p/p$, p is the number of processors, T_l is the execution time of the parallel algorithm implemented on one

CPU. It can be seen that the more the number of network components is, the higher speedup and efficiency are.

Table 1 The computation time of the sequential and parallel algorithm

B_n	T_s/s	T_p/s			
		1 CPU	2 CPUs	4 CPUs	8 CPUs
612	5.343	0.45	0.31	0.21	0.28
1472	63.39	5.53	3.1	1.6	1.08
3531	756.468	78.84	45.28	21.28	12.35
8624	10117.359	1047.19	584.4	275.49	158.82
22440	179384.71	17376.9	9581.61	4411.99	2510.24

Table 2 The speedup and efficiency of the parallel algorithm

B_n	2 CPUs		4 CPUs		8 CPUs	
	S_p	E_p/%	S_p	E_p/%	S_p	E_p/%
612	1.452	72.60	2.143	53.56	1.607	20.09
1472	1.784	89.20	3.456	86.40	5.12	64
3531	1.741	87.05	3.705	92.63	6.384	79.80
8624	1.792	89.60	3.801	95.03	6.594	82.43
22440	1.814	90.70	3.939	98.48	6.922	86.53

In order to manifest the validity of the parallel algorithm, the effective permittivities for 3D RC network model containing 612 (network size is $6\times6\times6$), 1472 ($8\times8\times8$), 3531 ($10\times11\times11$), 8624 ($14\times15\times14$), 22440 ($19\times20\times20$) and 32912 ($22\times22\times23$) components with the ratio 20% R − 80% C have been simulated. Results are shown in Fig. 3 and Fig. 4. It is seen that both the real part and imagery part of effective permittivity is nearly independent on the size of the network in the low frequency band. On the other hand, effective permittivity varies with network size in

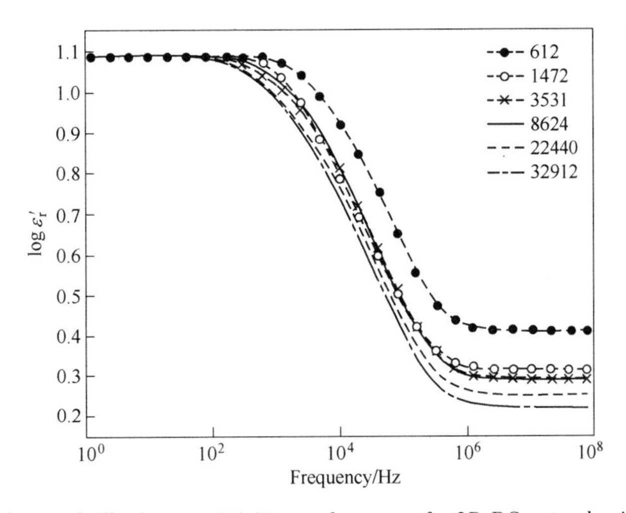

Fig. 3 The real part of effective permittivities vs frequency for 3D RC network with different size

high frequency band. Therefore, the simulation of larger RC network is a key issue for study the dielectric response of heterogeneous materials. In the simulation, the maximum network size is 32912 components[10], and is 37 times that of the network size reported in Ref. [11].

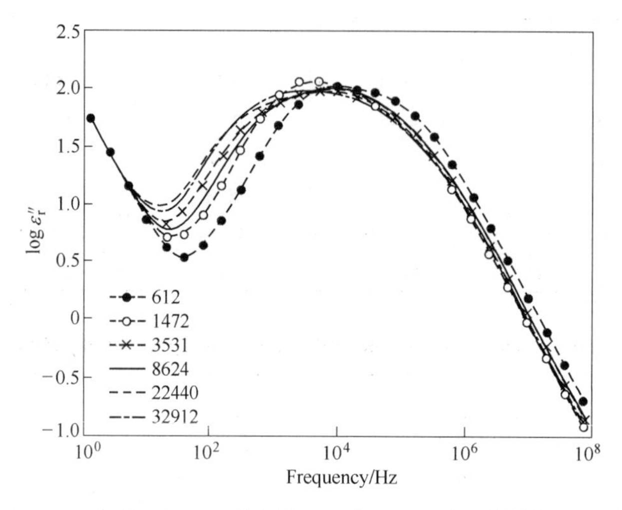

Fig. 4 The imaginary part of effective permittivities vs frequency for 3D RC network with different size

The effective conductance and capacitance for 3D RC networks containing 3531 (10 × 11 × 11) components with the ratios 10% R – 90% C, 20% R – 80% C, ⋯, and 90% R – 10% C have been simulated. Results are shown as Figs. 5 and 6. Both of them show that the linear variation region moves to higher frequency with increased ratio of resistors in the range of 10kHz to 1000kHz. It can be found from Fig. 5 that the slope of the linear variation region of the effective conductance decreases with the increase of resistor ratio in the network, which good agreement with the Eq. (2) in Ref. [12]. On the contrary, with the increase of the ratio of resistors, the slope of the linear variation region of the effective capacitance increases, which is consistent with the Eq. (3) in Ref. [13].

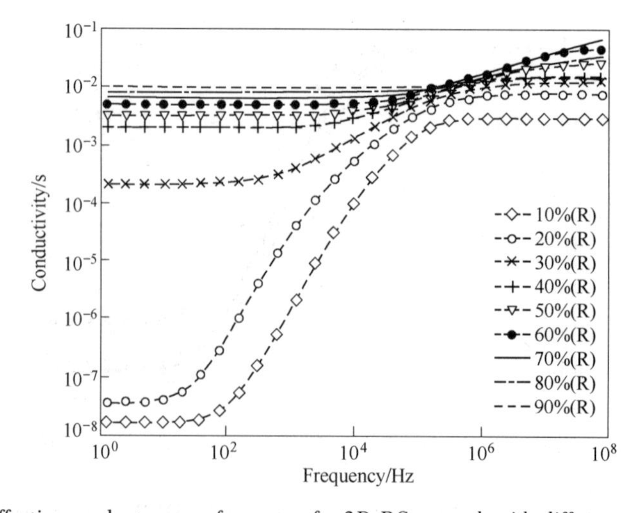

Fig. 5 Effective conductance vs frequency for 3D RC network with different resistor ratio

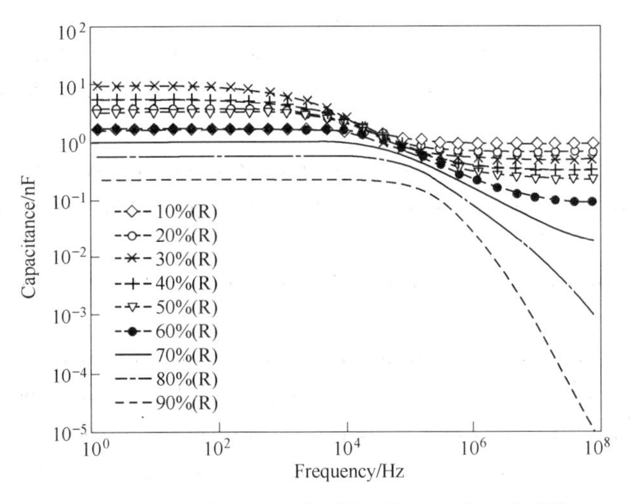

Fig. 6 Effective capacitance vs frequency for 3D RC network with different resistor ratio

4 Conclusions

A parallel algorithm for computing the effective electromagnetic properties of large - scale 3D random RC network model is presented in this paper. The dielectric response for 3D RC network with different components and different resistor ratio has been computed in the frequency band of 1Hz - 100MHz. The results show that the efficiency of the parallel algorithm gradually increases with the number of network components. The algorithm is valuable for researching the dielectric response of heterogeneous materials. The development of an online version based on this algorithm is a target of our lab in the future study. This would promote the application of the RC network model in the fields of microwave metallurgical engineering and materials design.

Acknowledgements

This work was supported by the National Natural Science Foundation of China (Grant No. 61161007), Scientific Research Fund Major Project of the Education Bureau of Yunnan Province (Grant No. ZD2011003), the Natural Science Foundation of Yunnan Province (project 2011FB018), the Scientific Research Foundation of Yunnan University (Grant No. 2010YB025).

References

[1] Olson G B. Computational design of hierarchically structured materials [J]. Science, 1997, 277 (5330): 1237 - 1242.

[2] Shelby R A, Smith D R, Schultz S. Experimental verification of a negative index of refraction [J]. Science, 2001, 292(5514): 77 - 79.

[3] Brosseau C. Modelling and simulation of dielectric heterostructures: a physical survey from an historical perspective [J]. Journal of Physics D: Applied Physics, 2006, 39(7): 1277.

[4] Sihvola A. Metamaterials in electromagnetics [J]. Metamaterials, 2007, 1(1): 2 - 11.

[5] Yang J J, Huang M, Yang C F, et al. Electromagnetic properties of heterogeneous materials and the local electri-

cal field enhancement effects[J]. Materials Review,2009,23:1 – 4.

[6]Sun J,Huang M,Peng J H,et al. The simulation of the frequency – dependent effective permittivity for composite materials [C]//Antennas Propagation and EM Theory (ISAPE), 2010 9th International Symposium on. IEEE,2010:701 – 704.

[7]Huang M Y. A Novel Technology and Theory Research for Preparing Titanium – rich Materials From Ilmenite Concentrate with High CaO and MgO Contents[D]. Kunming:Kunming University of Science and Technology, 2008.

[8] Nan C W. Heterogeneous Materials Physics: Microstructure and Property Link [M]. Beijing: Science Press,2005.

[9]Almond D P,Vainas B. The dielectric properties of random R – C networks as an explanation of the universal power law dielectric response of solids[J]. Journal of Physics:Condensed Matter,1999,11(46):9081.

[10]Bouamrane R,Almond D P. The "emergent scaling" phenomenon and the dielectric properties of random resistor – capacitor networks[J]. Journal of Physics:Condensed Matter,2003,15(24):4089.

[11] Xiao Z,Huang M,Wu Y F,et al. Modelling the universal dielectric response in heterogeneous materials using 3 – D RC networks[J]. Acta Physica Sinica,2008,57:957 – 961.

[12]Huang M,Yang J J,Xiao Z,et al. Modeling the dielectric response in heterogeneous materials using 3D RC networks[J]. Modern Physics Letters B,2009,23(25):3023 – 3033.

[13]Almond D P,Bowen C R,Rees D A S. Composite dielectrics and conductors:simulation,characterization and design[J]. Journal of Physics D:Applied Physics,2006,39(7):1295.

Chapter Ⅱ

New Technology of Microwave Applications in Material and Chemical Engineering

Microwave Plasma Sintering of Nanocrystalline Alumina

Jinhui Peng, Pinjie Hong, Shushan Dai, Dorothee Vinga Szabo

Abstract: Sintering of nanocrystalline alumina by microwave plasma has been studied. The relative density, microhardness of the samples with different grain sizes by microwave plasma and other sintering techniques have been compared.

Keywords: microwave plasma sintering; nanocrystalline alumina

1 Introduction

To reduce the final grain size and the densification temperature and to obtain good mechanical properties of nanocrystalline powders, some particular sintering techniques (such as sinter - forging[1], hot pressing sintering[2], vacuum sintering[3-6], microwave sintering[7-9] and spark plasma sntering or plasma - activated sintering[10-12]) have been studied. Plasma sintering can promote rapid densification and restrain grain growth in sintered materials[13,14], yet microwave plasma sintering of nanocrystalline alumina has scarcely been reported. It is the intention of this paper to prove whether it is possible to sintering nanocrystalline alumina using microwave plasma.

2 Experimental

The nanocrystalline alumina powders were produced by microwave plasma described previously[15-18]. The reaction was performed in a 50mm diameter vessel made of quartz, passing a TE_{01} mode cavity in succession connected to a 0.915GHz microwave generator. A mixture of argon with 20 vol. pct oxygen was used as plasma gas. The average grain sizes of nanocrystalline alumina were 4 - 8nm, 50 - 100nm alumina powders from Russia were also used.

The nanocrystalline alumina powders were pressed at 400MPa into 0.8cm diameter discs. The single mode (TE_{113}) tunable cylindrical resonant cavity with 2.45GHz was used as the microwave plasma sintering apparatus. The presintering step by adjusting the position of discs in the plasma zone was necessary to prevent the discs from exploding during rapid heating.

Microhardness was measured on polished surface by the Vickers microhardness tester (Fischerscope H100, Helmut Fischer Ltd. Germany), using loads of 100mN for 20s. Typical standard deviation was ± 6 GPa for Vickers microhardness. Generally not less than ten different regions were sampled to obtain the average hardness of each specimen.

TEM specimens were prepared by mechanical thinning and dimpling. Final thinning to electron transparency was done by argon ion - beam milling.

3 Results and discussions

3.1 Sintering behaviour

The densities of the samples with different grain sizes and in both microwave plasma and conventional furnaces are plotted in Fig. 1 as function of sintering time.

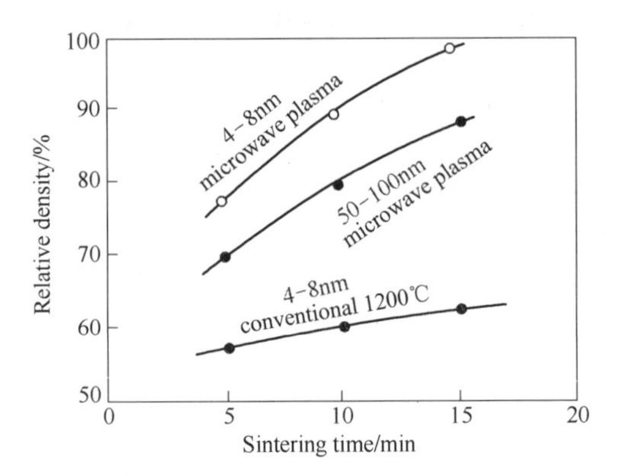

Fig. 1 Relative density of alumina as a function of sintering time

It was observed that 4 ~ 8nm nano - alumina could be sintered by microwave plasma to 99% of relative density in 15min whereas similar samples treated by a conventional furnace only to 63% in the same time at 1200℃.

Fig. 1 shows sintering curves of two nanoalumina samples with average green grain sizes of 4 ~ 8nm and 50 ~ 100nm by microwave plasma. It can be seen that densification of samples varies with grain sizes. The finer the green grain size, the faster the densification. Full density of the samples with 4 ~ 8nm green grain sizes can be achieved by microwave plasma for 15min, but the samples with 50 ~ 100nm green grain sizes only reached 89% of relative density under otherwise equal conditions.

The kinetics of densification in a monodispersed powder is described by the following equation

$$\partial\rho/\partial t = A\nu_n\Omega/KT(D_b\delta_b/d^4 + D_v/d^3) \qquad (1)$$

where ρ is the density; ν_n is the surface free energy; Ω is the atomic volume; D_b and D_v are the grainboundary and bulk diffusion coefficients respectively; δ_b is the grain boundary thickness. According to the Eq. (1), the densificatian rate is strongly dependent on the initial grain size d. The results of this paper are in good agreement with this equation.

3.2 Comparation of different synthesis and sintering techniques of alumina

A typical transmission electron micrograph reveals that the average grain size of alumina with 99% relative density exceeds 1000nm (see Fig. 2). This result is somewhat similar to what has been observed in the microwavc sintering and plasma activated sintering processes[7-9].

Table 1 compares the different synthesis and sintering techniques of alumina. It is also apparent that microwave plasma sintering of alumina with grain size less than 100nm is difficult, perhaps because of the γ phase to a phase transformation and associated texture[10-12].

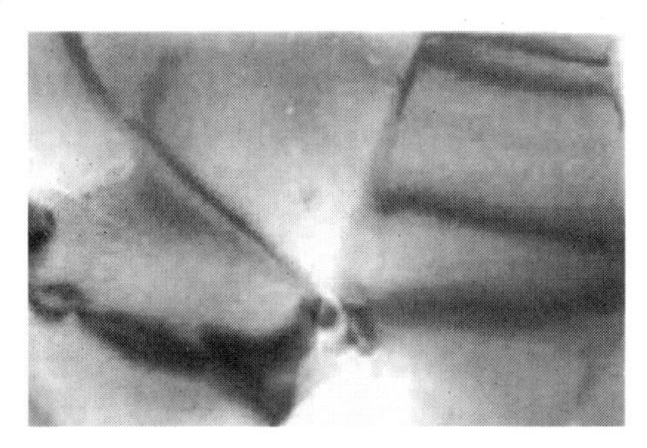

Fig. 2 Transmission electron micrograph of sinterted alumina with 99% relative density by microwave plasma

Table 1 Comparison of different synthesis and sintering techniques of alumina

Green alumina Synthesis technique	Phase	Grain size/nm	Sintering technique	Phase	Sinteredalumina relative density/%	Grain /nm	Ref.
Gas condftnsation synthesis	γ/α	5 ~ 20	Microwave sintering	α	89	> 1000	[7]
Gas condensation syotheais	γ	3 ~ 28	Plasma activated sintering	α	96	316	[12]
Gas coadensation synthesis	amorphous and γ	4 ~ 8	Microwave plasma sintering	α	99	> 1000	Thiswork

3.3 Microhardness

The average Vickers microhardness of sintered alumina in ambient conditions is shown in Fig. 3 as a function of the relative density for two samples with different initial grain sizes.

It can be seen that Vickers microhardness improves with increasing the relative densities. For example, Vickers microhardness of nano – alumina with 4 – 8nm initial grain sizes increases from 12GPa to 29GPa when the relative density increases from 78% to 99%. It is also clear from Fig. 2 that the better Vickers microhardness is yielded from the sample of smaller initial grain size.

Hardness data for alumina are compared in Table 2. Hardness values of present work are a little greater than the corresponding hardness values of sintered alumina and lower than bulk values. That could explain why hardness values are influenced by densification, grain size, indentation load and measurement methods.

Table 2 Comparison of hardness data for alumina

Handness/GPa	Load/N	Grain size/μm	Material	Ref.
29 ± 6	0.10	1	Sintered alumina	This work
19.96	3.0	0.45	Sintered alumina	[19]
25.51	3.0	3	Sintered alumina	[19]
18.64	3.0	10	Sintered alumina	[19]
21 ± 3	0.12		Polycrystalline bulk alumina	[20]
34 ± 3	0.12		Polycrystalline bulk alumina	[20]
40 ± 8	0.12		Sapphire (c – axis)	[20]

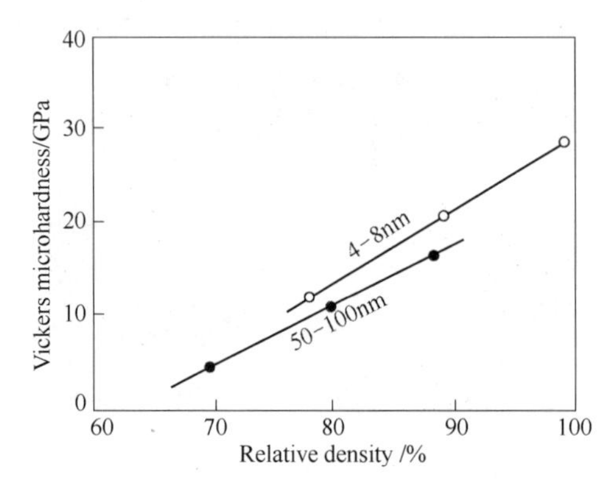

Fig. 3 Vickers microhardness of alumina as a function of reative density

4 Conclusion

Microwave plasma enables sintering of nanocrystalline alumina to be completed in a short time, but it is difficult to fully densify alumina without significant grain size growth.

References

[1] Hofler H J, Averback R S. Mater. Res. Soc. Symp. Proc, 1993, 286: 9.

[2] Hahn H, Logas J, Averback R S. Sintering characteristics of nanocrystalline TiO_2 [J]. Journal of Materials Research, 1990, 5(03): 609 - 614.

[3] Iijma S. Jpn. J Appl Phys, 1984, 23(6): 347.

[4] Sadangi R K, Kear B H, McCandlish L E. Nanostructured Materials, 1996, 6: 69.

[5] Skandan G. Nanostructured Materials, 1995, 5(2): 111.

[6] Bonerich J E, Marks L D. Mater. Res. Soc. Symp. Proc, 1993, 286: 3.

[7] Freim J, McKittrich J, Katsz J, Sickafus K. Nanostructured Materials, 1994, 4(4): 370.

[8] Zhang J S, Yang Y J, Oao L H. Mater. Res. Soc. Symp. Proc, 1994, 347: 591.

[9] Bykov Y, Gusev S, Eremeev A, Malygin N, et al. Nanostructured Materials, 1995, 5: 855.

[10] Nishimura T, Mitomo M, Hirosturu H, Kawahara M. J. of Mater. Sci. Lett, 1995, 14: 1046.

[11] Tracy M J, Groza J R. Nanostructured Materials, 1995, 14: 1046.

[12] Mishra R S, Schneider J A, Shackelford J F, et al. Nanostructured Materials, 1995, 5: 525.

[13] Bennett C E G, McKinnon N A, Williams L. S. Nature, 1968, 217: 1287.

[14] Bennett C E G, McKinnon N A. Glow Discharge of Alumina, in Kinetics of Reactions in Ionic Systems [M]. New York: Plenum Press, 1969: 408.

[15] Vollath D, Sickafus K E. Nanostructured Materials, 1992, 1: 427.

[16] Vollath D, Sickafus K E. Nanostructured Materials, 1993, 2: 451.

[17] Vollath D, Sickafus K E. Nanostructured Materials, 1994, 3: 927.

[18] Brook R J. Proc Brit. Ceram. So., 1982, 32: 7.

[19] Krell A J. Am. Ceram. Soc., 1995, 78: 1417.

[20] Nicholls J R, Hall D J, Tortorelli P F. Materials at High Temperature, 1994, 12(2): 141.

Influences of Temperatures on Tungsten Copper Alloy Prepared by Microwave Sintering

Lei Xu, Mi Yan, Jinhui Peng, C. Srinivasakannan,
Yi Xia, Libo Zhang, Guo Chen, Hongying Xia, Shixing Wang

Abstract: The CuW80 alloy was prepared by the method of microwave vacuum sintering. The effects of temperatures on the performance of CuW80 alloy was assessed based on the relative density and hardness. The micro structure of alloy was characterized using scanning electron microscopy, while XRD was utilized to identify the structure changes. Experimental results indicate CuW80 alloy with excellent performance under vacuum conditions prepared by microwave sintering. Density was found to increase with increase in sintering temperature linearly until 1200℃, while the rate of increase was found to reduce at higher temperatures, reaching an asymptote. The maximum relative density of the alloy was estimated to be 97.95% at 1300℃. At a sintering temperature of 1200℃, CuW80 alloy was more uniform with the main phase of alloy being $Cu_{0.4}W_{0.6}$ (PDF:50 - 1451) and maximum hardness being 222 HBS.
Keywords: Cu - W alloy; microwave sintering; microstructure; material properties

1 Introduction

Cu - W is an important alloy as it combines many fine features of tungsten and copper. The presence of W is favorable to enhance the properties such as high melting point, high - density, arc - erosion resistance, welding resistance, high temperature strength, while the presence of Cu is favorable to enhance the properties such as high electrical conductivity, thermal conductivity, ductility, and processing workability. Cu - W alloy is widely used in high - voltage switch, EDM electrodes and microelectronic materials. An increase in the demand for this alloy is expected with the development of the electronics industry and hence an improved method for preparation of this alloy is imperative[1-5].

Since tungsten and copper are immiscible, the conventional sintering methods will encounter many difficulties in manufacturing full density tungsten - copper composite material. Currently, powder metallurgy sintering method is commonly used in the preparation of Cu - W alloy[3-5]. In this method, compacted sintering is the last step of the production process, which demands improvement in the process such as lowering the sintering temperature and shortening the sintering time.

Microwave is an electromagnetic wave with wavelength ranging from 1mm to 1m, and the frequency at 300MHz to 300GHz[6-11]. Microwave sintering technology has a favorable effect in the heating and sintering the functional ceramics, magnetic materials, carbide and hard alloys fields[10-14]. Compared with conventional sintering techniques, it has the advantages of lower sintering temperature, shorter sintering duration, higher energy and heating efficiency. Application of microwave sintering resulting in a better quality in terms of high density, hardness, toughness and an excellent overall performance is well documented[15-17].

The present work attempts to utilize microwave vacuum sintering method to manufacture Cu – W alloy. The influence of temperature on the properties of the Cu – W alloy adopting microwave sintering is experimentally investigated. In addition, since metal tungsten powder with strong oxidizing, we must combine vacuum sintering technology to ensure the tungsten powder is not oxidized during the sintering process.

2 Experimental method

2.1 Materials

The tungsten powder (purity higher than 99.8%, average particle size, 9.63μm) was provided by Zigong Cemented Carbide Corp. Ltd., China, and electrolytic copper powder (purity higher than 99.7%, average particle size, 21.97μm) was provided by Shanghai Longxin Chemical Industry Co. Ltd., China.

2.2 Preparation of alloys

By using the method of microwave vacuum sintered Cu – W alloy powder compacts, was prepared with 20% copper content (mass percentage). The process steps include mixing – pressing – microwave vacuum sintering.

A mixture of Cu and W powder was prepared with copper content of 20%, with the help of a V – blender (V – 10, Wuxi Fu'an Powder Machinery Co. Ltd., China) by mixing it for 2 – 4h. The mixed powder was subjected to a pressure of 35MPa for die formation, to a sample diameter of 25mm.

The compacted of Cu – W alloy was placed in a silicon carbide crucible, and the microwave vacuum sintering procedure was initiated. The equipment adopted was 2.45GHz microwave vacuum high – temperature sintering furnace. Schematic illustration of the microwave sintering is shown in Fig. 1.

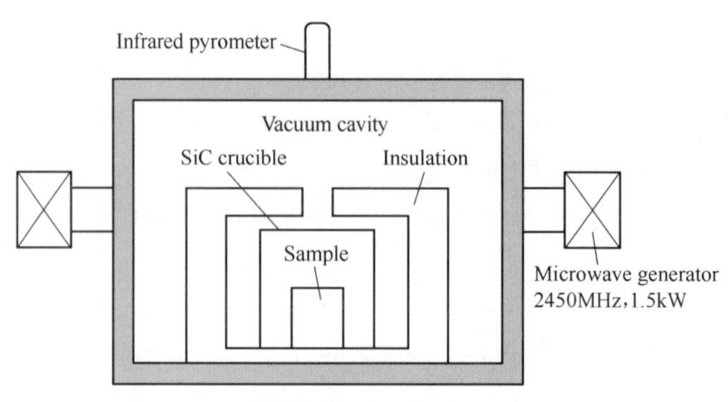

Fig. 1 Schematic of microwave sintering Cu – W alloy

The Cu – W alloy samples are prepared at different temperature. The sintering temperature is controlled at 1100℃, 1150℃, 1200℃, 1250℃ and 1300℃ respectively, for 1h duration. As is well – known, microwave sintering greatly reduces the duration as compared to conventional sintering. A typical comparison of the sintering between conventional and microwave heating schedules is as depicted in Fig. 2.

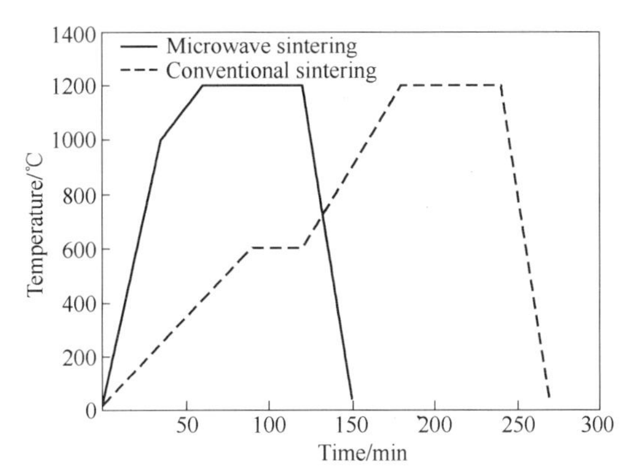

Fig. 2 Comparison of sintering Cu – W alloy using conventional and microwave processing

2.3 Material characterizations

After metallographic polishing of sample to a surface free of scratches, the microstructures of Cu – W alloy were observed by scanning electron microscopy (JSM – 5610LV, JEOL, Ltd., Tokyo, Japan). Sample analysis was conducted by XRD – 7000S diffractometer (Shimadzu, Ltd., Kyoto, Japan) using CuK_α radiation (λ = 0.154060nm). Hardness test was conducted by HB – 3000 hardness tester (Yantai Huayin Test Instrument Co. Ltd., China). The density of sintered body was characterized using Archimedes principle. The relative density of the alloy was calculated as detailed below.

$$\text{Relative Density} = \frac{\text{Sintered density}}{\text{Theoretical density}} \tag{1}$$

3 Results and discussion

3.1 Microstructure of copper – tungsten alloy

Fig. 3 shows the SEM images of CuW80 alloy prepared by microwave sintering at 1100℃, 1200℃ and 1300℃, respectively. As can be seen from the Fig. 3 (a), at a sintering temperature of 1100℃, the distribution of the alloy particles is uneven, tungsten particles aggregate and distributed irregularly. Coating of Cu component on W particle is limited, as well as dispersion of Cu and W. Furthermore, as the agglomerate W particles formed close pores, Cu couldn't fill the pores efficiently. Therefore, at a sintering temperature of 1100℃, more tiny pores exists in the alloy matrix, while at the sintering temperature of 1200℃ and 1300℃, the copper in alloy could migrate better and achieve a better liquid phase rearrangement. Thus, the tiny pores in the alloy matrix are relatively fewer.

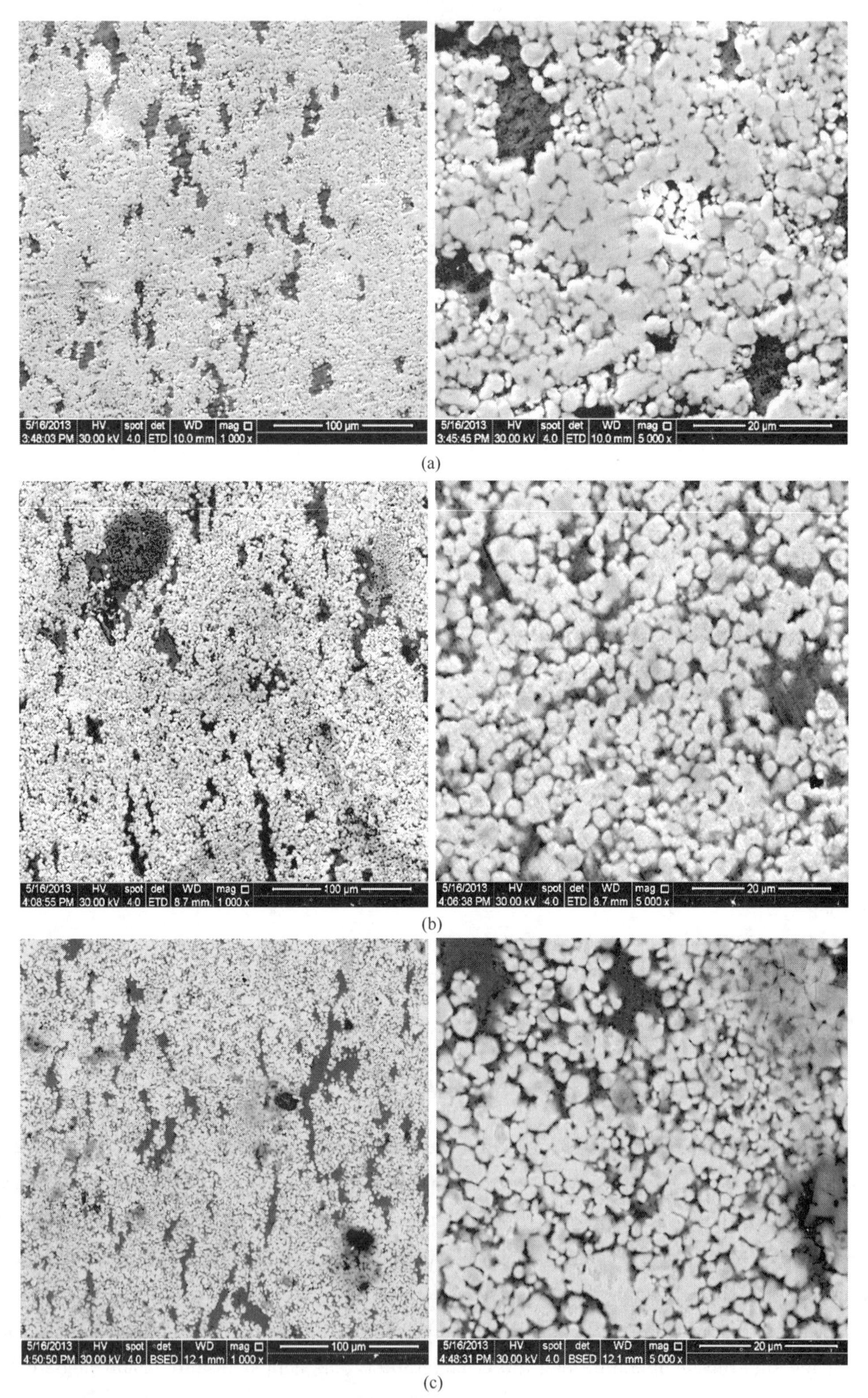

Fig. 3 SEM images of alloy microscopic structure

(a)1100℃; (b)1200℃; (c)1300℃

Fig. 3(b) shows that at a sintering temperature of 1200℃, W is uniform in terms of particle size distribution, with relatively dense structure and relatively better coating of Cu component on W. At a sintering temperature of 1300℃, the size of W was found to increase marginally, however with an uneven particle size distribution of W particle. The distance between W particles was observed to be less, while the contact between the interfaces of W grains was observed to increase marginally. This indicates good wettability of the copper tungsten surfaces. Copper has good mobility, and favorable for liquid phase diffusion and aggregation. With the improvement in the liquid phase rearrangement of the inferior alloy, an enrichment of the metal copper could be observed. Hence a sintering temperature of 1300℃ could be concluded to be better than 1100℃ and 1200℃.

3.2 Density of copper – tungsten alloy

The samples were prepared by wire cutting and were tested for the density. Fig. 4 shows density of copper tungsten alloy prepared by microwave sintering at different temperatures.

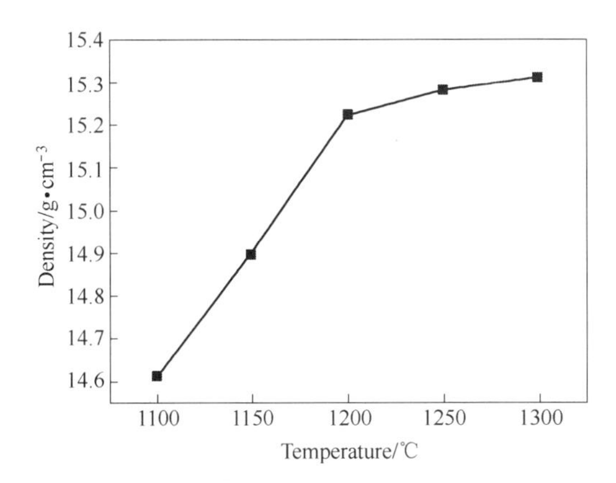

Fig. 4 Variation of the relative density of the CuW80 alloy with sintering temperature

As can be seen from Fig. 4, with the increase in sintering temperature, a linear increase in the density could be observed until a temperature of 1200℃, while it tends towards an asymptote at temperatures higher than 1200℃. Chen, et al. have reported Cu – W alloys with 20% Cu content with the relative density of 95.58% using the powder injection molding method[18]. In the present work the absolute density of Cu – W alloy was estimated to be 15.31g/cm^3 with the relative density being 97.95%, at the sintering temperature of 1300℃.

Luo, et al. have reported the maximum relative density of W – 20Cu alloys sintered at 1250℃ to be 95.7%, however, with a decrease in relative density at 1300℃[19]. Liu, et al. have reported an increase in relative density of the W/Cu FGM samples in the temperature range of 1150℃ to 1350℃, however at a relatively slower rate at temperature in excess of 1250℃[20]. The present work a rapid increase in the density was observed until a sintering temperature of 1200℃, while at a relatively slower rate at temperature in excess of 1200℃. Under the microwave field, copper – tungsten alloy is rapidly heated to a temperature above 1100℃, rather more uniformly, melting copper powder in the alloy, facilitating a better liquid flow driven by capillary force, resulting in more uniform filling of the pores by molten copper. Additionally the tungsten particles are rearranged such a way having lowest total surface area due to tight stuffing. With the increase in sintering temperature, an improvement in flow properties of Cu is evidenced, however, at temperatures near about 1200℃, the tungsten particle rearrangement and pore – filling is complete and hence at higher temperature a relatively slower increase in density.

3.3 Hardness of copper – tungsten alloy

Fig. 5 shows the plot of hardness with respect to microwave sintering temperature of copper tungsten alloy. A significant increase in the hardness of the alloy was observed with increase in sintering temperature until 1200℃, with the highest of hardness being 222 HBS. At sintering temperature in excess of 1200℃ the hardness of alloy was observed to decrease.

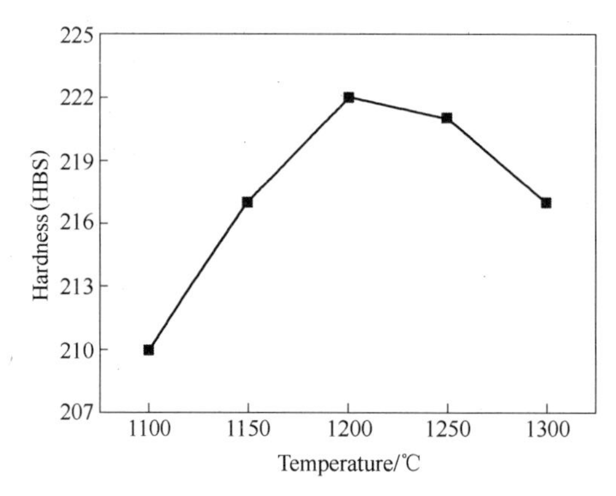

Fig. 5 Variation of the hardness of CuW80 alloy with sintering temperature

In the temperature range less than 1200℃, with rapid increase of temperature, the migration of molten copper and rearrangement of sample particles happen simultaneously in a small region, which fills the pore defects within the sample. At a relatively low temperature, migration of copper component to long range is not easy, also avoids large area of copper enrichment, and grain growth during solidification. All of which contribute to the increase in the hardness of alloy macro level at low temperatures. However, as the sintering temperature continues to rise, it contributes to an increase in both liquid migration of copper in the alloy and rearrangements of tungsten particles, leading to a large area enrichment of the copper and grain growth, which is consistent with the results observed in microstructure.

3.4 XRD analysis

Fig. 6 shows the XRD pattern of copper – tungsten alloy prepared by microwave vacuum sintering at different temperature. According to the Fig. 6, $Cu_{0.4}W_{0.6}$ (PDF: 50 – 1451) and Cu are the main phases in copper tungsten alloy. Liu, et al. have reported that the formation of $Cu_{0.4}W_{0.6}$ alloy phase will be conducive to improve the tissue distribution uniformity and alloy property[21].

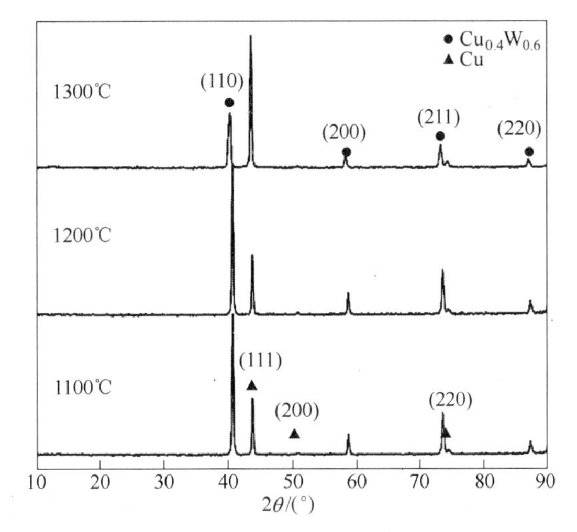

Fig. 6 X - ray diffraction patterns of CuW80 alloy(CuK_{α})

When sintered at 1300℃ ,the highest peak appears at phase Cu,and the lowest peak appears at phase $Cu_{0.4}W_{0.6}$ (PDF:50 - 1451). While sintered at 1100℃ and 1200℃ ,phase composition in the alloy is similar,and the alloy mainly exists in the phase of $Cu_{0.4}W_{0.6}$(PDF:50 - 1451),in the process of liquid phase sintering,part of Cu enters into W crystal,and stays stable,having favorable combination. When sintered at 1300℃ , diffusion of copper in the alloy is relatively smooth, and copper is easier to enrich,which is consistent with the results observed in microstructure.

4 Conclusion

Microwave vacuum sintering technology can be used to prepare CuW80 alloy with excellent performance,with the maximum relative density of alloy being 97.95% ,and hardness of 222 HBS.

At microwave sintering temperatures below 1200℃ ,a sharp increase in the density and hardness was observed,while at temperatures in excess of 1200℃ ,the rate of density increase reduced eventually attaining an asymptotic value at temperature around 1300℃ ,while the hardness was found to decrease beyond a temperature of 1200℃.

Acknowledgements

This work was financially supported by the National Natural Science Foundation of China(Grant No. 51204081),and by China Scholarship Council(No. 2011853521), and by Yunnan Provincial Science and Technology Innovation Talents Scheme - Technological Leading Talent of China (No. 2013HA002),and by Applied Basic Research Project of Yunnan Province of China(Grant No. 2013FZ008), and by Yunnan Provincial Department of Education Research Fund of China (Grant No. 2013Z118).

References

[1] Yang X H,Fan Z K,Liang S H,et al. Effects of TiC on properties and microstructures of Cu - W electrical con-

tact materials[J]. Rare Metal Mater. Eng. ,2007,36(5):817 - 821.
[2]Zhou Z J,Kwon Y S. Fabrication of W - Cu composite by resistance sintering under ultra - high pressure [J]. Journal of Materials Processing Technology,2005,168(1):107 - 111.
[3]Wu W A,Xiao C L,Wang G,Han H Q. Electrical Eng. Mater. ,2004,(4):7 - 11.
[4]Guo Y,Yi J,Luo S,et al. Fabrication of W - Cu composites by microwave infiltration[J]. Journal of Alloys and Compounds,2010,492(1):L75 - L78.
[5]Das J,Chakraborty A,Bagchi T P,et al. Improvement of machinability of tungsten by copper infiltration technique[J]. International Journal of Refractory Metals and Hard Materials,2008,26(6):530 - 539.
[6]Peng J H,Liu B G. Microwave Calcination Technology and Application[M]. Beijing:Science Press,2012.
[7] Zhang Y, Hao W H, Gao J H. Microwave Technologies and Applications [C]. Xi'an: XiDian University Press,2006.
[8]Roy R,Agrawal D,Cheng J,et al. Full sintering of powdered - metal bodies in a microwave field[J]. Nature, 1999,399(6737):668 - 670.
[9]Ernest C O. Microwave Power Engineering[M]. New York:Academic Press,1968.
[10]Agrawal D,Cheng J, Peng H, et al. Microwave energy applied to processing of high - temperature materials [J]. American Ceramic Society Bulletin,2008,87(3):39 - 44.
[11]Lekse J W,Stagger T J,Aitken J A. Microwave metallurgy:synthesis of intermetallic compounds via microwave irradiation[J]. Chemistry of Materials,2007,19(15):3601 - 3603.
[12]Agrawal D. Int. Symp. Adv. Process. Met. Mater. ,2006,4:183 - 192.
[13]Agrawal D,Cheng J,Seegopaul P,et al. Grain growth control in microwave sintering of ultrafine WC - Co composite powder compacts[J]. Powder Metallurgy,2000,43(1):15 - 16.
[14]Chandrasekaran S,Basak T,Ramanathan S. Experimental and theoretical investigation on microwave melting of metals[J]. Journal of Materials Processing Technology,2011,211(3):482 - 487.
[15]Zhou Y,Wang K,Liu R,et al. High performance tungsten synthesized by microwave sintering method [J]. International Journal of Refractory Metals and Hard Materials,2012,34:13 - 17.
[16]Zhu F X,PENG Y D. Sintering response of copper powder metal compact in microwave field[J]. Journal of Central South University:Science and Technology,2009,40(1):106 - 111.
[17]Zhou Y,Sun Q X,Liu R,et al. Microstructure and properties of fine grained W - 15wt. % Cu composite sintered by microwave from the sol - gel prepared powders [J]. Journal of Alloys and Compounds, 2013, 547:18 - 22.
[18]Surreddi K B,Scudino S,Sakaliyska M,et al. Crystallization behavior and consolidation of gas - atomized Al84 Gd6Ni7Co3 glassy powder[J]. Journal of Alloys and Compounds,2010,491(1):137 - 142.
[19]Shudong L,Jianhong Y,YingLi G,et al. Microwave sintering W - Cu composites:analyses of densification and microstructural homogenization[J]. Journal of Alloys and Compounds,2009,473(1):L5 - L9.
[20]Liu R,Hao T,Wang K,et al. Microwave sintering of W/Cu functionally graded materials[J]. Journal of Nuclear Materials,2012,431(1):196 - 201.
[21]Liu T,Fan J L,Cheng H C,Tian J M. Nano - processing Tech. ,4(2007):597 - 603.

Application of Response Surface Methodology(RSM) for Optimization of the Sintering Process of Preparation Calcia Partially Stabilized Zirconia(CaO - PSZ) Using Natural Baddeleyite

Jing Li, Jinhui Peng, Shenghui Guo, Libo Zhang

Abstract: Response surface methodology(RSM) was successfully applied to process of preparation calcia partially stabilized zirconia(CaO - PSZ). Besides that, natural baddeleyite was used as starting materials instead of chemical pure zirconia. The pressureless sintering process was optimized by the application of RSM. The independent variables, which had been found as the most effective variables on the relative density and bending strength by screening experiments, were determined as holding time, sintering temperature and heating rate. Two quadratic models were developed through RSM in terms of related independent variables to describe the relative density and bending strength as the responses. Based on contour plots and variance analysis, optimum operational conditions for maximizing relative density and bending strength, at cooling rate of 3℃/min, were 1540℃ of sintering temperature, 5h of holding time and heating rate of 3℃/min to obtain 98.57% for relative density and 165.72MPa for bending strength.

Keywords: CaO - PSZ; natural baddeleyite; RSM

1 Introduction

High - purity zirconia(ZrO_2) is a white crystalline powder. It exhibits three polymorphs depending on temperature. At room temperature pure zirconia can only be monoclinic[1,2]. As the temperature increase, monoclinic turns into a tetragonal phase at a certain temperature. Further increase in temperature leads to conversion of the tetragonal phase into cubic phase again, and it is a reversible process. The shift between tetragonal and monoclinic phase is of high interest in the transition of zirconia, and it is called martensitic transformation[3]. In view of pure ZrO_2 phase transformation characteristics, especially the volume expansion associated with transformation of tetragonal phase to monoclinic phase cannot be accommodated by zirconia grains, resulting in cracking of ZrO_2 material, unless some specific processes are used[4-6]. Otherwise, it would be difficult to prepare pure phase ZrO_2 material. In order to stabilize the zirconia phase, and ensure the existence of high - temperature phase at room temperature, it is common to use stabilizer.

Common zirconia stabilizers are rare earth or alkaline - earth oxides. It is possible to stabilize the ZrO_2 in the tetragonal and/or cubic forms at room temperature[7], by adding different stabilizers, such as MgO[8-11], CaO[12], Y_2O_3[13-15], CeO_2[16,17], Al_2O_3[18] and even a combination of them[19,20]. Different quantities of stabilizer can cause zirconia to stabilize in different phase composition. If only part of the t - ZrO_2 metastable to room temperature, partially stabilized zirconia (PSZ) is formed.

At present, the raw materials which are used to prepare PSZ ceramic are usually chemical pure or industrial pure. In this work, instead of chemical pure zirconia, natural baddeleyite was used as starting materials to prepare partially stabilized zirconia (PSZ) with calcia as stabilizer. Therefore, it not only decreased the cost, also reduced the social energy consumption and environmental pollution[21]. The production process of partially stabilized zirconia using natural baddeleyite, stabilized by the addition of calcia, has been optimized. The natural baddeleyite was obtained after the flotation of baddeleyite ore.

The traditional one - factor - at - a - time approach to optimization is time - consuming and incapable of reaching a true optimum because of taking no account of interaction among factors. On the contrary, statistical methods can take into account the interaction of variables in generating the process response. Therefore, a statistically designed experiment with minimum experimental runs is greatly desired. Response surface methods (RSM) consist of a group of empirical techniques devoted to the evolution of relations existing between a cluster of controlled experimental factors and the measured responses, useful for developing, improving and optimizing processes by carrying out a limited number of experiments. In this study, "central composite design" (CCD), was applied.

2 Experimental procedure and design

2.1 Experimental procedure

Instead of chemical pure zirconia, the natural baddeleyite (monoclinic ZrO_2) which was obtained by flotation of baddeleyite ore (the ZrO_2 content is not less than 99.30%) with the particle size of 152μm and high purity CaO were selected as the starting powders. Previously, the natural baddeleyite powder was milled to average particle size of 5.67μm using crusher. The experimented composition consisted of 96.2 wt% ZrO_2 (natural baddeleyite, 5.67μm) with additions of 3.8 wt% of CaO. The powders were rigorously wet mixed by means of planetary milling using agate ball (ball - feed weight ratio of (4 - 6) : 1) in ethanol for 12h (average particle size of 1.92μm), and oven - dried at 80℃ for 10h. The dried mixture blended with a determined quantity of binder, and was uniaxially pressed at 150MPa by single action at a constant strain for 8min. The size of samples was 3mm × 4mm × 40mm and U15mm × 3mm, respectively.

Subsequently, the samples were heat treated at 850℃ in air at a heating rate of 5℃/min during 4h in order to burn out the binder. After the discharge treatment, the furnace was allowed to cool at the cooling rate of 5℃/min. Finally, the samples were placed into a corundum board with ZrO_2 bed powder and sintered at the design experiment condition. The bulk densities of the sintered ceramics were determined geometrically and by Archimedes principle with distilled water as the medium. The bending strength of bars was measured by three point bending. In the tests crosshead speed was 0.5mm/min. Five measurements were averaged to minimize the error.

2.2 Experimental design

Response surface method(RSM) can describe the relationship between factors and the response accurately through using reasonable experiment design and fitting the relationship between multi - factor experiment factors and the level with the polynomial. RSM is a solution to the problem of multi - variable statistical methods. RSM usually contains three steps: (1) design and experiments; (2) response surface modelling through regression; (3) optimization.

Central composite design(CCD) is a kind of RSM. It is a test design method developed on the basis of a two level full factorial and partial experimental design. It is well suited for fitting a quadratic surface, which usually works well for process optimization. It is composed of a core factorial that forms a cube with sides that are two coded units in length(from -1 to $+1$). Therefore, it can evaluate the non - linear relationship between the assessment of indicators and factors.

The CCD was applied using the Design Expert software. The total number of experiments with three variables was 20($=2k+2k+6$), where k is the number of independent variables. Fourteen experiments were augmented with six replications at the center values(zero level) to evaluate the pure error. These data acquired from the experimental runs are then used to optimize sintering process. In this study, the response variables measured were relative density and bending strength.

3 Results and discussion

Experimental design along with the observed responses is shown in Table 1.

Table 1 Design matrix and responses

Run	X_1	A	X_2	B	X_3	C	Y_1	Y_2
18	—	1450	—	3	—	3	88.92	126.729
1	—	1450	—	3	+	7	92.21	136.568
17	—	1450	+	5	—	3	95.75	152.309
5	—	1450	+	5	+	7	93.87	151.82
3	+	1550	—	3	—	3	97.69	160.179
13	+	1550	—	3	+	7	96.46	153.373
19	+	1550	+	5	—	3	97.9	163.85
14	+	1550	+	5	+	7	97.09	159.99
15	-1.6818	1415.91	0	4	0	5	88.92	126.729
10	1.6818	1584.09	0	4	0	5	98.48	169.951
9	0	1500	-1.6818	2.3182	0	5	96.43	153.017
2	0	1500	1.6818	5.6818	0	5	97.71	160.256
20	0	1500	0	4	-1.6818	1.64	97.74	162.357
16	0	1500	0	4	1.6818	8.36	96.4	152.945
4	0	1500	0	4	0	5	96.49	155.234
6	0	1500	0	4	0	5	96.65	157.992

Continues Table 1

Run	X_1	A	X_2	B	X_3	C	Y_1	Y_2
7	0	1500	0	4	0	5	96.62	157.241
8	0	1500	0	4	0	5	96.58	156.976
11	0	1500	0	4	0	5	96.48	154.873
12	0	1500	0	4	0	5	96.5	156.014

Note: Variables: A—sintering temperature (℃); B—holding time (h); C—heating rate (℃/min); responses: Y_1—relative density (%); Y_2—bending strength (MPa).

Fitting the data to various models (linear, two factorial, quadratic and cubic) and their subsequent ANOVA showed that relative density and bending strength were most suitably described with quadratic polynomial model (Eqs. (1) and (2)):

$$Y_1 = 96.57 + 2.19A + 0.58B - 0.54C - 0.54AB + 0.20AC - 0.14BC - 1.23A^2 + 0.069B^2 + 0.20C^2 \quad (1)$$

$$Y_2 = 156.47 + 8.82A + 3.76B - 2.84C - 2.70AB - 1.01AC + 0.57BC - 4.48A^2 - 0.47B^2 + 1.24C^2 \quad (2)$$

The estimated model coefficients and their significance in relation to the experimental scatter are shown in Table 2. At a 90% (Table 2) confidence level, the model equation in terms of actual factors is (Eqs. (3) and (4)):

$$Y_1 = 96.57 + 2.19A + 0.58B - 0.54C - 0.54AB - 1.23A^2 \quad (3)$$

With the quadratic factors such as B^2 and C^2 and then the interaction factors AC and BC are not significant.

$$Y_2 = 156.47 + 8.82A + 3.76B - 2.84C - 4.48A^2 \quad (4)$$

With the quadratic factors such as B^2 and C^2 and then the interaction factors AB, AC and BC are not significant.

Table 2 Estimated model coefficients and their significance in relation to the experimental scatter

Coefficients of model	Coefficients estimate	Degrees of freedom	Sum of squares	F value	Signification
For Y_1					
b_0	96.57	—	—	—	—
A	2.19	1	65.39	152.51	S
B	0.58	1	4.54	10.59	S
C	-0.54	1	4.03	9.40	S
A^2	-1.23	1	21.64	50.48	S
B^2	0.069	1	0.068	0.16	NS
C^2	0.20	1	0.57	1.34	NS
AB	-0.54	1	2.38	5.54	S
AC	0.020	1	3.2E-003	7.464E-0003	NS
BC	-0.14	1	0.16	0.38	NS

Continues Table 2

Coefficients of model	Coefficients estimate	Degrees of freedom	Sum of squares	F value	Signification
			For Y_2		
b_0	156.47	—	—	—	—
A	8.82	1	1062.94	141.48	S
B	3.76	1	193.32	9.48	S
C	-2.84	1	110.48	5.89	S
A^2	-4.48	1	17.84	288.92	S
B^2	-0.47	1	0.09	3.13	NS
C^2	1.24	1	0.09	21.99	NS
AB	-2.70	1	2.2	58.42	NS
AC	-1.01	1	0	8.21	NS
BC	0.57	1	0.21	2.57	NS

The statistical significance of the model equation was evaluated by the F - test for analysis of variance(ANOVA). ANOVA evaluations of this model, shown in Table 3, imply that this model can describe the experiments. As can be seen in Table 3 the prob > F - values for relative density and bending strength are lower than 0.05 indicating that quadratic models were significant. The coefficient of determination(R^2) that was found to be close to 1(0.96 for both Y_1, and 0.92 for Y_2) also advocated a high correlation between observed and predicted values. The "lack of fit tests" compares the residual error to the "Pure Error" from replicated experimental design Mpoints. The p - values, greater than 0.05, for both the responses indicate that lack of fit for the model was insignificant. Adequate precision measures the signal to noise ratio and a ratio greater than 4 is desirable. The adequate precision for Y_1 and Y_2 were 19.598, and 13.438, respectively. These high values of adequate precision demonstrated that models are significant for the process.

Table 3 ANOVA analysis for responses Y_1(relative density(%)) and Y_2(bending strength(MPa))

Source	Sum of squares	DF	Mean square	F value	Prob > F	
			For Y_1			
Model	100.02	9	11.11	25.92	< 0.0001	Singificant
Residual	4.29	10	0.43			
Lack of Fit	4.26	5	0.85	159.37	< 0.0001	Significant
Pure error	0.027	5	5.35E-03			
			R^2 = 0.96			
			Adeq precision = 19.598			
			For Y_2			
Model	1766.36	9	196.26	12.97	0.0002	Significant
Residual	151.36	10	15.14			
Lack of fit	143.94	5	28.79	19.42	0.0027	Significant
Pure error	7.41	5	1.48			
			R^2 = 0.92			
			Adeq precision = 13.438			

On the model analysis of variance, the correlation coefficients of quadratic regression equations of relative density and bending strength are $R^2 = 0.96$ and $R^2 = 0.92$. And that indicate model fit very well the actual situation. F – values which were 25.92 and 12.97 imply that the models are significant.

The actual and predicted relative density and example illustrates the isoresponse curves are plotted in Figs. 1 – 3, respectively. Figs. 2 and 3 depicted the change of relative density and bending strength with sintering temperature, holding time and heating rate, plotted for the case where the cooling rate is 3℃/min. And the effect caused by factors on response variation was showed in Fig. 4.

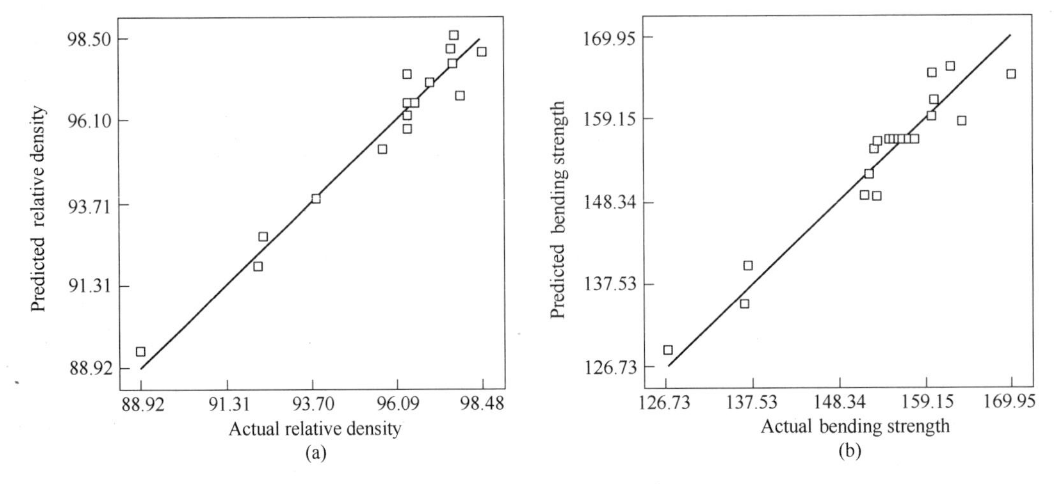

Fig. 1 Predicted response vs. actual response

The sintering temperature and heating rate have different effect on the responses (relative density and bending strength). Sintering temperature is positive, heating rate is negative. High temperature and low heating rate were profitably for reducing the rate of pore so as to increase the relative density of sintered body. The relative density and bending strength increase gradually with lengthened holding time and then become flat. Extending the holding time has good effect on the responses, but later nearly no effect when the transformation from monoclinic to tetragonal reaching equilibrium. Extension of holding time could help grain growth and densification of the sintered bodies. Because the sintering samples contain dozens of stomata and there exists only point contact between particles. On the influence of high temperature with low heating rate and extension of holding time, particles aggregate and connected pores turned into isolate and gradually reduced, and even disappeared to reach the final densification. But if densification was finished, increasing temperature could enter over burn state which would produce more fluxing holes and lower the density of sintered body. The results were obtained that the maximum relative density of 98.48% and bending strength of 169.951 were got at the 3℃/min of cooling rate.

The mathematical model generated during RSM implementation was validated by conducting experiment on given optimal medium setting. Process parameters of experimental optimization are shown in Table 4. The optimized parameters are sintering temperature of 1540℃, holding time of 5h, heating rate of 3℃/min.

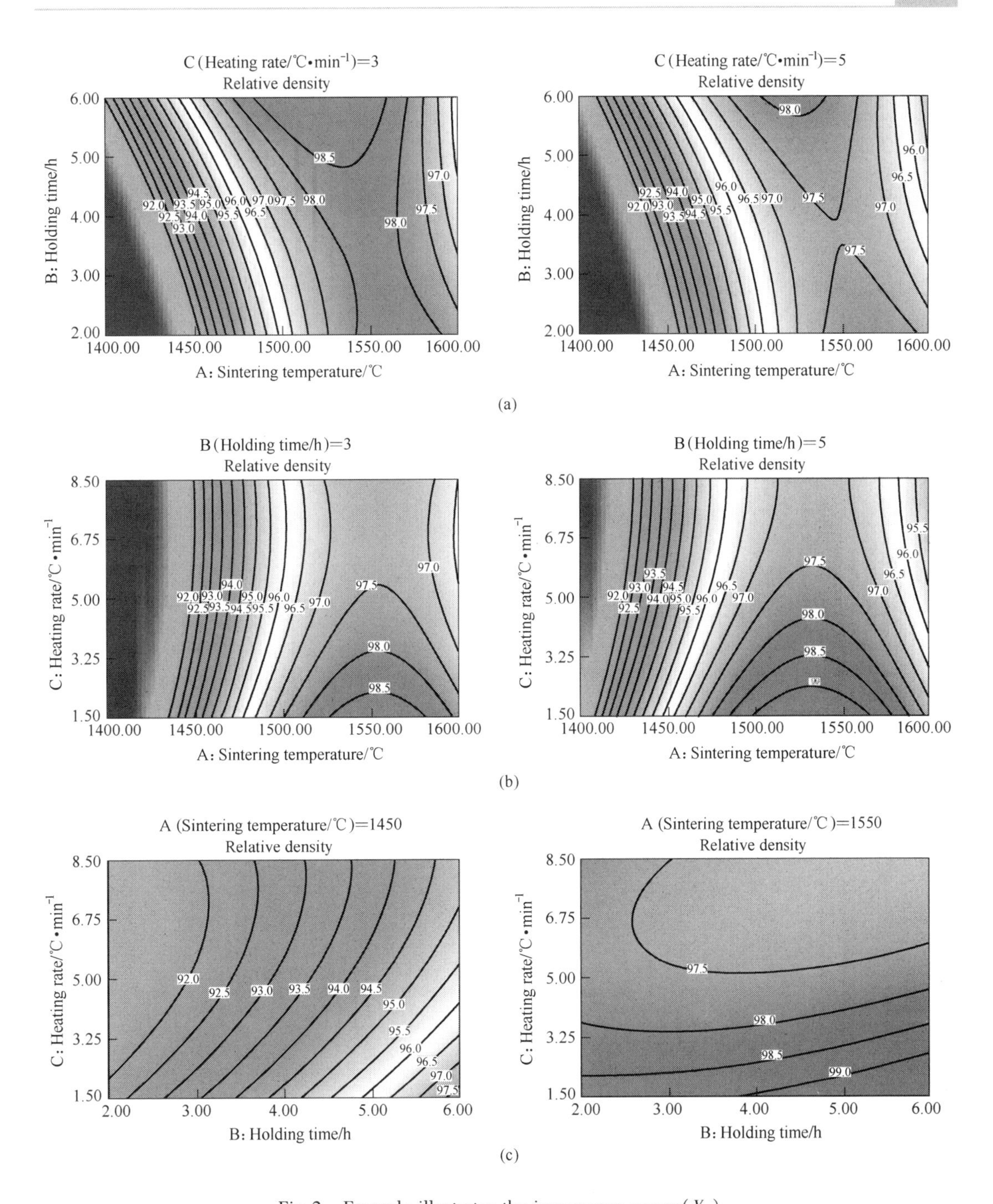

Fig. 2 Example illustrates the isoresponse curves(Y_1)

(a) Effect caused by sintering temperature and holding time on relative density variation with heating rate of 3℃/min and 5℃/min, respectively;

(b) Effect caused by sintering temperature and heating rate on relative density variation with holding time of 3h and 5h, respectively;

(c) Effect caused by holding time and heating rate on relative density variation with sintering temperature of 1450℃ and 1550℃, respectively

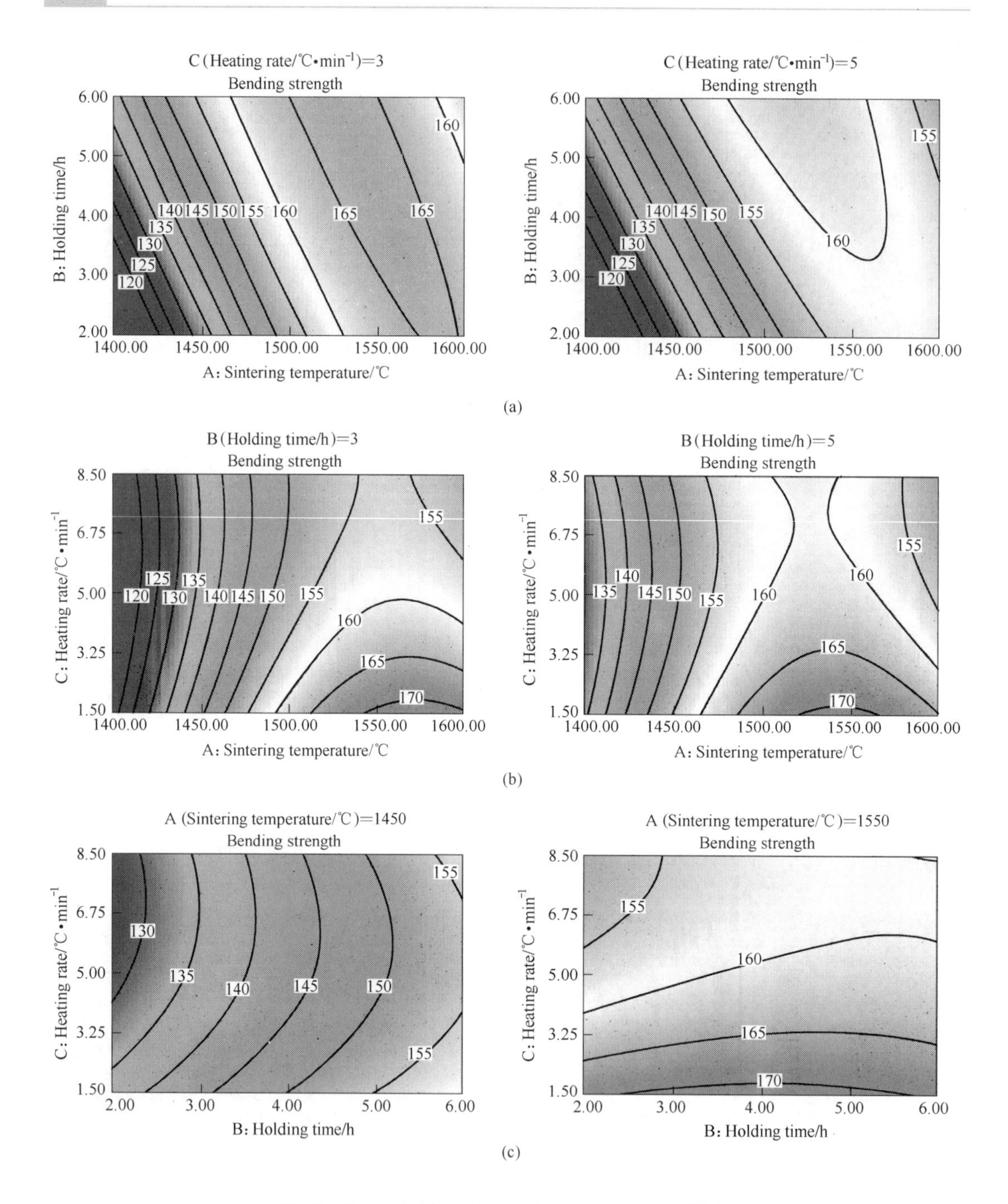

Fig. 3 Example illustrates the isoresponse curves(Y_2)

(a) Effect caused by sintering temperature and holding time on bending strength variation with heating rate of 3℃/min and 5℃/min, respectively;

(b) Effect caused by sintering temperature and heating rate on bending strength variation with holding time of 3h and 5h, respectively;

(c) Effect caused by holding time and heating rate on bending strength variation with sintering temperature of 1450℃ and 1550℃, respectively

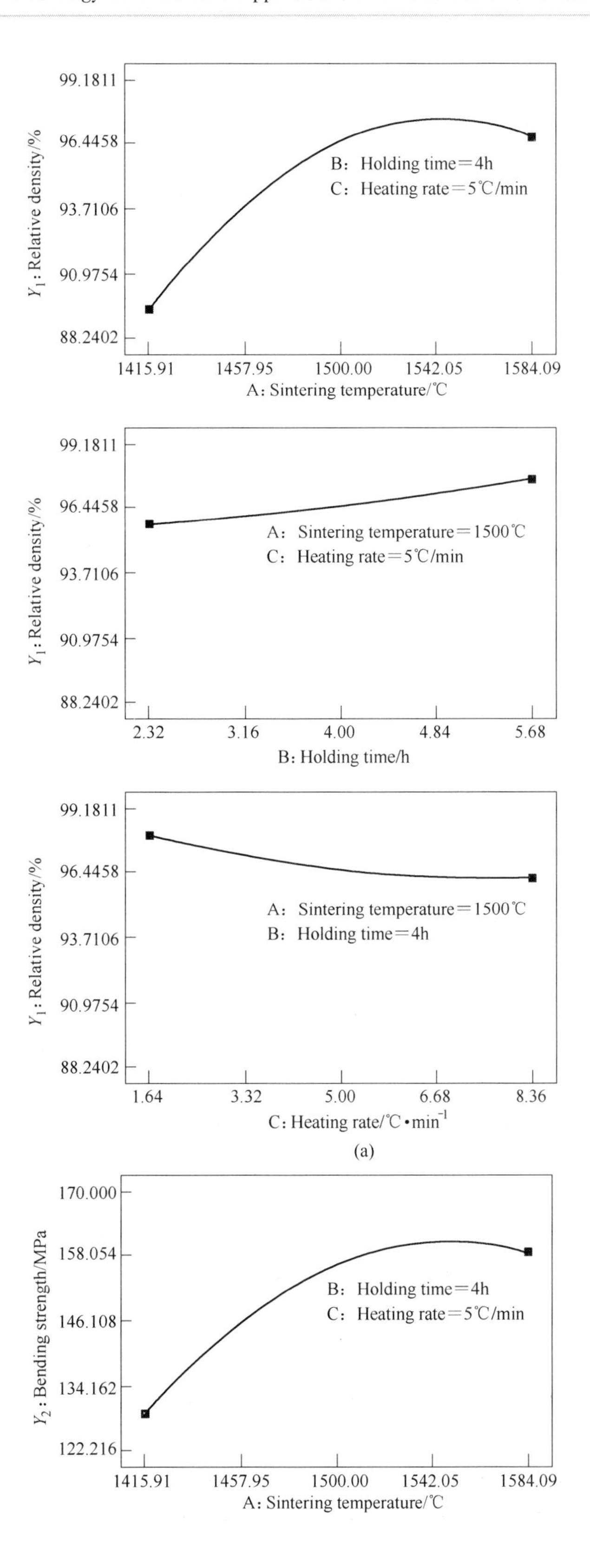

99.1811
96.4458
93.7106
90.9754
88.2402
Y_1: Relative density/%
B: Holding time=4h
C: Heating rate=5℃/min
1415.91
1457.95
1500.00
1542.05
1584.09
A: Sintering temperature/℃
A: Sintering temperature=1500℃
C: Heating rate=5℃/min
2.32
3.16
4.00
4.84
5.68
B: Holding time/h
A: Sintering temperature=1500℃
B: Holding time=4h
1.64
3.32
5.00
6.68
8.36
C: Heating rate/℃·min^{-1}
(a)
170.000
158.054
146.108
134.162
122.216
Y_2: Bending strength/MPa
B: Holding time=4h
C: Heating rate=5℃/min
A: Sintering temperature/℃

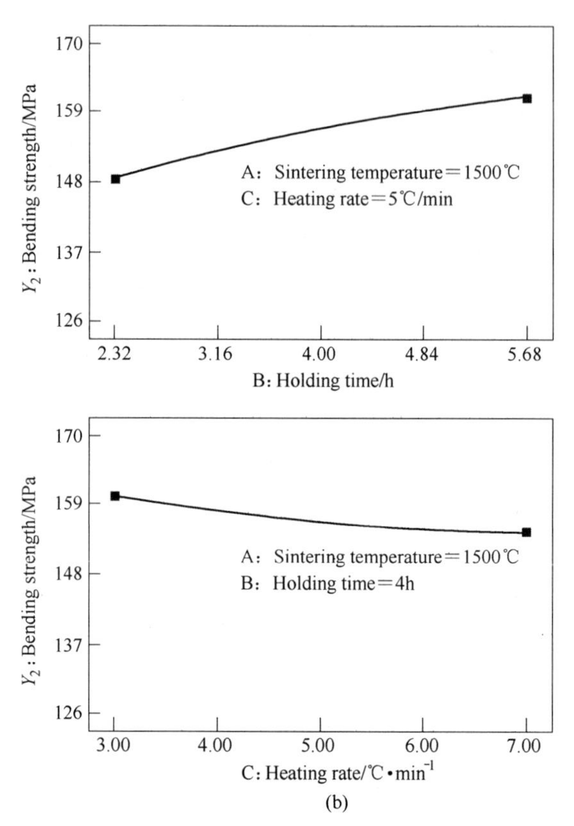

Fig. 4 The effects curves of three factors on responses(Y_1 and Y_2)

(a)The effects curves of three factors on Y_1; (b)The effects curves of three factors on Y_2

Table 4 Predicted values vs. validation experiment values

Sintering temperature x_1/℃	Holding time x_2/h	Heating rate x_3/℃ · min^{-1}	Relative density/%		Bending strength/MPa	
			Predicted value	Experiment value	Predicted value	Experiment value
1540	5	3	98. 61	98. 57	166. 124	165. 72

The experimental values are anastomosis with predicted value. The XRD spectrum and SEM picture of samples sintered at the optimized process are shown in Figs. 5 and 6.

Fig. 5 shows that partially stabilized zirconia ceramic was obtained after sintering process. The SEM image(Fig. 6)shows that the product obtained at the optimized process with the homogeneous structure and nearly no pore which result high relative density and good mechanical properties. That is because the samples before sintering typically contain dozens of stomatal, they are only point contact between particles. Strength of samples is very low. However, it will happen that the contact area expansion, particles gathering and volume contraction in high temperature. And grain boundaries forms with the shortening distance between particles. In the meantime, the stomatal become isolated from connected, gradually reduced, and escaped in the high temperature. Therefore, the densification of the PSZ was achieved eventually with fine homogeneous structure.

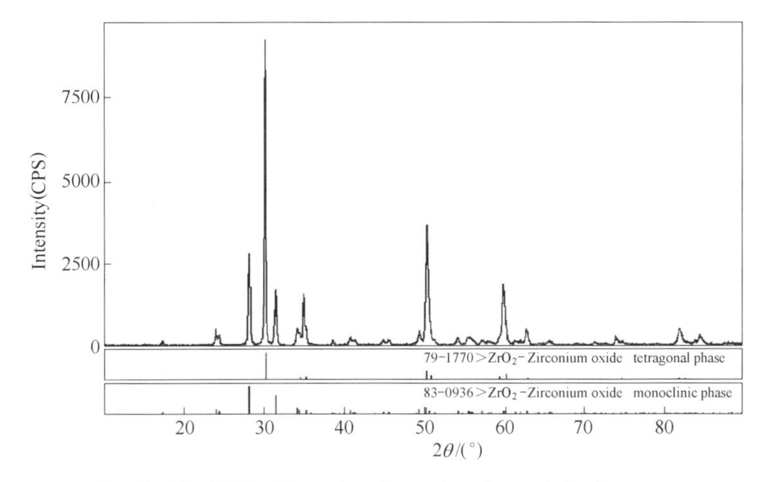

Fig. 5 The XRD of samples sintered at the optimized process

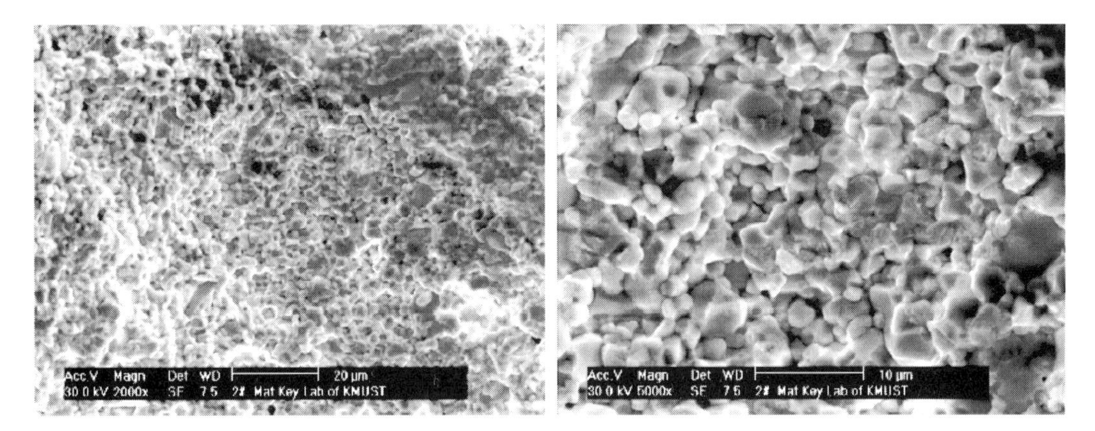

Fig. 6 SEM picture of samples sintered at the optimized process

4 Conclusions

(1) Optimization of the production of calcia partially stabilized zirconia (CaO – PSZ) by using natural baddeleyite shows that all the three reaction variables have their effects on the properties of CaO – PSZ (e. g. relative density and bending strength). The predicted model fits well with the experimental results. The temperature, 1540℃; holding time, 5h; and heating rate, 3℃/min were found to be the optimum conditions to achieve the maximum relative density of CaO – PSZ.

(2) In this paper the natural baddeleyite which was obtained by floating of baddeleyite ore, was used as raw material to prepare partially stabilized zirconia instead of chemical pure zirconia. Therefore, it can shorten the process and reduce energy consumption.

Acknowledgements

Project supported by the International Science & Technology Cooperation Program of China (No. 2012DFA70570), and the Yunnan Provincial International Cooperative Program (No. 2011IA004).

References

[1]Popović S,Grzˇeta B,Czakó – Nagy I,et al. Structural properties of the system m – ZrO_2 – α – Fe_2O_3 [J]. Journal of Alloys and Compounds,1996,241(1):10 – 15.

[2]Pazhani R,Padma K H,Varghese A,et al. Synthesis,vacuum sintering and dielectric characterization of zirconia (t – ZrO_2) nanopowder[J]. Journal of Alloys and Compounds,2011,509(24):6819 – 6823.

[3]Wolten G M. Diffusionless phase transformations in zirconia and hafnia[J]. Journal of the American Ceramic Society,1963,46(9):418 – 422.

[4]Sense K A. Journal of the American Ceramic Society,44(1963):465.

[5]Carniglia S C,Brown S D,Schroeder T F. Phase equilibria and physical properties of oxygen deficient zirconia and thoria[J]. Journal of the American Ceramic Society,1971,54:13 – 17.

[6]Garrett H J,Ruh R. Amer. Ceram. Soc. Bull. ,1968(47):578 – 579.

[7]Standard O C,Sorrell C C. Densification of zirconia – conventional methods[J]. Key Engineering Materials, 1998,153:251 – 300.

[8]Czeppe T,Zięba P,Pawlowski A. Erratum to Crystallographic and microchemical characterization of the early stages of eutectoid decomposition in MgO – partially stabilized ZrO_2[J]. Journal of the European Ceramic Society,2002,22:35 – 40.

[9]Sakuma T. Microstructural aspects on the cubic – tetragonal transformation in zirconia[J]. Key engineering materials,1998,153:75 – 96.

[10]Hughan R R,Hannnk R H J. J. Am. Ceram. Soc. ,1986,69:556 – 563.

[11]Luo H,Cai Q,Wei B,et al. Study on the microstructure and corrosion resistance of ZrO_2 – containing ceramic coatings formed on magnesium alloy by plasma electrolytic oxidation[J]. Journal of Alloys and Compounds, 2009,474(1):551 – 556.

[12]Kumar S,Pramamik P. Trans. J. Br. Ceram. Soc. ,1995,94:123 – 127.

[13]Basu B,Vleugels J,Van D B O. Microstructure – toughness – wear relationship of tetragonal zirconia ceramics [J]. Journal of the European Ceramic Society,2004,24(7):2031 – 2040.

[14]Borik M A,Lomonova E E,Osiko V V,et al. Partially stabilized zirconia single crystals:growth from the melt and investigation of the properties[J]. Journal of Crystal Growth,2005,275(1):e2173 – e2179.

[15]Yoshimura M,Oh S T,Sando M,et al. J. Alloys Comp. ,1999,290:284 – 289.

[16]Deville S,Attaoui H E,ChevalierJ. J. Eur. Ceram. Soc. ,2005,25:3089 – 3096.

[17]Wei Z,Li H,Zhang X,et al. Preparation and property investigation of CeO_2 – ZrO_2 – Al_2O_3 oxygen – storage compounds[J]. Journal of Alloys and Compounds,2008,455(1):322 – 326.

[18]Rittidech A,Tunkasiri T. J. Alloys Comp. ,2012,38:S125 – S129.

[19]Zhang Y L,Jin X J,Rong Y H,et al. On the t→m martensitic transformation in Ce – Y – TZP ceramics [J]. Acta materialia,2006,54(5):1289 – 1295.

[20]Moon J,Choi H,Kim H,et al. The effects of heat treatment on the phase transformation behavior of plasma – sprayed stabilized ZrO_2 coatings[J]. Surface and Coatings Technology,2002,155(1):1 – 10.

[21]Li J,Peng J,Guo S,et al. Application of response surface methodology (RSM) for optimization of sintering process for the preparation of magnesia partially stabilized zirconia (Mg – PSZ) using natural baddeleyite as starting material[J]. Ceramics International,2013,39(1):197 – 202.

Microwave Initiated Self – Propagating High – Temperature Synthesis of SiC

Jinhui Peng, Binner Jon, Bradshaw Steven

Abstract: The use of microwave energy to initiate self – propagating, high – temperature synthesis (SHS) of Si^{+} graphite mixtures has been investigated. The results indicate that, unlike with conventional ignition techniques, green densities in excess of 80% of theoretical can be ignited and the combustion wave front can be crudely controlled. It was found that the induction time for ignition increased with increasing green density and that a higher microwave power level was required with the denser green pellets to achieve the same ignition time. Combustion front velocity increased with green density. The degree of densification was found to decrease with increasing green density. For a given green density, the degree of densification increased with increasing microwave power. The product contained a significant proportion of ultrafine (36 – 72nm diameter) SiC whiskers; despite this, final densities as high as 83.6% of theoretical could be obtained without the use of applied pressure. This compares with the 50% densities obtained via conventional ignition techniques.

Keywords: microwave; self – propagating high – temperature synthesis; SiC; whiskers

1 Introduction

Self – propagating high – temperature synthesis (SHS), also known as combustion synthesis, is a process route that utilizes exothermic reactions to form materials[1-3]. It has several potential advantages over more conventional techniques, including high energy efficiency, a rapid production rate, and low processing costs[1,4]. To date, more than 500 different compounds have been synthesized using SHS[5].

However, the process can be limited by thermodynamic and kinetic considerations. The former limit occurs when a system has a low reaction enthalpy while the latter is dictated by low heat transfer and reaction rates. Examples of thermodynamic limitations occur in the SiC and B_4C systems, among others. For these ceramics, the adiabatic combustion temperature of the precursor materials is too low to effect a self – sustaining combustion wave[6]; for example, it has been demonstrated empirically that many SHS reactions will not be self – sustaining unless an adiabatic temperature 1800K is achieved[7]. This places materials, such as SiC, at the limit for the occurrence of an SHS reaction. Under normal conditions, the reaction is liable to be incomplete and thus several techniques have been used to enhance the combustion synthesis of SiC from elemental reactants. These include: preheating the reactants to raise the adiabatic temperature[8-12]; use of a high – pressure nitrogen atmosphere to initiate a silicon – nitrogen gas phase reaction first, followed by the Si – C reaction[13]; use of chemical or oxidative additives to change reaction mechanisms[14]; increasing the heating surface under high pressure (HPCS)[15], and application of an electric field to activate the combustion reaction[8,16,17].

It is well known that microwaves usually generate an inverse temperature profile during heating, i. e. ,the center of the body becomes hotter than the surface[18]. When used to initiate SHS reactions,the result is usually a combustion wavefront that propagates radially outward from the center[19]. This process can produce a completely different product morphology and lead to more complete conversion of reactants[20]. Although microwave heating has been used successfully to initiate several SHS processes[20-25],the technique has not been extensively investigated for silicon/carbon powder compacts,especially those with a high green density. The primary aims of the research reported here were to investigate the potential to use microwaves to:(1)achieve ignition in Si/graphite compacts with green densities greater than 80% of theoretical;(2)control the wavefront propagation;(3)to obtain products with improved microstructures.

2 Experimental

The precursors used were silicon and graphite powders,both being −325 mesh(<43mm),99.0% pure and from Strem Chemicals,UK. Stoichiometric amounts of the powders were weighed and dry mixed thoroughly using a ball mill for 24h and then uniaxially cold pressed to produce pellets 1.29cm in diameter and with heights ranging from 1.0cm to 1.5cm. This resulted in green densities in the range 52% −81% of theoretical for the Si + C powder mix. The pellets weighed between 2.3 −2.4g each.

The pressed pellets were placed individually in a vitreous silica crucible that was insulated using low dielectric loss aluminosilicate fiberboard to minimize heat losses. They were then heated in a controlled atmosphere,multimode microwave furnace operating at 2.45GHz with an argon atmosphere at ambient pressure. The furnace arrangement is shown schematically in Fig. 1;the maximum power available was 5kW.

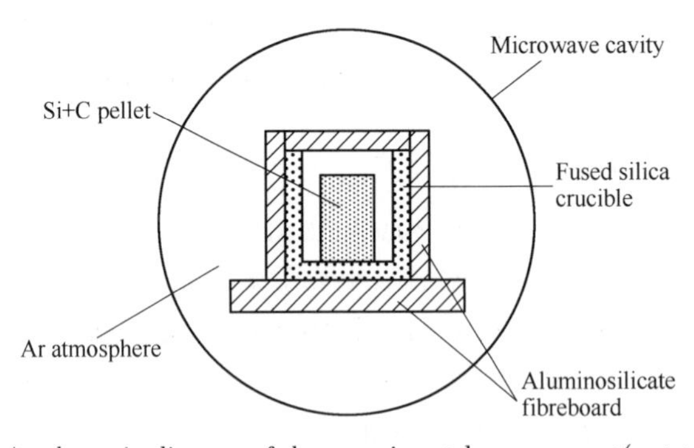

Fig. 1 A schematic diagram of the experimental arrangement(not to scale)

The samples were heated using different microwave power levels,through the point of ignition, until the SHS reaction was noted to have finished by observation through a viewport. After cooling, the density of each sample was measured by the Archimedes principle using the mercury − immersion method. The samples were then sectioned and photographed using optical microscopy and the product phases formed were characterized using X − ray diffraction(XRD) and scanning electron microscopy(SEM).

3 Results and discussion

Fig. 2 consists of a series of optical micrographs obtained from 52% dense green samples, heated using 0.85kW of microwave power, that were quenched at different times after ignition by turning off the microwave power and allowing the samples to rapidly cool. It can be seen that ignition occurred in the center of the samples and that the combustion wave front then propagated radially outward, consistent with the results of Dalton, et al.[19]. This contrasts with conventional SHS using localized ignition sources in which the wave front propagates from the point of ignition to the opposite surface. It was noticed that when the microwave radiation was turned off, the combustion reaction immediately halted; when the microwave power was restarted seconds later, ignition reoccurred, suggesting that the propagation of the combustion wave front in this weakly exothermic system could be controlled.

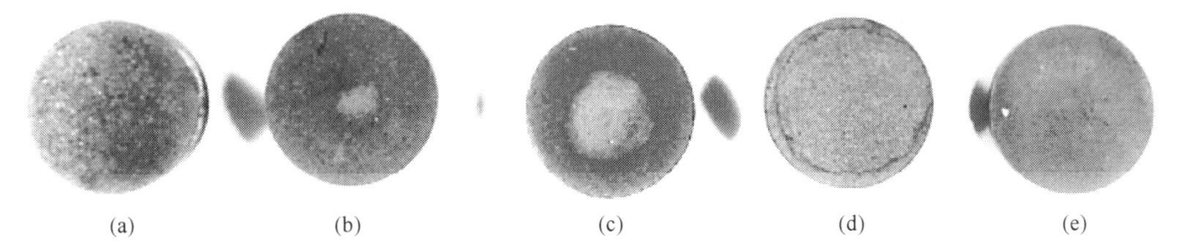

Fig. 2 Optical photographs of sample cross sections after exposure to microwave radiation for (a)0min, (b)1.6min, (c)4.0min, (d)7.3min and (e)8.1min after ignition

XRD analyses of the central and outer zones of the sample from Fig. 2(c) are shown in Fig. 3 (a) and (b), respectively. These indicate that the central zone consisted of a high-purity SiC product phase, while the outer zone retained the unreacted Si + C precursor mixture. This confirms that internal ignition occurred followed by outward propagation of the combustion front.

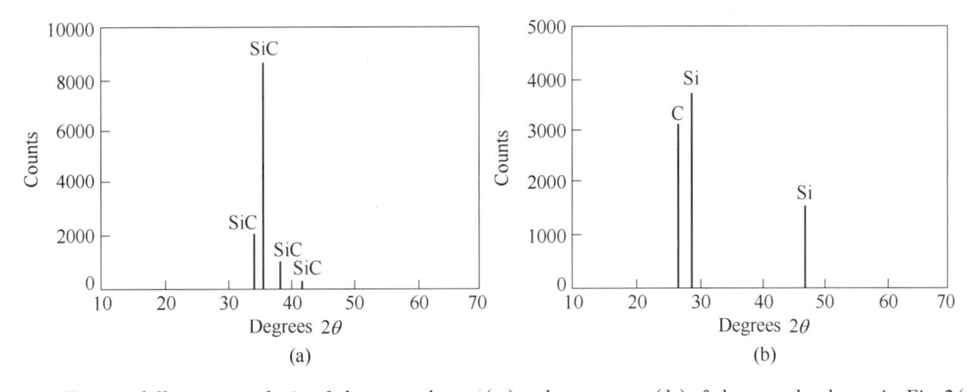

Fig. 3 X-ray diffraction analysis of the central zone(a) and outer zone(b) of the sample shown in Fig. 2(c)

The effects of green density and microwave power level on the induction time for ignition are shown in Fig. 4. It can be seen that the induction time increased with increasing green density, a result also found by Dalton, et al.[19] in the microwave ignition of TiO_2 - Al - C composites. This result matches that predicted for ignition by constant heat flux[26]. However, increasing the green density will increase the effective loss factor and, hence, reduce the penetration depth.

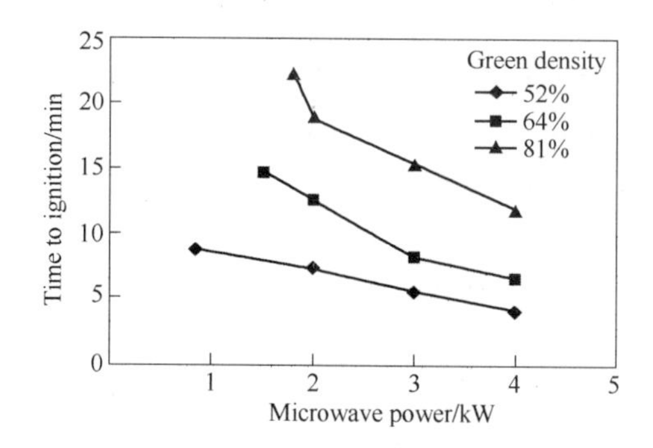

Fig. 4 Effect of microwave power and green density on induction time for ignition

The theoretical work on penetrative radiant heating reported by Merzhanov and Averson[26] also suggests that the time to ignition should increase with increasing penetration depth. Thus it appears that in the present work the density effect dominated the penetration depth effect. As would be expected, induction time decreased with increasing microwave power. It can also be seen that for the 81% dense sample, the induction time increased sharply at low microwave power. This trend implies that there is a minimum power below which ignition cannot be achieved.

Once ignition was achieved the wave front propagated faster through the denser samples, as can be seen in Fig. 5. This is the expected result for the case where excessive expansion caused by desorption of gases during combustion does not occur. It is interesting to note that in the work by Dalton, et al. [19] the combustion velocity was essentially constant across the radius of the samples. The increase in combustion velocity with increasing microwave power could also be expected, as it is known that higher initial temperatures result in higher velocities in conventional SHS[27]. For the 81% dense sample, it can be seen that the reaction time increased sharply at low microwave power. The effect can also be seen, but less noticeably, for the 64% sample. It is surmized that this would also have been true for the 54% dense sample, had the reaction been performed at even lower power than that reported here. These trends can be expected as they indicate that propagation can be achieved only above a minimum energy input. This is corroborated by the fact that switching off the microwave power caused propagation to cease.

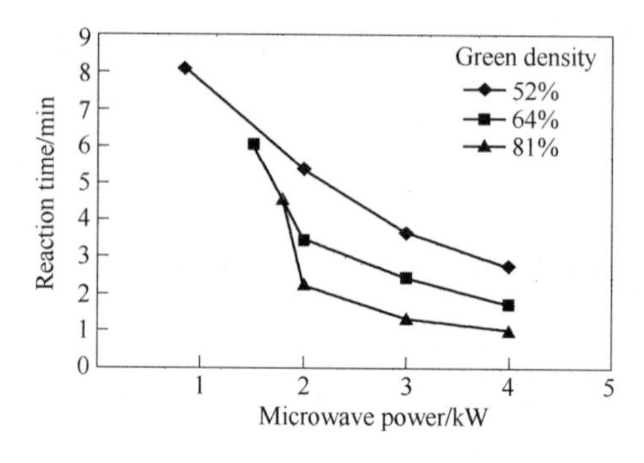

Fig. 5 Reaction time as a function microwave power and green density

The effects of microwave power and green density on final density are shown in Fig. 6. It can be seen that the degree of additional densification decreased with increasing green density. For the 52% green density samples, for example, the final SiC product density ranged from 60% to 66% of theoretical, with the higher densities being achieved via the use of higher microwave power levels. While the same general trend was observed with the 81% green density samples, additional densification was achieved only at the highest microwave power level used. Indeed, the final densities for 1.8kW and 2kW were less than the green density, which indicates that porosity was formed during the reaction and that this porosity could not be removed by any sintering achieved in the postreaction period. The trend of increasing densification with increasing microwave power for fixed green density may be because higher temperatures were reached, although lack of temperature measurement prevents corroboration of this. SiC products as dense as 83.6% of theoretical, the highest value obtained in the present work, have not been reported in the literature using conventional ignition methods without the simultaneous use of pressure. Typically, conventionally ignited SHS formed materials are very porous, about 50% of theoretical density[3]. Dalton, et al.[19] were also able to ignite and combust similarly dense samples in an investigation of microwave - assisted SHS.

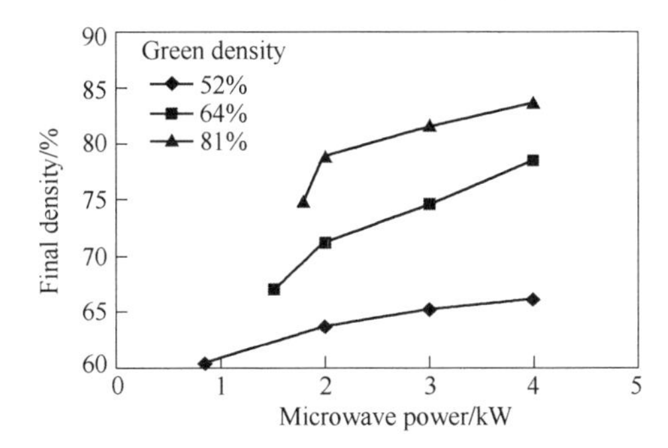

Fig. 6 Final density as a function of green density and microwave power

Fig. 7 shows that there was an increase in porosity at the interface between the unreacted and reacted zones after a short microwave heating time (52% dense sample, 0.85kW, 1.6min). This is likely to be the result of the decreased molar volume of the product compared with the reactants. The central zone of SiC has already started to sinter and densify. The generation of porosity at the interface between reactants and products creates the space known to be necessary for whisker formation; the initial creation of whiskers may be seen at the interface in Fig. 8. During the initial period of growth, the whiskers grew smoothly from the central zone in the direction of the combustion - front propagation. Growth of whiskers aligned in the direction of reaction propagation during SHS was also reported by Yi and Petric[28], who made Ti - Al - Nb matrix composites reinforced by TiB phases.

For samples where the reaction was not quenched, but allowed to reach completion, the whiskers exhibited an irregular shape as seen in Fig. 9(a). This result is typical for whisker formation where conditions (temperature, availability of material, etc.) change during growth.

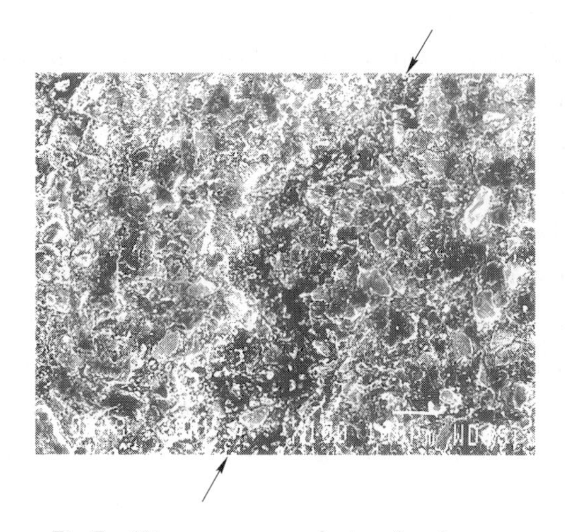

Fig. 7 Microstructures at the interface between unreacted and reacted zones after short microwave heating time (52% dense sample, 0.85kW, 1.6min)

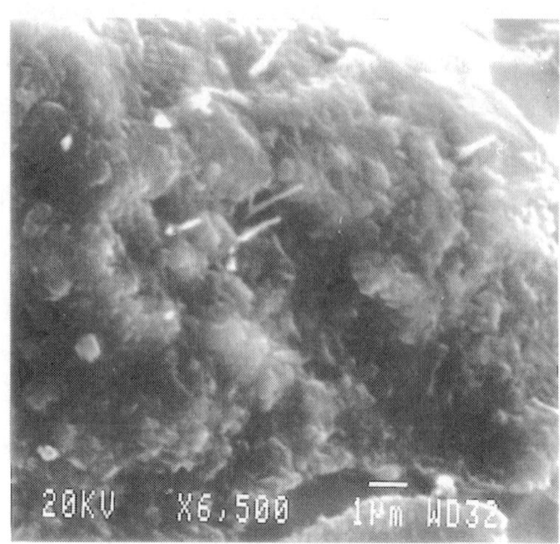

Fig. 8 SEM of the initial period of SiC whisker growth

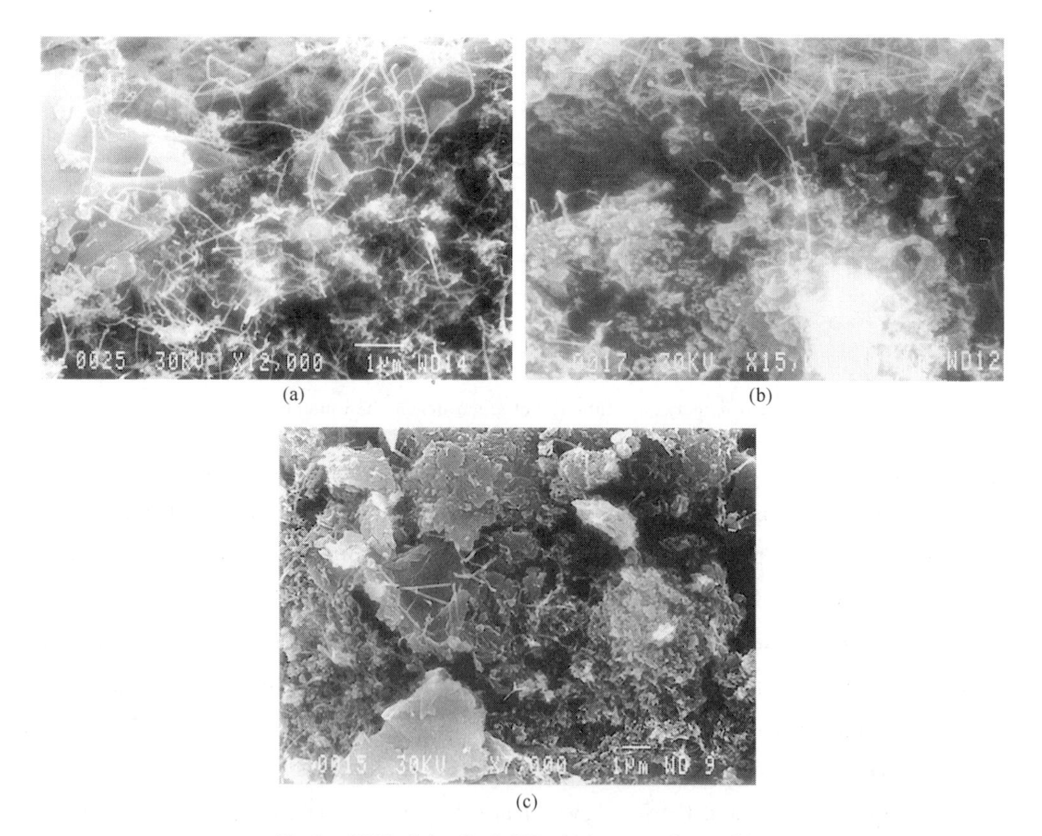

(a) (b) (c)

Fig. 9 SEM of the final SiC whiskers synthesized by the microwave initiated SHS process with different green densities (a) 52%; (b) 64%; (c) 81%

The effect of green density on the microstructures of the product is shown in Fig. 9(a) – (c). It was observed that there was a greater predominance of whiskers at lower green densities. This could be expected, since the lower the green density the greater the volume available for whisker growth. In addition, there will have been more time for whiskers to grow since the reaction times were longer.

The final densities of the products obtained in the present work are surprising given the fact that a significant fraction of the SiC was acicular. Typical whisker diameters lay in the range 36 – 72nm, while their lengths ranged from 360nm to 1500nm, yielding aspect ratios of 10 to 25. These whiskers are somewhat finer than commercially produced SiC whiskers. The formation of whiskers has not been reported during previous investigations into the conventional SHS processing of SiC.

This is not the first time that the formation of products with different morphologies to those obtained with conventional ignition techniques has been reported. For example, working with mixtures of titania, carbon, and aluminium, Yiin and Barmatz[23] observed solid Al_2O_3 whiskers in the microwave – initiated SHS product, while hollow whiskers with bulbous heads were observed when conventional techniques were used to ignite the SHS reaction under similar conditions. This indicated that there could be significant differences in the mechanistic path followed by microwave and conventionally ignited SHS reactions.

4 Conclusions

Microwave energy has been used to initiate SHS reactions in stoichiometric mixtures of Si^+ graphite powders. Ignition was achieved with green densities 81% of theoretical. The induction time for ignition increased with increasing green density. Once ignition was achieved, the wave front velocity was greater in the denser samples. It was found that the higher the green density, the lower the degree of additional densification achieved and that increasing microwave power resulted in a higher final density. The highest final density achieved was 83.6% of theoretical, which is considerably higher than is usually achieved with SHS. Reaction front propagation could be halted and restarted by switching on and off the microwave power during reaction. It was found that the product morphology consisted of particles and ultrafine whiskers. Whisker formation was more predominant at lower green density.

Acknowledgements

We are grateful to the Royal Society China Royal Fellowship Scheme of the United Kingdom for the financial support of this work. We are grateful to the Royal Society China Royal Fellowship Scheme of United Kingdom for the financial support of this work and to Professor Steven Bradshaw, University of Stellenbosch, S. Africa, for many useful discussions.

References

[1] Crider J F. Self – propagating high temperature synthesis—a Soviet method for producing ceramic materials [C]//Ceramic Engineering and Science Proceedings, 1982, 3: 519 – 528.

[2] Munir Z A, Anselmi – Tamburini U. Self – propagating exothermic reactions: the synthesis of high – temperature

materials by combustion[J]. Materials Science Reports, 1989, 3(7): 277 - 365.

[3] Bowen C R, Derby B. Selfpropagating high temperature synthesis of ceramic materials[J]. British Ceramic Transactions, 1997, 96(1): 25 - 31.

[4] Chung S L, Yu W L, Lin C N. A self - propagating high - temperature synthesis method for synthesis of AlN powder[J]. Journal of Materials Research, 1999, 14(05): 1928 - 1933.

[5] Merzhanov A G. History and recent developments in SHS[J]. Ceramics International, 1995, 21(5): 371 - 379.

[6] Feng A, Munir Z A. Effect of product conductivity on field—activated combustion synthesis[J]. Journal of the American Ceramic Society, 1997, 80(5): 1222 - 1230.

[7] Moore J J, Feng H J. Combustion synthesis of advanced materials: Part I. reaction parameters[J]. Progress in Materials Science, 1995, 39(4): 243 - 273.

[8] Feng A, Munir Z A. Field - assisted self - propagating synthesis of β - SiC[J]. Journal of Applied Physics, 1994, 76(3): 1927 - 1928.

[9] Yamada O, Miyamoto Y, Koizumi M. Self - propagating high - temperature synthesis of the SiC[J]. Journal of Materials Research, 1986, 1(02): 275 - 279.

[10] Pampuch R, Stobierski L, Lis J. Synthesis of sinterable β - SiC powders by a solid combustion method [J]. Journal of the American Ceramic Society, 1989, 72(8): 1434 - 1435.

[11] Pampuch R, Stobierski L, Lis J, et al. Solid combustion synthesis of β - SiC powders[J]. Materials Research Bulletin, 1987, 22(9): 1225 - 1231.

[12] Pampuch R. Advanced HT ceramic materials via solid combustion[J]. Journal of the European Ceramic Society, 1999, 19(13): 2395 - 2404.

[13] Maiti H S, Datta S, Basu R N. High - Tc superconductor coating on metal substrates by an electrophoretic technique[J]. Journal of the American Ceramic Society, 1989, 72(9): 1733 - 1735.

[14] Kharatyan S L, Nersisyan H H. Combustion synthesis of silicon carbide under oxidative activation conditions [J]. International Journal of Self - Propagating High - Temperature Synthesis, 1994, 3(1): 17 - 25.

[15] Yamada O, Miyamoto Y, Koizumi M. High pressure self - combustion sintering of silicon carbide[J]. American Ceramic Society Bulletin, 1985, 64(2): 319 - 321.

[16] Feng A, Munir Z A. The effect of an electric field on self - sustaining combustion synthesis: Part I. modeling studies[J]. Metallurgical and Materials Transactions B, 1995, 26(3): 581 - 586.

[17] Feng A, Munir Z A. The effect of an electric field on self - sustaining combustion synthesis: Part II. field - assisted synthesis of μ - SiC[J]. Metallurgical and Materials Transactions B, 1995, 26(3): 587 - 593.

[18] Binner J G, Cross T E. Applications for microwave heating in ceramic sintering: challenges and opportunities [J]. J. Hard Mater., 1993, 4: 177 - 185.

[19] Dalton R C, Ahmad I, Clark D E. Combustion synthesis using microwave energy[C]//14th Annual Conference on Composites and Advanced Ceramic Materials, Part 2 of 2: Ceramic Engineering and Science Proceedings, Volume 11, Issue 9/10. John Wiley & Sons, Inc., 1990: 1729 - 1742.

[20] Ahmad I, Dalton R, Clark D. Unique application of microwave energy to the processing of ceramic materials [J]. Journal of Microwave Power and Electromagnetic Energy, 1991, 26(3): 128 - 138.

[21] Komarenko P, Clark D E. Ceramic Engineering and Science Proceedings, 1994, 15(5): 1028 - 1035.

[22] Atong D, Clark D E. Synthesis of TiC - Al_2O_3 composites using microwave - induced self propagating high temperature synthesis(SHS) [C]//22nd Annual Conference on Composites, Advanced Ceramics, Materials, and Structures: B: Ceramic Engineering and Science Proceedings, Volume 19, Issue 4. John Wiley & Sons, Inc., 1998: 415 - 421.

[23] Yiin T, Barmatz M. in Microwaves: Theory and Application in Materials Processing Ⅲ, Ceramic Transactions,

59, D. E. Clark, D. C. Folz, S. J. Oda, and R. Silberglitt, eds. (American Ceramic Society, Westerville, OH) 1995:541 - 547.

[24] Clark D E, Ahmad I, Dalton R C. Microwave ignition and combustion synthesis of composites[J]. Materials Science and Engineering: A, 1991, 144(1):91 - 97.

[25] Willert - Porada M, Fisher B, Gerdes T. in Microwaves: Theory and Application in Materials Processing III, Ceramic, Transactions 36, D. E. Clark, W. R. Tinga, and J. R. Laia, eds. (American Ceramic Society, Westerville, OH) 1993:365 - 375.

[26] Merzhanov A G, Averson A E. The present state of the thermal ignition theory: an invited review [J]. Combustion and Flame, 1971, 16(1):89 - 124.

[27] Varma A, Rogachev A S, Mukasyan A S, et al. Combustion synthesis of advanced materials: principles and applications[J]. Advances in Chemical Engineering, 1998, 24:79 - 226.

[28] Yi H C, Petric A. Journal of Materials Synthesis and Processing, 1994, 2:355 - 366.

Microwave Ignited Combustion Synthesis of Aluminium Nitride

Jinhui Peng, J. Binner

Aluminum nitride, AlN, is a ceramic that has many attractive properties for electronics and refractory applications[1]. These include its high thermal conductivity, typically > 200W/(m · K), low thermal expansion coefficient and high electrical resistivity combined with its ability to be pressureless sintered using additives to theoretical density. AlN powders have also been used as fillers for polymer and glass compounds to enhance the heat transfer characteristics for electronic packaging applications[1]. Recently, the use of AlN whiskers in high – density substrates and functional films has also attracted interest[2].

AlN can be synthesized using a variety of methods[3], however two main processes have been in industrial operation for some time, the carbothermal reduction of aluminum oxide and the direct nitridation of aluminum. The former consists of heating a mix of alumina and carbon powders to above 1100℃ in nitrogen. Unreacted carbon is removed by controlled oxidation at ~650℃ in dry air while a second heat treatment at ≥1400℃ in vacuum reduces oxygen pickup and stabilizes the powder. Although this method has the advantages of readily available raw materials and the production of powders that are homogeneously sized and suffer low levels of agglomeration, avoiding carbon and oxygen impurities is difficult[1].

The second process also allows large quantities of powders to be produced with relatively low energy and raw materials' costs. Being highly exothermic, once the reaction starts it is able to proceed without any additional supply of external heat. However, it is difficult to produce fine, fully nitrided powders because during the reaction the Al powder can form large aggregates due to its melting point (660℃) being lower than the nitridation temperature (≥800℃)[4]. In addition, AlN formation occurs first on the surface of the Al particles inhibiting the diffusion of the nitrogen to the unreacted metal. Extended reaction times are therefore required to allow adequate diffusion while intermittent grinding steps are generally necessary to break the agglomerates and expose fresh surfaces of unreacted Al[5].

Microwaves are electromagnetic radiation in the approximate frequency range 0.3GHz to 300GHz. It is now well known that they usually generate an inverse temperature profile during heating, i. e., the center of the body becomes hotter than its surfaces[6]. Thus when microwaves are used to ignite the combustion synthesis process, ignition is initiated in the center of the body and the combustion wave front propagates radially outward[7]. This process has been shown to yield different product morphologies and more complete conversion of reactants in some systems[8]. Although only a few reports are available on the microwave assisted synthesis of aluminum nitride, previous work has revealed that both AlN nanoparticles[9] and AlN fibers have been observed[10] in the reaction products.

The precursors used in the present work were aluminum powder (Strem Chemicals, UK, > 99.7% purity and < 43μm diameter) and nitrogen (> 99.99% purity). Despite the fact that aluminum powder is normally a good absorber of microwave energy, it proved necessary to use a carbon powder susceptor in order to reachnthe required temperature to initiate the self - propagating synthesis reaction. It is possible that the relatively high flow rate of N_2 through the compact, 4L/min, provided a substantial cooling effect. The Al powder was therefore packed into the center of a double Al_2O_3 tube arrangement, Fig. 1, the inner tube measured 25mm id and the powder bed, supported by a layer of porous alumina fiberboard, was 50mm deep. The carbon powder microwave susceptor was placed in the annulus between the concentric alumina tubes. The gas tight, 6kW, 2.45GHz multimode microwave furnace was evacuated and then flushed with nitrogen prior to microwave heating being started. The temperature at the bottom, center and surface of the powder bed was measured using a moveable R - type sheathed thermocouple. The samples were heated using different microwave power levels, through the point of ignition, until the SHS reaction was noted to have finished by observation of the glow through a viewport. Typical heating rates were 70 - 140℃/min. The N_2 supply was maintained after the microwave power was turned off until the temperature dropped to below 300℃, then the product was allowed to cool to room temperature. After cooling, the product phases formed were characterized using X - ray diffraction (XRD) and scanning electron microscopy (SEM).

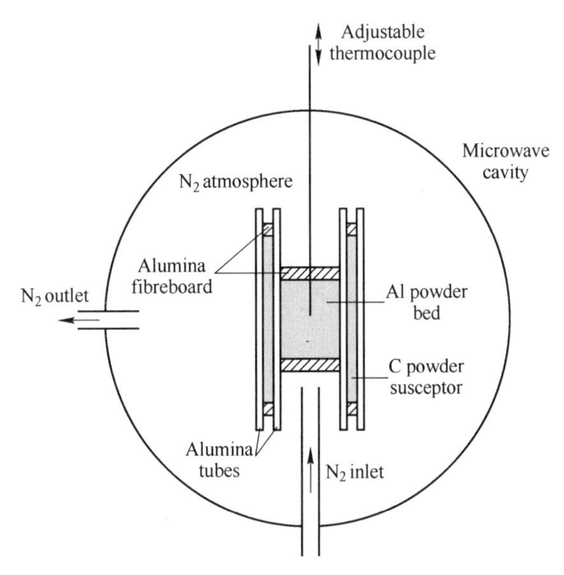

Fig. 1 A schematic diagram of the experimental arrangement (not to scale)

It was noted that there was a significant variation in peak temperature throughout the powder bed, e. g., 1166℃ at the bottom, 1280℃ in the center and 1214℃ at the top. It is believed that this temperature distribution reflects the competing processes of ignition at the center of the powder bed, cooling by the flowing N_2 up the bed and radiation losses from the surfaces of the bed. However, no attempt was made to generate a more uniform temperature distribution because it provided a useful set of "micro – climates" where local conditions varied. In addition, despite the short reaction durations, typically ≤15min, notionally complete conversion to AlN was often observed, as indicated by Fig. 2. This compares very favourably with results from more conventional processing in which complete conversion to AlN could not be achieved even after three passes through a transport flow reactor at up to 1650℃ and several hours processing[5]. However, after 2h post reaction ball milling, a further hour at temperatures as low as 900℃ was sufficient to complete the conversion[5]. This suggests that the unreacted Al had been encapsulated by an impermeable AlN coating that acted as a N_2 diffusion barrier, the effect of the ball milling being to fracture the coating.

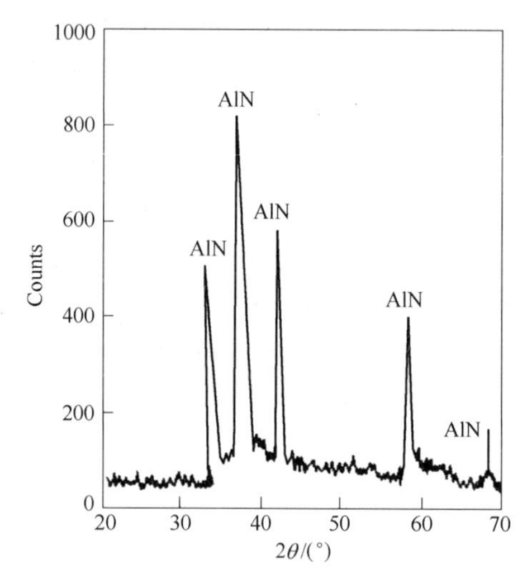

Fig. 2 XRD analysis of combustion products (temperature 1439 – 1553K; combustion time 15min)

Under certain circumstances however, probably when the nitridation reaction is rapid enough, then the encapsulating layer of AlN cracks because of molar volume and thermal expansion differences between the reactant and the product[11,12]. This is followed by the eruption of ultrafine aluminum droplets and vapor, which can be nitrided to form fine AlN particles, whiskers and/or fibers. It is believed that this mechanism also occurred in the present work, Figs. 3 and 4. It can be seen from the latter that both ultrafine AlN particles, 100 – 300nm, and whiskers, 60 – 200nm in diameter and several micrometers long, were observed in the reaction product. In general, the particles tended to dominate in the regions where the temperature was lowest, while the whiskers occurred more frequently in the center of the bed where the temperature was highest. Nearly all of whiskers in the coolest regions of the powder bed, i. e., at the bottom, also bore nodules on their tips; this might be indicative of the vapor – liquid – solid (VLS) mechanism. In contrast, fewer nodules were observed and the whiskers were much more kinked in the higher bed temperature regions. This might be indicative of a vapor – solid (VS) mechanism, though clearly further work is needed to define accurately the mechanism(s) responsible for these whisker morphologies.

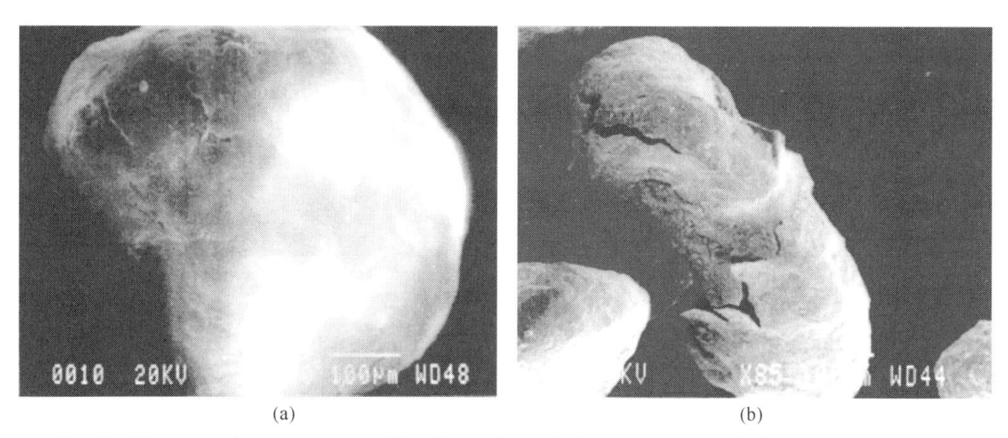

(a) (b)

Fig. 3 Scanning electron micrographs of an unheated aluminum particle(a) and after microwave heating to 1400K followed by termination prior to ignition of the combustion synthesis reaction (b)(the aluminum particle appears to be covered by a cracked nitride coating)

(a) (b)

(c)

Fig. 4 Scanning electron micrographs of AlN products synthesized using microwave ignited combustion synthesis

(a) At the bottom of the powder bed(1439K); (b) In the center region(1553K); (c) At the top(1487K)

Acknowledgements

The authors are grateful to the Royal Society China Royal Fellowship Scheme for supporting one of the authors(JP) and to Professor Steven Bradshaw, University of Stellenbosch, S. Africa, for many useful discussions.

References

[1]Sheppard L M. Aluminum nitride—a versatile but challenging material[J]. American Ceramic Society Bulletin, 1990,69(11):1801 - 1812.

[2]Jiang G J,Zhuang H R,Zhang J,et al. Morphologies and growth mechanisms of aluminum nitride whiskers by SHS method—Part 2[J]. Journal of Materials Science,2000,35(1):63 - 69.

[3]Haussonne F J M. Review of the synthesis methods for AlN[J]. Material and Manufacturing Process,1995,10(4):717 - 755.

[4]Chang A,Rhee S W,Baik S. Kinetics and mechanisms for nitridation of floating aluminum powder[J]. Journal of the American Ceramic Society,1995,78(1):33 - 40.

[5]Weimer A W,Cochran G A,Eisman G A,et al. Rapid process for manufacturing aluminum nitride powder [J]. Journal of the American Ceramic Society,1994,77(1):3 - 18.

[6]Binner J G P,Al - Dawery I A,Aneziris C,et al. Use of the inverse temperature profile in microwave processing of advanced ceramics[C]//MRS Proceedings. Cambridge University Press,1992,269:357.

[7]Dalton R C,Ahmad I,Clark D E. Combustion synthesis using microwave energy[C]//Ceramic Engineering and Science Proceedings,1990,11(9 - 10):1729 - 1742.

[8]Yiin T,Barmatz M,Feng H,et al. Microwave induced combustion synthesis of ceramic and ceramic - metal composites[J]. 1995.

[9]Ramesh P D,Rao K J. Microwave—assisted synthesis of aluminum nitride[J]. Advanced Materials,1995,7(2):177 - 179.

[10]Vaidhyanathan B,Agrawal D K,Roy R. Novel synthesis of nitride powders by microwave - assisted combustion [J]. Journal of Materials Research,2000,15(04):974 - 981.

[11]Hotta N,Kimura I,Ichiya K,et al. Continuous synthesis and properties of fine AlN powder by floating nitridation technique[J]. Nippon Seramikkusu Kyokai Gakujutsu Ronbunshi - Journal of the Ceramic Society of Japan,1988,96(7):731 - 735.

[12]Bradshaw S M,Spicer J L. Combustion synthesis of aluminum nitride particles and whiskers[J]. Journal of the American Ceramic Society,1999,82(9):2293 - 2300.

Microwave Initiated Self – Propagating High Temperature Synthesis of Materials: A Review

Jinhui Peng, J. Binner, S. Bradshaw

Abstract: The use of microwave energy to initiate self – propagating, high temperature synthesis (SHS) reactions has been reviewed. Microwave initiation usually results in ignition occurring at the centre of the body, with the combustion wavefront propagating radially outwards. This leads to a number of differences compared with conventionally ignited SHS reactions. These include ignition in both weakly exothermic systems and denser green bodies, crude control of the wavefront propagation, and the generation of different microstructures owing to dissimilar time temperature and spatial temperature profiles. The technology also extends the range of materials and compositions that can be produced in a self – propagating manner. Commercially important developments are likely to be those that utilise these features to produce tailored microstructures for niche applications.

1 Introduction

1.1 Self – propagating high temperature synthesis

Self – propagating, high temperature synthesis (SHS) is, in the simplest terms, the exploitation of a highly exothermic and usually very rapid chemical reaction to form useful materials[1-3]. The essential characteristic of the process is that the reaction will propagate as a narrow, high temperature front following local ignition. Advantages claimed for the process include higher product purity, a low energy requirement and the relative simplicity of the process. In addition, with the simultaneous application of external pressure, it is possible to make products in highly dense forms as has been demonstrated by investigations with TiC, TiB_2 and SiC[4-6]. Furthermore, recent studies on the combustion synthesis of intermetallic phases have shown that, depending on the mechanistic path followed by the combustion process, significant control of the porosity of the final product might be achievable[7].

Theoretical calculations have shown that several thousand compounds might be produced using SHS and over 500 different compounds have been synthesised. These include: refractory compounds such as borides, carbides, nitrides and silicides; oxides, including tantalates, niobates, ferrites and cuprates; intermetallics such as aluminides, germanides and nickelides; chalcogenides such as sulphides, selenides and tellurides; as well as other materials such as phosphides, hydrides and many others[8,9].

Many different techniques can be used to ignite the SHS reactions, including radiant flux[10], resistance heat coils[11], chemical ovens[12], sparks[13], direct resistive heating[14], and lasers[15] as well as, more recently, microwave radiation[16]. These methods yield three different modes of combustion, as illustrated in Fig. 1. Localised surface heating techniques result in a reaction wavefront that propagates from that surface through the bulk of the reactant mixture. In contrast, methods that provide volumetric heating have the potential to initiate volume combustion synthesis in which the reaction may occur more or less simultaneously throughout the sample. With microwave heating it is possible to achieve ignition at the centre of the sample with the wavefront propagating outwards to the surface.

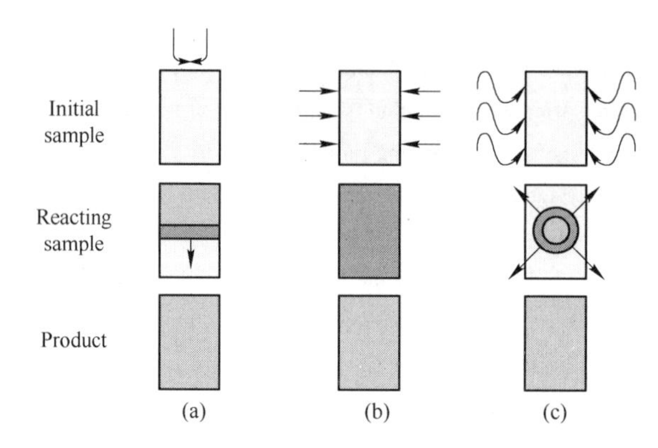

Fig. 1 Diagrams showing modes of combustion synthesis

(a) With surface ignition; (b) With volume ignition; (c) With internal igniton using microwaves

Although there are several examples of commercial production of materials by SHS[1], there remain a number of technical problems to be overcome[17]. These include control of product microstructure and the intrinsic creation of porosity owing to reactant/product molar volume differences when attempting to create dense bodies. Microwave initiation of SHS reactions is one of a number of techniques being studied which it is hoped will overcome some of the inherent limitations of SHS.

1.2 Microwave heating

Microwaves are the part of the electromagnetic radiation spectrum in the approximate frequency range 0.3GHz to 300GHz, with corresponding wavelengths ranging from 1m to 1mm in air. Dielectric materials can be heated when exposed to a high frequency electromagnetic field as a result of polarization and direct conduction effects. It is now well known that microwaves generate an inverse temperature profile during heating, i. e. the centre of a body becomes hotter than its surface[18]. This phenomenon results from the balance between heat loss from the surface, conduction of heat to/from the interior and volumetric dissipation of energy in the body. Initiating combustion synthesis with microwaves typically results in ignition in the centre of the body and the combustion wavefront propagating radially outward[16]. This can produce a product morphology completely different to that resulting from surface ignition, and may lead to a more complete conversion of reactants[19]. Microwave initiated SHS (MISHS) is thus an interesting and relatively new development. In this paper the latest developments are reviewed, the process advantages discussed and areas for further work suggested.

1.3 Characteristics of microwave initiated self – propagating high temperature synthesis process

The average microwave power P_{av} (W) dissipated within a material may be given by

$$P_{av} = \pi f \varepsilon_o \varepsilon''_{eff} \int_v (E^* \cdot E) \mathrm{d}V \quad (1)$$

where E is the magnitude of the internal field, V/m; f is the frequency, Hz; ε_o is the permittivity of free space (8.86×10^{-12} F/m); ε''_{eff} is the effective loss factor of the material. The equation shows that the ability to heat a material using microwaves depends on two primary factors, the intrinsic nature of the material (the effective loss factor) and the design and frequency of the microwave system. The frequency for microwave heating is limited to a relatively small range of frequencies reserved for international scientific and medical use. The most common are 2.45GHz and 915MHz (896MHz in the UK) although higher frequencies such as 24GHz and 83GHz are increasingly being investigated for microwave processing, particularly for sintering engineering and electro – ceramics. More power can be dissipated at higher frequencies, but the depth of power penetration is lower. This means that care must be taken to achieve uniform heating, especially with large objects. The magnitude of the internal field E is dependent on the size, geometry and location of the material within the microwave cavity and on the design and volume of the cavity itself; these are factors over which the designer has control.

For a given microwave system, however, the ability to heat materials using microwaves is primarily dependent on their intrinsic properties. For example, carbon (graphite) powder is an exceptionally good microwave absorber; small samples can be heated from ambient to greater than 1200℃ within 1min even at relatively low power levels such as 600W. On the other hand, SiO_2 is almost transparent to microwaves at ambient temperature and as a consequence it is very difficult to heat with microwaves[16,20,21]. Although bulk metals reflect microwaves (the basis behind radar), fine metal powders can be readily heated[22]. This is because metallic powders containing interparticle porosity act as dielectric materials. Microwave heating can then readily occur by ohmic loss (and possibly other loss mechanisms) if the metallic grain size is of the order of the skin depth in the metal.

Some empirical data for powders heated by microwaves at 2.45GHz are given in Table 1. The data have been collated from a variety of sources and involve the use of different power levels (although typically in the range 600W ± 1000W), for a range of times and in different microwave applicators and for samples of different sizes. The table should thus be used only as a general guide for a powder's ability to absorb microwaves.

Table 1 Data extracted from various sources[16,20,21,23-25] showing ability to heat a variety of different powders using microwaves

Material	Temperature reached/℃	Time taken/min	Material	Temperature reached/℃	Time taken/min
Al	600	1.5	Al_2O_3	78	4.5
B	1000	0.67	C(graphite)	1200	1
CaO	116	4	CeO_2	99	30

Continues Table 1

Material	Temperature reached/℃	Time taken/min	Material	Temperature reached/℃	Time taken/min
Co	697	3	Co_2O_3	1290	3
Cu	619	2	CuCl	619	13
CuO	701	0. 5	Fe	768	7
$FeCl_3$	41	4	Fe_2O_3	88	30
Fe_3O_4	510	2	La_2O_3	107	30
Mg	120	7	MgO	203	5. 5
$MnCl_2$	53	1. 75	MnO_2	1287	6
Mo	650	2. 5	MoO_3	69	5. 5
MoS_3	1106	7	NaCl	83	7
Nb	700	3. 17	Nb_2O_5	174	6
Ni	500	3. 25	NiO	1305	6. 25
PbO_2	182	7	Pb_3O_4	122	30
$SbCl_3$	224	1. 75	Si	1000	1. 17
SiO_2	79	7	$SnCl_2$	476	2
$SnCl_4$	49	80	SnO	102	30
Ta	700	2. 67	Ti	1150	1
TiB_2	843	7	TiO_2	122	30
V	557	1	V_2O_5	701	9
W	690	6. 25	WO_3	532	0. 5
Y_2O_3	115	7	Zn	581	3
$ZnCl_2$	609	7	ZnO	326	5. 5
Zr	710	4	ZrO_2	63	4

The temperatures indicated are generally not the maximum that could be reached, rather they are the values attained after the specified duration of heating.

It should be noted that since the effective loss factor generally increases with temperature, many materials that absorb microwaves poorly at ambient temperature will couple with them much better at higher temperatures. Microwave heating of such materials can thus be achieved as part of a hybrid approach in which conventional heating techniques are used to generate the required initial temperature[26-29]. There are two main methods by which this may be accomplished. In the first, a different energy source (e. g. a conventional furnace) is used to heat the material to the temperature at which it will couple with microwave energy on its own. The second technique places the sample to be heated in close proximity to a material that is a good microwave absorber at low temperature. This latter, known as a susceptor, heats the sample by radiation or conduction until the sample temperature is such that it couples directly with the microwave energy. This use of hybrid heating methods can be used to widen the scope of materials available for microwave initiated SHS still further.

The volumetric dissipation of power during microwave heating is fundamentally different from conventional heating methods and results in a qualitatively different time temperature History[18] (Fig. 2). Power dissipation in a material illuminated by a plane electromagnetic wave follows the exponential law

$$P = P_t e^{-2\alpha x} \tag{2}$$

where P_t is the power transmitted through the surface in the x direction; α is the attenuation constant. The latter is a function of frequency, dielectric and magnetic properties[30]. Since power is dissipated only in the material, the surrounding environment remains cooler than the body. Heat conduction into the body and heat loss from the surface result in the development of an inverse temperature profile with time, i. e. the interior becomes hotter than the surface. Fig. 2 illustrates the typical inverse temperature profile developed over time in a microwave irradiated dielectric. The quantitative aspects of the profile are dependent on many factors such as dielectric properties, thermal conductivity, duration of heating and boundary conditions, etc. [31] In the case of MISHS this inverse temperature profile typically leads to ignition occurring in the centre of the sample rather than at one of the surfaces. The potential advantages of this are discussed in a later section.

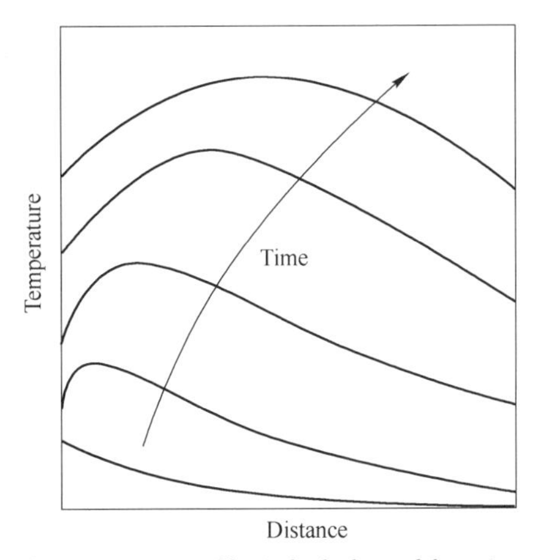

Fig. 2 Development of temperature profile in body heated by microwaves after Binner[18]

1.4 Experimental equipment used for microwave initiated, self - propagating, high temperature synthesis

A number of different microwave applicators have been used for the research, however they can generally be classified into two types: multimode cavities (of which the domestic microwave oven is an example) and single mode cavities. A multimode cavity is, in essence, a closed metallic box into which microwave energy is coupled. At least two of the dimensions of the box should be several wavelengths long. The multimode cavity is versatile and can, in principle, accept a large variety of loads to be heated although it can also suffer from the potential problem of poor heating uniformity, as a result of a non - homogeneous electric field, unless very well designed. In the single mode cavity, precise cavity dimensions allow a simple standing wave pattern to be established in which the microwave field is at a maximum at well defined points in space. This allows the load to be positioned appropriately for optimal absorption of electromagnetic energy. An example[19] is illustrated in Fig. 3. The single mode cavity generally has the advantage that higher electric field strengths can be achieved than with multimode cavities. This means that, in general, faster heating rates result and, depending on the microwave absorption characteristics of the material, higher temperatures can be reached. However, their use is restricted to the heating of much smaller samples than can be treated using a typical multimode applicator.

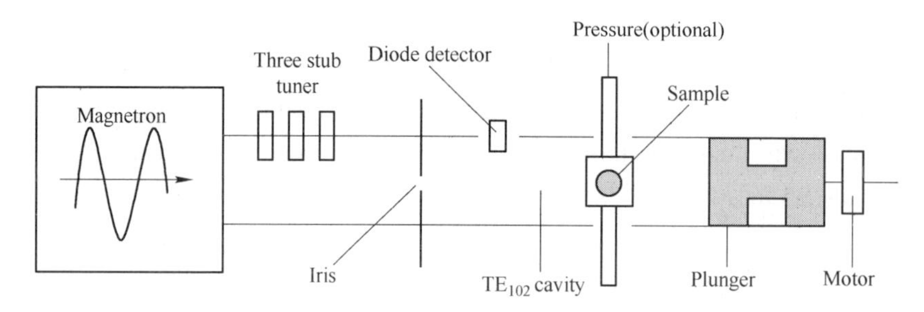

Fig. 3 Schematic single mode cavity sometimes used for microwave initiated, self - propagating, high temperature synthesis research after Yiin and Barmatz[19]

2 Developments in microwave initiated self - propagating high temperature synthesis processing

It is believed that the first studies which examined the potential for using microwaves as an ignition source for SHS were performed between 1989 - 1991 under the direction of Clark at the University of Florida, USA1[6,32,33] using modified domestic microwave ovens. Ten different materials were synthesised including carbides (TiC, SiC, B_4C), borides (TiB_2), silicides ($MoSi_2$, $TiSi_3$), nitrides (AlN, TiN), ceramic composites (Al_2O_3 - TiC) and cermets. It was found that microwave energy could heat readily both metal and nonmetal powder reactants used in combustion synthesis reactions that involved solid - solid and gas - solid reactions. The results showed that, under suitable conditions, internal ignition was achieved in most cases with a radially propagating combustion wavefront. Switching off the microwave power caused propagation to stop, suggesting that control of the front propagation might be possible. In addition, microwave energy could ignite compacted Ti + C samples with green densities greater than 80%, a possibility that had been reported as difficult with conventional ignition. Interestingly, Ti + C did not ignite internally, but close to the surface, presumably owing to a very low penetration depth at 2.45GHz.

Subsequent work by the Florida group[34] included the synthesis of cermets and composites such as AlN - Al, TiN - Ti, Al_2O_3 - TiC, Al_2O_3 - TiB_2, TiC - TiB_2, and the superconducting ceramic $YBa_2Cu_3O_7$. Once again the results indicated that the combination of microwave heating and combustion synthesis was especially useful for high green density compacts and the synthesised product showed little or no expansion. Frequently, conventional combustion synthesis of high density samples results either in unstable combustion, cracked or expanded products or fails to ignite the body. Potential appeared to have been offered by MISHS for the fabrication of dense ceramics and composites that were difficult to obtain with conventional SHS. Final densities were not reported however.

In 1993 Willert - Porada, et al. pursued this approach further by applying microwave heating to the combustion synthesis and sintering of Al_2O_3 - TiC ceramics[35]. As compared with conventional processing, pressureless microwave sintering of a combustible mixture composed of commercially available TiO_2 and Al powders using tar as the carbon source yielded "relatively dense" Al_2O_3 - TiC ceramics after only 20min at 1600℃. Conventional heating resulted in the formation of a rigid framework of TiC, which prevented high densities being achieved. The main difference between the conventional and microwave approaches was the heating profile generated; the use of microwaves allowed the temperature to be raised at the same rate as the exothermic reaction propagated. The research also highlighted the importance of the source of carbon used. When $Ti(O_2C_5H_7)_2$ - $(OC_3H_7)_2$, a metallorganic precursor to TiO_2, was used as an alternative to tar in a combustible mixture of Al - TiO_2 - C an unusual grain morphology was observed that was typical of crystal growth by condensation from the vapour. This indicated an excessive formation of volatile products for both the microwave and conventional sintering approaches. However, strong evidence was found from X - ray diffraction and EDX analysis that the whisker, tube or sphere like crystals were homogeneously composed of Al_2O_3 and TiC.

In 1994, Komarenko and Clark investigated both microwave and conventional ignition and heating of stoichiometric mixtures of Ti/Si/C powder compacts, with the intention of producing the Ti_3SiC_2 phase[36]. Results indicated that, at a processing temperature of 1400 - 1450℃, conventional ignition lead to a product composed primarily of TiC whilst microwave initiated SHS produced a product that was mainly Ti_3SiC_2 with only small amounts of TiC present. Thus the use of microwaves seems to have favoured the formation of the desired phase, though whether for thermodynamic or kinetic reasons is as yet unknown. It has been reported several times in conventional SHS studies that different ignition conditions result in different modes of combustion and in different products[37], although mechanistic explanations are still lacking.

In the same year, Bayya, et al. [38] investigated both conventional and microwave SHS techniques for synthesising Tl - 2212 and Tl - 2223 phase superconductors from their stoichiometric starting compositions. Because of the weakly exothermic nature of the reactions, it was necessary to preheat the reactants to above 300℃ in order to oxidise the copper metal powder and achieve stable combustion. However, even this did not lead to complete reactions being achieved owing to the long diffusion distances that exist in a four component system and the short time interval that the atoms have to diffuse and react as the combustion wave passes by. Nevertheless, the microwave initiated SHS technique achieved a higher degree of reaction than its conventional counterpart owing to the fact that the precursor materials were heated more uniformly and for a longer period of time. It was estimated that when using microwaves the reactants were at the combustion temperature for ~ 100s, as compared with just a few seconds with the conventionally ignited technique. The resultant powders were subsequently pressed into bars and sintered. The microwave SHS produced powders yielded essentially single phase Tl - 2212 and Tl - 2223 superconductors whilst the conventional SHS powders resulted in sintered bodies containing mixed phases. Thus the higher degree of reaction achieved when using microwave initiated SHS had resulted in superior powders that produced much better quality superconductors.

In 1995, Yiin and Barmatz[19] investigated the potential for applying uniaxial pressure during the microwave initiated combustion synthesis process. Working with mixtures of titania, carbon and aluminium, they used both slow (26K/s) and fast (82K/s) heating rates by controlling the microwave power. The reaction achieved may be stated as

$$3TiO_2 + 3C + (4 + x)Al \longrightarrow 3TiC + 2Al_2O_3 + xAl \tag{3}$$

where $x \sim 0$ or 4 was used. Pellets weighing 3g and with 50% of theoretical density were ignited using 50W and 75W respectively for slow and fast heating rates. This was achieved using the TE_{102} single mode microwave cavity illustrated in Fig. 3. Uniaxial pressures in the range of 1.4 – 9.8MPa (200 – 1400psi) were continuously applied along the vertical axis of some of the processed samples as the SHS reactions were initiated.

All the samples ignited with a slow heating rate resulted in product inhomogeneity. For the case of stoichiometric Al addition, slow heating resulted in circumferential cracking, indicating the development of strong radial stresses. The authors suspected this was due to melting of Al before ignition. They also felt that there might have been some electromagnetic shielding effect owing to the presence of a ring of molten Al. Compositional differences were suspected but not confirmed. Using the fast heating rate a uniform microstructure with 75% of theoretical density was achieved, without pressing, for $x \sim 0$, although radial cracking occurred. When $x \sim 4$, solid Al_2O_3 whiskers were observed in the microwave initiated SHS product. In contrast with cases for $x \sim 0$, ignition was found to have occurred near the surface of the pellet.

All the samples ignited with a slow heating rate resulted in product inhomogeneity. For the case of stoichiometric Al addition, slow heating resulted in circumferential cracking, indicating the development of strong radial stresses. The authors suspected this was due to melting of Al before ignition. They also felt that there might have been some electromagnetic shielding effect owing to the presence of a ring of molten Al. Compositional differences were suspected but not confirmed. Using the fast heating rate a uniform microstructure with 75% of theoretical density was achieved, without pressing, for $x \sim 0$, although radial cracking occurred. When $x \sim 4$, solid Al_2O_3 whiskers were observed in the microwave initiated SHS product. In contrast with cases for $x \sim 0$, ignition was found to have occurred near the surface of the pellet.

The application of the uniaxial force during the SHS process allowed the production of 85% dense homogeneous products for both microwave (fast heating) and conventional techniques under the optimum conditions. This suggests that, if dense ceramics are to be formed directly from SHS reactions, this will probably be achieved only through the use of high pressure. Rice[17] has also reached this conclusion, although not in the context of microwave ignition. It is also interesting to note that that there is an optimum delay time between the ignition of the reaction and the onset of pressure[1]; this must be such as to allow the expulsion of gases but be shorter than the cooling time.

Atong and Clark[39] reported in 1996 the effects of sample preparation and processing parameters on the synthesis of TiC – Al_2O_3 composites using conventional heating, microwave heating and hybrid microwave/conventional heating. Hybrid ignition was achieved using a susceptor placed around the material to preheat the sample to a critical temperature, above which microwaves coupled readily with the material. These samples had a higher density and more uniform microstructure compared with either microwave only heating or the conventional ignition technique because the susceptor reduced the temperature gradient in the sample thus allowing more uniform heating. In general, a laminar structure consisting of alternating layers of porous and dense regions, with an interval of ~290mm between them, was observed when either microwave or conventional heating were used, but this was not observed with the microwave hybrid heating. The laminar structure suggests the possibility of oscillating combustion, which could have been caused by heat loss. One would expect the hybrid system to have been closer to adiabatic and hence not to have exhibited unstable combustion. It was also found that the conventionally ignited samples required longer processing times to achieve complete conversion to the desired products compared with the samples ignited by the other two methods. At shorter processing times, both residual precursor materials and intermediate species were observed in the product.

In 1998, Zou, et al. ignited pellets consisting of a mixture of silica and aluminium powders using both a domestic microwave oven and a conventional furnace[40]. ASHS reaction occurred that produced silicon, aluminium, and alumina as the resultant phases. This product was subsequently heated in a nitrogen atmosphere without allowing the sample to cool after the end of the SHS process, i. e. effectively both the SHS process and the nitridation were completed as a one step process. The product was a mixture of silicon, aluminium nitride and alumina, that is, a form of sialon had been synthesised. Table 2 gives the conditions and products achieved using both the microwave and conventional processes.

Table 2 Comparison of results for production of sialon phases using microwave initiated, self – propagating, high temperature synthesis followed by microwave or conventional heating at 1450℃

Ignition method	Temperature /time	Reaction products
Microwave oven	1450℃/4h	Al_2O_3 and beta – sialon
Conventional furnace	1450℃/1h	beta – sialon > 15R – sialon > Al_2O_3

In the same year, Park, at al. synthesised $LaCrO_3$ powders by both hot plate and microwave induced combustion of metal nitrate – urea mixtures[41]. The product obtained by hot plate combustion was shown to consist of hard agglomerates. The particle size was found to be ~0. 12mm and the specific surface area was $14m^2/g$. When the product was ball milled for 6h the specific surface area only increased slightly, to $16m^2/g$. Although the product obtained using microwave induced combustion appeared to have a similar morphology, the size of the agglomerates was greatly reduced and the individual particle sizes were ~0. 02mm. This led to a much higher specific surface area of ~$25m^2/g$; after ball milling it increased significantly to ~$38m^2/g$, indicating that the agglomerates were more easily broken down. Hence the microwave process was considered to be a promising tool for the production of fine powders with high specific surface areas and thus high sinterability.

Recently, Gedevanishvili, et al. [42] synthesised, and in some cases also sintered, a wide variety of intermetallic compounds and alloys using metal compacts and either a pure microwave system or a hybrid heating system in which an external susceptor was used. The latter minimised surface heat losses. Though very few actual experimental details are provided, the authors claimed that the synthesis process involved similar diffusion mechanisms as in conventional powder metallurgy and that time and energy savings are predicted benefits. The work was also claimed to show that metal powders can couple effectively with microwave fields, although this has been known for many years.

Recent work by Peng, et al. [43] has further investigated the potential to produce dense ceramics using a 5kW, 2.45GHz, controlled atmosphere multimode microwave furnace to ignite SHS reactions involving Si^+ graphite compacts with green densities ranging from 52% - 81% of theoretical. The predominant product morphology was SiC whiskers 36 - 72nm in diameter and 0.36 - 1.5mm in length. The highest product density achieved was 83.6%. Although this was achieved without the use of pressure it is similar to that achieved with uniaxial pressing by Yiin and Barmatz[19] and is somewhat greater than that typically achieved with conventional ignition (~ 50% of theoretical). However, since the degree of additional densification decreased with increasing green density, the use of pressure is almost certainly essential if fully dense bodies are to be produced.

As part of the work, the effect of sample green density and microwave power level on wave velocity was studied[43]. The results showed that induction time before ignition was shorter the lower the green density but that the resultant combustion wave velocity was relatively slow, 0.8mm/min for 52% theoretical density. However, for samples with a high green density, such as 81.0% of theoretical, ignition required a higher level of microwave power (1.8kW), induction time was longer and the subsequent wave velocity was faster, 5.86mm/min. These results were in agreement with those of Dalton, et al. [16] In conventionally ignited SHS it is known that wave velocity usually increases with increasing initial temperature and also with green density, although for some systems there may be an optimum green density, probably owing to the effect of adsorbed gases[37]. The situation is more complex with MISHS, as green density affects both the dielectric properties and the thermal conductivity. The interplay between these parameters will determine the location of the ignition point and the time to ignition. It was also found that switching the microwave power off and on during combustion caused reaction to stop and then restart[43]. This suggested, as in the work of Dalton, et al. [16] that the use of microwaves might allow control of the propagation of the wavefront for either weakly exothermic systems or systems with very dense compacts.

In contrast, Chudoba, et al. [44,45] have examined the microwave initiated SHS of relatively large (200g) quantities of silicon/carbon black mixtures using a specially designed 850W, 2.45GHz microwave cavity with an argon atmosphere. Two antennas and the lack of a mode stirrer ensured that the microwave field was distributed inhomogeneously and that controllable hot spot configurations developed in the powder mixture samples. Three distinct morphologies were revealed by SEM. The most dominant were 5 - 20mm agglomerates of nanometre SiC particles (Fig. 4) that show some potential for being crushed into ultrafine SiC powder. The second were 1mm to 5mm highly faceted crystals almost certainly produced by the classical mechanism of dissolution - precipitation (Fig. 5); this was the dominant type of product formed during conventionally initiated SHS. The third, and very rarely found form, was empty "shells" composed of 0.5mm to 5mm SiC particles (Fig. 6). The reason for the different mechanisms is a matter for speculation at the present time, but it would suggest that less diffusion has occurred in the microwave case. Whether this is because the time at temperature was shorter or because the temperature of the reaction front was lower is not yet known.

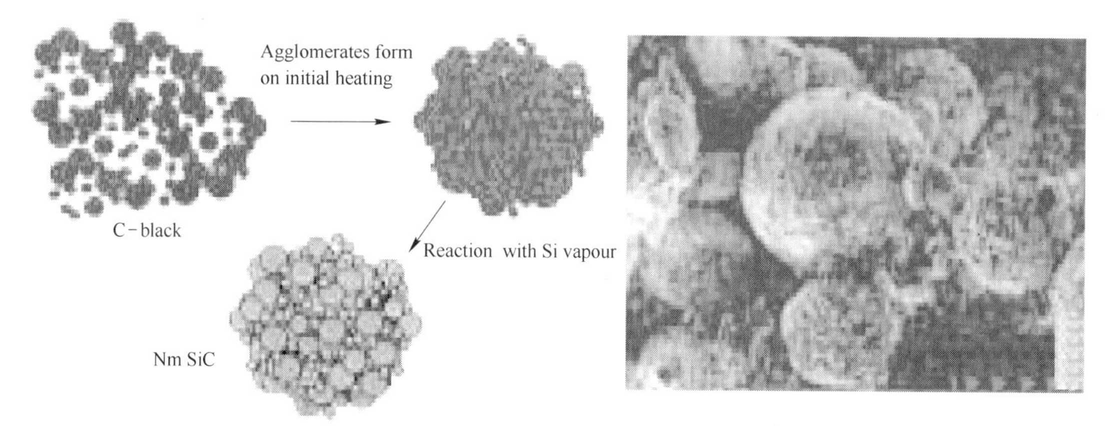

Fig. 4 Diagram and micrograph showing agglomerates of nanometre silicon carbide particles: it is possible that C black particles initially form into spherical agglomerates under action of surface tension of liquid Si formed at reaction temperatures; this may subsequently lead to formation of nanometre SiC grains from individual C black particles via reaction with Si vapour

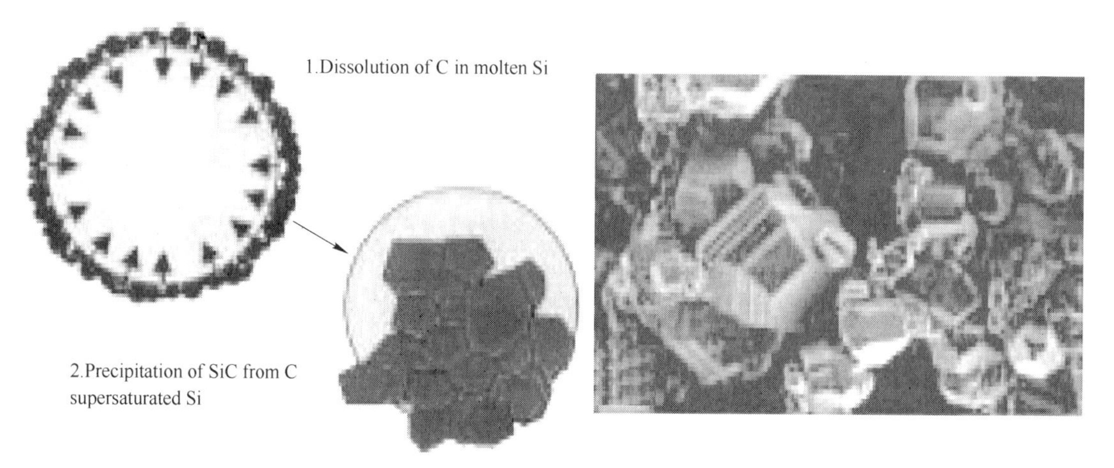

Fig. 5 Diagram and micrograph showing highly faceted crystals probably produced by classical mechanism of dissolution - precipitation: during heating Si particles melt and C is dissolved; on cooling SiC particles precipitate out of carbon supersaturated silicon to form agglomerates that are shaped much like original droplet of molten silicon

Peng and Binner[46] used microwaves to ignite aluminium powder beds in a 6kW multimode microwave furnace operating at 2.45GHz under flowing nitrogen. As a result the competing processes of ignition at the centre of the powder bed, cooling by N_2 flowing up the bed and radiation losses from the surfaces of the bed, there was a significant variation in peak temperature throughout the powder bed. For example, in one run 1166℃ was measured at the bottom, 1280℃ in the centre, and 1214℃ at the top. However, this provided a useful set of "microclimates" where local condi-

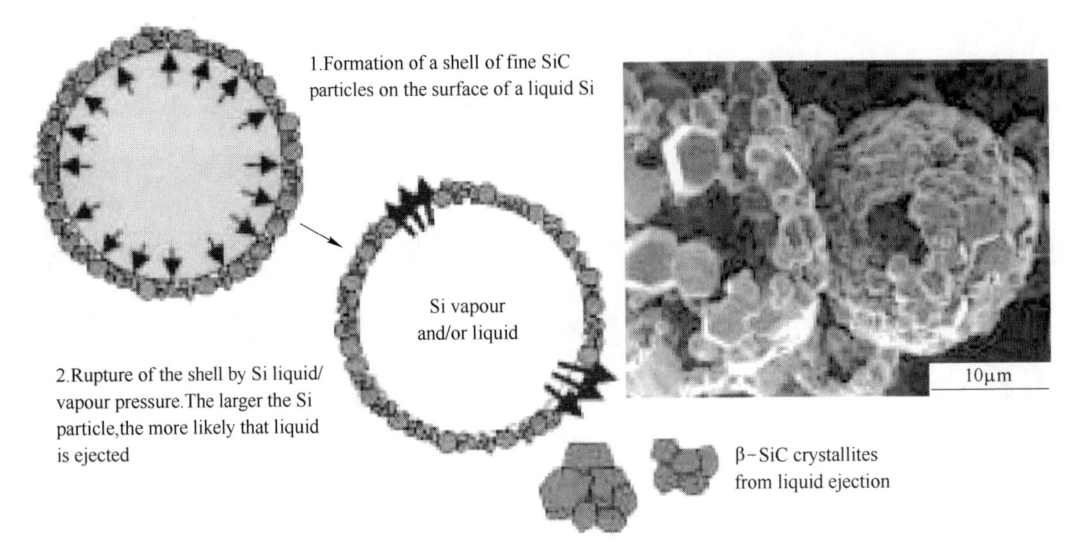

Fig. 6 Diagram and micrograph showing empty "shells" composed of fine SiC particles: it is possible that C reacts at surface of molten Si droplets to form SiC particles that create dense shell surrounding still molten Si core; at some point Si bursts out of shells, though whether as liquid or gas is unknown; almost complete absence of any SiC whiskers suggests that Si did not vaporise excessively during this reaction mechanism

tions varied. In general complete conversion from Al to AlN could be achieved in just 15min without any post – reaction processing. This compares very favourably with results from more conventional processing[47] in which complete conversion to AlN could not be achieved without post – reaction processing even after three passes through a transport flow reactor at up to 1650℃. The primary AlN product had a whisker morphology with diameters ranging from 60nm ± 200nm and several micrometers in length. Some ultrafine AlN particles (100nm ± 300nm) were also observed at the bottom of the powder bed where the temperature was lowest. The product morphology was similar to that found by Bradshaw and Spicer[48] who reported a number of interesting whisker morphologies during forced filtration combustion of AlN. The growth of whisker morphologies during either conventional or microwave initiated SHS has not yet received much attention. As one of the most important factors for whisker growth is the presence of free volume[49], it would be expected that SHS synthesis would favour whisker growth as many SHS reactions are accompanied by a molar volume reduction. Internal ignition of gas – solid reactions using microwaves provides an alternative means of ensuring that counter current filtration combustion can take place.

3 Modelling

Modelling of SHS has been a profitable field for mathematical physicists and considerable advances have been made. However it is only recently that attempts have been made to combine these models with descriptions of microwave heating.

In 1996, Bechtold, et al. [50] derived a simple mathematical model that described the localised heating and subsequent ignition in solid combustible materials owing to microwave heating. The asymptotic theory developed assumed a small Biot number and exploited the fact that the activation energy is large for these systems. The authors showed that the phenomena of ignition could be described by a single, localised, nonlinear differential equation. The solutions to this equation exhibited thermal runaway behaviour and explicit ignition times could be computed. Theses critical times were expressed in terms of various physical parameters in the system.

A one - dimensional model for the heating and ignition of a combustible solid by microwave energy was formulated and analysed in the limit of small inverse activation energy $\varepsilon = T_0/E_a$ and small Biot number (Bi) by Booty, et al.[51] in 1998. The limit of small inverse activation energy implies that the heating process is inert until the temperature within the material reaches a critical ignition value. The small Biot number (large thermal conductivity compared with convective heat loss) means that spatial temperature variations are small during the heat up stage. Analysis of the inert stage revealed the dynamics of inert hot spots, whilst first order corrections showed that hot spots would be located internally in the sample. Analysis of the system as the ignition temperature was approached showed that there were parameter regimes where the ignition site in a given experiment could not readily be predicted. It was shown that ignition would occur close to the surface on the incident field side for larger values of the ratio ε/Bi, and would move inwards as the ratio decreased. For materials with relatively low ignition temperatures, there were regimes of applied microwave power within which the ignition site could change abruptly in location as the power was varied. A way of avoiding this problem would be to seed the precursor sample with a high loss material that localised the conversion of electromagnetic energy into heat. The practical implications of this would have to be considered, as unwanted impurities would then be introduced.

3.1 Potential advantages of microwave initiated self - propagating high temperature synthesis process

In order to assess the potential of MISHS it is necessary to review the challenges faced by conventional SHS, these are summarised in Table 3[1,9 17,52] and then to see how the demonstrated benefits of MISHS can address some or all of these challenges. It should be noted that a number of nonmicrowave solutions have already been proposed, including dynamic compaction after SHS and variations on the directed oxidation (Lanxide type) process.

Table 3 Challenges facing self - propagating, high temperature synthesis (SHS) taken from Refs. [1,9,17,52]

Challenge	Comments
To produce dense bodies	Inherent molar volume changes generate porosity; simultaneous use of presses etc. can give thermal shock unless the presses are preheated; violent self propagating reactions yield rapid desorption of gases which breaks samples
Make composites from cheap starting materials	Use of expensive elemental reactants negates inherent cheaper energy and equipment costs
Producing ultrafine powders	Sintering and agglomeration need to be avoided, often by use of dilution with the product
Producing large single crystals	Microstructure control is inherently difficult
Igniting weakly exothermic systems	Chemical furnaces are sometimes used
Controlling the mode of propagation	SHS is inherently uncontrolled making a priori tailoring of processes necessary

It can be seen that most of the challenges involve establishing greater control over the process. In this respect, MISHS has been found to:

(1) allow ignition of denser green bodies than can be achieved with conventional ignition.

(2) allow ignition and propagation in weakly exothermic systems.

(3) allow some form of control of propagation (to date stopping and restarting the reaction) in certain systems, particularly those which are dense and/or weakly exothermic, though further work is required to determine precisely what level of control is available in different systems.

(4) generate different microstructures owing to different time ± temperature and spatial temperature profiles.

Thus it would appear that MISHS has the potential to address some of the challenges in SHS dense body production (e. g. ignition of denser green bodies, controlled propagation and possibly less violent desorption of gases). However, it appears unlikely that synthesis of fully dense parts by MISHS alone will be achieved, but that a profitable area for further research for the production of fully dense bodies would be either combining MISHS with pressing/hipping techniques or the construction and use of a high pressure microwave applicator. Such a system is currently being built at the High Pressure Research Institute in Warsaw, Poland[53]. Particular attention will have to be paid to prior desorption of adsorbed gases or controlled desorption reaction.

Microwave initiated SHS also successfully addresses the challenges of igniting weakly exothermic systems and is an alternative to chemical furnace technology etc. The technique can therefore be said to extend the range of materials and compositions that can be produced in a self - propagating manner. The dielectric loss factor of most materials increases with temperature and thus the reactants directly in front of the combustion wave during MISHS should absorb microwave energy very efficiently. This selective heating of a narrow zone adjacent to the reaction front might enable SHS reactions to proceed to completion, even when thermodynamics might not predict this. Clearly this would not be the case where melting of a metallic reactant occurred ahead of the reaction front.

In general the most successful applications of microwave heating have been those which have made use of the unique features of the technology. In the context of MISHS these would include volumetric preheating, internal ignition, and development of an inverse temperature profile. Commercially important developments of MISHS are likely to be those that utilise these features to produce tailored microstructures for niche applications. Further experimental work and modelling in combination with knowledge of market requirements will be required to find appropriate applications.

For microwave initiation of SHS to be successful, the dielectric properties of the precursors are of vital importance. As described in Eq. (1), the ability to heat a material using microwaves is primarily dependent on the effective loss factor, i. e. the ability to absorb microwave energy. On the basis of these characteristics and data from other, more conventional, SHS processes, over 100 different materials believed suitable for microwave ignition are proposed in Table 4. In some cases, previous work has already confirmed the suitability whilst others are the subject of current work by the present authors and others in the field.

Table 4 Materials predicted to be suitable for internal microwave ignition①

Borides	FeB, MoB, MoB_2, MoB_4, NbB, NbB_2, TaB, TaB_2, TiB, TiB_2[15,34], WB, WB_2, WB_4, W_2B, W_2B_5, VB_2, V_3B_2, ZrB, ZrB_2
Carbides	Al_3BC, Al_4C_3, B_4C[16,34], Mo_2C, Mo_2C_3, NbC, Nb_2C, SiC[16,34], TaC, Ta_2C, TiC[16,34], Ti_3SiC_2[36], VC, WC, W_2C, ZrC
Nitrides	AlN[15], BN, NbN, Nb_2N, NiN, Si_3N_4, Ta_2N, TaN, WN, ZrN
Silicides	MoSi, Mo_3Si, $MoSi_2$[16], FeSi, $NbSi_2$, $TaSi_2$, TiSi, $TiSi_2$, Ti_5Si_3[16,34], V_3Si, V_5Si, WSi_2, W_5Si_3, ZrSi, $ZrSi_2$, Zr_5Si_3, Ni_3Si
Intermetallics and alloys	Co – Al, Cr – Al, Cu – Al, Cu – Ti, Cu – Zn, Cu – Zn – Al, Fe – Al, Fe – Ni, Ge – Al, Mo – Al, Nb – Al, Nb – Ge, Nb – Ti – Al, Nb – V – Al, Ni – Al, Ni – Al – Ag, Ni – Al – Cr, Ni – Al – Cu, Ni – Al – Mo, Ni – Al – Ti, Ni – Al – V, Ni – Mg, Ni – Zr, Pt – Al, Sb – Al, Ta – Al, Ta – Al – Fe, Ti – Al, Ti – Al – Nb, Ti – Al – Sn, Ti – Co, Ti – Cu – Al, Ti – Fe, Ti – Ni, Ti – Ta – Al – W, W – Al, Zr – Al[42]
Composites	AlN – Al[34], B_4C – Al_2O_3, B_4C – TiB_2, B_4C – TiC, Cr_3C_2 – Al_2O_3, Fe – (W, Ti)C, MoB – Al_2O_3, $MoSi_2$ – SiC, Ni – ZrO_2 (Ni/YSZ), TiAl – TiB_2, TiB_2 – Al_2O_3[34], TiC – Al_2O_3[34,35], TiC – M(M = Ni, Co, Cr), TiC – Ni_3Al, TiC – TiB_2[34], TiC – TiB_2 – Al_2O_3, TiN – Ti[34], Ti_5Si_3 – Nb, ZrC – Al_2O_3
Functionally gradient materials	Cu – TiB_2, TiB_2 – Ti, TiC – Ni, TiC – NiAl[40]
Others	Beta – sialon[40]

①Materials listed in bold have been confirmed as suitable for internal microwave ignition. The numbers in square brackets provide the reference number(s).

3.2 Comparison with other novel self – propagation high temperature synthesls methods

Recently use has been made of a 50Hz electric field imposed across samples during conventionally ignited SHS[54]. This method is known as field activated combustion synthesis(FACS) and bears some similarities to MISHS. In particular, FACS allows reactions that would not be self – sustaining to become so. Modelling and measurements suggest formation of a preferential current path in the thin reaction front. A threshold voltage is required to establish and sustain the wavefront. Above a certain applied voltage no separate ignition source is required. Field activated combustion synthesis does not seem to change the reaction mechanism extends compositional limits particularly for the SHS of composites, e. g. $B_4C \pm TiB_2$. It was also found that application of the field had an effect on microstructure, in certain cases giving finer products than without the field. This was speculated to be due to higher temperatures and faster wave propagation and more rapid cooling(and hence limited grain growth). In essence this method is similar to that used by Yamada, et al.[14] who also found ignition in the centre of pellets. There are obvious paralles between FACS and MISHS, in that both methods extend the compositional limits of SHS, and that different time – temperature profiles(when compared with conventional ignition) give different product microstructures.

4 Conclusions

Research into the use of microwave energy to initiate self – propagating, high temperature synthesis (SHS) began a little over a decade ago. Since that time only a relatively small body of work has been performed compared with other microwave processing fields, with the work occurring in a small number of laboratories around the world. The results, however, have shown that the use of microwave energy results in a number of differences compared with conventionally ignited combustion synthesis reactions.

Research into the use of microwave energy to initiate self – propagating, high temperature synthesis (SHS) began a little over a decade ago. Since that time only a relatively small body of work has been performed compared with other microwave processing fields, with the work occurring in a small number of laboratories around the world. The results, however, have shown that the use of microwave energy results in a number of differences compared with conventionally ignited combustion synthesis reactions. greater than 80% of theoretical and allows higher product densities to be achieved compared with conventionally ignited processes. To date, however, the achievement of fully dense products remains elusive. Combination of microwave ignition with the application of high pressures during the reaction might offer a solution to this problem. Microwave initiated SHS can be used to ignite and sustain propagation in weakly exothermic systems, thus allowing a wider range of materials to be synthesised.

The use of microwaves results in different temperature profiles being followed, different product morphology being achieved and frequently yields different microstructures when compared with conventionally ignited SHS reactions. Further work is needed to quantify and understand these effects. Only then, can full advantage be taken of the unique feature of microwave heating, and the technology used to tailor specific microstructures for specific applications.

Acknowledgements

The authors are grateful to the Royal Society China Royal Fellowship Scheme of the United Kingdom for financially supporting Professor Peng.

References

[1] Merzhanov A G. Self – propagating high – temperature synthesis: twenty years of search and findings [J]. Combustion and Plasma Synthesis of High – Temperature Materials, 1990: 1 – 53.

[2] Munir Z A, Anselmi – Tamburini U. Self – propagating exothermic reactions: the synthesis of high – temperature materials by combustion [J]. Materials Science Reports, 1989, 3(7): 277 – 365.

[3] Golubjatnikov K A, Stangle G C, Spriggs R M. The economics of advanced self – propagating high – temperature synthesis materials fabrication [J]. American Ceramic Society Bulletin, 1993, 72(12): 96 – 102.

[4] Holt J B, Munir Z A. Combustion synthesis of titanium carbide: theory and experiment [J]. Journal of Materials Science, 1986, 21(1): 251 – 259.

[5] Miyamoto Y, Koizumi M, Yamada O. High – pressure self – combustion sintering for ceramics [J]. Journal of the American Ceramic Society, 1984, 67(11): c224 – c225.

[6] Yamada O, Suzuki T, Then J H, et al. The use of various metal carbonyls as catalyst precursors for coal hydroliquefaction[J]. Fuel Processing Technology, 1985, 11(3): 297 - 311.

[7] Munir Z A. Synthesis of high temperature materials by self - propagating combustion methods[J]. American Ceramic Society Bulletin, 1988, 67(2): 342 - 349.

[8] Merzhanov A G. Combustion processes that synthesize materials[J]. Journal of Materials Processing Technology, 1996, 56(1): 222 - 241.

[9] Merzhanov A G. History and recent developments in SHS[J]. Ceramics International, 1995, 21(5): 371 - 379.

[10] Escobar - Vargas S, Hernandez - Guerrero A, Baltazar - Cervantes J C. Ignition analysis of SHS processes subjected to a constant heat flux[J]. Asme - Publications - Htd, 1997, 352: 57 - 64.

[11] Merzhanov A G, shkiro V M, Borovinskaya I P. US Patent No. 37226643, US Patent Office, Washington, DC, 1973.

[12] Miyamoto Y, Ishikawa K, Fukao H, et al. Invivo setting behaviour of fast - setting calcium phosphate cement [J]. Biomaterials, 1995, 16(11): 855 - 860.

[13] Moore J J, Feng H J. Combustion synthesis of advanced materials: Part I. reaction parameters[J]. Progress in Materials Science, 1995, 39(4): 243 - 273.

[14] Yamada O, Miyamoto Y, Koizumi M. Self - propagating high - temperature synthesis of the SiC[J]. Journal of Materials Research, 1986, 1(02): 275 - 279.

[15] Shilyaev M I, Borzykh V É, Dorokhov A R. Laser ignition of nickel - aluminum powder systems [J]. Combustion, Explosion and Shock Waves, 1994, 30(2): 147 - 150.

[16] Dalton R C, Ahmad I, Clark D E. Combustion synthesis using microwave energy[C]//14th Annual Conference on Composites and Advanced Ceramic Materials, Part 2 of 2: Ceramic Engineering and Science Proceedings, Volume 11, Issue 9/10. John Wiley & Sons, Inc., 1990: 1729 - 1742.

[17] Combustion and Plasma Synthesis of High - temperature Materials[M]. New York etc.: VCH, 1990.

[18] Arai M, Binner J G P, Carr G E, et al. High temperature dielectric property measurements of engineering ceramics[J]. Ceram. Trans., 1993, 36: 483 - 492.

[19] Yiin T, Barmatz M, Feng H, et al. Microwave induced combustion synthesis of ceramic and ceramic - metal composites[J]. 1995.

[20] Sutton W H. Cer Soc[J]. Bull, 1989, 68(2): 376 - 386.

[21] Walkiewicz J W, Kazonich G, McGill S L. Microwave heating characteristics of selected minerals and compounds[J]. Minerals and Metallurgical Processing, 1988, 5(1): 39 - 42.

[22] Campisi I E, Summers L K, Finger K E, et al. Microwave absorption by lossy ceramic materials[C]//MRS Proceedings. Cambridge University Press, 1992, 269: 157.

[23] Peng J, Yang X. The New Applications of Microwave Power[M] Kunming: Yunnan Press of Science and Technology (in Chinese).

[24] Peng J. The Mechanism and Application of Microwave Heating to Non - ferrous Metallurgy[D]. Kunming: Kunming Institute of Technology, 1992: 22 - 58(in Chinese).

[25] Mingos D M P, Baghurst D R. Tilden Lecture. Applications of microwave dielectric heating effects to synthetic problems in chemistry[J]. Chem. Soc. Rev., 1991, 20(1): 1 - 47.

[26] Sutton W H. Microwave processing of ceramics - an overview[C]//MRS Proceedings. Cambridge University Press, 1992, 269: 3.

[27] Peng J, Ma J J. Microwaves, 1997, 5(13): 58 - 62(in Chinese).

[28] Meek T T. Proposed model for the sintering of a dielectric in a microwave field[J]. Journal of Materials Science Letters, 1987, 6(6): 638 - 640.

[29] De A, Ahmad I, Whiteney K, Clark D E. Ceram. Trans., 1991, 21: 319 – 328.

[30] Metaxas A C, Meredith R J. Industrial Microwave Heating[M]. IET, 1983.

[31] Dolande J, Datta A. Temperature profiles in microwave heating of solids: a systematic study[J]. Journal of Microwave Power and Electromagnetic Energy, 1993, 28(2): 58 – 67.

[32] Ahmad I, Chen C Y R. Post – processor for data path synthesis using multiport memories[C]//Computer – Aided Design, 1991. ICCAD – 91. Digest of Technical Papers, 1991 IEEE International Conference on. IEEE, 1991: 276 – 279.

[33] Clark D E, Ahmad I, Dalton R C. Combustion synthesis of materials using microwave energy[P]. International Patent Application WO 90/13513.

[34] Clark D E, Ahmad I, Dalton R C. Materials Science and Engineering A, 1991, A144: 91 – 97.

[35] Willert – Porada M, Fisher B, Gerdes T. Application of microwave heating to combustion synthesis and sintering of Al_2O_3 – TiC ceramics[M]//Ceramic Transactions: Microwaves. Theory and Application in Materials Processing II. Volume 36, 1993.

[36] Komorenko P, Clark D E. Ceramic Engineering and Science Proceedings. 1994, 15(5): 1028 – 1035.

[37] Varma A, Rogachev A S, Mukasyan A S, et al. Combustion synthesis of advanced materials: principles and applications[J]. Advances in Chemical Engineering, 1998, 24: 79 – 226.

[38] Bayya S S, Snyder R L. Self – propagating high – temperature synthesis (SHS) and microwave – assisted combustion synthesis (MACS) of the thallium superconducting phases[J]. Physica C: Superconductivity, 1994, 225 (1): 83 – 90.

[39] Atong D, Clark D E. Synthesis of TiC – Al_2O_3 composites using microwave – induced self – propagating high temperature synthesis (SHS) [C]//22nd Annual Conference on Composites, Advanced Ceramics, Materials, and Structures: B: Ceramic Engineering and Science Proceedings, Volume 19, Issue 4. John Wiley & Sons, Inc., 1998: 415 – 421.

[40] Ahn Z O U S, Lee H B. Study of the synthesis of sialon phases from silica – aluminium powder mixture using microwave energy[J]. Journal of Materials Science, 1998, 33(16): 4255 – 4259.

[41] Park H K, Han Y S, Kim D K, et al. Synthesis of $LaCrO_3$ powders by microwave induced combustion of metal nitrate – urea mixture solution[J]. Journal of Materials Science Letters, 1998, 17(9): 785 – 787.

[42] Gedevanishvili S, Agrawal D, Roy R. Microwave combustion synthesis and sintering of intermetallics and alloys [J]. Journal of Materials Science Letters, 1999, 18(9): 665 – 668.

[43] Peng J, Binner J G P, Bradshaw S. Journal of Materials Synthesis and Processing.

[44] Chudoba T, Kuzmenko D, Lojkowski W, Binner J, Cross T. Proc. Int. Conf. on "Microwave chemistry", Antibes, France, September Association for Microwave Power in Education and Research in Europe, 2000: 323 – 326.

[45] Chudoba T, Kuzmenko D, Lojkowski W, Binner J, Cross T. 8th Int. Conf. on Microwave and high frequency heating, Bayreuth, Germany, September 2001, Association for Microwave Power in Education and Research in Europe.

[46] Peng J, Binner J G P. Journal of Materials Science.

[47] Ravichandran G, Subhash G. Critical appraisal of limiting strain rates for compression testing of ceramics in a split Hopkinson pressure bar[J]. Journal of the American Ceramic Society, 1994, 77(1): 263 – 267.

[48] Villegas M, Caballero A C, Moure C, et al. Factors affecting the electrical conductivity of donor – doped $Bi_4Ti_3O_{12}$ piezoelectric ceramics[J]. Journal of the American Ceramic Society, 1999, 82(9): 2411 – 2416.

[49] Tiegs T N, Weaver S C. Carbide, Nitride and Boride Materials Synthesis and Processing[M]. London; Chapman and Hall, 1997: 411 – 432.

[50] Bechtold J K, Booty M R, Kriegsmann G A. Microwave – Assisted Ignition[C]//MRS Proceedings. Cambridge

University Press, 1996, 430: 369.

[51] Booty M R, Bechtold J K, Kriegsmann G A. Microwave - induced combustion: a one - dimensional model [J]. Combustion Theory and Modelling, 1998, 2(1): 57 - 80.

[52] Dunmead S D. Carbide, Nitride and Boride Materials Synthesis and Processing [M]. London: Chapman and Hall, 1997: 229 - 272.

[53] Lojkowski W. Private Communication.

[54] Munir Z A. The use of an electric field as a processing parameter in the combustion synthesis of ceramics and composites [J]. Metallurgical and Materials Transactions A, 1996, 27(8): 2080 - 2085.

Synthesis and Microwave Absorbing Properties of Corundum – mullite Refractories

Guo Chen, Jin Chen, Xiaojie Zhi, C. Srinivasakannan, Jinhui Peng

Abstract: A new technology for preparation of corundum – mullite refractories was proposed, and the microwave absorbing characteristics of prepared materials were systematically investigated, using the microwave cavity perturbation technique and the digital signal processing technique. The crystal structures of raw materials before and after sintering process were characterized using XRD. The analysis results indicated that corundum and mullite were mainly crystalline compounds in the sintering samples. All the X – ray diffraction peaks of sample matched well with those of the standard XRD pattern of corundum and mullite phase, respectively. The optimum conditions result of microwave absorbing properties in prepared corundum – mullite refractories with particle size of 75μm.

Keywords: microwave absorbing properties; corundum – mullite refractories; microwave cavity perturbation technique; digital signal processing technique

1 Introduction

Corundum – mullite is an attractive material for refractory. Compared with other materials, it has high purity, high density, high strength and good thermal shock resistance, is mainly used as high temperature (1350 – 1650℃) kiln linings or kiln furniture in ceramics and refractories industries, gas combustion furnace in petrochemical industry and other high temperature services where thermal shock happens frequently[1,2]. However, these processes pollute the environment[3]. To explore new method to produce corundum – mullite refractories with low energy consumption and less environment pollution is necessary.

Recently, many experimental measure methods have been developed and applied for the microwave absorbing characteristics of materials[4,5]. These methods involve using microwave technique and digital signal processing technique for measuring processes and analyzing the process results. Among them, microwave cavity perturbation technique stands out as a popular method utilized in fields of microwave absorbing characteristics of materials. Huang, et al. [6] invented a novel microwave sensor for measuring the properties of a liquid drop, and the technique can easily be extended to microwave processing materials. It was also found that the theory based on the microwave sensor was in good agreement with the experimental results. The moisture content of a sulphide mineral concentrate has been measured using the microwave cavity perturbation technique by Huang and Peng[7]. Guo, et al. [8] investigated microwave technique for the analysis of microwave absorbing characteristics of mixtures of different carbonaceous reducing agents and oxidized ilmenite. The results showed that the microwave – absorbing characteristic of these carbonaceous reducing agents(coconut - based activated carbon, coke and graphite) were all better than that of oxidized ilmenite under the conditions of particle size of 175 - 147μm. Maik, et al.[9] have developed a non - contact method to probe the electrical conductivity and complex permittivity of single and polycrystalline samples in a flow - through reactor in the temperature range of 20 - 500℃ and in various gas atmospheres.

Based on the concept mentioned above, analytically pure Al_2O_3, analytically pure SiO_2 and MgO powders were used to prepare corundum - mullite refractories. The microwave absorbing characteristics of corundum - mullite refractories were measured using microwave cavity perturbation technique. Effect of the particle size on attenuation/frequency shift of corundum - mullite refractories in the single resonant cavity was mainly researched. The crystal structures of corundum - mullite refractories were also analyzed using XRD.

2 Experimental

2.1 Materials

Analytically pure Al_2O_3, analytically pure SiO_2 and MgO powders with a median particle size less than 74μm and a purity of 99.9% were used to preparing the corundum - mullite refractories as raw materials, and the PVA were added as assistant adhesives for synthesis processing.

2.2 Characterization

The sintered samples were characterized by X - ray diffractometer(D/Max 2200, Rigaku, Japan) at a scanning rate of 0.25°/min with 2θ ranging from 5° to 100° using CuK_α radiation (λ = 0.15418nm) and a Ni filter. The voltage and anode current operated were 35kV and 20mA, respectively. A conical ball mill(XNQ - 67, Wuhan, China) was employed to mix different raw materials to achieve a certain composition. The morphological aspects of the sintered samples were also investigated by scanning electron microscope(SEM). The SEM instrument(XL30ESEM - TMP, Philips, Holland) was operated at 20kV in a low vacuum, while the energy dispersion scanner spectrometer(EDAX, USA) attached to the SEM was used for semi - quantitative chemical analysis.

2.3 Measuring instrumentation and principle

Microwave resonant cavity perturbation techniques were commonly used to measure the permittivity and permeability of samples at microwave frequencies[6]. Typical microwave resonant cavity perturbation system consists of microwave resonator, sweeping signal, detector and digital signal processor (DSP), multifunctional card, interface circuit and computer. The microwave signals were transmitted into the microwave sensor[8]. The computer controls the fast scanning microwave generator through multipurpose card. The output signals of the microwave sensor were picked up by the linear detector and DSP. Then they were fed into the low pass filter. After that, the output signals of the low pass filter were amplified and converted by the A/D converter. The data processing of the microwave resonant cavity perturbation system was finished on the computer. The software control of the set - up was performed by windows XP operating system, and programmed by Visual Basic 6.0[5]. The schematic diagram of the microwave resonant cavity perturbation system was illustrated in Fig. 1.

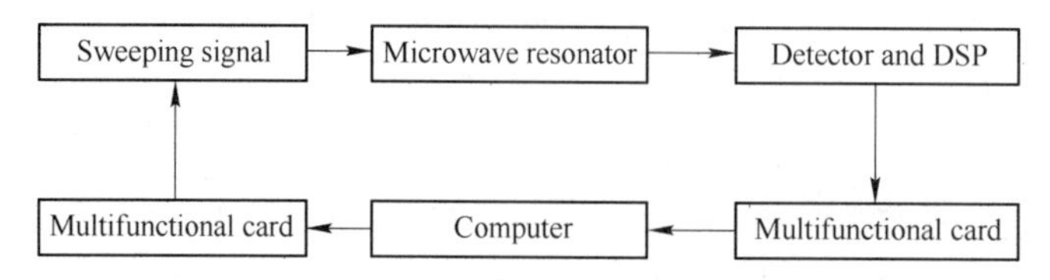

Fig. 1 Schematic diagram of microwave resonant cavity perturbation system

The microwave resonant cavity perturbation system was based on microwave cavity perturbation technique and digital signal processing technique. The measurement principle was that microwave was coupled with the microwave cavity resonator[10]. Derived from the theory of electric - magnetic field, the frequency shift and the output voltage of the microwave resonant cavity, which was given the following equations:

$$\frac{\Delta\omega}{\omega} = -\omega_0(\varepsilon'_r - 1)\int_{V_e} E_0^* \cdot E\mathrm{d}v/4W \tag{1}$$

$$\frac{1}{Q} - \frac{1}{Q_0} = 2\varepsilon_0\varepsilon''_r\int_{V_e} E_0^* \cdot E\mathrm{d}v/4W \tag{2}$$

$$W = \int_V [(E_0^* \cdot D_0 + H_0^* \cdot B_0) + (E_0^* \cdot D_1 + H_0^* \cdot B_1)]\mathrm{d}v \tag{3}$$

where W is the storage energy; ω is the angular frequency; ω_0 is the angular frequency without sample in resonant cavity; $\Delta\omega = \omega - \omega_0$ is the shift of angular frequency; E is the electric field intensity of the sample in the resonant sensor; E_0^*, H_0^* are the hetero conjugations of electric field intensity and electromagnetic field intensity in the resonant sensor before perturbation, respectively; D_0 and B_0 are the hetero conjugations of electric displacement and magnetic induction before perturbation, respectively; D_1 and B_1 are the increments of electric displacement and magnetic induction in samples after perturbation; v_c and v_e are the volumes of the sample and the resonant sensor, respectively; $\mathrm{d}v$ is the volume of the element; Q_0 and ω_0 are the quality factors (Q values) of the cavity unloaded (unperturbed condition) and loaded with the samples, respectively; ε_0 is the absolute permittivity of a vacuum (free space); ε'_r is the real part of the complex permittivity, which is related to the stored energy within the medium, in most cases, it is not a constant but it is a strong function of both the temperature and microwave frequency; ε''_r is the imaginary part of the complex permittivity, which is related to the dissipation (or loss) of energy within the medium and will also be dependent on the temperature and microwave frequency[10].

2.4 Procedure

Prior to the use, weighed a certain amount of analytically pure Al_2O_3, analytically pure SiO_2, MgO powders were put into the conical ball mill, and were ground separately and screened to particle size ranges of 1.11 - 20.61 μm. Then, the raw materials was loaded on a ceramics boat, which was placed in a drying oven and heated up to a drying temperature of 120℃, at a heating rate of 5℃/min. Upon reaching 120℃, the samples were held at the same conditions for 2h. After drying, the sample was cooled to room temperature. The dried material with various mass fractions of the mixtures was placed into the muffle furnace for each experimental run, and set to the desired temperature (1450℃) and holding times (120min). Finally, the microwave absorbing properties of the samples were obtained by putting the samples into the microwave resonant sensor in turn.

3 Results and discussion

The samples after sintering process is characterized by XRD and the results are illustrated in Fig. 2. It can be seen from Fig. 2 that corundum (JCPDS card No. 82 – 1541) and mullite (JCPDS card No. 15 – 0776) are mainly crystalline compounds in the sintering samples. All the X – ray diffraction peaks of sample matched well with those of the standard XRD pattern of corundum and mullite phase, respectively. The corundum has the strongest preferential orientation of (1 1 0) plane at $2\theta = 10.46°$, and the diffraction peaks of corundum gradually broadened and their intensities increased under roasting processing. The second strong preferential orientation of (1 1 2) planes of corundum prepared are observed at $2\theta = 21.74°$. The sintering samples has peaks at $2\theta = 26.34°$ and $26.00°$, where the strongest and the second strongest peaks of mullite phase occur, respectively. The third strongest peak at $2\theta = 16.44°$, remained and it. The XRD results confirmed that preparation of corundum – mullite refractories from analytically pure Al_2O_3, analytically pure SiO_2, and MgO powders was feasible.

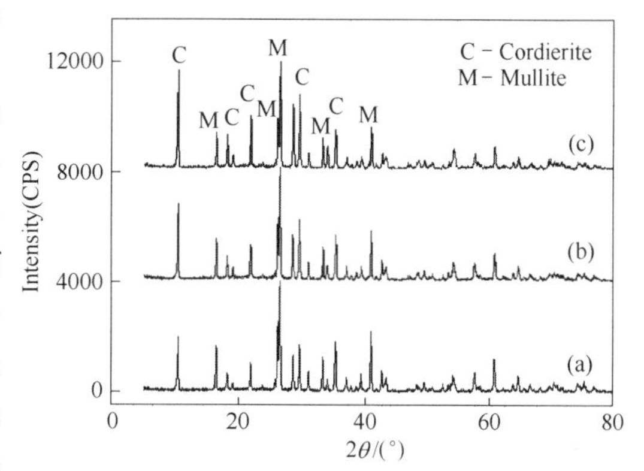

Fig. 2 X – ray diffraction pattern of sintering samples at 1450℃ for 120min
(a) corundum:30%, mullite:70%;
(b) corundum:40%, mullite:60%;
(c) corundum:50%, mullite:50%

The SEM images of fracture surface of sintering samples at 1450℃ for 120min with 50% corundum and 50% mullite were showed in Fig. 3. It can be seen from Fig. 3, numerous needle – like crystalline grains can be found on the fracture surface, and the length of these whiskers range from 1.74μm to 40.00μm. The density of corundum – mullite refractories increases with increase in the sintering temperature. The SEM results indicate that the whiskers of corundum – mullite have been synthesized.

The microwave absorbing properties of raw materials before and after sintering process are assessed using microwave resonant cavity perturbation techniques, based on the output voltage and resonator frequency of microwave sensor, by comparing the case with treated sample and raw materials in the microwave resonant cavity, and the results are shown in Figs. 4 and 5 respectively. The resonant curve of the microwave resonant cavity with empty chamber is shown in Fig. 4. It can be seen from Fig. 4 that the resonant curve of empty chamber has the highest resonant amplitude and the largest resonant frequency, and the other resonant curves indicate lower resonant amplitude and smaller resonant frequency, the results show that prepared corundum – mullite refractories has good microwave absorption property.

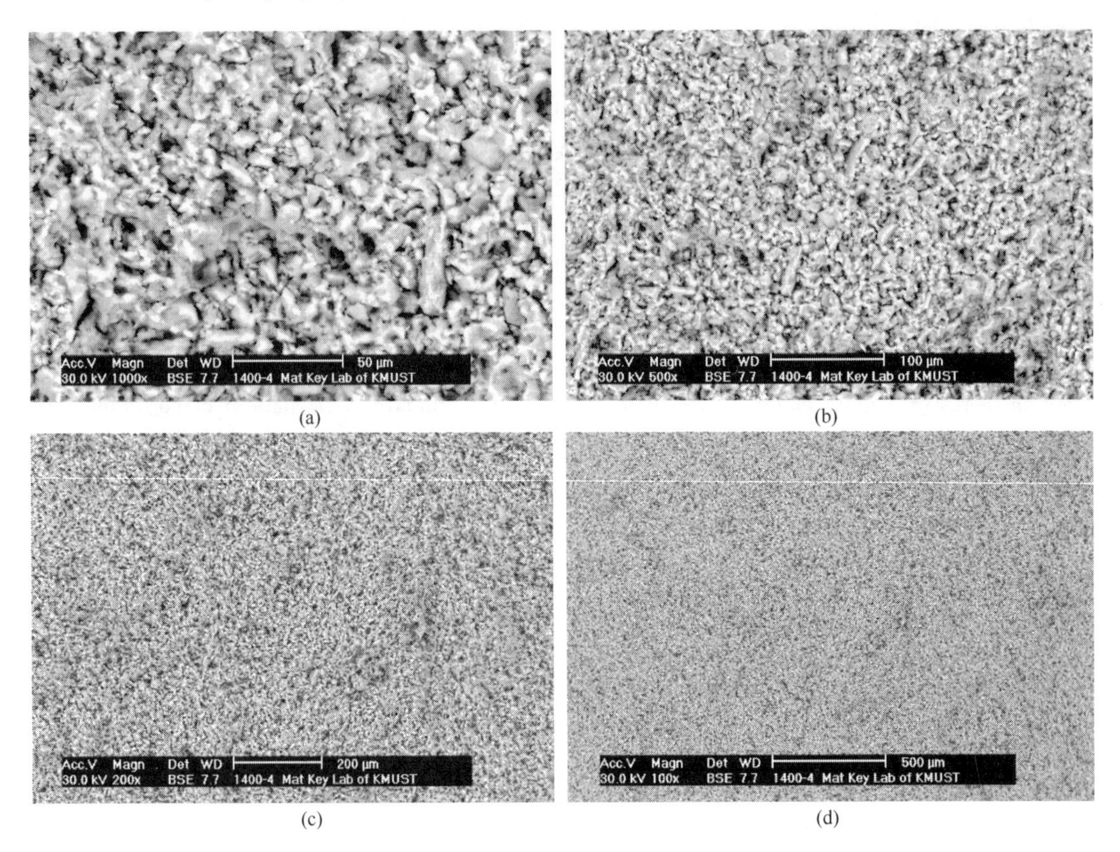

(a) (b) (c) (d)

Fig. 3 SEM images of sintering samples at 1450℃ for 120min with 50% corundum and 50% mullite
(a) 1000; (b) 500; (c), (d) 200

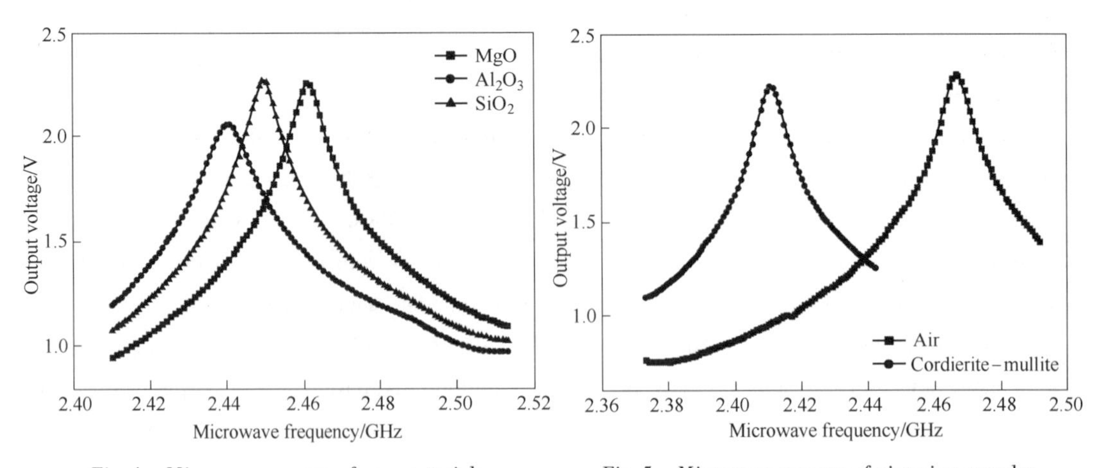

Fig. 4 Microwave spectra of raw materials

Fig. 5 Microwave spectra of sintering samples

Based on the above analysis, the real part of the permittivity of measured materials is directly proportional to microwave frequency shift in microwave spectra and is indicated in Eq. (1). And the imaginary part of the permittivity of measured materials is inversely proportional to amplitude of voltage in microwave spectra and is indicated in Eq. (2). The variation behavior of microwave spectra, such as the amplitude of voltage and the frequency shift of the first wave crest of microwave spectrum's are calculated and analyzed. The effects particle size of corundum – mullite refractories on microwave absorbing properties are obtained and obtained results are shown in Fig. 6 and Table 1. It can be observed in Fig. 6 and Table 1 that the frequency shift of corundum – mullite refractories increases gradually from about 0.0108GHz to 0.0149GHz with decreasing the particle size of corundum – mullite refractories from 150μm to 75μm. Similarly, the amplitude of voltage of corundum – mullite refractories also decreases from about 2.2857V to 2.2588V with decreasing the particle size of corundum – mullite refractories from 150μm to 106μm; it increases from 2.2588V to 2.2686V with decreasing the particle size of corundum – mullite refractories from 106μm to 75μm. From Fig. 6 and Table 1, it is found that frequency shift order for corundum – mullite refractories is $\Delta\omega_{75} > \Delta\omega_{106} > \Delta\omega_{150} > \Delta\omega_0$ ($\Delta\omega$ is frequency shift; subscript number is the particle size). Therefore, $\varepsilon'_{75} > \varepsilon'_{106} > \varepsilon'_{150} > \varepsilon'_0$. The amplitude of voltage order for corundum – mullite refractories is $U_0 > U_{75} > U_{150} > U_{106}$ (U is amplitude of voltage). Therefore, $\varepsilon''_{106} > \varepsilon''_{150} > \varepsilon''_{75} > \varepsilon''_0$. So, the optimum experiment parameters are found as follows: particle size of corundum – mullite refractories of 75μm.

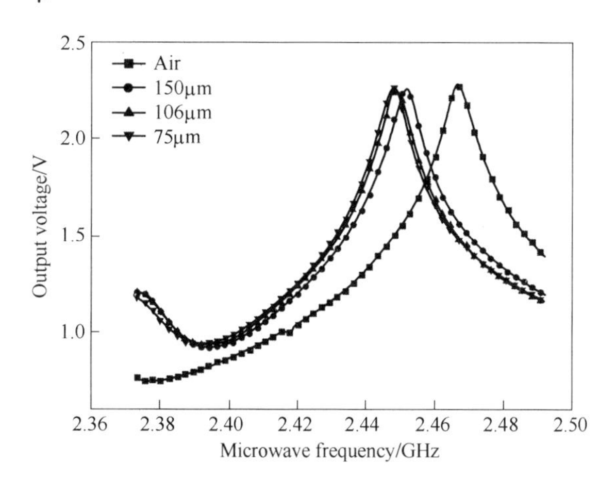

Fig. 6 Effect of different particle sizes on the microwave absorbing properties of corundum – mullite refractories

Table 1 Frequency shift and voltage attenuation of different particle sizes of corundum – mullite refractories

Particle size	Resonant frequency/GHz	Frequency shift	Amplitude of voltage/V
Air	f_0 = 2.4623	0	V_0 = 2.2857
150μm	f_1 = 2.4515	Δf_1 = 0.0108	V_1 = 2.2637
106μm	f_2 = 2.4484	Δf_2 = 0.0139	V_2 = 2.2588
75μm	f_3 = 2.4474	Δf_3 = 0.0149	V_3 = 2.2686

4 Conclusion

This study confirmed that preparation of corundum – mullite refractories from analytically pure Al_2O_3, analytically pure SiO_2, and MgO powders was feasible. The microwave absorbing characteristics of raw materials before and after sintering process were assessed using microwave resonant cavity perturbation techniques. The optimum conditions for preparation of corundum – mullite refractories have been identified to be an sintering temperature of 1450℃, and process time of 120min. The optimum conditions result of microwave absorbing properties in prepared corundum – mullite refractories with particle size of 75μm.

Acknowledgements

The authors acknowledge the financial supports from the National Scientific Foundation of China (No. 51102123, 51090385), the International S&T Cooperation Program of China (No. 2012DFA70570), the Doctoral Fund of Ministry of Education of China (No. 20125314120014, 20105314120002), the Yunnan Provincial International Cooperative Program (No. 2011IA004) and the Applied Foundation Fund of Yunnan Province of China (No. 2009ZC012M, 2012FD015).

References

[1] Medvedovski E. Alumina – mullite ceramics for structural applications[J]. Ceramics International, 2006, 32 (4): 369 – 375.

[2] Aksel C. The effect of mullite on the mechanical properties and thermal shock behaviour of alumina – mullite refractory materials[J]. Ceramics International, 2003, 29(2): 183 – 188.

[3] Meng B, Peng J. Effects of in situ synthesized mullite whiskers on flexural strength and fracture toughness of corundum – mullite refractory materials[J]. Ceramics International, 2013, 39(2): 1525 – 1531.

[4] Sheen J. Measurements of microwave dielectric properties by an amended cavity perturbation technique [J]. Measurement, 2009, 42(1): 57 – 61.

[5] Huang M, Peng J, Yang J, et al. A new equation for the description of dielectric losses under microwave irradiation[J]. Journal of Physics D: Applied Physics, 2006, 39(10): 2255.

[6] Huang M, Yang J, Wang J, et al. Microwave sensor for measuring the properties of a liquid drop [J]. Measurement Science and Technology, 2007, 18(7): 1934.

[7] Huang M, Peng J, Yang J, et al. Microwave cavity perturbation technique for measuring the moisture content of sulphide minerals concentrates[J]. Minerals Engineering, 2007, 20(1): 92 – 94.

[8] Guo S, Li W, Peng J, et al. Microwave – absorbing characteristics of mixtures of different carbonaceous reducing agents and oxidized ilmenite[J]. International Journal of Mineral Processing, 2009, 93(3): 289 – 293.

[9] Eichelbaum M, Stößer R, Karpov A, et al. The microwave cavity perturbation technique for contact – free and in situ electrical conductivity measurements in catalysis and materials science[J]. Physical Chemistry Chemical Physics, 2012, 14(3): 1302 – 1312.

[10] Chen G, Chen J, Peng J H. Microwave absorbing properties of mechanical activated ilmenite[J]. Metalurgia International, 2011, 16(8): 24 – 28.

Improvement of Electrochemical Properties of $LiNi_{1/3}Mn_{1/3}Co_{1/3}O_2$ by Coating with $La_{0.4}Ca_{0.6}CoO_3$

Jiang Du, Zhengfu Zhang, Jinhui Peng, Yamei Han, Yi Xia, Shenghui Guo, Shaohua Ju, Chongyan Leng, Guo Chen, Lei Xu, Junsai Sun, Hongge Yan

Abstract: In this paper, $La_{0.4}Ca_{0.6}CoO_3$ - coated $LiNi_{1/3}Mn_{1/3}Co_{1/3}O_2$ is successfully prepared by the sol - gel method associated with microwave pyrolysis method. The structure and electrochemical properties of the $La_{0.4}Ca_{0.6}CoO_3$ - coated $LiNi_{1/3}Co_{1/3}Mn_{1/3}O_2$ are investigated by using X - ray diffraction (XRD), electrochemical impedance spectroscopy (EIS), and charge/discharge tests. XRD analyses show that the $La_{0.4}Ca_{0.6}CoO_3$ coating does not change the structure of $LiNi_{1/3}Co_{1/3}Mn_{1/3}O_2$. The electrochemical performance studies demonstrate that 2wt. % $La_{0.4}Ca_{0.6}CoO_3$ - coated $LiNi_{1/3}Co_{1/3}Mn_{1/3}O_2$ powders exhibit the best electrochemical properties, with an initial discharge capacity of 156. 9mA · h/g and capacity retention of 98. 9% after 50 cycles when cycled at a current density of 0. 2C between 2. 75V and 4. 3V. $La_{0.4}Ca_{0.6}CoO_3$ coating can improve the rate performance because of the enhancement of the surface electronic/ionic transportation by the coating layer. EIS results suggest that the coating $La_{0.4}Ca_{0.6}CoO_3$ plays an important role in suppressing the increase of cell impedance with cycling especially for the increase of charge - transfer resistance.

Keywords: $LiNi_{1/3}Co_{1/3}Mn_{1/3}O_2$; coating; cathode; lithium ion battery; electrochemical properties

1 Introduction

For lithium ion battery, cathode material is one of the key factors to improve its performance. The $LiNi_{1/3}Co_{1/3}Mn_{1/3}O_2$ with layered structure has advantages of high specific capacity, good structure stability and high safety, and is considered to be one of the promising cathode materials for lithium ion batteries[1-4]. However, there is an obvious capacity fading of $LiNi_{1/3}Co_{1/3}Mn_{1/3}O_2$ at high cycling rate which widely prohibited the applications of these materials. The most plausible reasons are the lower ionic/electronic conductivity and side reactions at the interface of the electrode and electrolyte[5]. According to the literature[6-9], introduction of transition metal, rare earth elements, such as Co, La, can improve the ion conductivity. Moreover, as reported[6-8], a perovskite oxide(ABO_3) has high ionic conductivity because of the cation deficiency at the A - site and the tilting and/or the rotating of the BO_6 octahedra during the insertion and emerging process of Li^+. Recently, surface coating[10-13] has been proved to be a facile and effective strategy to improve the electrochemical performance of cathode materials[14-19]. However, the effects of the $La_{0.4}Ca_{0.6}CoO_3$ coating on $LiNi_{1/3}Co_{1/3}Mn_{1/3}O_2$ have still not been reported. In this study, we have explored the possibility of coating the $LiNi_{1/3}Co_{1/3}Mn_{1/3}O_2$ particles by $La_{0.4}Ca_{0.6}CoO_3$.

We adopted the sol – gel method associated with microwave pyrolysis method to get $La_{0.4}Ca_{0.6}$ – CoO_3 – coated $LiNi_{1/3}Co_{1/3}Mn_{1/3}O_2$ cathode material. The preparation, structure and electrochemical performance of the $La_{0.4}Ca_{0.6}CoO_3$ – coated $LiNi_{1/3}Co_{1/3}Mn_{1/3}O_2$ cathode material were discussed in comparison with the bare one.

2 Experimental

2.1 Preparation of $La_{0.4}Ca_{0.6}CoO_3$ – coated $LiNi_{1/3}Co_{1/3}Mn_{1/3}O_2$

Commercially available spherical $LiNi_{1/3}Co_{1/3}Mn_{1/3}O_2$ powder (Beijing Dangsheng, China) was utilized as bare sample. Coating $La_{0.4}Ca_{0.6}CoO_3$ on the $LiNi_{1/3}Co_{1/3}Mn_{1/3}O_2$ powders was formed by using a sol – gel method associated with microwave pyrolysis[20,21]. The $La_{0.4}Ca_{0.6}CoO_3$ coating sol was prepared from $Ca(NO_3)_2 \cdot 4H_2O$, $Co(NO_3)_2 \cdot 6H_2O$ and $La(NO_3)_3 \cdot nH_2O$ (Regent, Tianjin) in a stoichiometry which was dissolved in the distilled water. The proper conditions were determined as the following: pH = 1.5, temperature 40 – 90℃ and stirring rate 500r/min. The $LiNi_{1/3}$ – $Co_{1/3}Mn_{1/3}O_2$ powders were added into the coating sol and well – mixed in ethanol. The mixture was stirred in magnetic stirrer at 323K until the solvent was vaporized. The material was calcinated in a microwave reactor with a frequency of 2450MHz and a maximum power output of 1500W. The microwave power was adjusted according to the temperature of material. Temperature control was as follows: from room temperature heated up to 973K with a heating rate of 10℃/min and the temperature was maintained for 30min. Then the 2wt. % $La_{0.4}Ca_{0.6}CoO_3$ – coated $LiNi_{1/3}Co_{1/3}Mn_{1/3}O_2$ material was obtained. The schematic is shown in Fig. 1.

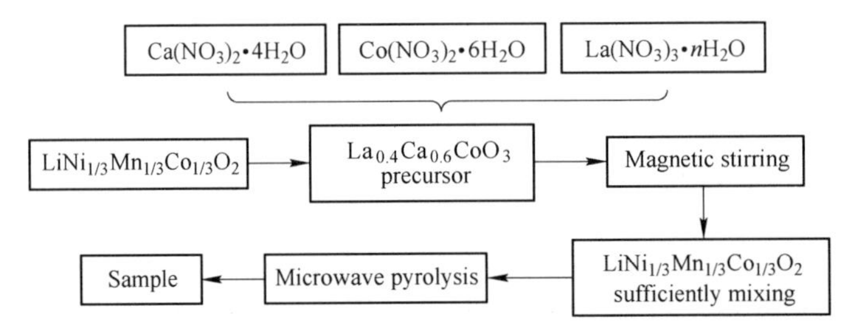

Fig. 1 Preparation of process flow diagram of electrode materials

2.2 Assembly of cells

The electrodes were fabricated from a 85∶10∶5 (mass percent) mixture of active materials, acetylene black as the current conductor and polyvinylidene fluoride (PVDF) as the binder. The PVDF was dissolved in N – methylpyrrolidinone, before the active material and the conductor mixture were added. After homogenization, the slurry was evacuated for 20min to remove the residual air. The slurry was then coated on a thin aluminum foil (20μm thick) and dried overnight at 60 – 70℃. The electrodes were pressed with a pressure of 10MPa and punched into 12mm – diameter disks. The electrochemical cells were prepared as 2025 coin – cell with lithium metal foil as both the counter and reference electrodes. The cells were assembled in an argon – filled glove box. The electrolyte used for analysis was 1mol $LiPF_6$ in ethylene carbonate/diethyl carbonate (1∶1).

2.3 X – Ray diffraction and electrochemical tests

X – ray diffraction (XRD) data for the finely ground samples were collected at 298K by using a Bruker D8 X – ray diffractometer with CuK_α radiation (λ = 0.15406nm). It was operated at 40kV and 300mA in the 2θ range of 10° to 90° in the continuous scan mode with the step size of 0.01° and the scan rate 1.0 °/min.

The cells were aged for 12h before being electrochemically cycled between 2.75 – 4.3V (versus Li/Li^+) by using Neware battery testing system and instrument after assembling. For the galvanostatic charge/discharge test at room temperature, the cells were charged at 0.2C and then discharged at 0.2C, 1C, 2C and 4C, respectively.

3 Results and discussion

3.1 Structure and surface morphology

Fig. 2 shows the XRD patterns and Miller indices of the bare and 2wt. % $La_{0.4}Ca_{0.6}CoO_3$ – coated $LiCo_{1/3}Ni_{1/3}Mn_{1/3}O_2$. In Fig. 2, the diffraction patterns of all the materials indicate hexagonal system and a single phase of well – defined α – $NaFeO_2$ structure with space group of R$\bar{3}$m. Both of the samples have a well – defined α – $NaFeO_2$ structure, and $La_{0.4}Ca_{0.6}CoO_3$ have no significant effects on $LiNi_{1/3}Co_{1/3}Mn_{1/3}O_2$. The splits in the (006/102) and (108/110) are around at 38° and 65° doublets, indicating the formation of a highly ordered layered structure[22,23]. The ratios of intensities of the I_{003}/I_{104} of the bare and $La_{0.4}Ca_{0.6}CoO_3$ – coated material are 1.666 and 1.694, respectively, well above the values reported for compounds like $LiNi_{1-x}Co_xO_2$ to deliver good electrochemical performance[24]. No other impurity phase is observed in XRD patterns for the $La_{0.4}Ca_{0.6}CoO_3$ – coated material, which suggests that the structure of $LiNi_{1/3}Co_{1/3}Mn_{1/3}O_2$ is not affected by the $La_{0.4}Ca_{0.6}CoO_3$ coating. Moreover, according to the XRD data and the Eq. (1), the structure parameters of all the samples are summarized in Table 1. The increase of structure parameters may be due to a little La and Ca enters the crystal lattice. After the high temperature treatment, one part of the La, Co and Ca exists on the surface of the $LiNi_{1/3}Co_{1/3}Mn_{1/3}O_2$ as $La_{0.4}Ca_{0.6}CoO_3$, and the other part of La, Co and Ca is diffused into the $LiNi_{1/3}Co_{1/3}Mn_{1/3}O_2$ bulk and modified its crystal structure.

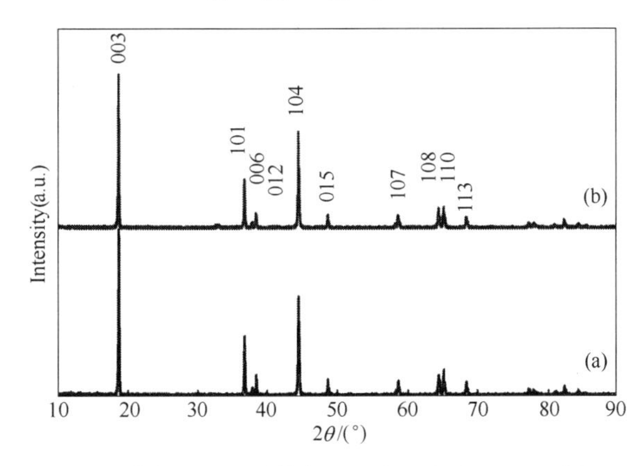

Fig. 2 XRD patterns of (a) bare and (b) $La_{0.4}Ca_{0.6}CoO_3$ – coated $LiNi_{1/3}Mn_{1/3}Co_{1/3}O_2$

Table 1 Lattice structural parameters of bare and $La_{0.4}Ca_{0.6}CoO_3$ - coated $LiNi_{1/3}Mn_{1/3}Co_{1/3}O_2$

	a/nm	c/nm	c/a	I_{003}/I_{104}
0wt. % —$La_{0.4}Ca_{0.6}CoO_3$	0.2858	1.4220	4.975	1.666
2wt. % —$La_{0.4}Ca_{0.6}CoO_3$	0.2862	1.4245	4.9775	1.694

The hexagonal crystal spacing can be estimated using Eq. (1)[25]:

$$\frac{1}{d^2} = \frac{4(h^2 + hk + k^2)}{3a^2} + \frac{2l}{c^2} \quad (1)$$

Scanning electron microscopy (SEM) observations were carried out in order to observe morphology of the $La_{0.4}Ca_{0.6}CoO_3$ - coated $LiNi_{1/3}Mn_{1/3}Co_{1/3}O_2$ powders. Fig. 3(a) and (b) shows the micromorphologies of $La_{0.4}Ca_{0.6}CoO_3$ - coated $LiNi_{1/3}Mn_{1/3}Co_{1/3}O_2$ materials in various magnifications, respectively. All samples have good spheral shape with diameters of about 8μm. The spheral particle is beneficial for achieving a high tap density and energy density. As can be seen in Fig. 3, the primary particle size is about 200 - 300nm in diameter and these small particles aggregated each other to form micro - sized spherical secondary particles. Hence, one can achieve both high rate capability and high tap density from the particles having the spherical morphology as shown in Fig. 3. According to Fig. 3, $La_{0.4}Ca_{0.6}CoO_3$ is distributed uniformly on the surface of particles of 2wt. % $La_{0.4}Ca_{0.6}CoO_3$ - coated $LiNi_{1/3}Mn_{1/3}Co_{1/3}O_2$. Furthermore, small white particles in Fig. 3 (b) are newly appeared and these can be attributed to crystalline $La_{0.4}Ca_{0.6}CoO_3$.

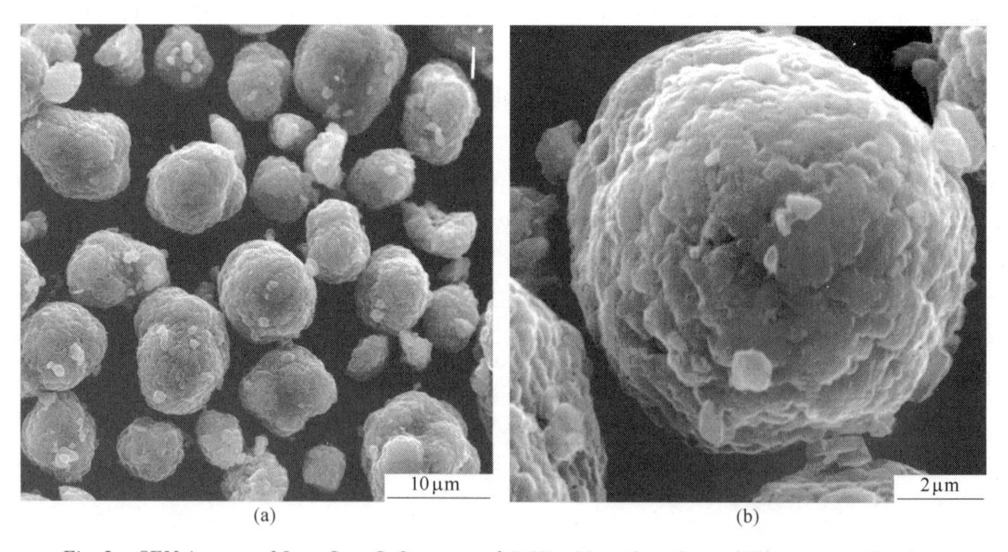

Fig. 3 SEM images of $La_{0.4}Ca_{0.6}CoO_3$ - coated $LiNi_{1/3}Mn_{1/3}Co_{1/3}O_2$ at different magnifications (a) 3000; (b) 12000

3.2 Elemental analysis

The X - ray fluorescence results of $LiNi_{1/3}Mn_{1/3}Co_{1/3}O_2$ and 2wt. % $La_{0.4}Ca_{0.6}CoO_3$ - coated $LiNi_{1/3}Mn_{1/3}Co_{1/3}O_2$ material are shown in Table 2. It reveals that the content of La and Ca are 0.927wt. % and 0.675wt. % respectively and pyrolytic products are stoichiometric $La_{0.4}Ca_{0.6}CoO_3$ -

coated $LiNi_{1/3}Mn_{1/3}Co_{1/3}O_2$.

Table 2 X - ray fluorescence results of $LiNi_{1/3}Mn_{1/3}Co_{1/3}O_2$ - and $La_{0.4}Ca_{0.6}CoO_3$ - coated (2wt. %) $LiNi_{1/3}Mn_{1/3}Co_{1/3}O_2$

	O	Ni	Co	Mn	La	Ca
0wt. % —$La_{0.4}Ca_{0.6}CoO_3$	43.00	18.59	18.25	17.64	0	0
2wt. % —$La_{0.4}Ca_{0.6}CoO_3$	41.53	18.89	18.55	17.89	0.927	0.675

3.3 Electrochemical properties of $La_{0.4}Ca_{0.6}CoO_3$ - coated $LiNi_{1/3}Mn_{1/3}Co_{1/3}O_2$

Electrochemical characterization was carried out with a coin - type cell. The discharge curves of the bare and $La_{0.4}Ca_{0.6}CoO_3$ - coated $LiNi_{1/3}Mn_{1/3}Co_{1/3}O_2$ between 2.75V and 4.3V at 0.2C, 1C, 2C rates are shown in Fig. 4. The first discharge - specific capacity for material without the

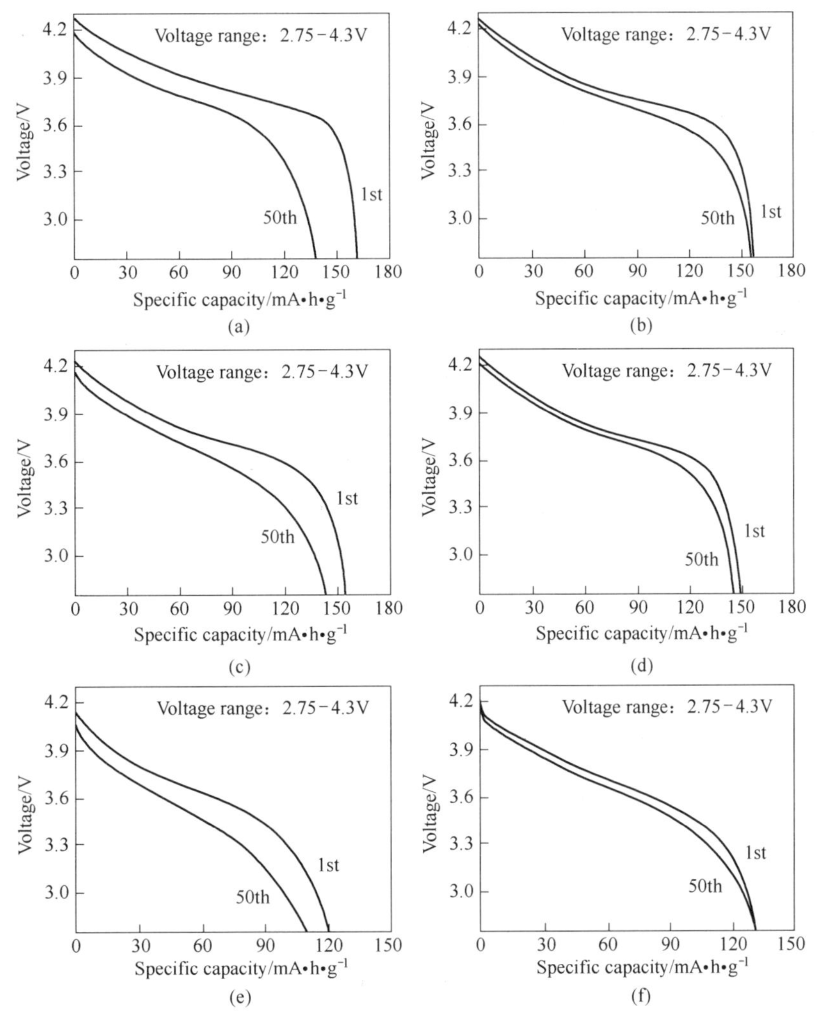

Fig. 4 Discharge curves of bare at 0.2C(a), coated at 0.2C(b), bare at 1C(c), coated at 1C(d), bare at 2C(e), and coated at 2C in 2.75 - 4.3V(f)

$La_{0.4}Ca_{0.6}CoO_3$ coating is 161.8mA · h/g at first time and 137.3mA · h/g at 50th times at 0.2C rates, while it is 156.9mA · h/g at first time and 155.1mA · h/g at 50th times at 0.2C rates for the $La_{0.4}Ca_{0.6}CoO_3$ – coated $LiNi_{1/3}Mn_{1/3}Co_{1/3}O_2$ material. As can be seen, the first discharge capacity at 0.2C rates reduces slightly, because $La_{0.4}Ca_{0.6}CoO_3$ is covered in the positive electrode material surface. The first discharge specific capacity for the bare material is 154.72mA · h/g at first time and 143.05mA · h/g at 50th times at 1C rate, while it is 148.66mA · h/g at first time and 144.88mA · h/g at 50th times at 1C rate for the $La_{0.4}Ca_{0.6}CoO_3$ – coated $LiNi_{1/3}Mn_{1/3}Co_{1/3}O_2$ material. Capacity retention of the $La_{0.4}Ca_{0.6}CoO_3$ – coated $LiNi_{1/3}Mn_{1/3}Co_{1/3}O_2$ material is better than bare $LiNi_{1/3}Mn_{1/3}Co_{1/3}O_2$ at 1C rate, which corresponds to 97.5% and 92.5%, respectively. For the bare material, the first discharge capacity is 120.08mA · h/g, the capacity retention is 91.31%; however, for the coated one, its first discharge capacity and capacity retention reach 132.10mA · h/g and 99.99%, both larger than those of the bare $LiNi_{1/3}Mn_{1/3}Co_{1/3}O_2$. Such improved discharge capacity suggests that the coating $La_{0.4}Ca_{0.6}CoO_3$ on active material particles help assure good electronic contact within the composite cathode and do not impede Li^+ transport across the $LiNi_{1/3}Mn_{1/3}Co_{1/3}O_2$/electrolyte interface[26,27]. Clearly, $La_{0.4}Ca_{0.6}CoO_3$ coating on $LiNi_{1/3}Mn_{1/3}Co_{1/3}O_2$ can improve the cyclic stability and capability of $LiNi_{1/3}Mn_{1/3}Co_{1/3}O_2$ during the charge – discharge process. The result is similar to Lee, et al.[28] and Wang, et al.[29].

Fig. 5 shows the rate capabilities and cyclic performance of bare and $La_{0.4}Ca_{0.6}CoO_3$ coated $LiNi_{1/3}Mn_{1/3}Co_{1/3}O_2$ electrodes at 0.2C, 1C, 2C and 4C rates in the voltage range of 2.75 – 4.3V. For instance, the capacity retention for the bare electrode at the 0.2C rate is only 84.8%. However, the $La_{0.4}Ca_{0.6}CoO_3$ coated sample exhibits as high as 98.9% capacity retention under the same measurement condition. Similarly at 0.2C discharge rates, after 50th cycles capacity retention of the $La_{0.4}Ca_{0.6}CoO_3$ – coated $LiNi_{1/3}Mn_{1/3}Co_{1/3}O_2$ is 97.5%, while the bare $LiNi_{1/3}Mn_{1/3}Co_{1/3}O_2$ is 92.5%. After 50th cycles capacity retention of the bare $LiNi_{1/3}Mn_{1/3}Co_{1/3}O_2$ and $La_{0.4}Ca_{0.6}CoO_3$ – coated $LiNi_{1/3}Mn_{1/3}Co_{1/3}O_2$ material are 91.31% and 99.99% at 2C rates, respectively. At 4C rates, after 50th cycles the capacity retention of the bare $LiNi_{1/3}Mn_{1/3}Co_{1/3}O_2$ and

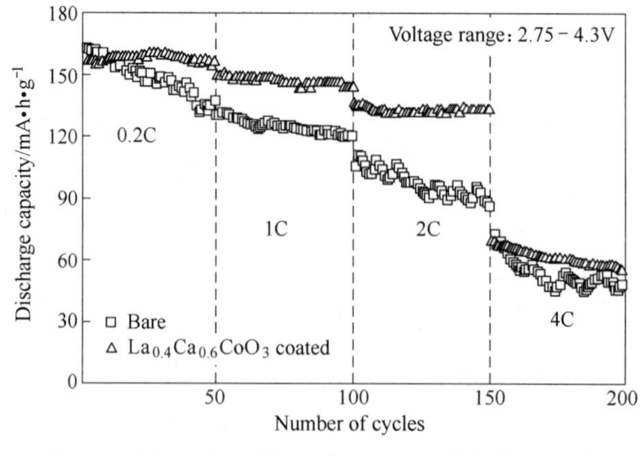

Fig. 5 Discharge specific capacities and cyclic performances of the bare and $La_{0.4}Ca_{0.6}CoO_3$ – coated $LiNi_{1/3}Mn_{1/3}Co_{1/3}O_2$ electrodes in the voltage of 2.75 – 4.3V at 0.2, 1, 2, and 4C rates

$La_{0.4}Ca_{0.6}CoO_3$ – coated $LiNi_{1/3}Mn_{1/3}Co_{1/3}O_2$ material are 69.9% and 80.1%, respectively. These results indicate that the 2wt.% $La_{0.4}Ca_{0.6}CoO_3$ coating can improve the capacity retention and cycle stability of $LiNi_{1/3}Mn_{1/3}Co_{1/3}O_2$. Moreover, the phenomenon of improved rate capability and cyclic performance of the coated material become more obvious with the increase of current density. The improved performance may be due to the protection of the electrode surface by the $La_{0.4}Ca_{0.6}CoO_3$ coating from electrolyte corrosion[30,31]. Moreover, $La_{0.4}Ca_{0.6}CoO_3$ coatings may also have acted as an efficient transportation medium of Li^+ between cathode and electrolyte. These properties led to increased rate capability of the $La_{0.4}Ca_{0.6}CoO_3$ coated cathode.

3.4 AC impedance analysis

To explore the reason for the enhanced electrochemical performance, alternating current (AC) impedance measurements were carried out after the 2nd and 50th charge – discharge cycles. Fig. 6 shows impedance profiles of the cells using the bare and the $La_{0.4}Ca_{0.6}CoO_3$ – coated $LiNi_{1/3}Mn_{1/3}$-$Co_{1/3}O_2$ as positive electrode materials after the 2nd and 50th charge – discharge cycles at 0.2C at a room temperature. The Nyquist plots for both electrodes present two semicircles, one in the

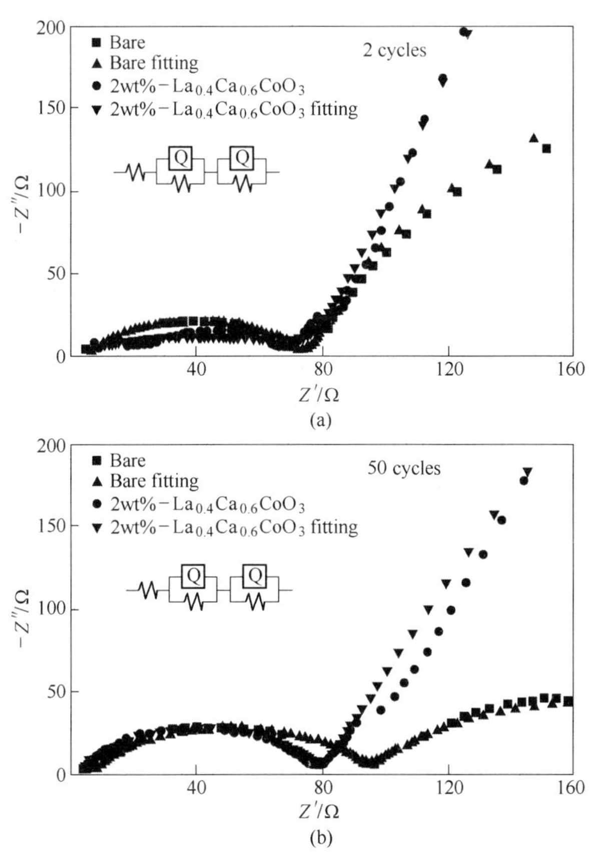

Fig. 6 The AC impedance diagram of bare and $La_{0.4}Ca_{0.6}CoO_3$ – coated $LiNi_{1/3}Mn_{1/3}Co_{1/3}O_2$ material after charge and discharge two times (a) and 50 times (b)

high – to – medium frequency region and the other in the low frequency region. The first semicircle is attributed the resistance of the surface film and the second one is attributed to the charge transfer resistance[32-34]. As shown in Fig. 6, we find that the charge transfer resistance is reduced indistinctively by the $La_{0.4}Ca_{0.6}CoO_3$ coating after the $La_{0.4}Ca_{0.6}CoO_3$ coating. The charge – transfer resistance in the positive electrode with the bare and $La_{0.4}Ca_{0.6}CoO_3$ – coated samples increases after 50th charge – discharge cycles. The resistances of the cells with the bare and the $La_{0.4}Ca_{0.6}CoO_3$ – coated $LiNi_{1/3}Mn_{1/3}Co_{1/3}O_2$ are about 95Ω and 78Ω, respectively. The increasing of the charge – transfer resistance in the positive electrode causes the discharge capacity fading of the cells with the bare and $La_{0.4}Ca_{0.6}CoO_3$ – coated samples as shown in Fig. 4. Basically, the capacity loss can be ascribed to the side reaction between electrode and electrolyte, but the coated $La_{0.4}Ca_{0.6}CoO_3$ may act as an obstacle to the contact of electrode and electrolyte. Moreover, the interfacial resistance of the cell with the $La_{0.4}Ca_{0.6}CoO_3$ – coated sample is smaller than that with the bare sample. This result suggests that $La_{0.4}Ca_{0.6}CoO_3$ coating is effective to decrease the interfacial resistance[35].

4 Conclusions

$LiNi_{1/3}Mn_{1/3}Co_{1/3}O_2$ particles coated with $La_{0.4}Ca_{0.6}CoO_3$ were successfully prepared by the sol – gel method associated with microwave pyrolysis method. The capacity, rate capability and cycling performance of $La_{0.4}Ca_{0.6}CoO_3$ – coated $LiNi_{1/3}Mn_{1/3}Co_{1/3}O_2$ were significantly better than that of $LiNi_{1/3}Mn_{1/3}Co_{1/3}O_2$. At high rates, the 2wt. % $La_{0.4}Ca_{0.6}CoO_3$ – coated $LiNi_{1/3}Mn_{1/3}Co_{1/3}O_2$ cathode exhibited much enhanced rate capability and cycling performance compared to the bare sample at room temperature. The charge – discharge performance of the cells was improved by the $La_{0.4}Ca_{0.6}CoO_3$ coating because of the decrease in the interfacial resistance. Thus, an appropriate $La_{0.4}Ca_{0.6}CoO_3$ coating was a promising method to overcome the existing problems of $LiNi_{1/3}Mn_{1/3}Co_{1/3}O_2$ cathode for high power applications.

Acknowledgements

This work was supported by the Research Projects of China National NSFC(No. U1202272).

References

[1] Liu Z L, Yu A S, Lee J Y. Synthesis and characterization of $LiNi_{1-x-y}Co_xMn_yO_2$ as the cathode materials of secondary lithium batteries[J]. Journal of Power Sources, 1999, 81: 416 – 419.

[2] Ohzuku T, Makimura Y. Layered lithium insertion material of $LiCo_{1/3}Ni_{1/3}Mn_{1/3}O_2$ for lithium – ion batteries [J]. Chemistry Letters, 2001, (7): 642 – 643.

[3] Zheng J M, Li J, Zhang Z R, et al. The effects of TiO_2 coating on the electrochemical performance of $Li[Li_{0.2}Mn_{0.54}Ni_{0.13}Co_{0.13}]O_2$ cathode material for lithium – ion battery[J]. Solid State Ionics, 2008, 179(27 – 32): 1794 – 1799.

[4] Dahbi M, Wikberg J M, Saadoune I, et al. A delithiated $LiNi_{0.65}Co_{0.25}Mn_{0.10}O_2$ electrode material: a structural, magnetic and electrochemical study[J]. Electrochimica Acta, 2009, 54(11): 3211 – 3217.

[5] Vetter J, Novak P, Wagner M R, et al. Ageing mechanisms in lithium - ion batteries[J]. Journal of Power Sources, 2005, 147(1): 269 - 281.

[6] Gao J, Ying J, Jiang C, et al. Preparation and characterization of spherical La - doped $Li_4Ti_5O_{12}$ anode material for lithium ion batteries[J]. Ionics, 2009, 15(5): 597 - 601.

[7] Ishihara T, Akbay T, Furutani H, et al. Improved oxide ion conductivity of Co doped $La_{0.8}Sr_{0.2}Ga_{0.8}Mg_{0.2}O_3$ perovskite type oxiden[J]. Solid State Ionics, 1998, 113: 585 - 591.

[8] Bohnke O, Bohnke C, Fourquet J L. Mechanism of ionic conduction and electrochemical intercalation of lithium into the perovskite lanthanum lithium titanate[J]. Solid State Ionics, 1996, 91(1): 21 - 31.

[9] Li Z, Zhang Z, Jiang W, et al. Direct measurement of lanthanum uptake and distribution in internodal cells of Chara[J]. Plant Science, 2008, 174(5): 496 - 501.

[10] Alva G, Kim C, Yi T, et al. Improving the stability of $LiNi_{0.5}Mn_{1.5}O_4$ as a high - power cathode material for Li - ion batteries by MgO coating[C]//Meeting Abstracts. The Electrochemical Society, 2013(14): 993.

[11] Sinhal N N, Munichandraiah N. Synthesis and characterization of carbon - coated $LiNi_{1/3}Co_{1/3}Mn_{1/3}O_2$ in a single step by an inverse microemulsion route[J]. ACS Applied Materials & Interfaces, 2009, 1(6): 1241 - 1249.

[12] Kim H S, Kim K, Moon S I, et al. A study on carbon - coated $LiNi_{1/3}Mn_{1/3}Co_{1/3}O_2$ cathode material for lithium secondary batteries[J]. Journal of Solid State Electrochemistry, 2008, 12(7 - 8): 867 - 872.

[13] Hu S K, Cheng G H, Cheng M Y, et al. Cycle life improvement of ZrO_2 - coated spherical $LiNi_{1/3}Co_{1/3}Mn_{1/3}O_2$ cathode material for lithium ion batteries[J]. Journal of Power Sources, 2009, 188(2): 564 - 569.

[14] Kim Y, Kim H S, Martin S W. Synthesis and electrochemical characteristics of Al_2O_3 - coated $LiNi_{1/3}Co_{1/3}Mn_{1/3}O_2$ cathode materials for lithium ion batteries[J]. Electrochimica Acta, 2006, 52(3): 1316 - 1322.

[15] Guo R, Shi P, Cheng X, et al. Effect of ZnO modification on the performance of $LiNi_{0.5}Co_{0.25}Mn_{0.25}O_2$ cathode material[J]. Electrochimica Acta, 2009, 54(24): 5796 - 5803.

[16] Park B C, Kim H B, Myung S T, et al. Improvement of structural and electrochemical properties of AlF_3 - coated $Li[Ni_{1/3}Co_{1/3}Mn_{1/3}]O_2$ cathode materials on high voltage region[J]. Journal of Power Sources, 2008, 178(2): 826 - 831.

[17] Kim H S, Kim Y, Kim S I, et al. Enhanced electrochemical properties of $LiNi_{1/3}Co_{1/3}Mn_{1/3}O_2$ cathode material by coating with $LiAlO_2$ nanoparticles[J]. Journal of Power Sources, 2006, 161(1): 623 - 627.

[18] Li D, Sasaki Y, Kobayakawa K, et al. Preparation, morphology and electrochemical characteristics of $LiNi_{1/3}Mn_{1/3}Co_{1/3}O_2$ with LiF addition[J]. Electrochimica Acta, 2006, 52(2): 643 - 648.

[19] West W C, Soler J, Smart M C, et al. Electrochemical behavior of layered solid solution Li_2MnO_3 - $LiMO_2$ (M = Ni, Mn, Co) Li - ion cathodes with and without alumina coatings[J]. Journal of the Electrochemical Society, 2011, 158(8): A883 - A889.

[20] Zhang Z F, Ma Q B, Chen Q H. Transactions of Nonferrous Metals Society of China, 2001, 21: 1111 - 1117.

[21] Han Y M, Zhang Z F, Zhang L B, et al. Influence of carbon coating prepared by microwave pyrolysis on properties of $LiNi_{1/3}Mn_{1/3}Co_{1/3}O_2$[J]. Transactions of Nonferrous Metals Society of China, 2013, 23: 2971 - 2976.

[22] Zhang H H, Qiao Q Q, Li G R, et al. Surface nitridation of Li - rich layered $Li(Li_{0.17}Ni_{0.25}Mn_{0.58})O_2$ oxide as cathode material for lithium - ion battery[J]. Journal of Materials Chemistry, 2012, 22: 13104 - 13109.

[23] Wang J, Qiu B, Cao H, et al. Electrochemical properties of $0.6Li[Li_{1/3}Mn_{2/3}]O_2 - 0.4LiNi_xMn_yCo_{1-x-y}O_2$ cathode materials for lithium - ion batteries[J]. Journal of Power Sources, 2012, 218: 128 - 133.

[24] Shaju K M, Subba Rao G V, Chowdari B V R. X - ray photoelectron spectroscopy and electrochemical behaviour of 4V cathode, $Li(Ni_{1/2}Mn_{1/2})O_2$[J]. Electrochima Acta, 2003, 48: 1505 - 1514.

[25] Fan Q. Material Sciences, 2012, 2: 68 - 71.

[26] Marcinek M L, Wilcox J W, Doeff M M, et al. Microwave plasma chemical vapor deposition of carbon coatings on $LiNi_{1/3}Co_{1/3}Mn_{1/3}O_2$ for Li - ion battery composite cathodes[J]. Journal of the Electrochemical Society, 2009, 156: A48 - A51.

[27] Kerlau M, Marcinek M, Srinivasan V, et al. Studies of local degradation phenomena in composite cathodes for lithium - ion batteries[J]. Electrochima Acta, 2007, 52: 5422 - 5429.

[28] Li D, Kato Y, Kobayakawa K, et al. Preparation and electrochemical characteristics of $LiNi_{1/3}Mn_{1/3}Co_{1/3}O_2$ coated with metal oxides coating[J]. Journal of Power Sources, 2006, 160: 1342 - 1348.

[29] Wang M, Wu F, Su Y, et al. Modification of $LiCo_{1/3}Ni_{1/3}Mn_{1/3}O_2$ cathode material by CeO_2 - coating [J]. Science in China Series E: Technological Sciences, 2009, 52: 2737 - 2741.

[30] Liu T, Zhao S X, Wang K, et al. CuO - coated $Li[Ni_{0.5}Co_{0.2}Mn_{0.3}]O_2$ cathode material with improved cycling performance at high rates[J]. Electrochima Acta, 85: 605 - 611.

[31] Li T, Ai X P, Yang H X. Article previous article next article table of contents reversible electrochemical conversion reaction of Li_2O/CuO nanocomposites and their application as high - capacity cathode materials for Li - ion batteries[J]. The Journal of Physical Chemistry C, 2011, 115: 6167 - 6174.

[32] Liu J, Wang Q, Reeja - Jayan B, et al. Carbon - coated high capacity layered $Li[Li_{0.2}Mn_{0.54}Ni_{0.13}Co_{0.13}]O_2$ cathodes[J]. Electrochemistry Communications, 2010, 12: 750 - 753.

[33] Seki S, Kobayashi Y, Miyashiro H, et al. Degradation mechanism analysis of all - solid - state lithium polymer secondary batteries by using the impedance measurement[J]. Journal of Power Sources, 2005, 146: 741 - 744.

[34] Riley L A, Atta A V, Cavanagh A S, et al. Electrochemical effects of ALD surface modification on combustion synthesized $LiNi_{1/3}Mn_{1/3}Co_{1/3}O_2$ as a layered - cathode material[J]. Journal of Power Sources, 2011, 196: 3317 - 3324.

[35] Machida N, Kashiwagi J, Naito M, et al. Electrochemical properties of all - solid - state batteries with ZrO_2 - coated $LiNi_{1/3}Mn_{1/3}Co_{1/3}O_2$ as cathode materials[J]. Solid State Ionics, 2012, 225: 354 - 358.

Influence of Carbon Coating Prepared by Microwave Pyrolysis on Properties of $LiNi_{1/3}Mn_{1/3}Co_{1/3}O_2$

Yamei Han, Zhengfu Zhang, Libo Zhang, Jinhui Peng, Mengbi Fu, C. Srinivasakannan, Jiang Du

Abstract: A novel synthesis method of carbon – coated $LiNi_{1/3}Mn_{1/3}Co_{1/3}O_2$ cathode material for lithiumion battery was reported. The carbon coating was produced from a precursor, glucose, by microwave – pyrolysis method. The prepared powders were characterized by canning electron microscopy (SEM), X – ray diffraction (XRD), X – ray fluorescence (XRF) and charge/discharge tests. XRD results indicated that the carbon coating does not change the phase structure of $LiNi_{1/3}Mn_{1/3}Co_{1/3}O_2$ material. SEM results show that the surface of spherical carbon – coated material becomes rough. Electrochemical performance results show that the carbon coating can improve the cycling performance of $LiNi_{1/3}Mn_{1/3}Co_{1/3}O_2$. The specific discharge capacity retention of the carbon – coated $LiNi_{1/3}Mn_{1/3}Co_{1/3}O_2$ reached 85.0% – 96.0% at the 50th cycle at 0.2C rate, and the specific discharge capacity retention is improved at a high rate.

Keywords: lithium – ion battery; cathode material; carbon coating; microwave pyrolysis method; electrochemical performance

1 Introduction

Lithium ion battery has become the state – of – power sources for portable appliances such as cellular phones, notebook computers, digital cameras and also a prime candidate for hybrid electric vehicles (HEVs), plug – in hybrid electric vehicles (PHEVs) and electric vehicles (EVs) for reasons of high power density, high energy density, safety and cycling performance[1-3]. Literature related to lithium transition metal oxides and their derivatives, such as $LiMn_2O_4$, $LiCoO_2$ and $Li(Ni,Co)O_2$ have been extensively reported, which are commercially available as 4V class cathode materials for lithium ion batteries. Among them, $LiCoO_2$ is widely used as positive electrode in commercial lithium secondary batteries. However, utilization of $LiCoO_2$ has the drawback due to scarcity of cobalt, being expensive and toxic. The cathode material synthesized by Ohzuku and Makimura[4], $LiNi_{1/3}Mn_{1/3}Co_{1/3}O_2$ attracts lots of attentions for its lower cost, less toxic and higher capacity, which is superior to the commercial material of $LiCoO_2$. It was reported that even 200mA · h/g can be attainable by charging up to 2.8V and 4.6V, with superior cycle performance[5]. Although $LiNi_{1/3}Mn_{1/3}Co_{1/3}O_2$ exhibits excellent performance comparatively, it has drawbacks such as low electronic conductivity, low tap density and relatively low cycling performance at high – rate, which impedes its industrial acceptance. Some measures have been reported to improve the cycle performance of $LiNi_{1/3}Mn_{1/3}Co_{1/3}O_2$, such as doping small amounts of additional ions[6-8], and modification of the surfaces[9-11]. In particular, a thin carbon coating on particles is known to be effective not only in enhancing the conductivity of the cathode material but also in protecting the particles from chemical attack by the electrolyte. However, only a few studies on the carbon - coated $LiNi_{1/3}Co_{1/3}Mn_{1/3}O_2$ material were reported. Diverse methods of carbon coatings, such as carbon black compounds or pyrolysis of adsorbed organic compounds[12,13], were employed. Nevertheless, all of these methods need some hours for providing carbon coating on the particles at an elevated temperature. It is well known that microwave process widely used to synthesize many inorganic materials and advanced sinter material[14,15], such as $LiMn_2O_4$[16-18], $LiNiO_2$[19], $LiFePO_4$[20] and $LiCoO_2$[21], due to its short reaction time, low energy consumption, high efficiency of synthesis and product crystal structure[22].

In this study, in order to decrease the process time, reduce the synthesis cost and improve the cycle performance of cathode material, carbon was coated on the surface of the $LiNi_{1/3}Mn_{1/3}Co_{1/3}O_2$ powder by microwave pyrolysis method in a novel microwave reactor (as shown in Fig. 1). Coin cells were prepared with the carbon - coated $LiNi_{1/3}Mn_{1/3}Co_{1/3}O_2$ powder and lithium metal, and their electrochemical performance was measured.

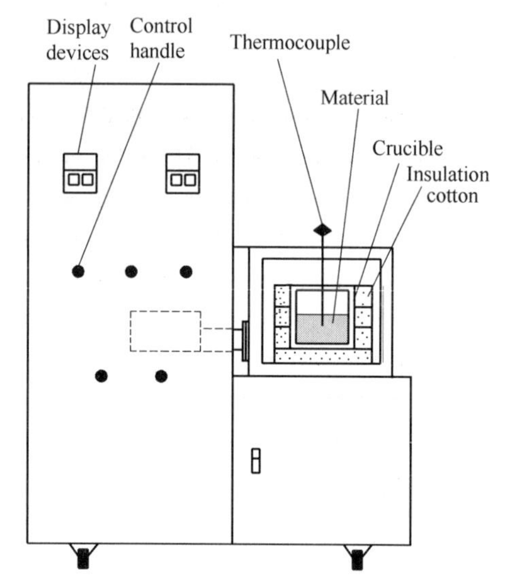

Fig. 1 Schematic of microwave reactor

2 Experimental

2.1 Sample preparation

The spherical $(Ni_{1/3}Mn_{1/3}Co_{1/3})(OH)_2$ (Lanzhou Jinchuan) and Li_2CO_3 (China lithium, 99.9%) powders with molar ratio 1:1.06 were ball - milled thoroughly. The excess amount of Li was used to compensate for the loss of Li during calcinations. The powder after drying at 323K was sintered in the microwave reactor at 1243K for 30min. $LiNi_{1/3}Mn_{1/3}Co_{1/3}O_2$ powder was mixed with glucose of 2% and 3% in mass fraction by mechanically mixing for 16h, respectively, and then pyrolyzed in flowing nitrogen in a microwave reactor at 673K for 15min.

2.2 Experimental equipment

Experiments were carried out using a novel microwave reactor, the schematic of which is shown in Fig. 1. The microwave reactor with a frequency of 2450MHz and a maximum power output of 1500W was made by Kunming University of Science and Technology, China.

2.3 Characterization

X - ray diffraction (XRD) data for the finely ground samples were collected at 298K using a Bruker D8 X - ray diffractometer with CuK_α radiation ($\lambda = 0.15406nm$). It was operated at 40kV and

300mA in the 2θ range of 10° to 80° in the continuous scan mode with the step size of 0. 01° and the scan rate 1. 0°/min. The particle shapes and morphologies of the bare and carbon – coated $LiNi_{1/3}Mn_{1/3}Co_{1/3}O_2$ materials were obtained using scanning electron microscopy (SEM, Philips XL – 30E). Rigaku ZSX100e X – ray fluorescence spectrometer was applied to multi – element determination in the bare and carbon – coated $LiNi_{1/3}Mn_{1/3}Co_{1/3}O_2$ materials. The electrodes were fabricated from a 85∶10∶5 (mass percent) mixture of active materials, acetylene black as the current conductor and polyvinylidene difluoride (PVDF) as the binder. The PVDF was dissolved in N – methylpyrrolidinone, before the active material and the conductor mixture were added. After homogenization, the slurry was evacuated for 20min to remove the residual air. The slurry was then coated on a thin aluminum foil (20μm thick) and dried overnight at 60 – 70℃. The electrode was pressed with a pressure of 10MPa and punched into 12mm diameter disks. The electrochemical cells were prepared as 2025 – coin – cell hardware with lithium metal foil as both the counter and reference electrodes. The cells were assembled in a vacuum glove box. The electrolyte used for analysis was 1mol/L $LiPF_6$ in ethylene carbonate/diethyl carbonate (1∶1).

3 Results and discussion

3.1 Structure

Fig. 2 shows the XRD patterns and miller indices of the bare and carbon – coated $LiCo_{1/3}Ni_{1/3}Mn_{1/3}O_2$. In Fig. 2, the XRD patterns of the two materials indicated a single phase of α – $NaFeO_2$ structure with space group of $R\bar{3}m$. The splits in the (006/012) and (108/110) are around at 38° and 65° doublets, indicating the formation of a highly ordered layered structure. The ratios of intensities of the $I_{(003)}/I_{(104)}$ of the bare and carbon – coated material are 1. 60759 and 1. 83147, respectively, well above the values reported for compounds like $LiNi_{1-x}Co_xO_2$ to deliver good electrochemical performance[23]. No other impurity phase was observed in XRD patterns for the carbon – coated material, which suggests that the structure of $LiNi_{1/3}Co_{1/3}Mn_{1/3}O_2$ is not affected by the carbon coating. Moreover, the structure parameters of the two samples are summarized in Table 1 according to the XRD data and the hexagonal crystal spacing calculation equation:

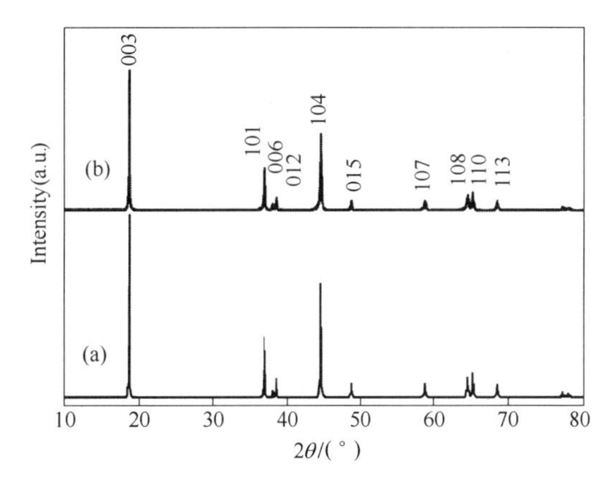

Fig. 2 XRD patterns of bare $LiNi_{1/3}Mn_{1/3}Co_{1/3}O_2$ (a) and 2% carbon – coated $LiNi_{1/3}Mn_{1/3}Co_{1/3}O_2$ (b)

$$\frac{1}{d^2} = \frac{4(h^2 + hk + k^2)}{3a^2} + \frac{2l}{c^2} \tag{1}$$

Table 1 Lattice structural parameters of bare and 2% carbon - coated $LiNi_{1/3}Mn_{1/3}Co_{1/3}O_2$

	a/nm	c/nm	c /a	$I_{(003)}/I_{(104)}$
Bare	0. 2858193	1. 422028	4. 975269	1. 60759
Carbon - coated	0. 2859432	1. 419277	4. 963491	1. 83147

3. 2 Morphology

The SEM images of the bare and carbon - coated $LiNi_{1/3}Mn_{1/3}Co_{1/3}O_2$ particles are shown in Fig. 3. For the bare $LiNi_{1/3}Mn_{1/3}Co_{1/3}O_2$, the rod - shaped particles with a submicron size were agglomerated to form a sphere - shaped particle with diameter of 10 - 14μm. A smooth and clean surface is observed in the bare particles. After coating carbon, the surface of $LiNi_{1/3}Mn_{1/3}Co_{1/3}O_2$ particles is rough. Similar SEM images were also observed in ZrF_x - coated $Li[Ni_{1/3}Co_{1/3}Mn_{1/3}]O_2$[24] and the carbon - coated $LiNi_{1/3}Mn_{1/3}Co_{1/3}O_2$[25]. Compared with the average length of the primary grain of bare, that of the carbon - coated material increased from 540nm to 750nm. The primary grain of the carbon - coated $LiNi_{1/3}Mn_{1/3}Co_{1/3}O_2$ particles treated by microwave is more distinct than the bare one.

Fig. 3 SEM images of bare $LiNi_{1/3}Mn_{1/3}Co_{1/3}O_2$ (a) and 2% carbon - coated $LiNi_{1/3}Mn_{1/3}Co_{1/3}O_2$ (b)

3. 3 Elemental analysis

Table 2 shows the X - ray fluorescence results of 2% carbon coated $LiNi_{1/3}Mn_{1/3}Co_{1/3}O_2$ and 3% carbon coated $LiNi_{1/3}Mn_{1/3}Co_{1/3}O_2$. The XRF analysis indicates that O, Ni, Co, Mn and C are in the carbon - coated $LiNi_{1/3}Mn_{1/3}Co_{1/3}O_2$ particle. The practical contents of carbon are 1. 9106% and 2. 9968% after microwave pyrolysis of 2% carbon - coated $LiNi_{1/3}Mn_{1/3}Co_{1/3}O_2$ and 3% carbon -

coated $LiNi_{1/3}Mn_{1/3}Co_{1/3}O_2$, respectively.

Table 2 XRF results of 2% carbon and 3% carbon coated $LiNi_{1/3}Mn_{1/3}Co_{1/3}O_2$

Sample	w(O)/%	w(Ni)/%	w(Co)/%	w(Mn)/%	w(C)/%
2% carbon	41.5288	18.8922	18.5518	17.8886	1.9106
3% carbon	41.6463	18.8708	18.4789	17.8127	2.9968

3.4 Electrochemical performance

Fig. 4 shows the discharge curves of that bare and 2% carbon – coated $LiNi_{1/3}Mn_{1/3}Co_{1/3}O_2$ electrodes cycled between 2.75V and 4.3V at 0.2C rate. The discharge specific capacity for the $LiNi_{1/3}Mn_{1/3}Co_{1/3}O_2$ electrode without carbon coating were 161.7mA · h/g at the first time and 137.4mA · h/g at fiftieth times at 0.2C rate, while it were 159.3mA · h/g at first time and 152.9mA · h/g at fiftieth times at 0.2C rate for the 2% carbon – coated $LiNi_{1/3}Mn_{1/3}Co_{1/3}O_2$ electrode. As can be seen, the first discharge specific capacity after carbon – coated at 0.2C reduces slightly because a layer of barrier, carbon, is covered on the positive electrode material surface, which blocks cathode materials to contact with electrolyte.

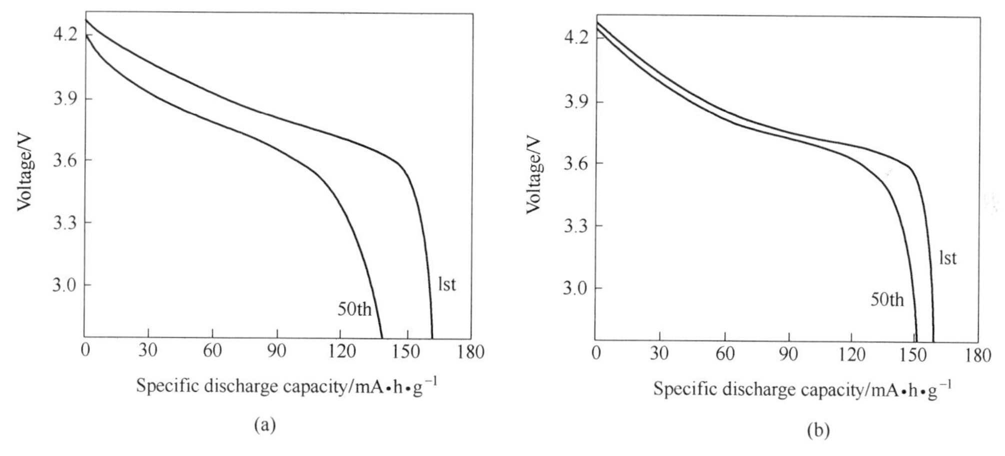

Fig. 4 Discharge curves of bare(a) and 2% carbon – coated (b) $LiNi_{1/3}Mn_{1/3}Co_{1/3}O_2$ electrodes tested at 0.2C rate in 2.75 – 4.3V

Fig. 5 shows the discharge curves of bare and 2% carbon – coated $LiNi_{1/3}Mn_{1/3}Co_{1/3}O_2$ electrodes cycled between 2.75V and 4.3V at 2C rate. The discharge specific capacity for the $LiNi_{1/3}$-$Mn_{1/3}Co_{1/3}O_2$ electrode without carbon coating were 105.5mA · h/g at the first time and 86.4mA · h/g at the 50th time at 2C rate, while they are 130.8mA · h/g at the first time and 112.6mA · h/g at 50th time at 2C rate for the 2% carbon – coated $LiNi_{1/3}Mn_{1/3}Co_{1/3}O_2$ electrode.

Comparison of the cycling performances between the bare and 2% carbon – coated $LiNi_{1/3}Mn_{1/3}Co_{1/3}O_2$ electrodes at 0.2C rate are shown in Fig. 6. The specific discharge capacity of the bare $LiNi_{1/3}Mn_{1/3}Co_{1/3}O_2$ electrode at 0.2C rate was about 161.7mA · h/g, while the 2% carbon – coated $LiNi_{1/3}Mn_{1/3}Co_{1/3}O_2$ electrode exhibites comparable discharge specific capacity of 152.9mA · h/g.

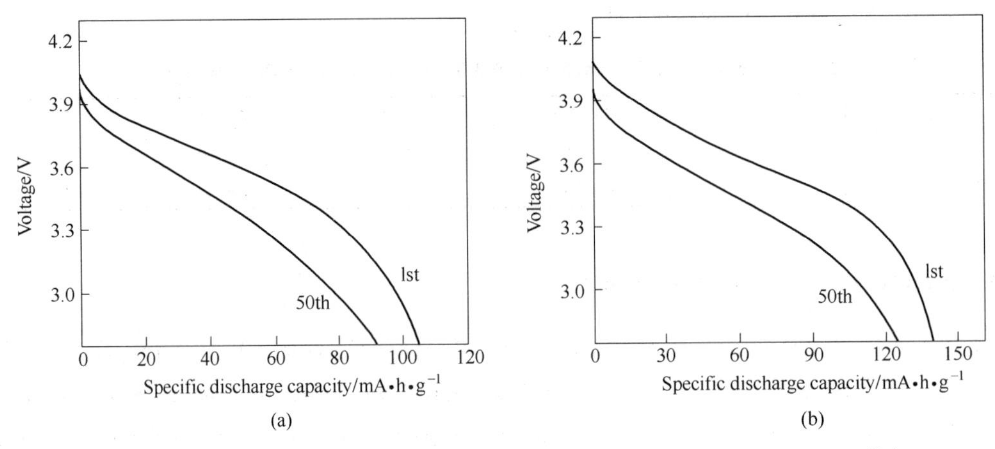

Fig. 5 Discharge curves of bare(a) and 2% carbon - coated $LiNi_{1/3}Mn_{1/3}Co_{1/3}O_2$(b) $LiNi_{1/3}Mn_{1/3}Co_{1/3}O_2$ electrodes tested at 0.2C rate in 2.75 - 4.3V

However, after the 50th cycles the specific capacity retention of the bare and 2% carbon - coated $LiNi_{1/3}Mn_{1/3}Co_{1/3}O_2$ electrodes are 84.8% and 95.5%, respectively.

Fig. 7 shows the discharge specific capacities and cyclic properties of the bare and 2% carbon - coated $LiNi_{1/3}Mn_{1/3}Co_{1/3}O_2$ electrodes at 0.2C, 2C, 4C and 8C rates in the voltage range of 2.75 - 4.3V. At 2C rate, after the 50th cycle the specific capacity retention of 2% carbon - coated $LiNi_{1/3}$-$Mn_{1/3}Co_{1/3}O_2$ electrode is 86.1%, while the bare $LiNi_{1/3}Mn_{1/3}Co_{1/3}O_2$ electrode is 81.9%.

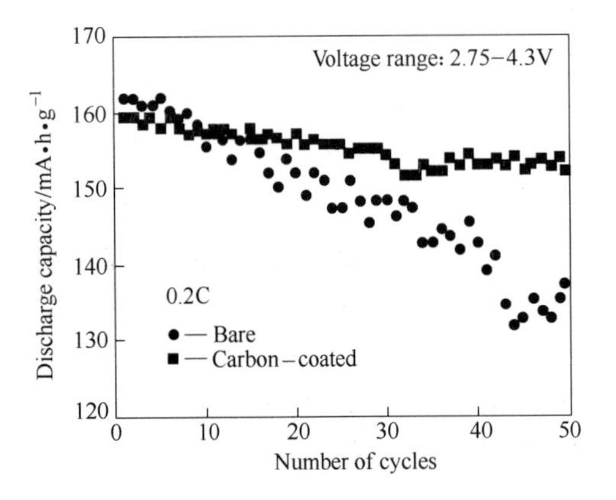

Fig. 6 Cyclic performance of bare and 2% carbon - coated $LiNi_{1/3}Mn_{1/3}Co_{1/3}O_2$ electrodes in 2.75 - 4.3V at 0.2C rate

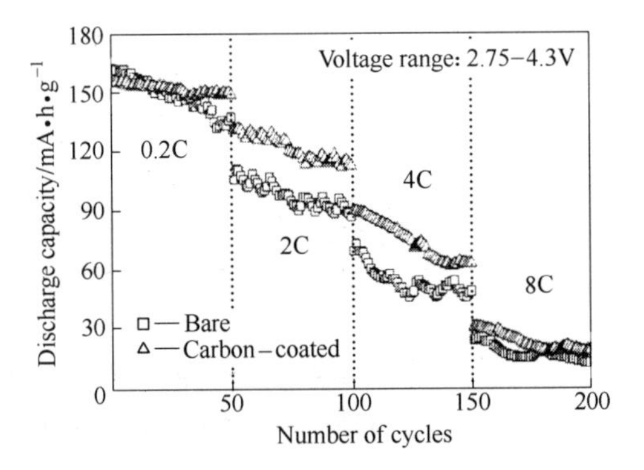

Fig. 7 Specific discharge capacities and cyclic performance of bare and 2% carbon - coated $LiNi_{1/3}Mn_{1/3}Co_{1/3}O_2$ electrodes in 2.75 - 4.3V at 0.2C, 2C, 4C and 8 C rates

Fig. 8 displays the capacity retention (ratio of the discharge capacity at an assigned discharge rate to the discharge capacity at a 0.2C rate) as a function of the rate. After 50 cycles the specific capacity retention of the bare and 2% carbon - coated $LiNi_{1/3}Mn_{1/3}Co_{1/3}O_2$ electrodes are 69.7% and 69.9% at 4C rate, respectively. At 8C rate, after 50 cycles the capacity retention of the bare and 2% carbon - coated $LiNi_{1/3}Mn_{1/3}Co_{1/3}O_2$ electrodes are 50.0% and 59.8%, respectively. The specific capacity retention of the carbon - coated $LiNi_{1/3}Mn_{1/3}Co_{1/3}O_2$ electrode is better than bare $LiNi_{1/3}Mn_{1/3}Co_{1/3}O_2$ electrode at any discharge rate, because the carbon coating blocks the contact between cathode materials and electrolyte, which reduces the corrosion of positive electrode material.

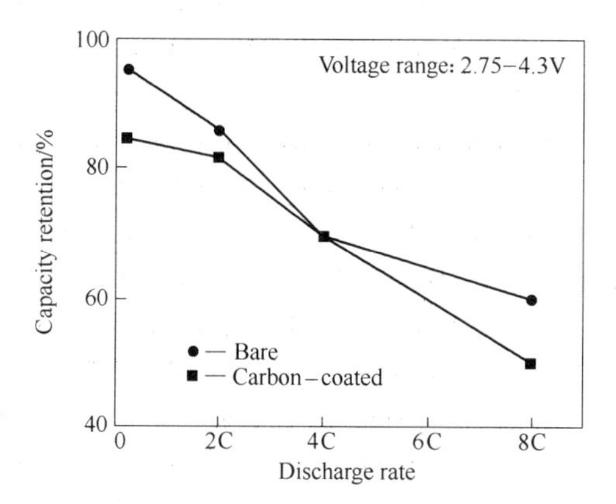

Fig. 8 Specific capacity retentions of bare and 2% carbon - coated $LiNi_{1/3}Mn_{1/3}Co_{1/3}O_2$ electrodes as function of discharge rate

Although the specific discharge capacities of the bare and 2% carbon - coated $LiNi_{1/3}Mn_{1/3}Co_{1/3}O_2$ electrodes decreased with the increase of charge - discharge cycles, the specific capacity retention of the 2% carbon - coated $LiNi_{1/3}Mn_{1/3}Co_{1/3}O_2$ electrode remains better than the bare $LiNi_{1/3}Mn_{1/3}Co_{1/3}O_2$ electrode. The results affirms that the carbon - coated $LiNi_{1/3}Mn_{1/3}Co_{1/3}O_2$ electrode exhibits good discharge rate capability and cycle performance. Perhaps, it could be attributed to the increase in electronic conductivity of the carbon - coated material, reducing the cell polarization and oxygen evolution from the cathodes at the end of charge[25]. The carbon coating could act as a protection layer of positive material, which restrains the dissolution of metal ions and reduces the impedance of the charge transfer[26]. The improved electronic conductivity of the carbon - coated material surface may contribute to the improvement in electrical contact of the material among particles, thereby improving the electrochemical performance of the material[27].

4 Conclusions

(1) The carbon - coated $LiNi_{1/3}Mn_{1/3}Co_{1/3}O_2$ material was successfully synthesized by microwave - pyrolysis method using glucose as carbon source. The carbon coating does not change the phase structure of $LiNi_{1/3}Mn_{1/3}Co_{1/3}O_2$ material.

(2) The experimental results show that the carbon coating enhances the rate capability and cycling performance of the $LiNi_{1/3}Mn_{1/3}Co_{1/3}O_2$ electrode.

References

[1] Liu H K, Wang G X, Guo Z P, et al. Nanomaterials for lithium - ion rechargeable batteries[J]. Journal of Nanoscience and Nanotechnology, 2006, 6(1): 1 - 15.

[2] Whittingham M S. Lithium batteries and cathode materials[J]. Chemical Reviews, 2004, 104(10): 4271 - 4302.

[3]Tarascon J M, Armand M. Issues and challenges facing rechargeable lithium batteries[J]. Nature, 2001, 414(6861): 359 - 367.

[4]Ohzuku T, Makimura Y. Layered lithium insertion material of $LiCo_{1/3}Ni_{1/3}Mn_{1/3}O_2$ for lithium - ion batteries [J]. Chemistry Letters, 2001(7): 642 - 643.

[5]Choi J, Manthiram A. Comparison of the electrochemical behaviors of stoichiometric $LiNi_{1/3}Co_{1/3}Mn_{1/3}O_2$ and lithium excess $Li_{1.03}(Ni_{1/3}Co_{1/3}Mn_{1/3})_{0.79}O_2$[J]. Electrochemical and Solid State Letters A, 2004, 7(10): 365 - 368.

[6]Guo J, Jiao L F, Yuan H T, et al. Effect of structural and electrochemical properties of different Cr - doped contents of $Li(Ni_{1/3}Mn_{1/3}Co_{1/3})O_2$[J]. Electrochimica Acta, 2006, 51: 6275 - 6280.

[7]Zhai J, Zhao M, Wang D D. Effect of Mn - doping on performance of $Li_3V_2(PO_4)_3/C$ cathode material for lithium ion batteries[J]. Transactions of Nonferrous Metals Society of China, 2011, 21: 523 - 528.

[8]Ye S Y, Xia Y Y, Zhang P W, et al. Al, B, and F doped $LiNi_{1/3}Co_{1/3}Mn_{1/3}O_2$ as cathode material of lithium - ion batteries[J]. Journal of Solid State Electrochemistry, 2007, 11: 805 - 810.

[9]Yang Z, Li X H, Wang Z X, et al. Surface modification of spherical $LiNi_{1/3}Co_{1/3}Mn_{1/3}O_2$ with Al_2O_3 using heterogeneous nucleation process[J]. Transactions of Nonferrous Metals Society of China, 2007, 17: 1319 - 1323.

[10]Li J G, Wang L, Zhang Q, et al. Electrochemical performance of SrF_2 - coated $LiNi_{1/3}Co_{1/3}Mn_{1/3}O_2$ cathode materials for Li - ion batteries[J]. Journal of Power Sources, 2009, 190: 149 - 153.

[11]Wang H Y, Tang A D, Huang K L, et al. Uniform AlF_3 thin layer to improve rate capability of $LiNi_{1/3}Co_{1/3}Mn_{1/3}O_2$ material for Li - ion batteries[J]. Transactions of Nonferrous Metals Society of China, 2010, 20: 803 - 808.

[12]Kim H S, Kim K, Moon S I, et al. A study on carbon - coated $LiNi_{1/3}Mn_{1/3}Co_{1/3}O_2$ cathode material for lithium secondary batteries[J]. Journal of Solid State Electrochemistry, 2008, 12: 867 - 872.

[13]Guo R, Shi P F, Cheng X Q, et al. Synthesis and characterization of carbon - coated $LiNi_{1/3}Co_{1/3}Mn_{1/3}O_2$ cathode material prepared by polyvinyl alcohol pyrolysis route[J]. Journal of Alloys and Compounds, 2009, 473: 53 - 59.

[14]Yang Y J, Sheu C I, Cheng S Y, et al. Si - Ca species modification and microwave sintering for NiZn ferrites [J]. Journal of Magnetism and Magnetic Materials, 2004, 284: 220 - 226.

[15]Bhat M H, Chakravarthy B P, Ramakrishnan P A, et al. Microwave synthesis of electrode materials for lithium batteries[J]. Bulletin of Materials Science, 2000, 23(6): 461 - 466.

[16]Nakayama M, Watanabe K, Ikuta H, et al. Grain size control of $LiMn_2O_4$ cathode material using microwave synthesis method[J]. Solid State Ionics, 2003, 164: 35 - 42.

[17]Yang S T, Jia J H, Ding L, et al. Studies of structure and cycleability of $LiMn_2O_4$ and $LiNd_{0.01}Mn_{1.99}O_4$ as cathode for Li - ion batteries[J]. Electrochimica Acta, 2003, 48: 569 - 573.

[18]Yan H W, Huang X J, Chen L Q. Microwave synthesis of $LiMn_2O_4$ cathode material[J]. Journal of Power Sources, 1999, 81: 647 - 650.

[19]Kalyani P, Kalaiselvi N, Renganathan N G. Microwave - assisted synthesis of $LiNiO_2$—a preliminary investigation[J]. Journal of Power Sources, 2003, 123: 53 - 60.

[20]Zhang Y, Feng H, Wu X B, et al. One - step microwave synthesis and characterization of carbon - modified nanocrystalline $LiFePO_4$[J]. Electrochimica Acta, 2009, 54(11): 3206 - 3210.

[21]Subramanian V, Chen C L, Chou H S, et al. Microwave - assisted solid - state synthesis of $LiCoO_2$ and its electrochemical properties as a cathode material for lithium batteries[J]. Journal of Materials Chemistry, 2001, 11: 3348 - 3353.

[22]Shen B J, Ma J S, Wu H C, et al. Microwave - mediated hydrothermal synthesis and electrochemical properties of $LiNi_{1/3}Co_{1/3}Mn_{1/3}O_2$ powders[J]. Materials Letters, 2008, 62: 4075 - 4077.

[23] Shaju K M, Subba Rao G V, Chowdari B V R. X - ray photoelectron spectroscopy and electrochemical behaviour of 4V cathode, Li($Ni_{1/2}Mn_{1/2}$)O_2[J]. Electrochimica Acta, 2003, 48: 1505 - 1514.

[24] Yun S H, Park K S, Park Y J. The electrochemical property of ZrF x - coated Li[$Ni_{1/3}Co_{1/3}Mn_{1/3}$]O_2 cathode material[J]. Journal of Power Sources, 2010, 195: 6108 - 6115.

[25] Lin B, Wen Z Y, Han J, et al. Electrochemical properties of carbon - coated Li[$Ni_{1/3}Co_{1/3}Mn_{1/3}$]O_2 cathode material for lithium - ion batteries[J]. Solid State Ionics, 2008, 179: 1750 - 1753.

[26] Shin H C, Cho W I, Jang H. Electrochemical properties of the carbon - coated $LiFePO_4$ as a cathode material for lithium ion secondary batteries[J]. Journal of Power Sources, 2006, 159: 1383 - 1388.

[27] Shin H C, Cho W I, Jang H. Electrochemical properties of carbon - coated $LiFePO_4$ cathode using graphite carbon black, and acetylene black[J]. Electrochimica Acta, 2006, 52: 1472 - 1476.

Preparation of High Surface Area Activated Carbon from Coconut Shells Using Microwave Heating

Kunbin Yang, Jinhui Peng, C. Srinivasakannan,
Libo Zhang, Hongying Xia, Xinhui Duan

Abstract: The present study attempts to utilize coconut shell to prepare activated carbon using agents such as steam, CO_2 and a mixture of steam – CO_2 with microwave heating. Experimental results show that BET surface area of activated carbons irrespective of the activation agent resulted in surface area in excess of 2000m/g. The activation time using microwave heating is very much shorter, while the yield of the activated carbon compares well with the conventional heating methods. The activated carbon prepared using CO_2 activation has the largest BET surface area, however the activation time is approximately 2.5 times higher than the activation using steam or mixture of steam – CO_2. The chemical structure of activated carbons examined using Fourier transformed infra – red spectra(FTIR) did not show any variation in the surface functional groups of the activated carbon prepared using different activation agents.

Keywords: activated carbon; high surface area; pore size distribution; micropore volume; coconut shells

1 Introduction

Activated carbons with highly developed surface area are widely used in a variety of industries for applications which included separation/purification of liquids and gases, removal of toxic substances, as catalysts and catalyst support[1,2]. With the development of technology, the applications of activated carbons keep expanding, with newer applications such as super – capacitors, electrodes, gas storage, and so on[3,4]. Activated carbons have been traditionally produced by the partial gasification of the char either with steam or CO_2 or a combination of both. The gasification reaction results in removal of carbon atoms and in the process simultaneously produce a wide range of pores (predominantly micropores), resulting in porous activated carbon. Precursors to activated carbons are either of botanical origin(e. g. wood, coconut shells and nut shells) or of de – graded and coalified plant matter(e. g. peat, lignite and all ranks of coal). Agricultural by – products are considered as very important feedstock as they are renewable and low – cost materials[5,6].

In general, there are two main steps for the preparation of activated carbon: (1) the carbonization of carbonaceous precursor below 800℃, in the absence of oxygen; (2) the activation of carbonized product(char), either using physical or chemical activation methods. Generally physical activation is a two – step process which involves carbonization of a carbonaceous material followed by activation of the resulting char at elevated temperature in presence of suitable oxidizing gases such as carbon dioxide, steam, air or their mixtures. In the chemical activation process the precursors are impregnated with dehydrating chemicals such as H_3PO_4, $ZnCl_2$, K_2CO_3, NaOH or KOH and carbonized at desired conditions in a single step. Chemical activation offers several advantages which include single step activation, low activation temperatures, low activation time, higher yields and better porous structure. However the process involves a complex recovery and recycle of the activating agent, which generates liquid discharge that demands effluent treatment. Physical activation process is widely adopted industrially for commercial production owing to the simplicity of process and the ability to produce activated carbons with well developed micro porosity and desirable physical characteristics such as the good physical strength.

The conventional heating methods results in surface heating from the hearth wall, which do not ensure a uniform temperature for different shapes and sizes of samples. This generates a temperature gradient from the hot surface of the sample particle to its interior and impedes the effective removal of gaseous products to its surroundings, thereby resulting in long activation time and higher energy consumption. Recently, microwave heating is being used in various technological and scientific fields for variety of applications. The main difference between microwave heating and conventional heating systems is in the way the heat is generated. Energy transfer is not by conduction or conventional heating, but is readily transformed into heat inside the particles by dipole rotation and ionic conduction. When high frequency voltages are applied to a material, the response of the molecules with a permanent dipole to the applied potential field is to change their orientation in the direction opposite to that of the applied field. The synchronized agitation of molecules then generates heat[7-9]. Therefore, the tremendous temperature gradient from the interior of the char particle to its cool surface allows the microwave induced reaction to proceed more quickly and effectively, resulting in energy saving and shorter reaction time.

Although microwave heating has been used to produce and regenerate activated carbon, the relevant literature is very limited. The studies pertaining to preparation of activated carbon using physical activation has been limited to Guo and Lua and Williams and Parkes[10], while the regeneration studies were limited to Ania, et al. [7], Ania, et al. [11] and Coss and Cha[12] Nabais, et al. ,[8] have reported the surface chemistry modification of activated carbon fibers by means of microwave heating. None of the previous attempts have reported comparison of activated carbon prepared with steam, CO_2 and a mixture of steam - CO_2 by microwave heating for coconut shells. The present study attempts to investigate the effect of activation time on the yield and porous structure of activated carbons produced from coconut shells employing different activation agents such as steam, CO_2 and combination of team - CO_2.

2 Experiment

2.1 Materials

Coconut shells utilized in the present study were obtained from Xishuangbanna in Yunnan province of China. The starting materials were manually chosen, cleaned with deionized water, dried at 110℃ for 48h and ground using a roller mill and sieved to a size range of 3.35 - 4.75mm and stored in sealed containers for experimentation. The proximate analysis of the coconut shell is shown in Table 1.

Table 1 Proximate analysis results of coconut shell

Sample	Moisture(wt.)/%	Volatile matter(wt.)/%	Fixed carbon(wt.)/%	Ash(wt.)/%
Coconut shells	10. 53	70. 06	18. 75	0. 66

2. 2 Preparation of activated carbons

2. 2. 1 Carbonization of coconut shells

Carbonization was performed in a horizontal tube furnace by electric heating and coconut shells were placed inside a stainless steel reactor. The coconut shells were heated up to a carbonization temperature of 1000℃ at a heating rate of 10℃/min with conventional heating and were held for 2h at the carbonization temperature under N_2 gas flow(100cm^3/min). After carbonization, the samples were cooled to room temperature under N_2 flow(100cm^3/min).

2. 2. 2 Activation procedure

The activation experiments were carried out in a self - made microwave tubular furnace, which has a single - mode continuous controllable power and is shown in Fig. 1. The microwave frequency was 2. 45GHz, while the output power could be set to a maximum of 3000W. The temperature of sample in the microwave equipment was monitored using a type of K(chromel - alumel) thermocouple, placed at closest proximity to the sample.

Fig. 1 Microwave reactor of multi - mode with continuous controllable power

Approximately 25g of the pre carbonized material was placed into reactor and set to the desired temperature along with the N_2 flow rate at 200cm^3/min. Upon reaching the desired temperature the gasifying agent was allowed into the reactor at a desired flow rate. Approximately 5 to 7min of time was required to raise the temperature of sample to the desired temperature of 900℃. Three experimental series were conducted employing the flowing conditions:

(1) Steam activation(1. 35g/min): to assess the effect of activation time at 900℃;

(2) CO_2 activation(600cm^3/min): to assess the effect of activation time at 900℃;

(3) CO_2 - steam mixture activation(600cm^3/min + 1. 35g/min): to assess the effect of activation time at 900℃.

Selected sets of experiments were repeated to ensure the reprotducibility of experimental data.

2.3 Structure characterization

The pore structure of the sample is characterized by nitrogen adsorption at 77K with an accelerated surface area and porosimetry system (Autosorb - 1 - C, Quantachrome). Prior to gas adsorption measurements, the carbon was degassed at 300℃ in a vacuum condition for a period of at least 2h. Nitrogen adsorption isotherm was measured over a relative pressure (p/p_0) range from approximately 10^{-7} to 1. The BET surface area was calculated from the isotherms by using the Brunauer - Emmett - Teller (BET) equation[13]. The cross - sectional area for nitrogen molecule was assumed to be 0.162nm. The Dubinin - Radushkevich (DR) method was used to calculate the micropore volume[14]. The micropore size distribution was ascertained by Non - local Density Functional Theory (NLDFT)[15] by minimizing the grand potential as a function of the fluid density profile. The total volume[13] was estimated by converting the amount of N_2 gas adsorbed at a relative pressure of 0.95 to equivalent liquid volume of the adsorbate (N_2). The mesopore volume was estimated by the subtracting the micropore volume from the total volume.

3 Results and discussion

3.1 Characteristics of porosity in coconut shell chars

The carbonization process prior to activation enriches the carbon content with removal of volatile matter which creates the initial porosity in the char. The pore structure of the char was estimated using the nitrogen adsorption isotherm and is shown in Fig. 2. The total BET surface area and the pore volume distribution is listed in Table 2. The nitrogen adsorption isotherm (Fig. 2) of coconut shell chars correspond an intermediate between type Ⅰ and Ⅱ of the referred IUPAC classification (Rouquerol, et al., 1999). A BET surface area of 702m^2/g is significant, with micro pore volume accounting to 66%. This type of isotherm is usually exhibited by microporous solids containing a well developed mesopore structure. This indicates that chars produced at 1000℃ offer higher potential to produce activated carbon of greater adsorption capacity. The porosity is generated due to the conversion of hemicellulose, cellulose and lignin in coconut shells by the process of dehydrating, linkage breaking reactions, the structural ordering process of the residual carbon and finally polymerization reaction[16]. At a carbonization temperature of 1000℃, a significantly higher rate of polymerization reaction results in faster size reduction of the sample contributing to the formation of micropores in the samples[17-19]. The high porous char produced at 1000℃ forms a good basis for further enhancing the porous nature of the char, by promoting the reaction between the char and the reacting agents such as steam and CO_2[20].

Table 2 Characteristics of porosity in coconut shell chars prepared at temperature 1000℃

Sample	BET surface area/$m^2 \cdot g^{-1}$	$V_{tot}/cm^3 \cdot g^{-1}$	$V_{micro}/cm^3 \cdot g^{-1}$	$V_{meso}/cm^3 \cdot g^{-1}$
Coconut shell chars	702	0.5319	0.3482	0.1837

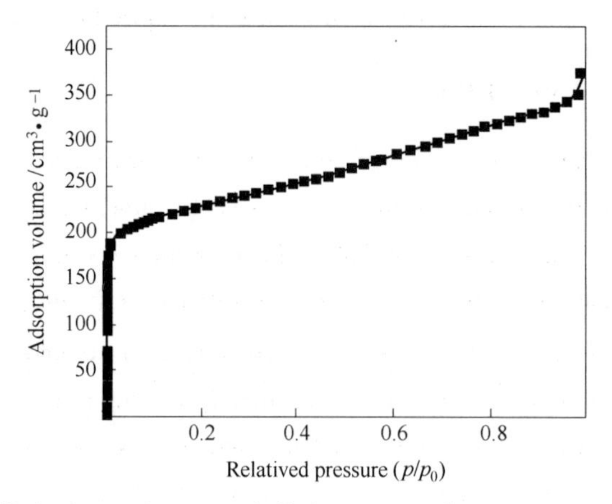

Fig. 2 N_2 isotherm of coconut shell chars prepared at temperature 1000℃

3.2 Characteristics of activated carbons

3.2.1 Effects of activation agent and activation time on yield of activated carbons

The yield of activated carbon is an important factor in the process of activated carbon preparation as it has direct bearing on the process economics. The yield is defined as the % ratio of weight of activated carbon produced to the weight of carbonized char utilized for activation. It should be noted that the yield in the process of conversion of coconut shell into carbonized char is 22%. Fig. 3 illustrates the relationship between the activation time and the yield of activated carbon corresponding to different activation agent at an activation temperature of 900℃. It can be seen that the yield of activated carbon decreases progressively with the activation time which could be attributed to the reaction between carbon and activation agent. The rate of reduction of yield with respect to activation time is the lowest for the CO_2 activation.

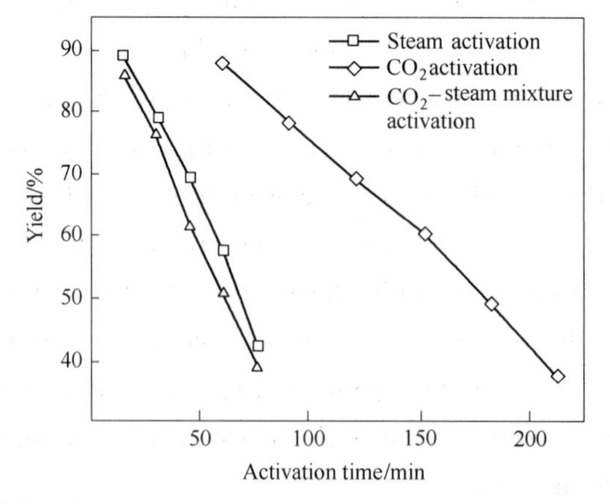

Fig. 3 Effects of activation agent and hold time on yield of activated carbons

While it is an order of magnitude higher using steam activation. The rate of reaction of the steam with carbon is well establelished to be higher than the CO_2 by several other earlier reported works[21,22], which accounts for the faster reduction in yield using steam activation as compared to

CO_2 activation. However reduction in yield for combination of mixture of steam and CO_2 was found to be higher than steam alone, which could be attributed the higher net flow of the activation agent and occurrence of simultaneous reaction of steam and CO_2 with carbon.

3.2.2 N_2 isotherms of activated carbon

It is well known that the adsorption capacity of an adsorbent largely depends on the amount of micropores and surface area. The activation process develops porosity in the carbon by creating a more orderly porous structure. There are usually four stages in the pore development during the activation process: (1) opening of previously inaccessibly pores; (2) creation of new pores by selective activation; (3) widening of the existing pores; (4) merger of the existing pores due to pore wall breakage[23].

The most popular and widely followed method utilized to estimate the porous nature of adsorbents is based on the nitrogen adsorption isotherm. Fig. 4 shows the nitrogen adsorption isotherms exhibited by activated carbons prepared at 900℃ for different activation time with steam, CO_2 and CO_2 - steam mixture by microwave heating. It can be ascertained from Fig. 4 that the isotherms of all acti-

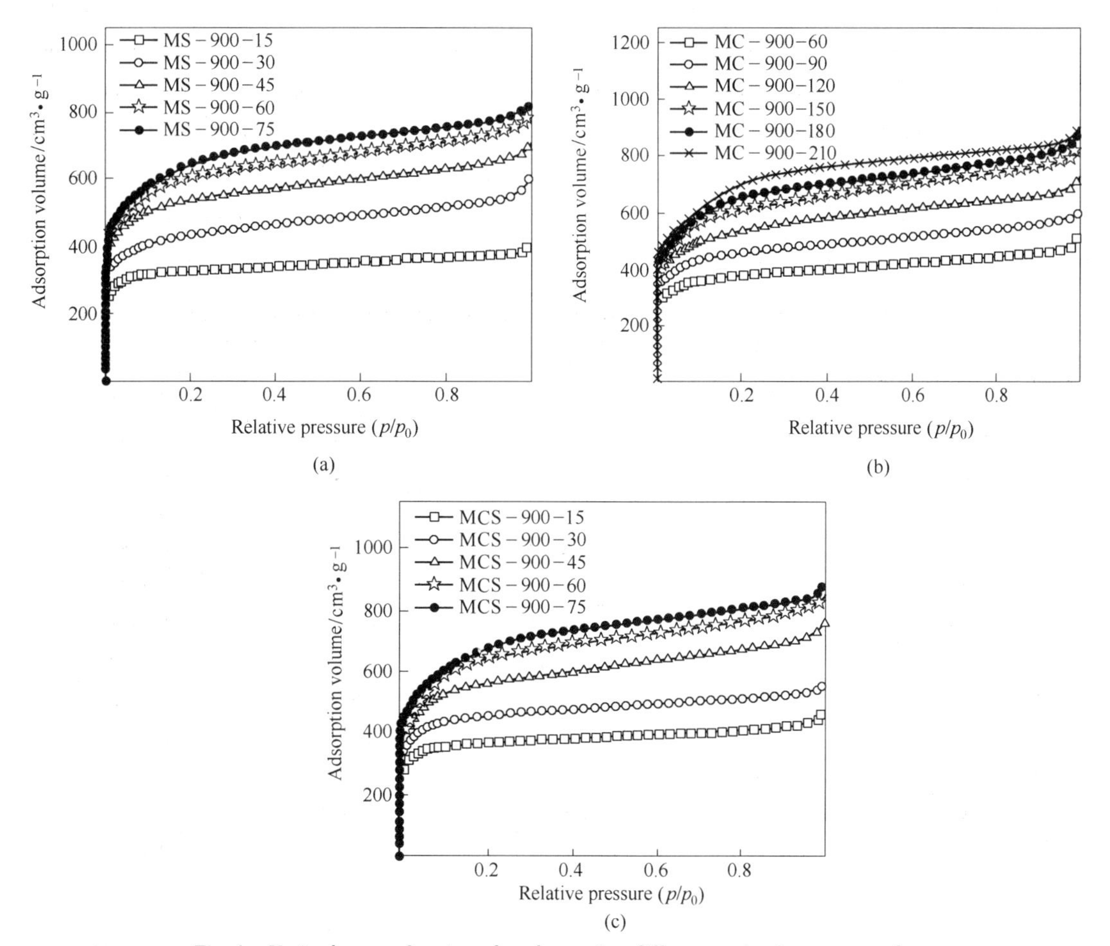

Fig. 4 N_2 isotherms of activated carbon using different activation agent under different hold time at activation temperature of 900℃

Activation agents: (a) steam activation; (b) CO_2 activation; (c) CO_2 - steam mixture activation

vated carbons pertain to intermediate between type Ⅰ and Ⅱ of the referred IUPAC classification[24]. This type of isotherm is usually exhibited by microporous solids that include a well developed mesopore structure. An in – crease in the activation time produces a positive effect on the amount of nitrogen adsorbed indicating development of additional pores in the activated carbon. From the micropore and mesopore volumes provided in Table 3, it can inferred that the mesopores account approximately about 20% of the total pore volume irrespective of the activating agent for carbons with well developed porous structure.

A comparison of the nitrogen isotherms of activated carbon with different activation agent reveal that activation using CO_2 – steam mixture produces a higher nitrogen adsorption capacity compared to activation using CO_2 or steam, at identical activation times. It should however be noted that the total flow rate of the mixture of CO_2 – steam is much higher than the individual flow of either CO_2 or steam. The increased porosity using CO_2 – steam mixture could possibly be due to higher reactivity of steam which gen – erates and widens micropores at a faster rate, which aids the diffusion of CO_2 to the interior parts of the carbon contributing to additional development of micropores. Further the higher net flow of activation agent could possibly enhance the reaction rate of the steam and CO_2 with carbon. However the nitrogen adsorption capacity of the activated carbon prepared using CO_2 activation has the highest surface area but at a much higher activation time, owing to the lower rate of reaction of CO_2 with carbon. It can be concluded that the CO_2 based activation demands a higher activation time to the order of 2.5 times as compared to steam or combination of steam – CO_2 to produce activated carbon with comparable porous nature.

Table 3 Characteristics of porosity of activated carbons prepared with different activation agent

Sample	SBET/$m^2 \cdot g^{-1}$	V_{tot}/$cm^3 \cdot g^{-1}$	V_{micro}/$cm^3 \cdot g^{-1}$	V_{meso}/$cm^3 \cdot g^{-1}$
MS – 900 – 15	1011	0.585	0.5179	0.0671
MS – 900 – 30	1363	0.8454	0.6727	0.1727
MS – 900 – 45	1677	1.0250	0.8339	0.1911
MS – 900 – 60	1888	1.1570	0.9483	0.2087
MS – 900 – 75	2079	1.2120	0.9735	0.2385
MC – 900 – 60	1162	0.7159	0.5703	0.1456
MC – 900 – 90	1425	0.8820	0.7022	0.1798
MC – 900 – 120	1703	1.0320	0.8153	0.2167
MC – 900 – 150	1905	1.204	0.9365	0.2675
MC – 900 – 180	2080	1.2700	0.9974	0.2726
MC – 900 – 210	2288	1.2990	100120	0.287
MCS – 900 – 15	1139	0.6911	0.5869	0.1042
MCS – 900 – 30	1424	0.8276	0.7229	0.1047
MCS – 900 – 45	1761	1.1020	0.8773	0.2247
MCS – 900 – 60	2020	1.2480	1.0080	0.2400
MCS – 900 – 75	2194	1.2930	1.0100	0.2830

Note: M—microwave heating; S—steam activation; C—CO_2 activation; CS—CO_2 + steam activation; 900—activation temperature (℃); 15 – 210—activation time (min).

Table 3 lists the BET surface area, total pore volume, micropore volume and mesopore volume of activated carbon corresponding to various activation agents at different activation times. It can be observed from the data that BET surface area and pore volume increases with increase in the activation time irrespective of the activation agent. The magnitude of increase in porosity reduces at higher activation time, possibly indicating proximity to the optimum activation time. The improvement in porous structure coupled with decrease in yield of the activated carbon, with increase in the activation time indicates the increase in extent of the reaction between the activation agent and carbon. The increased extent of the reaction aids enhancement and generation of new pores in the carbon resulting in higher surface area and pore volume. A BET surface area of activated carbon in excess of 2000m^2/g could be produced with all the three different combinations of activation agents, however with different activation times.

A comparison of the pore characteristics of the activated carbons due to the present study with other literature reported values for the coconut shell based precursor, under optimum activation conditions are listed in Table 4. It can be observed from the data provided in Table 4 that the BET surface area due to the present study is much higher than those of reported using physical activation. The yield of activated carbon well compares with the conventional heating methods reported in literature. A yield of 40% , with high surface area of 2200m^2/g could be attributed to the higher carbonization temperature (1000℃) than the conventional carbonization temperatures (550 - 700℃)[25] and to the microwave heating. At high carbonization temperatures, a more ordered structure is likely to be developed in the char that leads to a slower rate of gasification in the interior of the particle[20,23].

Table 4 Comparison of the characteristics of porosity in activated carbons of present work with other literature date under optimum conditions

References	Activation and heating method	Activation time/min	SBET /m^2·g^{-1}	V_{tot} /cm^3·g^{-1}	V_{micro} /cm^3·g^{-1}	Yield /%
Present work	Physical activation(steam), microwave heating	75	2079	1.212	0.9735	42.2
	Physical activation(CO_2), microwave heating	210	2288	1.299	1.012	37.5
	Physical activation(CO_2 + steam), microwave heating	75	2194	1.293	1.010	39.2
Li, et al. (2008)[20]	Physical activation(steam), conventional heating	120	1926	1.260	0.931	39.1
Singh, et al. (2008)[25]	Physical(inter atmosphere), conventional heating	60	378	0.26	0.12	
Sarkar and Bose(1997)	Physical activation(steam), conventional heating	120	1018		0.4008	
Su, et al. (2003)[26]	Physical(without activation agent), conventional heating		663	0.23		23.2
Achaw and Afrane(2008)[27]	Physical activation (35% steam, 65% N_2), conventional heating	120	524	0.226	0.210	
Cagnon, et al. (2009)[28]	Physical activation(N_2 + H_2O), conventional heating	120		0.39	0.35	76.3

Continues Table 4

References	Activation and heating method	Activation time/min	SBET $/m^2 \cdot g^{-1}$	V_{tot} $/cm^3 \cdot g^{-1}$	V_{micro} $/cm^3 \cdot g^{-1}$	Yield /%
Cagnon, et al. (2003)[13]	Physical activation($N_2 + H_2O$), conventional heating	210		0.9812	0.52	44.5
Su, et al. (2007)[29]	Physical activation(CO_2), conventional heating	2880	1964	0.5768		
Din, et al. (2009)[30]	Chemical and physical activation(with KOH followed by CO_2 activation), conventional heating	120	1026	0.5768		
Azevedo, et al. (2007)[31]	Chemical activation($ZnCl_2$), conventional heating		1266	0.731	0.676	
	Chemical and physical activation(with $ZnCl_2$ followed by physical activation), conventional heating		2114	1.307	1.142	
Hu and Srinivasan(1999)[32]	Chemical(KOH), conventional heating	120	2451	1.210		23.6

The activation time due to present work is much shorter owing to the thermal efficiency of microwave heating system. The uniform, faster heating rate due to microwave heating increases the reaction rate between the carbon and the activation agent, which leads to more active sites taking part in reaction. It can be concluded that the higher carbonization temperature of 1000℃ coupled with microwave heating plays a key role for the enhanced porous nature of the carbon and for higher yields. The micropore volume of the activated carbon corresponding to high surface area samples contributes more than 80%. Activated carbons prepared from char for other precursors at high carbonization temperature have been reported to have higher micropore volume. These precursors include almond shell[23], prune pit[33], olive stone[6] and palm shell[34].

3.2.3 Pore size distributions of activated carbons

Fig. 5(a) shows the pore size distributions of activated carbons prepared with steam, CO_2 and CO_2 – steam mixture corresponding to different degrees of activation. The degree of activation indicated in % corresponds to the yield of activated carbon. Label "Steam 88.9%" indicate the pore size distribution of the activated carbon sample with 88.9% yield or in other words the activated carbon produced at low activation time resulting in high yield. At low degrees of activation(Fig. 5(a)), the pore development is mainly in the ultra micropore(0.5 – 0.7nm) range with the steam having higher degree of pore development. The quantum of pore development using steam – CO_2 is one half as compared to steam alone. Similarly, the pore development using CO_2 is one half as compared to the steam – CO_2 mixture. This is well in agreement with the low rate of activation of carbon – CO_2 reaction in comparison with the carbon – steam reaction. The quantum of pores higher than ultra micropores is insignificant at lower degree of activation, evidenced from Fig. 5(b) which do not show significant rise in the cumulative pore volume beyond the ultra micropore range, in addition, non existence of pores beyond a pore size of 1.4nm is insignificant. The comparison of the pore development with degree of activation well indicates the formation of micropores initially, which progressively widens with increase in the degrees of activation. At higher degrees of activation, the higher total pore volume is due to the continued generation of the ultra micropores and widening the existing pores progressively in tune with the extent of the carbon conversion. The difference in the quantum of ultra micropores is marginal at higher degree of activation among the different activation agents considering the variation in the degree of activation in the data presented in Fig. 5(a). These observations are in concurrence with the earlier reports due to Arenas and Chejine[35].

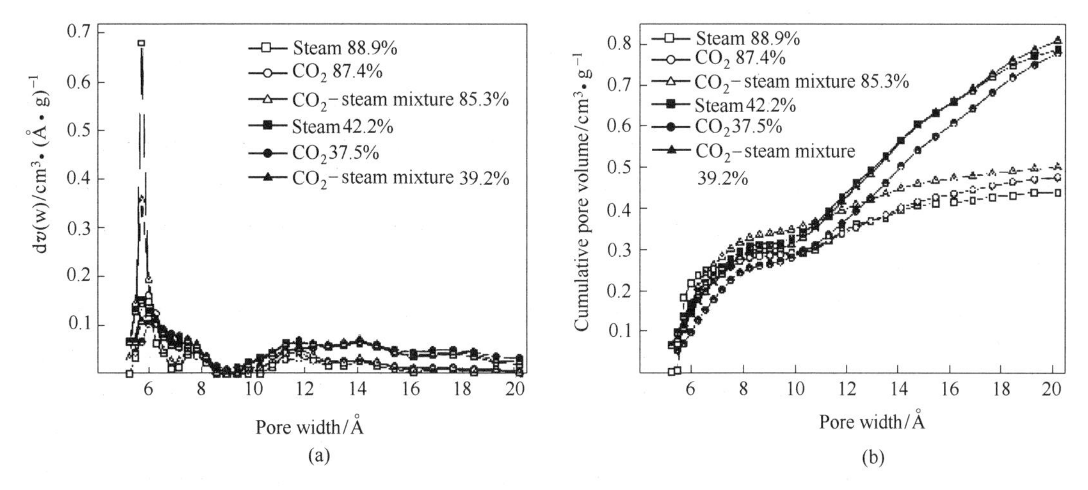

Fig. 5 Pore size distribution of activated carbons prepared under 900℃ with steam, CO_2 and CO_2 – steam mixture activation by microwave heating method

(a) Pore size distribution; (b) Cumulative pore volume distribution (1Å = 0.1nm)

3.3 FTIR analysis

The Fourier transformed infra – red spectra of the activated car – bons prepared using microwave heating with different activation agent is shown in Fig. 6. The FTIR spectroscopy provides information on the chemical structure of the materials. It can be observed from the figure that irrespective of the activation agent utilized the overall shapes of the spectra are very similar. The band at around 3430cm^{-1} can be assigned to the O – H stretching vibration mode of hydroxyl functional

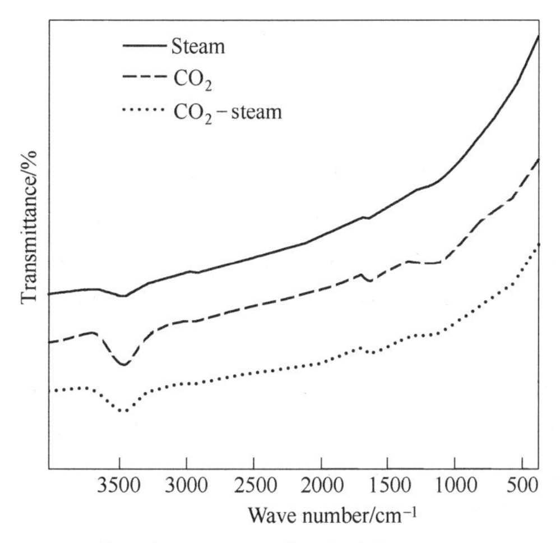

Fig. 6 FTIR spectra of the activated carbons prepared with different activation agent by microwave heating

groups, while the band at around 2917cm^{-1} can be assigned to the C – H symmetric and asymmetric vibration mode of methyl and methylene groups. The band at around 1631cm^{-1} can be assigned to C = C stretching vibration mode of olefinic C = C bonds[36].

4 Conclusions

Coconut shell is the most popular and industrially widely adopted precursor for preparation of activated carbon. The presen study utilizes coconut shell to prepare activated carbon using different activating agents such as steam, CO_2 and combination of steam – CO_2 with microwave heating, in order to compare the porous structure and yield of activated carbon. A two – stage activation process with carbonization at 1000℃ under N_2 atmosphere with conventional heating followed by activation at 900℃ with microwave heating has been utilized for preparation of activated carbon. The BET surface area of the activated carbons irrespective of the activation agent resulted in surface area in excess of 2000m^2/g. The activation time using microwave heating is very much shorter, while the yield of the activated carbon well compares with the conventional heating methods. The activated carbons are primarily micropores with the mesopores contributing to 20% of the pore volume, for activated carbons with well developed pores. The process of pore formation is faster with steam as the activating agent compared to CO_2 owing to the higher rate of reaction of steam with carbon. The ultra micropores formed initially activation which con – tinues to widen to higher pore diameter in tune with the extent of the reaction. The activated carbon prepared using CO_2 activation has the largest BET surface area; however the activation time is approximately 2. 5 times higher than the activation using steam or mixture of steam – CO_2. The chemical structure of activated carbons examined using Fourier transformed infra – red spectra did not show any variation in the surface functional groups of the activated carbon prepared using different activation agents.

Acknowledgements

The authors would like to express their gratitude to the China International Science and Technology Cooperation Program(No. 2008DFA91500) , the International Collaboration Project of Yunnan Provincial Science and Technology Department(No. 2006GH01).

References

[1] Moon, Seung H, Shim, et al. A novel process for CO_2/CH_4 gas separation on activated carbon fibers – electric swing adsorption[J]. Journal of Colloid and Interface Science, 2006, 298: 523 – 528.

[2] Fuente A M, Pulgar G, González F, et al. Activated carbon supported Pt catalysts: effect of support texture and metal precursor on activity of acetone hydrogenation [J]. Applied Catalysis A: General, 2001, 208 (1 – 2): 35 – 46.

[3] Yuan A B, Zhang Q L. A novel hybrid manganese dioxide/activated carbon supercapacitor using lithium hydroxide electrolyte[J]. Electrochemistry Communications, 2006, 8(7): 1173 – 1178.

[4] Biloé S, Goetz V, Guillot A, et al. Optimal design of an activated carbon for an adsorbed natural gas storage system[J]. Carbon, 2002, 40(8): 1295 – 1308.

[5] Ioannidou O, Zabaniotou A. Agricultural residues as precursors for activated carbon production—A review [J].

Renewable and Sustainable Energy Reviews,2007,11(9):1966 - 2005.

[6]Gonzalez J D L,Vilchez F M,Rodríguez - Reinoso F. Characterization of active carbons from olive stones [J]. Carbon,1970,8:117 - 124.

[7]Ania C O,Parra J B,Menéndez J A,et al. Effect of microwave and conventional regeneration on the microporous and mesoporous network and on the adsorptive capacity of activated carbons[J]. Microporous and Mesoporous Materials,2005,85(1 - 2):7 - 15.

[8]Nabais J M V,Carrott P J M,Carrott M M L. Preparation and modification of activated carbon fibers by microwave heating[J]. Carbon,2004,42(7):1315 - 1320.

[9]Jones D A,Lelyveld T P,Mavrofidis S D. Microwave heating applications in environmental engineering—a review[J]. Resource,Conservation and Recycling,2002,34(2):75 - 90.

[10]Williams,Howard M,Parkes,et al. Activation of a phenolic resin - derived carbon in air using microwave thermogravimetry[J]. Carbon,2008,46(8):1169 - 1172.

[11]Ania C O,Menéndez J A,Parra J B,et al. Microwave - induced regeneration activated carbons polluted with phenol. a comparison with conventional thermal regeneration[J]. Carbon,2004,42(7):1377 - 1381.

[12]Coss P M,Cha C Y. Microwave regeneration of activated carbon used for the removal of solvents from vented air[J]. Journal of the Air and Waste Management Association,2000,50:529 - 535.

[13]Gregg S J,Sing K S W. Adsorption,Surface Area and Porosity[M]. New York:Academic Press,1983.

[14]Dubinin M M. Progress in Surface and Membrane[M]. New York:Academic press,1975.

[15]Lastoskie C,Gubbins K E,Quirke N. Pore size distribution analysis and networking:studies of microporous sorbents[J]. Studies in Surface Science and Catalysis,1994,87:51 - 60.

[16]Byrne C E,Nagle D C. Carbonization of wood for advanced materials applications[J]. Carbon,1997,35(2):259 - 266.

[17]Benoît C,Py X,et al. The effect of the carbonization/activation procedure on the microporous texture of the subsequent chars and active carbons[J]. Microporous and Mesporous Materials,2003,57(3):273 - 282.

[18]Inagaki M,Nishikawa T,et al. Carbonization of kenaf to prepare highly - microporous carbons[J]. Carbon,2004,42(4):890 - 893.

[19]Tan J S,Ani F N,et al. Diffusional behavior and adsorption capacity of palm shell chars for oxygen and nitrogen - the effect of carbonization temperature[J]. Carbon,2003,41(4):840 - 842.

[20]Li W,Yang K B,Peng J H. Effects of carbonization temperatures on characteristics of porosity in coconut shell chars and activated carbons derives from carbonized coconut shell chars[J]. Industrial Crops and Products,2008,28(2):190 - 198.

[21]González,Juan F,Encinar,et al. Preparation of activated carbons from used tyres by gasification with steam and carbon dioxide[J]. Applied Surface Science,2006,252(17):5999 - 6004.

[22]Román S,González J F,González - García C M,et al. Control of pore development during CO_2 and steam activation of olive stones[J]. Fuel Processing Technology,2008,89(8):715 - 720.

[23]Rodríguez - Reinoso F,Lahaye J,Ehrburger P. Fundamental issues in control of carbon gasification reactivity [J]. Kluwer Academic,1991,533 - 571.

[24]Rouquerol F,Rouquerol J,Sing K. Adsorption by Powders and Porous Solids,Principles,Methodology and Applications[M]. London:Academic Press,1999.

[25]Singh,Kunwar P,Malik,et al. Liquid - phase adsorption of phenols using activated carbons derived from agricultural waste material[J]. Journal of Hazardous Materials,2008,150(3):626 - 641.

[26]Su W,Zhou L,Zhou Y P. Preparation of microporous activated carbon from coconut shells without activating agents[J]. Carbon,2003,41:861 - 863.

[27] Achaw O W, Afrane G. The evolution of the pore structure of coconut shells during the preparation of coconut shell – based activated carbons[J]. Microporous and Mesoporous Materials, 2008, 112(1 – 3): 284 – 290.

[28] Cagnon, Benoît, Py, et al. Contributions of hemicellulose, cellulose and lignin to the mass and the porous properties of chars and steam activated carbons from various lignocellulosic precursors[J]. Bioresource Technology, 2009, 100(1): 292 – 298.

[29] Su W, Zhou Y P, Wei L F. Effect of microstructure and surface modification on the hydrogen adsorption active carbons[J]. New Carbon Materials, 2007, 22(2): 135 – 140.

[30] Din, Azam T. Mohd, Hameed, et al. Batch adsorption of phenol onto physiochemical – activated coconut shell [J]. Journal of Hazardous Materials, 2009, 161(2 – 3): 1522 – 1529.

[31] Azevedo D C S, Araújo J C S, Bastos – Neto M, et al. Microporous activated carbon prepared from coconut shells using chemical activation with zinc chloride[J]. Microporous and Mesoporous Materials, 2007, 100(1 – 3): 361 – 364.

[32] Hu Z H, Srinivasan M P. Preparation of high – surface – area activated carbons from coconut shell [J]. Microporous and Mesoporous Materials, 1999, 27(1): 11 – 18.

[33] Gergova K, Petrov N, Minkova V. A comparison of adsorption characteristics of various activated carbon [J]. Journal of Chemical Technology and Biotechnology, 1993, 56(1): 77 – 82.

[34] Daud, Wan M A W, Ali, et al. The effects of carbonization temperature on pore development in palm – shell – based activated carbon[J]. Carbon, 2000, 38(14): 1925 – 1932.

[35] Arenas E, Chejine F. The Effect of the activating agent and temperature on the porosity development of physically activated coal chars[J]. Carbon, 2004, 42(12 – 13): 2451 – 2455.

[36] Nakanishi K. Infra – red Absorption Spectroscopy – Practical[M]. San Francisco CA: Holden – Day, 1962.

Comparison of Activated Carbon Prepared from Jatropha Hull by Conventional Heating and Microwave Heating

Xinhui Duan, C. Srinivasakannan, Jinhui Peng, Libo Zhang, Zhengyong Zhang

Abstract: An attempt to compare the yield and porous nature of the activated carbon prepared using the conventional and microwave assisted heating, is the focus of the present work. Towards this Jatropha hull (a biomass precursor) is activated using the popular activating agents, steam and CO_2 to assess the relative merit of activating agents and the heating methods. The process optimization exercise is carried out with the minimum number of experiments following the standard full factorial statistical design of experiments (RSM). The activated carbon prepared under the optimized conditions is compared based on the yield and porous nature. The yield of activated carbon is not found to vary significantly for the steam activation, irrespective of the heating method, while it is found to double using CO_2 activation with microwave heating as compared to conventional heating. The pore volume and the surface area is found to double using the microwave heating with steam, while it is found to be of the same order of magnitude using CO_2 activation. Although the porosity of carbon is of the same order of magnitude using CO_2 activation, the activation temperature, the activation time, CO_2 flow rate are significantly lower than the conventional heating rendering the process more economical than the conventional heating. The steam - carbon reaction rate is significantly higher than the carbon - CO_2 reaction rate, rendering the time requiring for activation lesser using steam activation as compared to CO_2 activation.

Keywords: activated carbon; jatropha hull; microwave heating; pore size distribution; comparison

1 Introduction

The increasing appeals for environment protection, the global depletion of fossil oil reserves forces more and more countries to focus on renewable fuels. Signifificant efforts are made globally to develop biomass energy, of which the primary focus is on biodiesel, which is a combustible fuel manufactured by the esterification of renewable oils, fats and fatty acids[1]. At present, Jatropha curcas is widely accepted to be a promising plant for production of biodiesel which has the potential to reduce our dependence on the fossil fuel, as well as a renewable source of energy. The International Energy Agency (IEA) has forecasted the global energy demand to grow by more than 50% from the current consumption by 2030 with China and India alone contributing for 45% of the anticipated demand[2]. The production of biodiesel from Jatropha curcas would generate large quantities of by - products like Jatropha hull (extracted meat). It is estimated that the Jatropha curcas is currently being cultivated in China in area excess of $30000hm^2$. With the estimated yield of plant being 65t per hectare, the total output is estimated to be approximately around 2Mt. With an optimistic estimate of 30% waste biomass generation, the projected Jatropha hull can be 0.6Mt[3-7]. Raw Jatropha hull is toxic[8,9] and demands appropriate treatment in order not be harmful to human and the environment. One of the possible ways of utilization of this large quantity of toxic waste rich in lignin, is to utilize it as a possible precursor for preparation of activated carbon, a popular adsorbent widely used in industries for variety of applications. Activated carbons are widely prepared from variety of biomass precursors and the industry demands availability of cheap source of biomass that produce good quality activated carbons.

Generally, activated carbons are prepared either by physical or chemical activation. The advantage of chemical activation lies in the shorter treatment time and lower activation temperature[10,11], while this craft would induce extra elements into the carbon materials to some degree. This process contaminates the activated carbon, and the environment in the process of production as well. On the other hand, physical activation usually utilizes steam or CO_2 as the activation agent which is environmental friendly, but this craft demands higher activation temperature and longer activation time, which leads to relatively lower yield and higher energy cost. With the development of modern technology, microwave heating has recently received increasing attention from the scientific community for preparation of activated carbon[12], which has the potential to become a viable alternative for the conventional activation methods. Compared with conventional heating method, the microwave heating considerably reduces the activation time[13], as it provides advantages such as uniform interior heating, high heating rate, selective heating, greater control of the heating process, no direct contact between the heating source and heated materials and reduced equipment size and waste[14-17].

Ania, et al. [18] used microwave - assisted technology for the reactivation of activated carbon exhausted with phenol and salicylic acid and reported a significantly higher adsorption capacity as compared to conventional thermal reactivation. Li and Peng[13] reported a significant reduction in the activation time for the preparation of activated carbon from coconut shell, at a pilot plant level with microwave heating. Guo and Lua[19] have reported preparation of activated carbon from oil - palm - stone with microwave heating, with the resultant carbon being relatively high density and predominantly microporous. Although microwave heating in the preparation of activated carbon is being utilized at laboratory scale level, it shows a great potential for industrial production due to its number of advantages over the conventional heating methods.

The response surface methodology (RSM) is widely used to study the effects of multiply process variables on product properties as well as for process optimization. RSM is a combination of mathematical and statistical techniques[20]. The optimization of RSM is particularly useful when all the independent variables and their levels and responses are not clearly known. Utilization of RSM for process development and optimization is widely reported in literatures[21-27].

It is well known that carbon can be converted into high porous activated carbon either using steam or CO_2 as the activating agent and either using conventional heating or microwave heating. The present paper attempts to compare the quality of the best of the activated carbons prepared by microwave heating using steam and CO_2. The resultant activated carbon is characterized using the iodine adsorption capacity, yield, surface area (BET), pore structure distribution and scanning electron microscope (SEM) in order to assess the relative merits of the activation agents and heating method. The present paper recommends economical and efficient way for production of activated carbon from Jatropha hull, which otherwise would be a huge quantity of solid waste of toxic nature, without any utility.

2 Experimental

2.1 Materials

Jatropha hull received from ShenYu Company, Kunming, Yunnan province of China is washed with deionized water to remove the foreign materials and dried in an air oven at 105℃. The dry materials are sieved to the particle size of 2 – 3mm and stored in airtight container for further experimentation. The proximate analysis of the Jatropha hull is shown in Table 1.

Table 1 Proximate analysis of Jatropha hull

Sample	Moisture (wt.)/%	Volatile matter (wt.)/%	Fixed carbon (wt.)/%	Ash (wt.)/%
Jatropha hull	9.50	60.78	25.48	3.80

2.2 Activation craft

Experiments on activation of Jatropha hull has been conducted with conventional and microwave heating using steam and CO_2 as activation agents. Prior to activation, carbonization of Jatropha hull is carried out by loading 100g of dried material in to a muffie furnace, under N_2 gas flow ($100cm^3$/min) and heated up to a carbonization temperature of 600℃, at a heating rate of 10℃/min. Upon reaching 600℃ the samples is held at the same conditions for 1h. After carbonization the sample is cooled to room temperature under N_2 flow ($100cm^3$/min). The yield of char is found to be around 40%. The literature pertaining to other biomass indicates that the pyrolysis temperature, heating rate and the pyrolysis time vary widely. Preliminary experiments in this study revealed that there is no significant weight loss on pyrolysis mass, beyond 1h time at 600℃, 10℃ heating rate and hence the conditions were chosen for pyrolysis.

The carbonized char is activated using a tubular furnace and microwave heating device by varying activation temperature, activation time and the flow rate of activation agent. Range of the process parameters covered in the present study is shown in Table 2, which were decided based on the optimum conditions reported widely in literature for other precursors. The temperature control unit of the microwave heating device varies the power input to maintain the furnace temperature at the desired set point, which is measured by nichrome – nickel silicon armor type thermo – element, placed such that it touched the material. The thermo – element has the dimension of 8mm diameter and a length of 450mm, with the temperature range of 0 – 1250℃ and a measurement precision of ±0.5℃. The completion of activation process under the set of desired activation conditions is marked by terminating the flow of activation agent, by switching to the nitrogen flow until the activated carbon cooled to the room temperature. The scheme of the experimental apparatus using microwave heating is shown in Fig. 1. The power input of the microwave device could be increased to a maximum of 3kW at a frequency of 2450Hz. The sample is loaded in a quartz pipe and placed at centre of the microwave device. In the case of conventional heating system, conventional tubular furnace is utilized, which has electronic control system to control the furnace temperature at desired temperature.

Table 2 Upper and lower limits of the process variables studied

Sample	Conventional heating with steam	Conventional heating with CO_2	Microwave heating with steam	Microwave heating with CO_2
Activation temperature/℃	800 - 1000	850 - 950	800 - 1000	850 - 950
Activation time/min	10 - 35	25 - 40	10 - 19	10 - 30
Flow of steam (g/min) and CO_2 (mL/min)	1.5 - 5.5	200 - 500	2 - 5	150 - 450

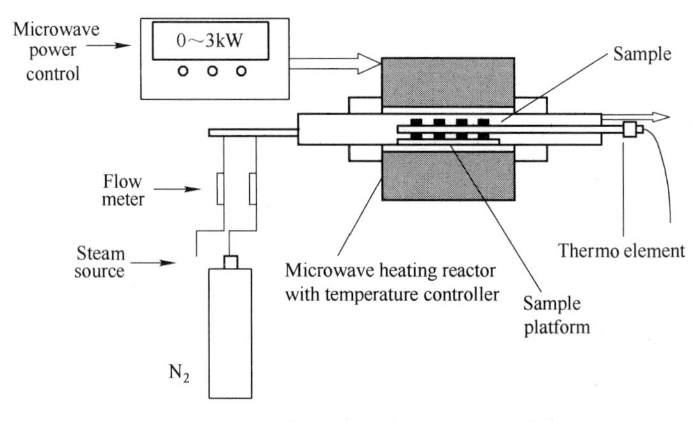

Fig. 1 The diagram of apparatus for the preparation of activated carbons with microwave

2.3 Product characterization

The product is subject to characterization for iodine number, yield and the BET surface area. The yield is defined as grams of activated carbon per gram char utilized for activation. The iodine adsorption capacity is represented as iodine number which indicates milligrams of iodine adsorbed by a gram of activated carbon (mg/g), by using the Standard Testing Methods of PR China (GB/T 12496. 8—1999) for testing acting activated carbons. The BET surface area, average pore size distribution are estimated using the surface area analyzer, Autosorbe 1 - C made by Quantachrome Instruments, USA. The microstructures were analyzed by the scanning electron microscope (SEM, Philips XL30ESEM - TMP). The optimized conditions for the preparation of activated with the result of iodine number and yield are shown in Table 3, all the experimental designs are three level designs, details of the optimization process, discussions on the influences of different preparation conditions on the property of activated carbons obtained and characterization of the products are reported in our previous papers[28,29].

Table 3 Preparation conditions of activated carbon and the experimental result

Sample	Conventional heating with steam	Conventional heating with CO_2	Microwave heating with steam	Microwave heating with CO_2
Activation temperature/℃	900	950	900	900
Activation time/min	22	40	19	30
Flow of steam(g/min) and CO_2(mL/min) Sample	5.5	500	5	300
Iodine number/mg · g^{-1}	950	1008	988	998
Yield/%	13.32	18.02	16.56	36.60

3 Results and discussions

Process optimization of the Jatropha hull precursor using conventional and microwave heating with steam and CO_2 as the activating agents has been attempted in the present study covering the parameters such as activation agent flow rate, the activation temperature and the activation time. A normal process optimization exercise involves varying one parameter at a time, which leads to large number of experiments. Since the present work attempts to optimize the process conditions of four processes, a statistical process optimization tool Response Surface Methodology (RSM)[30,31] has been utilized to estimate the optimum conditions with minimum number of experiments. The objective of the present paper is not to detail the process of optimization, for the four different processes conditions, but to compare and contrast the optimized activated carbon samples prepared under the optimized conditions. The process was optimized based on the iodine number and the yield of the activated carbon. The process of pore creation is well establelished and is due to the reaction of activation agent with the carbon. Either of the processes involving heating the samples conventionally or using microwave using steam or CO_2 as the activating agent, the iodine absorption capacity increases with increase in the extent of reaction. The increase in iodine absorption capacity indicates increase in porous nature of the carbon, while the increase in extent of the reaction is characterized by the reduction in the yield of the activated carbon. The yield of the activated carbon decreases with increase in any of the process parameters such as the increase in activation temperature, activation time and the flow rate of the activation medium, while it is found to increase the iodine absorption capacity. The iodine absorption capacity increases until an optimum value beyond which it is found to decrease with further decrease in yield of the carbon which was widely attributed to increase in pore size leading to merger of the pores. The effects of process parameters are in concurrence with the existing knowledge as reported in literature[32-34]. The rate of reaction of the steam with carbon is well established to be higher as compared to the with carbon is well established to be higher as compared to the rate of CO_2 with carbon and hence the extent of reaction with steam activation is much higher compared to the CO_2 activation for specified activation time[35,36].

The optimized process conditions are listed in Table 3 and it can be found that the optimized temperature remains around 900℃ with the exception of conventional heating with CO_2, while the flow rate of the activating agents are higher in conventional heating process compared to microwave heating. The optimized activation time is lower for microwave heating while it is higher for conventional heating. A lower flow rate of the activating agent and lower activation time in comparison with the conventional heating could be attributed to the faster and uniform heating of the carbon in the micro - wave field. The microwave energy is delivered to materials directly through molecular interaction in the electromagnetic field and the ability of conversion is decided by the dielectric properties of materials[37]. Generally, in conventional heating system, energy is transferred to materials from the surface to inside through convection, conduction, and radiation of heat[38], on the contrary in microwave heating system, the selective heating method of microwave energy lead to a heating inside of the materials first, then to the surface. The activated carbons are excellent microwave adsorptive materials[39] and hence heats up to the desired temperature much faster than the conventional heating method.

A comparison of the yield activated carbon with conventional and microwave heated systems, show close proximity to each other with steam as the activating agent. This indicates that the rate of carbon - steam reaction is not significantly altered, although the heating using microwave could be uniform and faster. This could be due to the fact that the rate limiting step could be the diffusion of the steam to the active sites in the carbon. Although the yields are in similar range, a significant increase in the pore volume is observed with microwave heating as compared to conventional heating. The pore volume as well as the surface area is found to almost double in as compared to conventional heating (Table 4).

Table 4 Pore size distribution, total pore volume and average pore size of activated carbon

Sample	Surface area/$m^2 \cdot g^{-1}$	Pore volume/$cm^3 \cdot g^{-1}$	Average pore width/nm
Char	480	0.42	3.5
Conventional heating with steam	748	0.53	2.85
Conventional heating with CO_2	1207	0.86	2.86
Microwave heating with steam	1350	1.07	3.10
Microwave heating with CO_2	1284	0.87	2.71

A similar exercise to compare the conventional and microwave assisted activation using CO_2 reveal that the optimum yields using microwave assisted heating is double that of conventional heating, while the activation temperature, time of activation, and CO_2 flow rates significantly lower than the conventional heating. This is highly favorable considering the fact that the pore volumes are in the same range in both the cases. However, the average pore size is found to be smaller in the case of microwave heating as compared to conventional, leading to a marginally higher surface area with microwave heating (Table 4).

Comparing with the other combinations showed in Table 4, the process of microwave heating with steam seems to be the most effective way in the production of activated carbon, although the yield of the production is comparable, however, the activation time is less than the other processes and the surface area, total pore volume and average pore size are significant higher than the others. The reason is probably be, on one hand, regardless of the activation agents, comparing with conventional heating process, the activation time using microwave is reduced because of the advantages of microwave heating of higher heating rates (inner heating), no direct contact between heating source and the heated materials, possibility of selective heating; On the other hand, comparing with microwave heating with CO_2, the molecular diameter of H_2O is smaller than that of CO_2, thus, the diffusion resistance is much smaller, which makes it more easier to diffuse into the micro structure of the activated carbon for the development of micropore, which contributes to the better properties of the productions. However, it should be indicated that the Jatropha hull is some kind of slice naturally, that makes it very easy to react with the activation agent, unable to get along with long period of activation process because of the little thickness, which also makes the yield of the activated carbon obtained not high enough after reaching the desire iodine number. Thus, the activation time of conventional heating is not very long, and comparing with the activation time of microwave heating, the advantage of less activation time by microwave heating is not significant enough.

3.1 Characterization of activated carbon

The nitrogen adsorption isotherms of the activated carbons prepared in conventional heating system and microwave heating system with steam and CO_2 as activation agent respectively is shown in Fig. 2. The samples are analyzed at 77K with an accelerated surface area and porosimetry system (Autosorb - 1 - C, Quantachrome). Prior to gas adsorption measurements, the carbon is degassed at 300℃ in a vacuum condition for a period of at least 2h. Nitrogen adsorption isotherm is measured over a relative pressure (p/p_0) range from approximately 10^{-7} to 1. The BET surface area is calculated from the isotherms by using the Brunauere - Emmette - Teller (BET) equation[40]. The cross - sectional area for nitrogen molecule is assumed to be 0. 162nm. The Dubinine - Radushkevich (DR) method is used to calculate the micropore volume[41]. The micropore size distribution is ascertained by Non - local Density Functional Theory (NLDFT)[42] by minimizing the grand potential as a function of the fluid density profile. The total volume is estimated by con - verting the amount of N_2

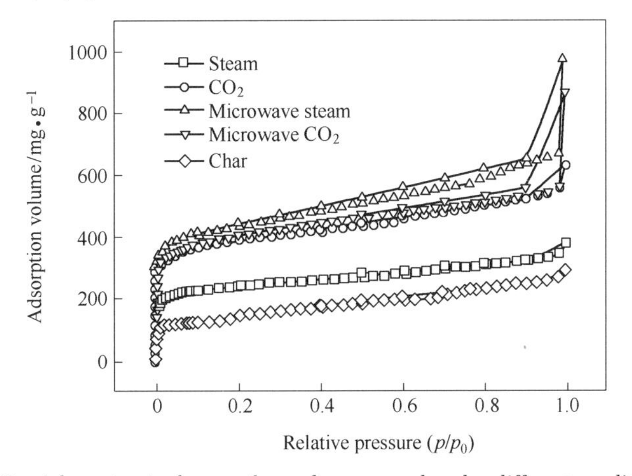

Fig. 2 Adsorption isotherms of samples prepared under different conditions

gas adsorbed at a relative pressure of 0.95 to equivalent liquid volume of the adsorbate(N_2)[43].

It can be seen from Fig. 2 that the volumes adsorbed increase sharply at low relative pressure which indicate the filling of micropores and reach a plateau. The adsorption capacity continues to increase with increase in the relative pressure up to a p/p_0 value of 1, which pertains to intermediate between type Ⅰ and Ⅱ of the referred IUPAC classification. This type of isotherm is usually exhibited by microporous solids that include a well developed mesopore structure. The surface area, pore volume, average pore size are shown in Table 4. The NLDFT analysis of the activated carbons prepared under different conditions are shown in Fig. 3 in form of the cumulative pore volume distribution, which compare the pore size distribution curves over an extended range of micro - and mesopores regions[44].

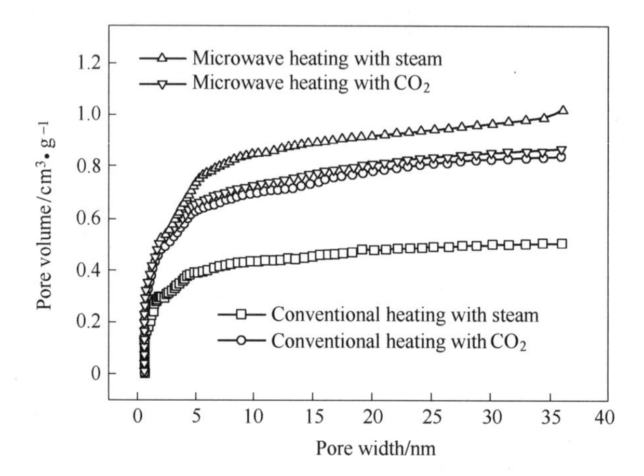

Fig. 3 Cumulative pore volume distribution chart for Jatropha hull activated carbon prepared under different conditions

3.2 Microscopic structure analysis

Fig. 4 shows the SEM images of the char and activated carbons prepared under different conditions. As can be seen from Fig. 4(a), the surface of the primary char utilized for activation is planar with deeper hole structure, which indicated that the carbonization stage mainly creates macro and mesoporous carbon[45,46]. Upon activation, as can be seen from the rest of the images, there are lots of small crevices on the surface, which forms small pores over the surface, having well developed pore network. The reaction of carbon with activation agent generates large number of micropores which significantly increases the surface area of the activated carbon.

4 Conclusions

Conventional and microwave heating systems for the preparation of activated carbon form Jatropha hull with different activation agents (steam and CO_2) are compared in this paper and the results are as follows:

(1) The yield of activated carbon is not found to vary significantly for the steam activation, irrespective of the heating method, while it is found to double using CO_2 activation with microwave

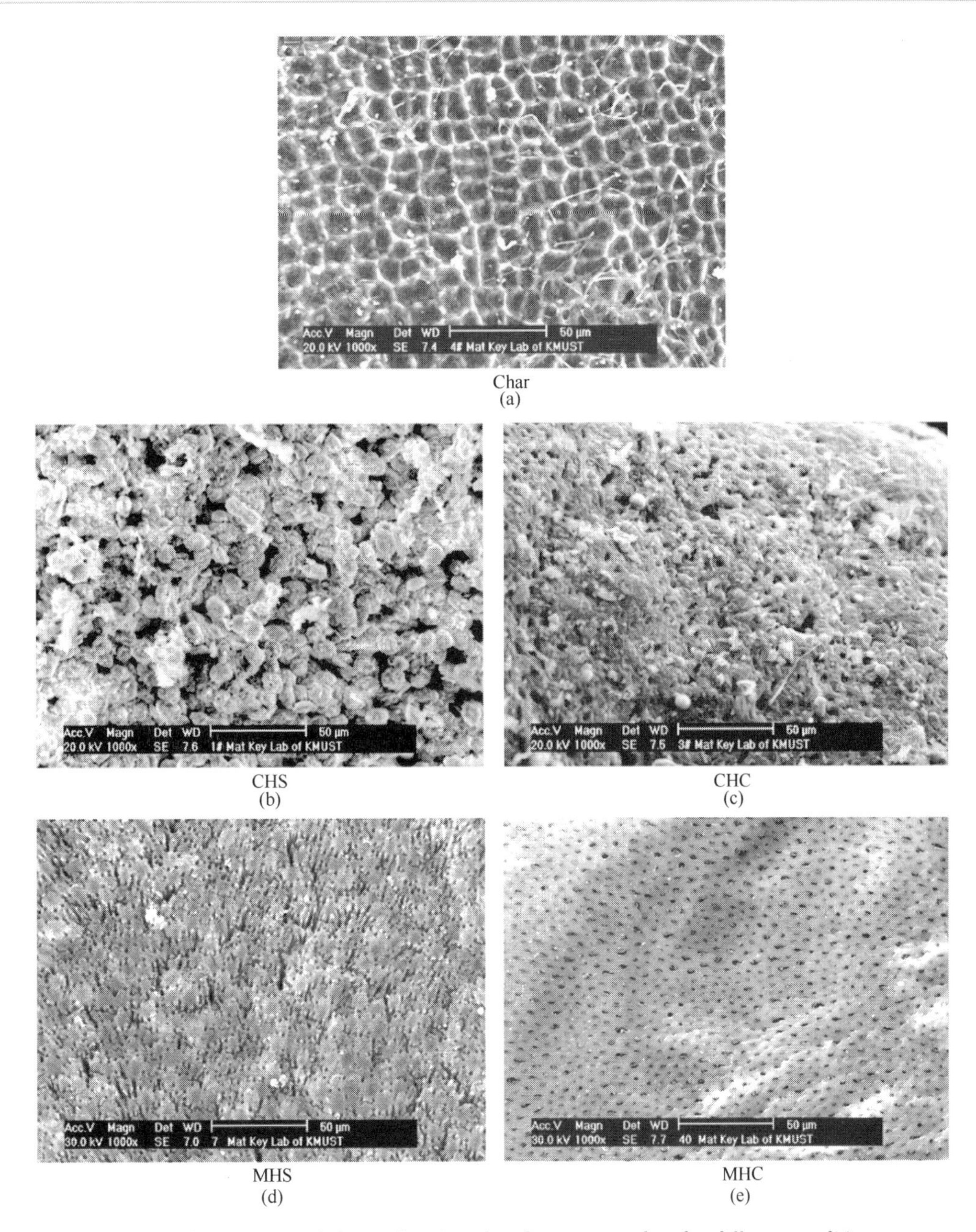

Char
(a)

CHS
(b)

CHC
(c)

MHS
(d)

MHC
(e)

Fig. 4 SEM images of char and activated carbons prepared under different condition

CHS—conventional heating with steam; CHC—conventional heating with CO_2;

MHS—microwave heating with steam; MHC—microwave heating with CO_2

heating as compared to conventional heating.

(2) The pore volume and the surface area is found to double using the microwave heating with steam, while it is found to be of the same order of magnitude using CO_2 activation.

(3) Although the porosity of carbon is of the same order of magnitude using CO_2 activation, the activation temperature, the activation time, CO_2 flow rate are significantly lower than the conventional heating rendering the process more economical than the conventional heating.

Acknowledgements

The authors would like to express their gratitude to the National Key Technology R&D Program of China(No. 2007BAD32B04)for financial support.

References

[1]Oliveira L S, Franca A S, Camargos R R S. Coffee oil as a potential feedstock for biodiesel production [J]. Bioresour Technol, 2008, 99(8): 3244 - 3250.

[2]World energy outlook - China and India insights. 9 rue de la Fe'de'ration, 75739 Paris Cedex 15, France: International Energy Agency, Head of Communication and Information Office, 2007.

[3]Openshaw K. A review of Jatropha curcas: an oil plant of unfulfilled promise[J]. Biomass Bioenerg, 2000, 19 (1): 1 - 15.

[4]Tang J R, Guo R C. About Jatropha curcas L. and relevant research progress[J]. Forest Inventory Plann, 2007, 32: 36 - 39(In Chinese).

[5]Wu G J, Liu J. Development of energy plant: progress and suggestions[J]. Bull Chinese Acad Imp. Sci., 2006, 21: 51 - 57(In Chinese).

[6]Wang Y, Long C L. Utilization of Jatropha cuscas L. An "Energy Plant" for biodiesel[J]. Anhui Agri. Sci., 2007, 35: 426 - 429(In Chinese).

[7]Zhao R F, Hua J. Preparation of activated carbon from Jatropha curcas L shell by chemical activation [J]. Sichuan Environ., 2007, 26: 16 - 18(In Chinese).

[8]Sharma D K, Pandey A K. Use of Jatropha curcas hull biomass for bioactive compost production[J]. Biomass Bioenerg, 2009, 33(1): 159 - 162.

[9]Sirisomboon P, Kitchaiya P, Pholpho T. Physical and mechanical properties of Jatropha curcas L. fruits, nuts and kernels[J]. Biosystems Engineering, 2007, 97(2): 201 - 207.

[10]Gratuito M K B, Panyathanmaporn T, Chumnanklang R A. Production of activated carbon from coconut shell: optimization using response surface methodology[J]. Bioresource Technology, 2008, 99(11): 4887 - 4895.

[11]Taya T, Ucarb S, Karagoz S. Preparation and characterization of activated carbon from waste biomass [J]. Hazard Mater., 2009, 165(1 - 3): 481 - 485.

[12]Yuen F K, Hameed B H. Recent developments in the preparation and regeneration of activated carbons by microwaves[J]. Advances in Colloid and Interface Science, 2009, 149(1 - 2): 19 - 27.

[13]Li W, Peng J H. Preparation of activated carbon from coconut shell chars in pilot - scale microwave heating equipment at 60kW[J]. Waste Management, 2009, 29(2): 756 - 760.

[14]Jones D A, Lelyveld T P, Mavrofidis S D, et al. Microwave heating applications in environmental engineering—a review[J]. Resources, Conservation and Recycling, 2002, 34(2): 75 - 90.

[15]Appleton T J, Colder R I, Kingman S W, et al. Microwave technology for energy - efficient processing of waste [J]. Applied Energy, 2005, 81(1): 85 - 113.

[16]Thostenson E T, Chou T W. Microwave processing: fundamentals and applications[J]. Composites Part A: Applied Science and Manufacturing, 1999, 30(9): 1055 - 1071.

[17]Venkatesh M S, Raghavan G S V. An overview of microwave processing and dielectric properties of agri - food materials[J]. Biosystems Engineering, 2004, 88(1): 1 - 18.

[18]Ania C O, Menéndez J A, Parra J B, et al. Microwave - induced regeneration of activated carbons polluted with phenol. A comparison with conventional thermal regeneration[J]. Carbon, 2004, 42(7): 1383 - 1387.

[19] Guo J, Lua A C. Preparation of activated carbons from oil - palm - stone chars by microwave - induced carbon dioxide activation[J]. Carbon, 2000, 38(14): 1985 - 1993.

[20] Khuri A Z, Cornell J A. Response Surface Designs and Analysis[M]. New York: Marcel Dekker, 1987.

[21] Sacchetti G, Pinnavaia G G, Guidolin E, et al. Effects of extrusion temperature and feed composition on the functional, physical and sensory properties of chestnut and rice flour - based snack - like products[J]. Food Research International, 2004, 37(5): 527 - 534.

[22] Jaya Shankar T, Bandyopadhyay S. Optimization of extrusion process variables using a genetic algorithm [J]. Food and Bioproducts Processing, 2004, 82(2): 143 - 150.

[23] Jaya Shankar T, Bandyopadhyay S. Process variables during single - screw extrusion of fish and rice - flour blends[J]. Journal of Food Processing and Preservation, 2005, 29(2): 151 - 164.

[24] Bhattacharya S, Prakash M. Extrusion of blends of rice and chick pea flours: a response surface analysis [J]. Journal of Food Engineering, 1994, 21(3): 315 - 330.

[25] Shankar T J, Sokhansanj S, Bandyopadhyay S, et al. A case study on optimization of biomass flow during single - screw extrusion cooking using genetic algorithm(GA) and response surface method(RSM) [J]. Food and Bioprocess Technology, 2010, 3(4): 498 - 510.

[26] Rout R K, Bandyopadhyay S. A comparative study of shrimp feed pellets processed through cooking extruder and meat mincer[J]. Aquacultural Engineering, 1999, 19(2): 71 - 79.

[27] Giri S K, Bandyopadhyay S. Effect of extrusion variables on extrudate characteristics of fish muscle - rice flour blend in a single - screw extruder[J]. Journal of Food Processing and Preservation, 2000, 24(3): 177 - 190.

[28] Duan X H, Srinivasakannan C, Peng J H, et al. Preparation of activated carbon from Jatropha hull with microwave heating: optimization using response surface methodology[J]. Fuel Processing Technology, 2011, 92(3): 394 - 400.

[29] Duan X H, Peng J H, Srinivasakannan C, et al. Process optimization for the preparation of activated carbon from Jatropha hull using response surface methodology[J]. Energy Sources, Part A: Recovery, Utilization, and Environmental Effects, 2011, 33(21): 2005 - 2017.

[30] Baçaoui A, Yaacoubi A, Dahbi A, et al. Optimization of conditions for the preparation of activated carbons from olive - waste cakes[J]. Carbon, 2001, 39(3): 425 - 432.

[31] Azargohar R, Dalai A K. Production of activated carbon from Luscar char: experimental and modeling studies [J]. Microporous and Mesoporous Materials, 2005, 85(3): 219 - 225.

[32] Zhang Z, Qu W, Peng J, et al. Comparison between microwave and conventional thermal reactivations of spent activated carbon generated from vinyl acetate synthesis[J]. Desalination, 2009, 249(1): 247 - 252.

[33] Minkova V, Razvigorova M, Bjornbom E, et al. Effect of water vapour and biomass nature on the yield and quality of the pyrolysis products from biomass[J]. Fuel Processing Technology, 2001, 70(1): 53 - 61.

[34] Toles C A, Marshall W E, Wartelle L H, et al. Steam - or carbon dioxide - activated carbons from almond shells: physical, chemical and adsorptive properties and estimated cost of production[J]. Bioresource Technology, 2000, 75(3): 197 - 203.

[35] González J F, Encinar J M, González - García C M, et al. Preparation of activated carbons from used tyres by gasification with steam and carbon dioxide[J]. Applied Surface Science, 2006, 252(17): 5999 - 6004.

[36] Román S, González J F, González - García C M, et al. Control of pre development during CO_2 and steam activation of olive stones[J]. Fuel Proc. Tech., 2008, 89: 715 - 720.

[37] Haque K E. Microwave energy for mineral treatment processes—a brief review[J]. International Journal of Mineral Processing, 1999, 57(1): 1 - 24.

[38] Thostenson E T, Chou T W. Microwave processing: fundamentals and applications[J]. Composites Part A: Ap-

plied Science and Manufacturing,1999,30(9):1055 - 1071.

[39] Jou G. Application of activated carbon in a microwave radiation field to treat trichloroethylene[J]. Carbon, 1998,36(11):1643 - 1648.

[40] Xie Q. Study on Control over Coal Carbonization and Preparation of Activation Carbon[D]. Xuzhou: China Univ. of Mining & Tech. ,1996.

[41] Xie Q. Principles of Control over Coal Carbonization and its Application in Preparation of Activated Carbon [M]. Xuzhou: China Univ. of Mining & Tech. Press,2002.

[42] Smaiések M, éCernây S. Active Carbon: Manufacture, Properties and Applications[M]. Amsterdam and New York: Elsevier Pub. Co. ,1970.

[43] Lyubchik S B, Benoit R, Beguin F. Influence of chemical modification of anthracite on the porosity of the resulting activated carbons[J]. Carbon,2002,40(8):1287 - 1294.

[44] Bisio C, Gatti G, Boccaleri E, et al. Understanding physico - chemical properties of saponite synthetic clays [J]. Microporous and Mesoporous Materials,2008,107(1):90 - 101.

[45] Lastoskie C, Gubbins K E, Quirke N. Pore size distribution analysis of microporous carbons: a density functional theory approach[J]. The Journal of Physical Chemistry,1993,97(18):4786 - 4796.

[46] Arriagada R, Garcia R, Molina - Sabio M, et al. Effect of steam activation on the porosity and chemical nature of activated carbons from (Eucalyptus globules) and peach stones [J]. Microporous Materials, 1997, 8 (3):123 - 130.

Comparison between Microwave and Conventional Thermal Reactivations of Spent Activated Carbon Generated from Vinyl Acetate Synthesis

Zhengyong Zhang, Wenwen Qu, Jinhui Peng, Libo Zhang, Xiangyuan Ma, Zebiao Zhang, Wei Li

Abstract: Microwave and traditional thermal reactivations of activated carbon (AC) used as catalyst support in vinyl acetate synthesis have been investigated. Experiments have been carried out by using a single mode microwave device (MW) operating at 2450MHz and a conventional electric furnace (CF) under steam and CO_2 atmosphere, respectively. The surface properties of the spent AC and the reactivated samples were characterized by means of N_2 adsorption and SEM, and compared the effects of different heating mechanisms and activating agents on the adsorption capacities and pore structures of the reactivated AC. These results indicated that the AC obtained by microwave irradiation showed higher adsorption capacities for iodine, methylene blue (MB) and acetate acid, higher BET surface areas and mesoporosity than those obtained by conventional thermal heating. The reactivated samples activated by steam had a narrower and more extensive microporosity as well as higher BET than those activated by carbon dioxide under the same heating equipment. From the results, it was concluded that microwave heating combined with steam as an activating agent could remarkably increase the reactivating efficiency compared to the traditional thermal heating.

Keywords: spent activated; carbon; microwave heating; reactivation; pore structure; adsorption capacity

1 Introduction

Activated carbon (AC) has been employed in a large number of applications due to their pronounced textural properties and the inert nature in extreme pH conditions. For instance, they can be used as catalysts, and what is even more important, as catalyst supports[1-4]. However, the AC easily becomes saturated after sustained use. So the spent AC becomes a waste that contains a great amount of zinc occurring in acetate, oxide or a metallic form as well as organic compounds which are by-products of vinyl acetate synthesis[5]. In China, around 9000t of the spent AC was generated from vinyl acetate synthesis per year as solid wastes. They were usually taken to landfill or dumped. But the hazardous natures of the spent AC bring about additional pollution to the environment and a large amount of reusable materials was wasted while the price of AC keeps rising. Therefore, how to resolve these problems and reuse the valuable secondary materials has become an important issue.

Lately, our research group has developed a new process to deal with large quantities of spent catalyst supports from vinyl acetate synthesis, which includes two steps: leaching of zinc and reactivation of spent AC. The spent catalyst pretreated by MW irradiation and leached by a mixture of ammonia, ammonium bicarbonate and water exhibits a high leaching rate of zinc(~93%), achieving the effective separation of activated carbon and zinc. The experiment detail of leaching zinc from spent catalyst has been described elsewhere[6]. However, the obtained spent AC could not be directly reused as catalyst supports due to the low adsorption capacity and BET surface area. Hence, the subsequently reactivating process of spent AC plays a key role to improve its adsorption capacity and efficiency from the modification of pore size distribution(PSD) and surface chemistry by different reactivation techniques.

Over these years, a wide variety of reactivation techniques have been suggested and applied to deal with spent AC[7,8]. A wide variety of regeneration techniques have been suggested and applied to deal with spent AC, such as thermal regeneration, ultrasonic regeneration and chemical regeneration[9,10]. One of the most widespread used methods is thermal treatment in a given atmosphere (steam, carbon dioxide or inert gas)[11,12]. Unfortunately, the conventional thermal reactivation technique is characterized by time and energy consumption in order to keep the relatively high temperature of the heated materials. After successive heating and cooling cycles, the carbon becomes damaged and the exhaust gases need further treatment[13]. Hence, there has been considerable interest in search of alternative reactivation methods in recent years.

Microwave heating has the potential to become a viable alternative for the conventional reactivation method. The main applications of microwave heating with respect to activated carbon are production, reactivation and treatment of compounds adsorbed onto the carbon matrix. Compared with traditional heating techniques, microwave heating offers additional advantages, including higher heating rates, no direct contact between heating source and the heated materials, possibility of selective heating, precise control of temperature, small equipment size and reduced waste[14]. This method has been applied to regenerate AC exhausted with a series of organic pollutants[15-17]. A good example is the investigation of using microwave - assisted technology for the reactivation of AC exhausted with phenol[16] and salicylic acid[17] by C. O. Ania, et al. Their work revealed that AC reactivated by microwave irradiation shows an outstanding adsorption capacity and efficiency compared to conventional thermal reactivation[16]. Interesting results by X. Liu, et al. presented that the adsorption rate and capacities of the acid orange exhausted - 7ACs could be even higher than those of virgin AC after several microwave reactivation cycles[15]. More important, microwave technology has also been reported to be used as an assistant method for recovering activated carbon from spent catalysts of vinyl synthesis[6]. In a word, microwave technology is considered to be a promising technology for future applications of carbon materials.

The work described here was performed to compare the reactivation of spent AC using microwave heating and conventional thermal method under steam and carbon dioxide atmosphere, respectively. The spent AC raw AC(non exhausted) and final product were characterized by acetic acid adsorption, Brunauer - Emmett - Teller(BET) surface area and pore structure distribution(PSD). The purpose of the present work from the comparison of the four reactivation processes of spent AC, is to find out the most economical and efficient process to achieve industrial production of AC from the spent catalyst.

2 Experimental

2.1 Materials

The spent AC was obtained from the spent catalyst of vinyl synthesis pretreated by microwave irradiation and then leaching of zinc by an alkaline solution. The iodine number, MB adsorption, BET surface area and pore volume of spent AC were shown in Tables 1 and 2. The XRD pattern of spent AC was shown in Fig. 1. It appears that the spent AC was mainly composed of activated carbon, and only carbon peaks appeared. Prior to experiments, the carbon was washed several times with distilled water to remove soluble components, dried in anoven at 120℃ until reaching a constant weight, and stored in a desiccator for use.

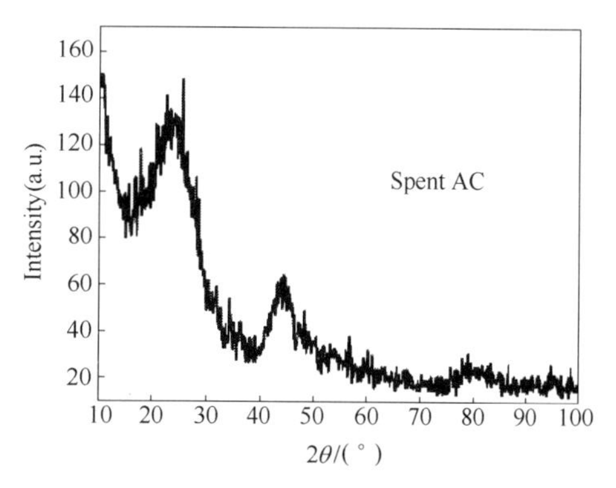

Fig. 1 XRD pattern of spent activated carbon

Table 1 The adsorption properties of the spent AC and reactivated samples

Samples	Iodine number /mg · g^{-1}	Methylene blue adsorption /mg · g^{-1}	Adsorption capacity of acetic acid/mg · g^{-1}
Spent AC	321	150	95
Raw AC	1050	260	565
MWW	1194	285	581
MWC	1158	240	564
CFW	1181	240	571
CFC	1091	210	518

Note: MWW—the sample reactivated by microwave device under steam atmosphere. MWC—the sample reactivated by microwave device under carbon dioxide atmosphere. CFW—the sample reactivated by conventional furnace under steam atmosphere. CFC—the sample reactivated by conventional furnace under carbon dioxide atmosphere. All samples were obtained in the optimum condition.

Table 2 The BET surface area and pore volume distribution of the spent AC and reactivated samples

Samples	SBET/m^2 · g^{-1}	DFT method/cm^3 · g^{-1}		Total pore volume	
		$V_{narrow-micropores}$	$V_{medium-micropores}$	$V_{mesopores}$	cm^3/g
Spent AC	271	0.02	0.07	0.13	0.22
Raw AC	932	0.13	0.22	0.24	0.59
MWW	1548	0.06	0.52	0.20	0.81
MWC	1255	0.06	0.44	0.09	0.59
CFW	1308	0.06	0.44	0.25	0.76
CFC	1178	0.05	0.42	0.08	0.55

Reactivation of the spent AC was carried out in a conventional furnace(CF) and a microwave device (MW), respectively. During the reactivating process in the MW, the exhausted AC was weighed 15g and situated inside a quartz reactor(40dia.) which is placed into the microwave reactor cavity in steam or CO_2 atmosphere and exposed to microwave radiation under controlled operating conditions. For a given AC, the temperature can be accurately controlled by adjusting the input power. The activation temperature(1000℃) of the sample with MW treatment was measured by using pyrometer. The scheme of the experimental apparatus is shown in Fig. 2. An 800W MW at the frequency of 2450MHz was used in this study. Distillate and the contaminant vapor were collected after passing through two bottles both containing 30mL of 0.1mol/L NaOH solution. Similarly, the reactivation in a conventional electric furnace was performed with the same weight of 15g spent AC by using an optical pyrometer to control the temperature and using steam and CO_2 as activating agents, respectively.

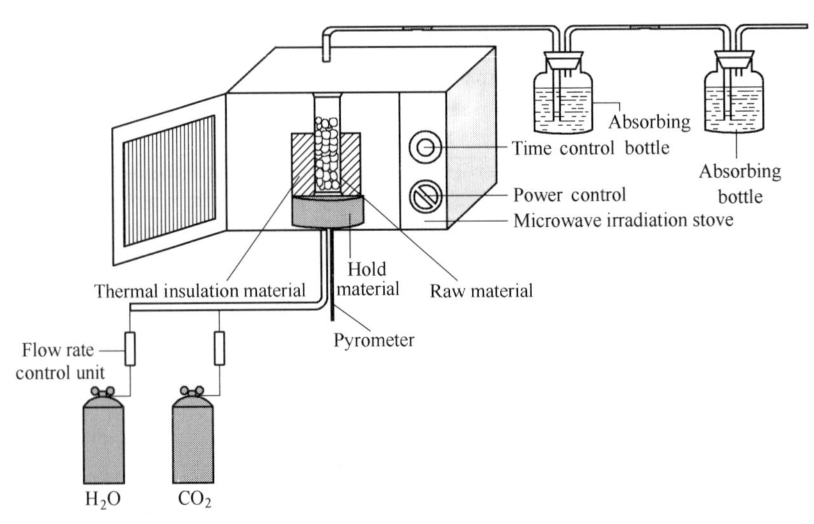

Fig. 2 The diagram of apparatus for the reactivation of activated carbons using physical activation

The four reactivation process can be described as follows: reactivation experiments were carried out at different temperature. Initially, the sample was heated in a nitrogen atmosphere ($120cm^3$/min) and once the desired temperature was reached, the nitrogen was rapidly switched to steam or CO_2 with a fixed flow rate, and reactivation was held for a different period of time. The reactivated product was then cooled to ambient temperature and washed with distilled water until the pH of the washing solution reached[6,7] Then, the washed activated carbon samples were dried in an electric oven in which the temperature was set at 105℃. The detailed reactivation procedure has been described in our previous work[18]. The optimum conditions of each reactivation method are shown in Table 3. For convenience, the resultant samples prepared under optimum conditions were labeled MWW, MWC, CFW and CFC, respectively. The burn - off was calculated by $\frac{W_2 - W_1}{W_2} \times 100\%$, where W_2 is the original weight of spent AC, and W_1 is the weight of reactivated AC.

Table 3 The preparation conditions of MWW, MWC, CFW and CFC

Samples factors	MWW	MWC	CFW	CFC
Activated temperature/℃	1000	1000	1000	1000
Microwave power/W	Adjust automatically	Adjust automatically	—	—
Activated times/min	40	25	90	100
The rate of steam/$g \cdot min^{-1}$	2.03	—	3.95	—
The rate of carbon dioxide/$m^3 \cdot h^{-1}$	—	0.2	—	0.5
Burn - off(wt.)/%	36	26	37	21

2.2 Analysis

Specific surface areas of reactivated AC samples were determined by BET method using N_2 as the adsorbent at -196℃ by an automatic apparatus(Nova2). Prior to gas adsorption measurement, the samples were outgassed at 300℃ in a vacuum condition for 12h to reach the degree of vacuum of around 10 -5 Torr. The pore volume distribution was evaluated by applying the DFT model[19,20]. Moreover, the XRD pattern of spent AC was obtained with a Rigaku diffractometer(D/max2500) using CuK_α radiation, and the microstructures were analyzed by the Scanning Electron Microscope (SEM, Philips XL30ESEM - TMP). Finally, an adsorption test of acetic acid onto the reactivated AC was carried out in a glycerine bath at 130℃ for 30min. Dried samples weighing 2.0000 ± 0.0002g were placed in a glass ware(diameter 30mm × 30mm), which in turn was placed inside a glycerine bath at 120℃. The temperature continued to rise to 130℃ and was held for 30min, then the glass ware was taken out and cooled - down in a desiccator until it reached room temperature. The adsorption capacity of acetic acid can be calculated by $(m_1 - m) \times 1000/m$, where m_1 is the weight of reactivated AC after adsorption test; and m is the original weight of spent AC used for the adsorption test.

3 Results and discussions

When exposed to the microwave, the temperature of the spent AC reached in the carbon bed depends on the nature of the AC(dielectric properties, chemical composition, particle size, moisture content, etc.)[21]. All samples employed in this study were uniformed with the same weight of 15g for comparison. When reactivating spent AC by steam activation, the activation time in CF was 90min, while the activation time in MW was 40min subjected to similar burn - off values. When the samples were reactivated with CO_2, the reactivation time in microwave energy(25min) was found to be 3 times shorter than that in the conventional furnace(100min). It is indicated that microwave heating is more efficient than the conventional furnace. Remarkably, the sample can rapidly reach the required temperature(1000℃) within 7 -9min from room temperature when heated by microwave energy, whereas nearly 25min were necessary for the samples treated in CF. This is probably because microwave energy is delivered directly to materials through molecular interaction in the electromagnetic field. The ability of conversion is based on the dielectric properties of materials. In contrast, with conventional thermal treatment, energy is transferred to materials through convection, conduction, and radiation of heat from the surface to inside[22]. Thus, the preferable energy transfer mechanism of microwave heating leads to homogeneous and rapid thermal reactions. It should be pointed out that activated carbon is an excellent microwave adsorptive material[23]. In general, only in a few minutes the temperature of activated carbon can rise from room temperature to 1000℃.

3.1 Characterization of the reactivated activated carbon

Fig. 3 shows the nitrogen adsorption isotherms of the spent AC treated in CF and MW as well as in the presence of the activating gas of steam and CO_2, respectively. According to IUPAC classifications, all nitrogen adsorption isotherms indicate that the shapes of these isotherms are intermediate between types Ⅰ and Ⅱ, pointing to microporous AC with a considerable development of mesoporosity[24]. The point about $p/p_0 = 0.1$ indicates that monolayer coverage was completed and multilayer adsorption begins on this stage. It means that the PSD was broad. This can be further testified by BET method. According to the definition from IUPAC, adsorbent pores are classified into three groups: micropore (size <2nm), mesopore (2 – 50nm), and macropore (>50nm)[25]. Micropores can be further divided into two subgroups: narrow – micropores (<0.7nm) and medium – micropores (width from 0.7nm to 2nm). The narrow – micropores, medium – micropores and mesopores volumewere evaluated by the DFT model applied to the nitrogen adsorption isotherms. The pore volume distribution pattern and the main textural parameters of the samples treated in MW and CF were shown in Fig. 4 and Table 2, respectively.

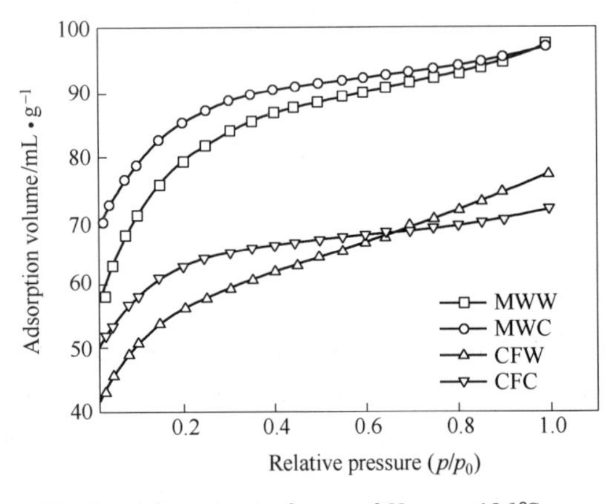

Fig. 3 Adsorption isotherms of N_2 at –196℃ on the samples reactivated under different conditions

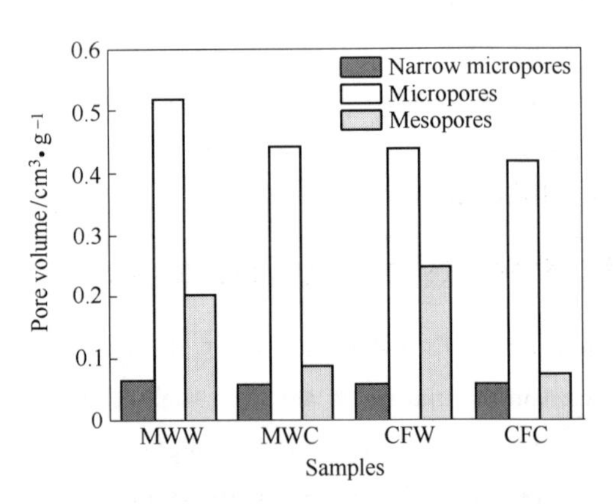

Fig. 4 Pore volume distributions of the samples evaluated by DFT model

Many experimental findings proved that the reactivation efficiencies of AC depend largely on the operating conditions especially temperature and activation atmosphere[26]. F. Rodriguez – Reinoso, et al. once reported that both CO_2 and steam initially develop the microporosity in the spent AC, but with different effects[27,28]. In this work, one can easily observe that steam activation produced a higher BET surface area than CO_2 activation with the same heating device (Table 2), attributing to the deeper burn – off degree under steam activation atmosphere.

In both devices, although it presents a different proportion in micro/meso distributions, the SBET and the mesoporosity of AC activated by steam is higher by about 24% and 57% than the samples reactivated under CO_2 atmosphere, respectively. Similarly, microporosity is higher in the samples reactivated in steam atmosphere than those by CO_2 activation. Mesoporosity may permit a better connection between the microporous pattern of the activated samples and the exterior of the carbon, permitting the retained molecules to evolve more rapidly from the carbon. The pore size distribution of the samples in the case of the steam activation illustrates that steam as an activation agent is more reactive than CO_2. This means that the samples under steam at mosphere activated deeper, which induce a larger loss of carbon. Thus, the burn - off of AC by steam activation is higher than that by CO_2 activation. Moreover, this phenomenon was reinforced in MW. For instance, the burns - off of MWC and CFC are 26% and 21%, respectively.

However, under the same atmosphere (steam or carbon dioxide), reactivation of the spent AC in MW is essentially microporous, which is the main contribution to activated carbon adsorption capacity. During the activating process, the increase of the micropores can contribute to the strong increase of the mesoporosity detected in the activated carbon. This is further confirmed by N_2 adsorption isotherm shown in Fig. 3. Nitrogen adsorption is a standard procedure for the determination of porosity of carbonaceous adsorbents. The adsorption isotherm includes the information about the porous structure of the adsorbent[29]. I. e., in Fig. 3 it is observed that reactivation by MW produced a higher N_2 adsorption than reactivation by CF with the same activation agent. Thus, the SBET of the samples reactivated in a microwave device is larger than those in conventional furnace (see Table 2). For example, the SBET of MWW is $1548m^2/g$ against $1308m^2/g$ in the CFW.

When reactivating spent AC using a different device in steam atmosphere, the adsorption capacity (Table 1) in the MW after reactivation was also significantly high. But in the case of the CF, the lack of a well developed mesoporosity delayed the migration of the desorbed molecules. Consequently, the formation of adsorbent deposits and damage to porosity was enhanced. For instance, after reactivation, the iodine number and the MB adsorption were 1194mg/g and 285mg/g in the MW, respectively, compared to 1181mg/g and 240mg/g in the conventional heating oven. But the time was found to be 44% shorter when the samples were reactivated by means of microwave energy (40min) compared to the conventional furnace (90min). The same results were gained under carbon dioxide atmosphere by a different device. This result might be due to the larger porous structure of the samples reactivated in the MW. Thus, it is inferred that microwave heating better retains the original porous framework of samples than CF treatment.

In order to evaluate the reactivation efficiency, the adsorption capacity of spent AC and raw AC were shown in Table 1. All the adsorption capacities of MWW, MWC, CFW and CFC were far exceeding the spent AC, but only the adsorption capacity of MWW was larger than the raw AC, attributing the cause to the higher BET surface area and total pore volume of reactivated AC in the MW using steam as an activating agent. Moreover, the PSD of the spent AC and raw AC (non exhausted) obtained by DFT analysis of nitrogen adsorption is otherm was shown in Fig. 5. As can be seen in Fig. 5, the raw AC had a predominant micropore content of about 59. 32%, and still had a small fraction of mesopore in the range 2 – 5nm. The sample reactivated in MW under steam atmosphere had more predominant microporosity by about 71.60% owing to renew the saturated micropores and produce newpores again in the process of reactivation.

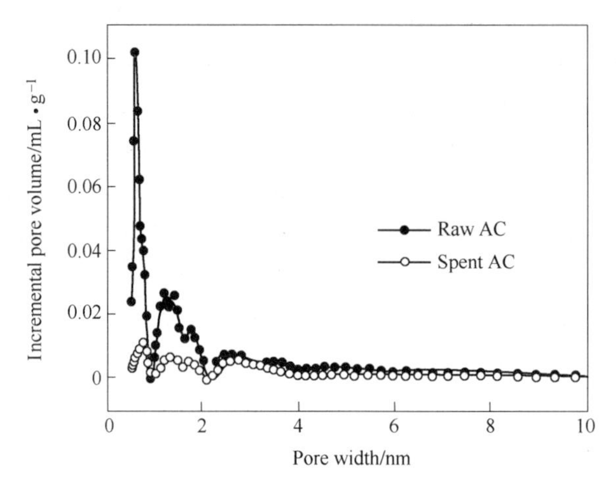

Fig. 5 Pore size distribution of raw activated carbon and spent activated carbon using DFT method

The results of the adsorption of acetic acid test were shown in Table 1, the adsorption capacity of acetic acid of reactivated AC was beyond the national standard of China (grade – A > 530mg/g, catalyst support activated carbon for vinyl acetate synthesis: GB/T 13803.5—1999) regardless of reactivation devices using steam as the activating agent. Reactivation of the samples by MW has higher adsorption capacity of acetic acid than raw AC ascribing to their PSD.

3.2 Microstructures analysis

The SEM of spent AC, MWW, MWC, CFW and CFC are shown in Fig. 6. Observing the SEM image

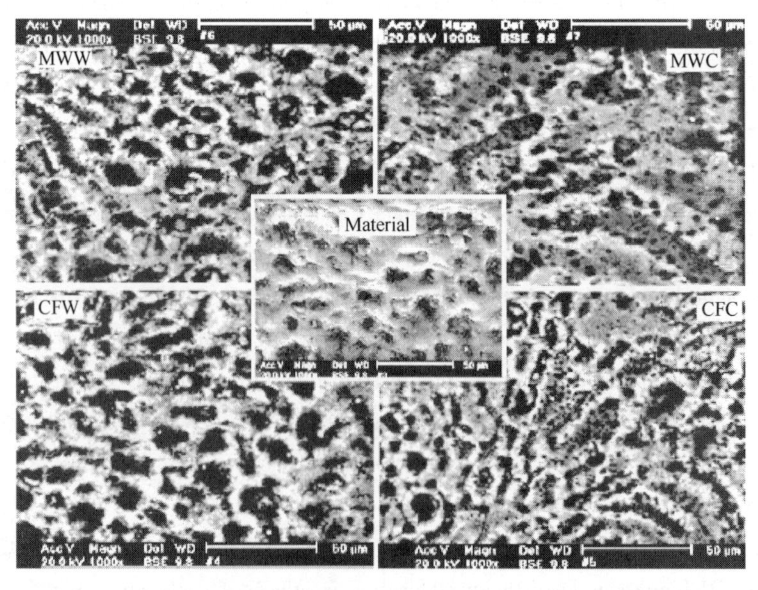

Fig. 6 SEM images of the spent AC and the samples reactivated under different conditions

of spent AC, we can find that the pores of spent activated carbon were still stuffed with some organic by – products, which may have remained inside the pore network of the spent AC and did not decompose until reaching a high temperature. In addition, the SEM photographs of MWW and MWC show more micropores than those of CFW and CFC.

4 Conclusions

By comparing microwave and conventional thermal, four reactivated technologies for multi – purpose utilization of spent catalyst support have been attempted. The following conclusions were obtained.

The results of BET and DFT analyses indicated that the reactivated AC had a microporous structure with a considerable mesoporous structure. Under the same activation atmosphere, microwave heating was found to better retain the original porous framework of samples. The SBET and the total pore volume were found to be much higher compared to samples treated in the conventional furnace. Moreover, experimental results showed that microwave heating could shorten the processing time remarkably, which implied a lower consumption of gas and energy. The microstructures of the AC characterized by SEM images further confirmed the results from the analysis of nitrogen adsorption isotherm.

After microwave treatment the iodine number, MB adsorption and acetic acid adsorption capacity were found to be much higher compared to those of the conventional furnace under the same activation atmosphere, owing to the changes induced in the textural properties and surface characterization of the reactivated samples. In this study, the method of reactivation of spent AC in microwave device under steam atmosphere has the most prospects towards industrialization. This is attributed mainly to the BET surface area and the total pore volume of MWW was 1.7 and 1.4 times of raw AC, respectively. The high performance activated carbon can be more widely applied. In the future work, it would be interesting to study the pilot – scale experiment of reactivation of spent AC of vinyl acetate synthesis.

Acknowledgements

The authors would like to express their gratitude for the financial support through the International Collaboration Project of Science and Technology of China (No. 2008DFA91500), the International Collaboration Project of Yunnan Provincial Science and Technology Department (No. 2006GH01) and the science and technology development program of Yunnan Environmental Protection Bureau.

References

[1] Liu X, Quan X, Bo L, et al. Temperature measurement of GAC and decomposition of PCP loaded on GAC and GAC – supported copper catalyst in microwave irradiation [J]. Applied Catalysis A: General, 2004, 264 (1): 53 – 58.

[2] Kang M, Bae Y S, Lee C H. Effect of heat treatment of activated carbon supports on the loading and activity of Pt catalyst [J]. Carbon, 2005, 43 (7): 1512 – 1516.

[3]Kim P,Kim H,Joo J B,et al. Preparation and application of nanoporous carbon templated by silica particle for use as a catalyst support for direct methanol fuel cell[J]. Journal of Power Sources,2005,145(2):139 - 146.

[4]Galvez M E,Boyano A,Lazaro M J,et al. A study of the mechanisms of NO reduction over vanadium loaded activated carbon catalysts[J]. Chemical Engineering Journal,2008,144(1):10 - 20.

[5]Dabek L. Sorption of zinc ions from aqueous solutions on regenerated activated carbons[J]. Journal of Hazardous Materials,2003,101(2):191 - 201.

[6]Li W,Peng J,Zhang L,et al. Pilot - scale extraction of zinc from the spent catalyst of vinyl acetate synthesis by microwave irradiation[J]. Hydrometallurgy,2008,92(1):79 - 85.

[7]Bagreev A,Rahman H,Bandosz T J. Thermal regeneration of a spent activated carbon previously used as hydrogen sulfide adsorbent[J]. Carbon,2001,39(9):1319 - 1326.

[8]Okoniewska E,Lach J,Kacprzak M,et al. The trial of regeneration of used impregnated activated carbons after manganese sorption[J]. Desalination,2008,223(1):256 - 263.

[9]Hamdaoui O,Naffrechoux E,Suptil J,et al. Ultrasonic desorption of p - chlorophenol from granular activated carbon[J]. Chemical Engineering Journal,2005,106(2):153 - 161.

[10]Zhang H. Regeneration of exhausted activated carbon by electrochemical method[J]. Chemical Engineering Journal,2002,85(1):81 - 85.

[11]Sabio E,Gonzalez E,Gonzalez J F,et al. Thermal regeneration of activated carbon saturated with p - nitrophenol[J]. Carbon,2004,42(11):2285 - 2293.

[12]Maroto - Valer M M,Dranca I,Clifford D,et al. Thermal regeneration of activated carbons saturated with ortho - and meta - chlorophenols[J]. Thermochimica Acta,2006,444(2):148 - 156.

[13]Ania C O,Parra J B,Menendez J A,et al. Effect of microwave and conventional regeneration on the microporous and mesoporous network and on the adsorptive capacity of activated carbons[J]. Microporous and Mesoporous Materials,2005,85(1):7 - 15.

[14]Jones D A,Lelyveld T P,Mavrofidis S D,et al. Microwave heating applications in environmental engineering—a review[J]. Resources,Conservation and Recycling,2002,34(2):75 - 90.

[15]Quan X,Liu X,Bo L,et al. Regeneration of acid orange 7 - exhausted granular activated carbons with microwave irradiation[J]. Water Research,2004,38(20):4484 - 4490.

[16]Ania C O,Menéndez J A,Parra J B,et al. Microwave - induced regeneration of activated carbons polluted with phenol. A comparison with conventional thermal regeneration[J]. Carbon,2004,42(7):1383 - 1387.

[17]Ania C O,Parra J B,Menendez J A,et al. Microwave - assisted regeneration of activated carbons loaded with pharmaceuticals[J]. Water Research,2007,41(15):3299 - 3306.

[18]Xia H Y,Peng J H,Liu X H,et al. Regeneration of catalyst support activated carbon for vinyl acetate synthesis with water vapor activation[J]. Chemical Engineering(China),2007,4:015.

[19]Valladares D L,Rodríguez Reinoso F,Zgrablich G. Characterization of active carbons:the influence of the method in the determination of the pore size distribution[J]. Carbon,1998,36(10):1491 - 1499.

[20]Ismadji S,Bhatia S K. A modified pore - filling isotherm for liquid - phase adsorption in activated carbon [J]. Langmuir,2001,17(5):1488 - 1498.

[21]Haque K E. Microwave energy for mineral treatment processes—a brief review[J]. International Journal of Mineral Processing,1999,57(1):1 - 24.

[22]Thostenson E T,Chou T W. Microwave processing:fundamentals and applications[J]. Composites Part A:Applied Science and Manufacturing,1999,30(9):1055 - 1071.

[23]Jou G. Application of activated carbon in a microwave radiation field to treat trichloroethylene[J]. Carbon,1998,36(11):1643 - 1648.

[24] Rouquerol J, Rouquerol F, Llewellyn P, et al. Adsorption by Powders and Porous Solids: Principles, Methodology and Applications[M]. Academic Press, 2013: 12.

[25] Sing K S W. Reporting physisorption data for gas/solid systems with special reference to the determination of surface area and porosity (provisional) [J]. Pure and Applied Chemistry, 1982, 54(11): 2201 – 2218.

[26] González J F, Encinar J M, González – García C M, et al. Preparation of activated carbons from used tyres by gasification with steam and carbon dioxide[J]. Applied Surface Science, 2006, 252(17): 5999 – 6004.

[27] Rodriguez – Reinoso F, Molina – Sabio M, Gonzalez M T. The use of steam and CO_2 as activating agents in the preparation of activated carbons[J]. Carbon, 1995, 33(1): 15 – 23.

[28] Molina – Sabio M, Gonzalez M T, Rodriguez – Reinoso F, et al. Effect of steam and carbon dioxide activation in the micropore size distribution of activated carbon[J]. Carbon, 1996, 34(4): 505 – 509.

[29] Ustinov E A, Do D D, Fenelonov V B. Pore size distribution analysis of activated carbons: Application of density functional theory using nongraphitized carbon black as a reference system [J]. Carbon, 2006, 44(4): 653 – 663.

Utilization of Crofton Weed for Preparation of Activated Carbon by Microwave Induced CO_2 Activation

Zhaoqiang Zheng, Hongying Xia, C. Srinivasakannan, Jinhui Peng, Libo Zhang

Abstract: Crofton weed was converted into a high - quality activated carbon (CWAC) via microwave - induced CO_2 physical activation. The operational variables including activation temperature, activation duration and CO_2 flow rate on the adsorption capability and activated carbon yield were identified. Additionally the surface characteristics of CWAC were characterized by nitrogen adsorption isotherms, FTIR and SEM. The operating variables were optimized utilizing the response surface methodology and were identified to be an activation temperature of 980℃, an activation duration of 90min and a CO_2 flow rate of 300mL/min with a iodine adsorption capacity of 972mg/g and yield of 18.03%. The key parameters that characterize quality of the porous carbon such as the BET surface area, total pore volume and average pore diameter were estimated to be 1036m^2/g, 0.71mL/g and 2.75nm, respectively. The findings strongly support the feasibility of microwave heating for preparation of high surface area porous carbon from crofton weed via CO_2 activation.

Keywords: crofton weed, activated carbon, microwave, response surface methodology

1 Introduction

Activated carbons (AC) are important kinds of porous carbon material with abundantly developed pore structure, strong adsorption ability, high surface area and thermo stability, they are widely used in many different industries, such as separation and purification processes, including both gas and aqueous media[1-3], catalyst supports[4] and removal of organic dyes and pollutants from industrial wastewater[5] and from other aqueous media[6]. The selection of an appropriate precursor plays an important role deciding the characteristics of the AC as well as the economics of the manufacturing plant. To identify new precursors that are cheap, accessible and available in large quantity has been a perennial challenge in commercial manufacture for economic benefits[7]. Towards which, different biomass based feedstock such as rice bran, coconut shell and waste materials were used as the raw materials since they are sustainable sources having high fixed carbon content[8-10].

Croton weed, a kind of global exotic weeds originated from Mexico, and which has spread extensively in many countries around the world such as America, Australia and the countries in Southeast Asia due to its strong ability to adapt to different environmental conditions[11]. Since 1940s, Croton weed has spread extensively in south and western of China. Lots of the farm lands, pasture fields and forests have been destroyed causing huge economic losses. This has drawn the attention of the society and many methods have been developed to control it, such as manual, chemical and biological control, no obvious progress is made. In 2003, the Chinese Ministry of Environmental Protection released a list of "The First Batch of Exotic Invasive Species" and croton weed was rated the first[12,13]. According to the published literatures, croton weed can be used as bio - pesticide[14], organic fertilizer and feedstuff, feedstock for production of marsh gas[15]. Although, croton weed can be utilized as a biomass resource to prepare the AC, the relevant literature is very limited. The attempts pertaining to preparation of CWAC has been limited to Xia, et al. and Wu, et al. [16,17].

Activated carbons have been traditionally produced by the partial gasification of the char either with steam or CO_2 or a combination of both. The gasification reaction results in removal of most reactive carbon atoms and in the process simultaneously produce a wide range of pores (predominantly micropores), resulting in porous activated carbon. In general, the methods for AC production are divided into two classes: physical activation and chemical activation. Physical activation is essentially a two - step process, where the carbonization of a carbonaceous material forms the first step, while the second step involves the activation of the resulting char at elevated temperature in the presence of suitable oxidizing gases such as carbon dioxide, steam, air or their mixtures. Chemical activation involves the impregnation of a carbonaceous material with an activation agent and heat treatment of the impregnated material under inert atmosphere. Physical activation is widely adopted industrially for commercial production owing to the simplicity of process and the ability to produce AC with well developed micro porosity and desirable physical characteristics such as the good physical strength.

Referring to the heating methods for the preparing of AC, interests are growing in the application of microwave (MW) heating. The conventional heating methods do not ensure a uniform temperature of the precursor owing to their variation in the size and shape as the mode of heating is through conduction and convection. This conventional heating mode generates a temperature gradient from the hot surface of the sample particle to its interior and impedes the effective removal of gaseous products to its surroundings, demanding higher processing time and energy consumption. Recently, microwave heating is being increasingly utilized for variety of applications, as heating is uniform, where the absorbed microwave readily transforms into heat inside the particles by dipole rotation and ionic conduction[18-20]. Recently MW heating has been widely used to produces as well as to regenerate AC, the relevant literature is very limited[21-23]. There is no study regarding AC prepared from the Crofton weed with MW heating in the presence of physical activation.

This urged research towards upgrading and utilization of the harmful biomass crofton weed. In this regard, the objective of this work is to evaluate the operational conditions for improving the porosity and adsorption capacity of CWAC using MW heating. Effects of the activation temperature, activation duration and CO_2 rate on the adsorption capacity and yield of CWAC were investigated systematically. The resultant products were characterized using the nitrogen adsorption isotherm, FTIR and SEM analysis.

2 Materials and methods

2.1 Materials

Crofton weed were collected from Kunming, Yunnan Province of China. The raw materials were crushed, sieved into a uniform size of 5 – 7mm, then were washed thoroughly with distilled water to remove foreign material and then oven – dried at 105℃ overnight and stored in a moisture free environment for utilization in the experiment. The proximate analyses of the Crofton weed were as follow: volatile 76.41%, ash 1.90% and fixed carbon 21.69%.

2.2 Carbonization of crofton weed

The carbonization of raw precursors was carried out by loading 1000g into a muffle furnace, under the nitrogen flow atmosphere (100mL/min) and heating in to a carbonization temperature of 500℃ at a heating rate of 20℃/min and was held for 1.5h at the carbonization temperature. The yield of char was found to be around 35% and the proximate analyses of the char were as follow: volatile 15.84%, ash 8.86% and fixed carbon 75.30%. The carbon content of the char was found to have increased significantly upon carbonization.

2.3 MW heating system and preparation of CWAC

A self – made MW tube furnace, which was employed in the activation experiments, was shown in Fig. 1. It consist of a microwave power control system (with the output power of 0 – 3kW and the microwave frequency of 2.45GHz), the temperature control system (temperature controlled by the input microwave power during the activation process, measured by the thermoelement type of nichrome – nickel silicon armor, placed such that it touched raw, with a measurement precision of ±0.5℃), and flow control system (controlled by the flow meter with CO_2 and N_2). A known amount of the carbonized materials were placed in the platform of quartz reactor fixed in the MW tube furnace. Prior to use, N_2 was used to purge any air in the furnace at a pre – set flow rate of 100mL/min. Subsequently, the materials were heated from the room temperature to the desired temperature (850 – 1000℃). The desired temperature could be achieved in duration of approximately 10 – 12min that correspond to heating rate of 120℃/min, which was extremely faster than the conventional heating. Once the carbonized materials reached the desired temperature the N_2 flow was terminated and replaced by CO_2 into the reactor at the desired flow rate for desired activation duration.

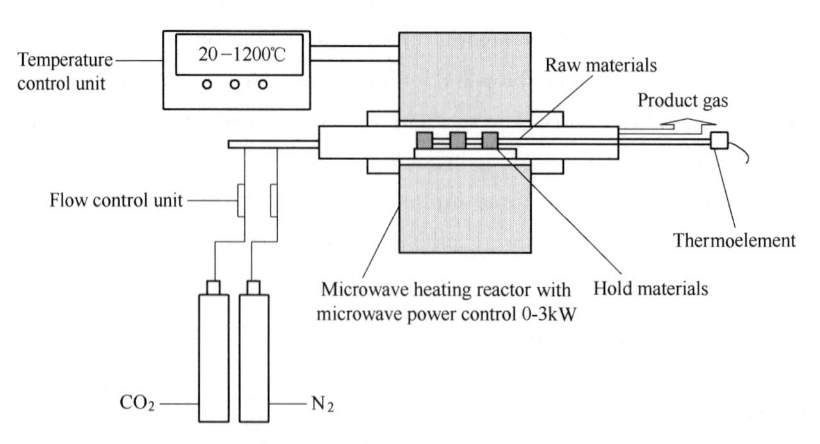

Fig. 1 The diagram of apparatus for the preparation of microwave heating equipment

The completion of activation process was marked by termination of the CO_2 supply and starting with N_2 flow until the CWAC were cooled to the room temperature, which were then dried at 105℃ and stored for further characterization.

2.4 Measurements of iodine adsorption capacity and yield

The iodine adsorption capacity is represented as iodine number, which is important parameter widely used to characterize micro porous material. Iodine number was tested for the CWAC according to the National Standard Testing Methods of P. R. China (GB/T 12496.8—1999). The pH measurements were carried out by the method of the National Standard Testing Methods of P. R. China (GB/T 12496.7—1999). The CWAC yield was defined as the ratio of weight of activated carbon to the weight of carbonized materials utilized for activation.

2.5 Experimental design

RSM was utilized to optimize the activation process as it is a popular statistical tool for modeling and analysis of multi parameter processes, and central composite design (CCD) was employed to design the activation experiments[24]. In this study, the effects of three independent variables, X_1 (activation temperature), X_2 (activation duration), X_3 (CO_2 flow rate), at five levels were investigated (Table 1). The iodine number and yield of CWAC were taken as the two responses of the designed experiments. A total of 20 experiments consisting of 8 factorial points, 6 axial points and 6 replicates at the central points were employed. The experimental design matrix is provided in Table 2. The experimental data were analyzed using the Design Expert software version 7.1.5 (Stat – Ease Inc., Minneapilis, USA).

Table 1 Independent variables and their levels for central composite design

Independent variables	Symbol	Coded variable levels				
		-1.68179	-1	0	+1	+1.68179
activation temperature/℃	X_1	899.55	920	950	980	100.45
activation duration/min	X_2	9.55	30	60	90	110.45
CO_2 flow rate/mL · min^{-1}	X_3	131.82	200	300	400	468.18

2.6 Characterization of activated carbon

The specific surface areas of CWAC produced by the optimum conditions as well as the carbonized char were determined by the BET method using nitrogen as the adsorbate at 77K utilizing the commonly used Autosorb instrument. Prior to the measurements, the samples were out gassed at 300℃ under nitrogen for at least 12h. Nitrogen adsorption isotherms of CWAC and char were tested over a relative pressure (p/p_0) range from 10^{-7} to 1. The total pore volumes were estimated to be the equivalent liquid volume of the adsorbate (N_2) at a relative pressure of 0.99. The BET surface of the samples was calculated by the Brunauer – Emmett – Teller (BET) equation. The micropore size distribution was analyzed by using the Non – local Density Functional Theory(NLDFT). Scanning electron microscopy(SEM, Philips XL30ESEM – TMP) analysis was carried out assess the surface morphology.

The Fourier transform infrared spectroscopy (FTIR) was applied to qualitatively identify the chemical function groups present in the CWAC. FTIR spectra were operating in the range of 4000 – 400cm^{-1} by using AVATAR 330(Thermo Nicolet CO., USA) Spectrophotometer. The transmission spectra of the samples were prepared by mixing up with KBr crystals and pressed into a pellet.

3 Results and discussion

Table 2 shows the experimental conditions for preparation of CWAC generated by the Design Expert software covering the parameters such as activation temperature, activation duration and CO_2 flow rate. The CWAC are characterized for iodine number, yield and results are listed as well in Table 2.

Table 2 Experimental design matrix and results by microwave heating with CO_2

Run	X_1/℃	X_2/min	X_3/mL · min^{-1}	Y_1/mg · g^{-1}	Y_2/%
1	920.00	30.00	200.00	729	66.83
2	980.00	30.00	200.00	818	50.36
3	920.00	90.00	200.00	833	39.35
4	980.00	90.00	200.00	890	30.42
5	920.00	30.00	400.00	801	49.95
6	980.00	30.00	400.00	905	26.16
7	920.00	90.00	400.00	879	22.55
8	980.00	90.00	400.00	989	16.41
9	899.55	60.00	300.00	789	50.88
10	1000.45	60.00	300.00	929	18.67
11	950.00	9.55	300.00	701	58.19
12	950.00	110.45	300.00	933	19.45
13	950.00	60.00	131.82	765	53.56
14	950.00	60.00	468.18	946	17.52
15	950.00	60.00	300.00	900	26.35
16	950.00	60.00	300.00	910	26.27
17	950.00	60.00	300.00	908	26.28
18	950.00	60.00	300.00	905	26.31
19	950.00	60.00	300.00	901	26.34
20	950.00	60.00	300.00	907	26.26

3.1 Response analysis and verification of the regression model

The popular design method, Central composite design (CCD) was used to develop correlation between the dependent and independent variables and to identify the significant factors contribution to the regression model. The independent variables selected were activation temperature (X_1), activation duration (X_2), CO_2 flow rate (X_3), at five levels were investigated by central composite design (Table 1), while the dependent variables are iodine number (Y_1) and yield (Y_2). As can be seen from Table 2, the iodine number was found to range from 701mg/g to 989mg/g, while the yield of the activation yield found range from 17.52% to 66.83%. Experimental runs performed at the center point of all the variables (15 – 20runs) were utilized to determine the experimental error. For iodine number and yield, the two – factor interaction model was selected as suggested by the software. The final empirical models in terms of coded factors (excluding the insignificant terms) for iodine number (Y_1) and yield (Y_2) are shown in Eqs. (1) and (2), respectively:

$$Y_1 = +904.72 + 43.60X_1 + 52.21X_1 + 44.55X_3 - 13.41X_1^2 - 26.67X_2^2 - 14.65X_3^2 \quad (1)$$

$$Y_2 = +26.28 - 8.02X_1 - 10.96X_2 - 9.7X_3 + 3.15X_1X_2 + 3.16X_1^2 + 4.59X_2^2 + 3.43X_3^2 \quad (2)$$

The quality of the model developed was always evaluated using the correlation coefficient values (R^2), which were 0.9663 for Eq. (1) and 0.9886 for Eq. (2), respectively. Both the R^2 values were of high proximity to unity, indicating the suitability of the model equation, evidencing good agreement between experimental data and the prediction of iodine number and yield using the model[25]. The results obtained from the analysis of variance (ANOVA) were also carried out to prove the validity of the model. Validating the model adequacy is an import part of the data analysis, since it would lead to poor or misleading results if it is an inadequate fit. The ANOVA for the quadratic model of iodine number was presented in Table 3. The model F – value of 31.87 implied the model was significant. There was only a 0.01% chance that a "Model F – Value" of this large could occur due to noise. Values of "Prob > F" less than 0.05 indicated that the model terms were significant. In this case X_1, X_2, X_3 and the interaction terms (X_1^2, X_2^2, X_3^2) were found to be significant model terms based on the low "P" values. Values greater than 0.1 indicated the model terms were not significant. Adequate precision measures of the signal to noise ratio. A ratio greater than 4 is desirable, which was observed to be 20.78 for the present experiments.

Table 3 Analysis of variance for the iodine number

Source	Sum of squares	Degree of freedom	Mean square	F – value	Prob > F
Model	104700	9	11633.63	31.87	<0.0001
X_1	25962.19	1	25962.19	71.13	<0.0001
X_2	37228.61	1	37228.61	103.00	<0.0001
X_3	27105.05	1	27105.05	74.26	<0.0001
X_1X_2	84.50	1	84.50	0.23	0.6408
X_1X_3	578.00	1	578.00	1.58	0.2368

Continues Table 3

Source	Sum of squares	Degree of freedom	Mean square	F - value	Prob > F
X_2X_3	24.50	1	24.50	0.067	0.8008
X_1^2	25911.96	1	25911.96	7.10	0.0237
X_2^2	10250.06	1	10250.06	28.08	0.0003
X_3^2	3092.35	1	3092.35	8.47	0.0155

Note: R^2 = 0.9663; R^2_{adj} = 0.9360; Adeq Precision = 20.78 (> 4).

The ANOVA result of quadratic model corresponding to yield was shown in Table 4. A model F - value of 96.34 and $P > F$ of < 0.0001 indicated the suitability of model. In this case, X_1, X_2, X_3 along with the interaction parameters (X_1X_2, X_1^2, X_2^2, X_3^2) were found to be significant. In conclusion, from the ANOVA results obtained, it was shown that the model was appropriateness to be utilized within the range of the variables covered in the present work.

Table 4 Analysis of variance for the yield

Source	Sum of squares	Degree of freedom	Mean square	F - value	Prob > F
model	4418.72	9	490.97	96.34	< 0.0001
X_1	877.97	1	877.97	172.29	< 0.0001
X_2	1641.44	1	1641.44	322.11	< 0.0001
X_3	1285.56	1	1285.56	252.27	< 0.0001
X_1X_2	79.32	1	79.32	15.56	0.0028
X_1X_3	2.57	1	2.57	0.50	0.4942
X_2X_3	13.18	1	13.18	2.59	0.1388
X_1^2	143.64	1	143.64	28.19	0.0003
X_2^2	303.24	1	303.24	59.51	< 0.0001
X_3^2	169.30	1	169.30	33.22	0.0002

Note: R^2 = 0.9886; R^2_{adj} = 0.9783; Adeq Precision = 35.939 (> 4).

3.2 Iodine adsorption capacity of CWAC

Fig. 2 shows the comparison of predicted versus the experimental data for iodine number of CWAC. Experimental values were the measured response data for a particular run while the predicted values were evaluated using the model Eq. (1)[26]. As seen in the figure, the experimental data were evenly distributed on the both sides of the model prediction, indicating the suitability of the model developed in capturing the correlation between the process and response variables. Based on the F values (Table 3), activation duration sowed the highest of F Value (103.0) indicating it to be the most significant parameters as compared to activation temperature and CO_2 flow rate.

Three - dimensional response surfaces were created to show the effects of the independent variables on iodine number were shown in Figs. 3 and 4. Fig. 3 shows the combined effect of activation temperature and activation duration on the iodine number (CO_2 flow rate is held constant at 300mL/min) while Fig. 4 shows the combined effect of activation temperature and CO_2 flow rate

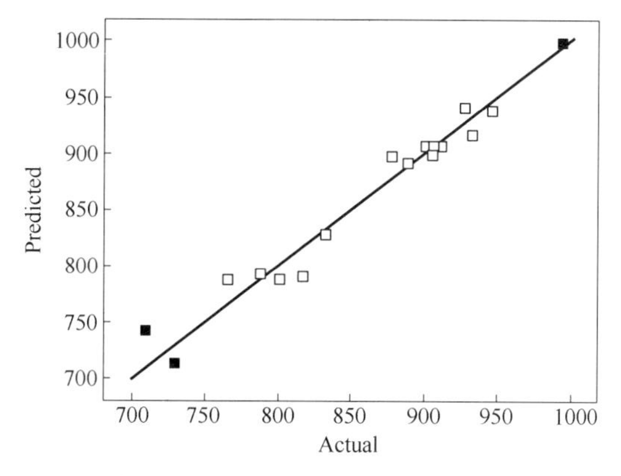

Fig. 2 Predicted vs. experimental iodine number of CWAC

(activation duration was fixed at 60 min). As can be seen, the iodine number gradually increases with the increase in activation temperature, activation duration and CO_2 flow rate, with the maximum iodine number of 989mg/g, corresponding to the maximum of all the three major factors within the range studied. However the rate of increase tends to slow down at high values reaching an asymptote, indicating the possible maximum.

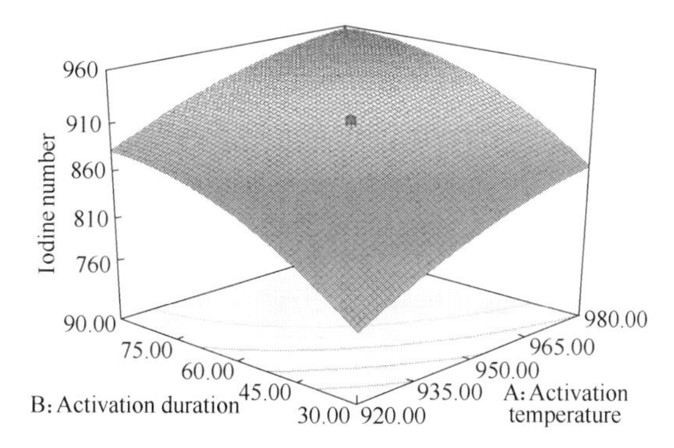

Fig. 3 Three - dimensional response surface plot of iodine number: effect of activation duration and activation temperature on iodine number (carbon dioxide flow: 300mL/min)

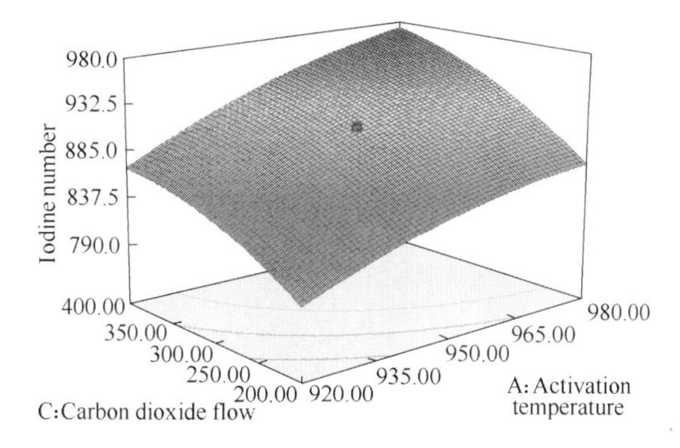

Fig. 4 Three - dimensional response surface plot of iodine number: effect of carbon dioxide flow and activation temperature on iodine number (activation duration: 60min)

It is well known that the adsorption depends on the pore structure. It is reported that the pore diameter should be at least 1.7 times of the molecular widest dimension in order to be good adsorption site to capture a molecule[27]. The iodine molecule is strongly adsorbed due to its smaller size of 0.27nm permitting its penetration into micropores under the size of 1nm[28]. The development of porosity was associated with gasification according to the reaction:

$$C + CO_2 \longrightarrow 2CO \quad (3)$$

The carbon dioxide would diffuse into the internal structure of char matrix, react with active sites in the carbon molecule, eventually developing a pore. The continuation of the reaction would generate new pores as well as widen the existing pores. The size and shape of the pores are bound to vary with the variation in the process conditions. The extent of the reaction would increase with an increase in the activation temperature, activation duration and CO_2 flow rate. The increase in porosity is expected to be directly proportional to the increase in the carbon conversion. The increase in the micropores can be assessed based on the increase in the iodine adsorption capacity of the carbon. The effect of activation temperature along with the activation duration and carbon dioxide flow on iodine number of CWAC is shown in Figs. 3 and 4. The rate of increase in iodine number is higher at the low temperature and activation duration while the rate of increase was found to be lower at higher temperature and duration, indicating the possible maximum being reached within the range of temperature and duration covered in the present study. Table 5 lists the comparison of maximum iodine number of various AC derived from different precursors reported in the literature. As can be seen, the CWAC prepared in the present work has relatively high iodine number as compared with the values reported in literature.

Table 5 Comparison iodine adsorption of AC prepared from different biomass

Precursors	Heating method	Activating agent	Activation duration/min	Iodine adsorption /$mg \cdot g^{-1}$	Reference
Crofton weed	Microwave heating	CO_2	90	989	Present study
Edible fungi residue	Microwave heating	K_2CO_3	16	732	[29]
Oil sands coke	Conventional heating	CO_2	360	670	[30]
Polyethyleneterephthalate wastes	Conventional heating	CO_2	240	630	[31]
Rice bran	Conventional heating	Steam	90	220	[8]
Paper mill sewage sludge	Conventional heating	Steam	40	180	[32]
Olive - waste cakes	Conventional heating	H_3PO_4	120	583	[33]

3.3 Yield of CWAC

The model prediction against the actual experimental data on the yield of CWAC is shown in Fig. 5. As can be seen, the experimental data are evenly distributed on the both sides of the model prediction, and the predicted value are more close to the actual values than Fig. 2. This was due to the high R^2 and low standard deviation values of the model equation as compared with Eq. (1) for iodine number. Based on the F values (Table 4), X_2 was found to have the highest of F value of 322.11, implying its significant effect on yield as compared to X_1 and X_3.

Table 2 compiles the yield of CWAC for each experiment. Three-dimensional response surfaces plot of yield with respect to the activation temperature and activation duration is shown in Fig. 5, while Fig. 6 shows the effect of activation temperature and CO_2 flow rate on the yield of CWAC. Figs. 5 and 6 were generated with the other parameter held at the center point, typically at a CO_2 flow rate of 300mL/min and a activation duration of 60min. From both the figures, the yield of CWAC is found to decrease with the increase in all the three parameters. Also the results indicate the minimum of yield corresponds to the maximum of the three parameters. An increase in any of the three parameters effectively contributes to an increase in the extent of C - CO_2 reaction, contributing to the reduction in the yield of CWAC.

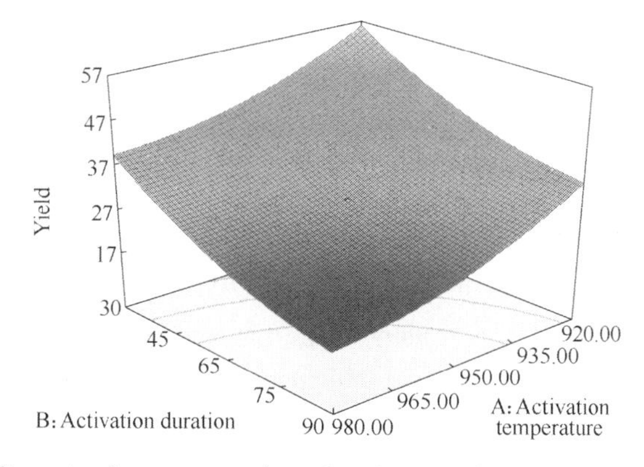

Fig. 5 Three-dimensional response surface plot of activated carbon yield: effect of activation temperature and activation duration on yield (carbon dioxide flow: 300mL/min)

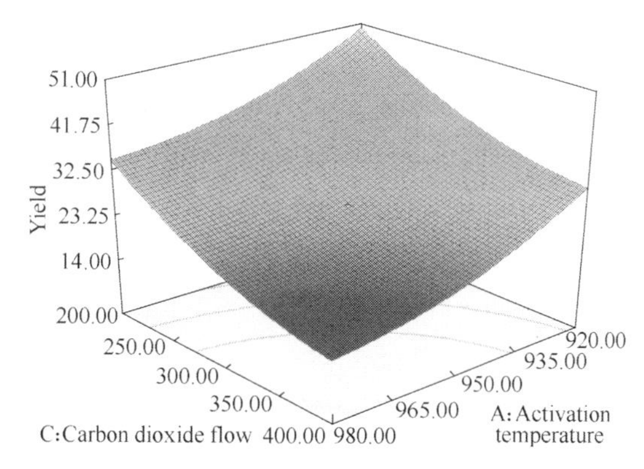

Fig. 6 Three-dimensional response surface plot of activated carbon yield: effect of stem flow rate and activation temperature on yield (activation duration: 60min)

3.4 Process optimization

Industrial production of AC augur a high iodine number and yield as both of which contribute to improve the economics of commercial manufacture. However, both the response variables the iodine adsorption capacity(Y_1) and yield(Y_2) respond opposite to each other, demanding identification of an optimum combination of the parameters that maximize the iodine number and the yield of carbon. The optimum condition was identified by invoking the optimization tool available with the Design Expert Software. The optimized process conditions along with the validation of the optimized process conditions are presented in Table 6. In order to authenticate the identified optimized experimental conditions, three repeat runs were conducted and the average of the three runs is posted in Table 6. The proximity in iodine number and the yield between the repeat runs and the optimized conditions validate the success of the optimization process.

Table 6 Validation of Process Optimization

Activation temperature, X_1/℃	Activation duration, X_2/min	Carbon dioxide flow, X_3/mL · min^{-1}	Iodine number/mg · g^{-1}		Yield/%	
			Predicted	Experimental	Predicted	Experimental
980.00	90.00	300	957	972	18.20	18.03

3.5 Characterizations of CWAC

Nitrogen adsorption is a standard procedure for determination of porosity of the carbonaceous adsorbents. The nitrogen adsorption isotherm of char and CWAC under the optimum condition estimated using the Autosorb instrument at 77K is shown Fig. 7. It was found that these isotherms present Ⅳ hybrid shapes as defined by the IUPAC classification[34]. Fig. 7 compares the nitrogen adsorption capacity of the char with respect to the CWAC. As can be seen, the nitrogen adsorption isotherm of CWAC is exceedingly higher than that of char, clearly indicating the higher amount of pores in CWAC. A continued increase in the nitrogen adsorption capacity beyond a p/p_0 value of 0.1, are typical for mesoporous material, which is exhibited by the char as well as the CWAC. Microporous materials don't exhibit an increase in adsorption capacity beyond a p/p_0 value of 0.1. The cumulative pore volume plots shown in Fig. 8, while Fig. 9 shows the pore size distribution, both of which substantiate the amount of pores in the mesoporous range, with the average pore diameter estimated to be 2.75nm.

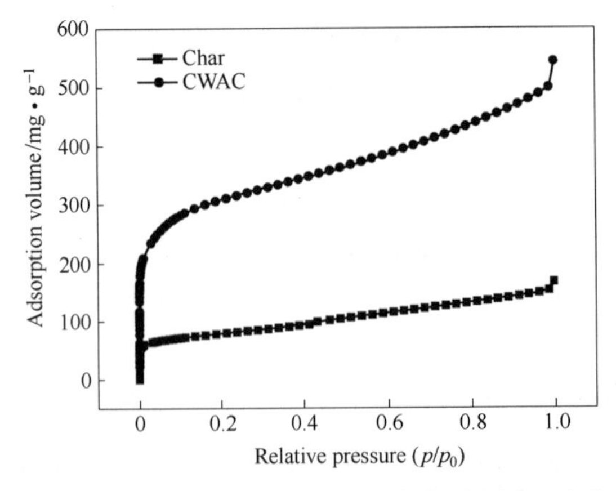

Fig. 7 Nitrogen adsorption isotherm of the CWAC and char

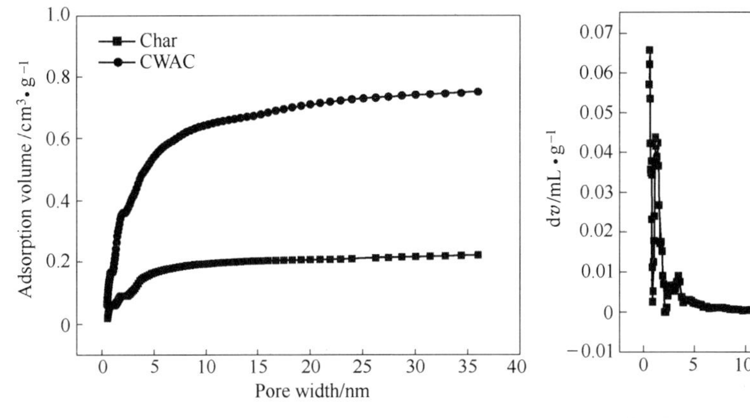

Fig. 8 Cumulative pore volume distribution chart for CWAC and char

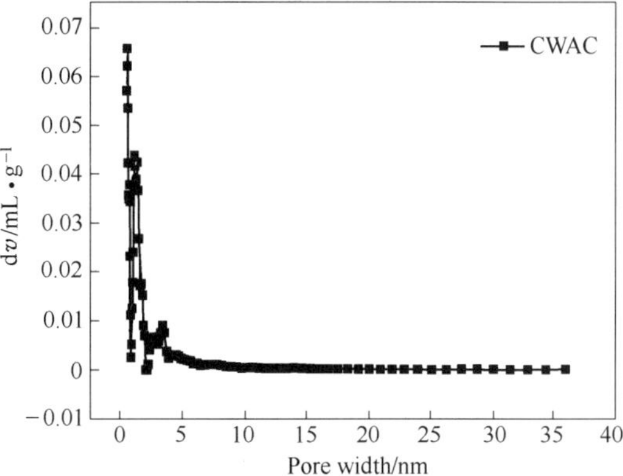

Fig. 9 Pore size distribution chart for CWAC

The pore structural parameters are summarized in Table 7. A comparison of the quality of CWAC with the char exhibits a significant increase in the pore volume, the micropore volume and the surface area attributed to the activation process. Table 8 shows the pH, chemical composition and ash composition analyses of CWAC. It is clear from Table 8, that CWAC has a neutral pH, high carbon value and low sulphur content. The main components of ash are silicon dioxide, calcium oxide, aluminum oxide, magnesium oxide and iron oxide.

Table 7 Pore structural parameters of CWAC vs. chars

Properties		CWAC	Chars
Pore volume	mL/g	0.71	0.23
Average pore diameter	nm	2.75	3.6
Micropore volume	%	35	27.1
Mesopore volume	%	65	72.9
BET surface area	m^2/g	1036	260

Table 8 pH, chemical composition and ash composition analyses of CWAC

pH	7.02								
Chemical Composition(wt.)/%									
content	C	O	H	N	S				
value	79.54	15.25	3.31	1.82	0.08				
Ash Composition(wt.)/%									
content	Fe_2O_3	Al_2O_3	CaO	TiO_2	SiO_2	MgO	Na_2O	CuO	K_2O
value	3.02	9.54	27.14	0.52	31.51	4.62	2.38	0.23	1.38

3.6 SEM analysis of microstructure

The microscopic structure of the crofton weed char(before activation) and the CWAC are shown in Fig. 10, Fig. 10(a) shows the SEM microstructure of char while Fig. 10(b) shows the SEM microstructure of CWAC. We can found the surface of the precursor was devoid of any tangible pores since it is covered by impurities. However, as shown in Fig. 10(b), the surface of CWAC has large number of pores of irregular and heterogeneous morphology, which attests a significant development of pore structure. A comparison of the microstructure of CWAC with the char indicates that the activation process plays an important role in removing surface impurities and contributing to pore - formation.

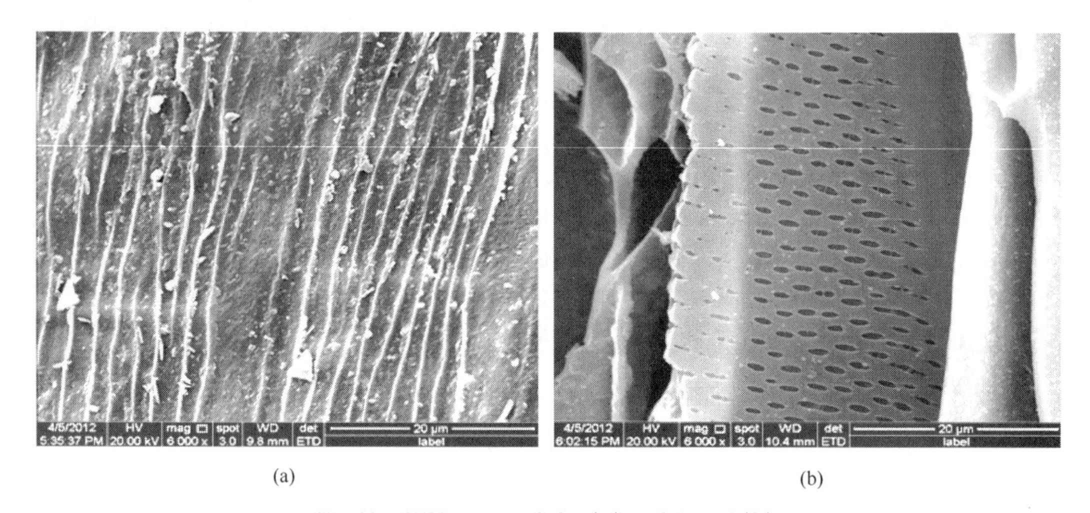

(a) (b)

Fig. 10 SEM images of char(a) and CWAC(b)

4 Conclusions

Crofton weed, a harmful biomass is utilized for preparing AC with microwave heating exhibit well developed pore structure. The effects of three vital process parameters, activation temperature, activation duration and CO_2 flow rate on the adsorption capacity and yield of AC were investigated systematically. The process parameters were optimized utilizing the Design Expert Software and were identified to be an activation duration of 90min, an activation temperature of 980℃ and a CO_2 flow rate of 300mL/min, with the resultant iodine number and yield being 972mg/g and 18.03% respectively. The key parameters that characterize quality of the porous carbon such as the BET surface area, total pore volume and average pore diameter were estimated to be $1036m^2/g$, 0.71mL/g and 2.75nm respectively, for the sample corresponding to the optimized process conditions. Additionally the surface chemistry groups and pore structure is characterized using FTIR and SEM, respectively.

Acknowledgements

The authors would like to express their gratitude to the Specialized Research Fund for the Doctoral

Program of Higher Education of China(No. 20115314120014) and the Kunming University of Science and Technology Personnel Training Fund(No. KKSY201252077) for financial support.

References

[1]Li Y,Du Q,Liu T,et al. Preparation of activated carbon from Enteromorpha prolifera and its use on cationic red X - GRL removal[J]. Applied Surface Science,2011,257(24):10621 - 10627.

[2]Nieto - Delgado C,Terrones M,Rangel - Mendez J R. Development of highly microporous activated carbon from the alcoholic beverage industry organic by - products[J]. Biomass and Bioenergy,2011,35(1):103 - 112.

[3]Ahmadpour A,Do D D,The preparation of active carbons from coal by chemical and physical activation [J]. Carbon. ,1996(34):471 - 479.

[4]Mudoga H L,Yucel H,Kincal N S. Decolorization of sugar syrups using commercial and sugar beet pulp based activated carbons[J]. Bioresource Technology,2008,99(9):3528 - 3533.

[5]Hameed B H,Ahmad A L,Latiff K N A. Adsorption of basic dye(methylene blue) onto activated carbon prepared from rattan sawdust[J]. Dyes and Pigments,2007,75(1):143 - 149.

[6]Hejazifar M,Azizian S,Sarikhani H,et al. Microwave assisted preparation of efficient activated carbon from grapevine rhytidome for the removal of methyl violet from aqueous solution[J]. Journal of Analytical and Applied Pyrolysis,2011,92(1):258 - 266.

[7]Duan X H, Srinivasakannan C, Qu W W, et al. Regeneration of microwave assisted spent activated carbon: process optimization,adsorption isotherms and kinetics[J]. Chemical Engineering and Processing: Process Intensification,2012,53:53 - 62.

[8]Suzuki R M,Andrade A D,Sousa J C,et al. Preparation and characterization of activated carbon from rice bran [J]. Bioresource Technology,2007,98(10):1985 - 1991.

[9]Su W,Zhou L,Zhou Y. Preparation of microporous activated carbon from raw coconut shell by two - step procedure[J]. Chinese Journal of Chemical Engineering,2006,14(2):266 - 269.

[10]Dias J M,Alvim - Ferraz M,Almeida M F,et al. Waste materials for activated carbon preparation and its use in aqueous - phase treatment: a review[J]. Journal of Environmental Management,2007,85(4):833 - 846.

[11]Sang W,Zhu L,Axmacher J C. Invasion pattern of Eupatorium adenophorum spreng in southern China [J]. Biological Invasions,2010,12(6):1721 - 1730.

[12]Weyerstahl P,Marschall H,Seelmann I,et al. Constituents of the flower essential oil of Ageratina adenophora (Spreng.) K. et R. from India[J]. Flavour and Fragrance Journal,1997,12(6):387 - 396.

[13]Guo S,Li W,Zhang L,et al. Kinetics and equilibrium adsorption study of lead(Ⅱ) onto the low cost adsorbent—Eupatorium adenophorum spreng [J] . Process Safety and Environmental Protection, 2009, 87 (5):343 - 351.

[14]Sahoo A,Singh B,Sharma O P. Evaluation of feeding value of Eupatorium adenophorum in combination with mulberry leaves[J]. Livestock Science,2011,136(2):175 - 183.

[15]Madan S P M. An alternative resource for biogas production[J]. Energy Sources,2000,22(8):713 - 721.

[16]Xia H,Peng J,Yang K. Study on the preparation of activated carbon from Eupatorium adenophorum spreng by microwave radiation[J]. Ion Exchange and Adsorption,2008,24(1):16.

[17]Wu C H,Qin Y J,Zhang J Y,Wu W,Xu H B. Preparation for activated carbon from the stem of Eupatorium adenophorum spreng by microwave radiation[J]. Journal of Fujian Agriculture and Forestry University(Natural Science Edition),2009,38(4):428 - 430.

[18]Yuen F K,Hameed B H. Recent developments in the preparation and regeneration of activated carbons by mi-

crowaves[J]. Advances in Colloid and Interface Science,2009,149(1):19 - 27.

[19] Yagmur E, Ozmak M, Aktas Z. A novel method for production of activated carbon from waste tea by chemical activation with microwave energy[J]. Fuel,2008,87(15):3278 - 3285.

[20] Venkatesh M S, Raghavan G S V. An overview of microwave processing and dielectric properties of agri - food materials[J]. Biosystems Engineering,2004,88(1):1 - 18.

[21] Williams H M, Parkes G. Activation of a phenolic resin - derived carbon in air using microwave thermogravimetry[J]. Carbon,2008,46(8):1169 - 1172.

[22] Coss P M, Cha C Y. Microwave regeneration of activated carbon used for removal of solvents from vented air [J]. Journal of the Air & Waste Management Association,2000,50(4):529 - 535.

[23] Duan X H, Srinivasakannan C, Peng J H, et al. Preparation of activated carbon from Jatropha hull with microwave heating: optimization using response surface methodology[J]. Fuel Processing Technology,2011,92(3): 394 - 400.

[24] Azargohar R, Dalai A K. Production of activated carbon from Luscar char: experimental and modeling studies [J]. Microporous and Mesoporous Materials,2005,85(3):219 - 225.

[25] Novak N, Majcen L M A, Bogataj M. Determination of cost optimal operating conditions for decoloration and mineralization of CI Reactive Blue 268 by UV/H_2O_2 process[J]. Chemical Engineering Journal,2009,151 (1):209 - 219.

[26] Körbahti B K, Rauf M A. Response surface methodology(RSM) analysis of photoinduced decoloration of toludine blue[J]. Chemical Engineering Journal,2008,136(1):25 - 30.

[27] Kasaoka S, Sakata Y, Tanaka E, et al. Design of molecular - sieve carbon. Studies on the adsorption of various dyes in the liquid phase[J]. Int. Chem. Eng,1989,29(4):734 - 742.

[28] Baçaoui A, Yaacoubi A, Dahbi A, et al. Optimization of conditions for the preparation of activated carbons from olive - waste cakes[J]. Carbon,2001,39(3):425 - 432.

[29] Xiao H, Peng H, Deng S, et al. Preparation of activated carbon from edible fungi residue by microwave assisted K_2CO_3 activation——application in reactive black 5 adsorption from aqueous solution[J]. Bioresource Technology,2012,111:127 - 133.

[30] Small C C, Hashisho Z, Ulrich A C. Preparation and characterization of activated carbon from oil sands coke [J]. Fuel,2012,92(1):69 - 76.

[31] Esfandiari A, Kaghazchi T, Soleimani M. Preparation and evaluation of activated carbons obtained by physical activation of polyethyleneterephthalate(PET) wastes[J]. Journal of the Taiwan Institute of Chemical Engineers,2012,43(4):631 - 637.

[32] Li W H, Yue Q Y, Gao B Y, et al. Preparation of sludge - based activated carbon made from paper mill sewage sludge by steam activation for dye wastewater treatment[J]. Desalination,2011,278(1):179 - 185.

[33] Baccar R, Bouzid J, Feki M, et al. Preparation of activated carbon from Tunisian olive - waste cakes and its application for adsorption of heavy metal ions[J]. Journal of Hazardous Materials,2009,162(2):1522 - 1529.

[34] Ravikovitch P I, Neimark A V. Characterization of nanoporous materials from adsorption and desorption isotherms[J]. Colloids and Surfaces A: Physicochemical and Engineering Aspects,2001,187:11 - 21.

Regeneration of Microwave Assisted Spent Activated Carbon: Process Optimization, Adsorption Isotherms and Kinetics

Xinhui Duan, C. Srinivasakannan, Wenwen Qu, Xin Wang, Jinhui Peng, Libo Zhang

Abstract: Microwave assisted regeneration of spent coal based activated carbon from the silicon industry has been attempted using steam as the regenerating agent. The response surface methodology (RSM) technique was utilized to optimize the process conditions and the optimum conditions have been identified to be a regeneration temperature of 950℃, regeneration time of 60min and steam flow rate of 2.5g/min. The optimum conditions result in an activated carbon with iodine number of 1103mg/g and a yield of 68.5% respectively. The BET surface area correspond to 1302m^2/g, with the pore volume of 0.86cm^3/g. The activated carbon is heteroporous with a micropore volume of 69.27%. The regenerated carbon is tested for its suitability for adsorption of methylene blue dye molecule. The adsorption isotherms were generated and the maximum adsorption capacity was found to be 385mg/g, with the isotherm adhering to Langmuir isotherm model. The kinetic of adsorption was found to match pseudo-second-order kinetic model. The results indicate potential application of the regenerated activated carbon for liquid phase adsorption involving high molecular weight compounds.

Keywords: microwave assisted regeneration; spent activated carbon; optimization; basic dye; adsorption isotherm and kinetic

1 Introduction

Activated carbons are porous material with extremely high surface area, they have been widely used in a variety of industrial applications such as separation/purification of liquids and gases, removal of toxic substances, catalysts and catalyst support, super-capacitors, electrodes and gas storage[1-7]. The popularity of activated carbon as absorbent can be evidenced from the quantum of the activated carbon being manufactured and traded across the globe[8-10].

Generally, the adsorption capacity is directly influenced by the porous nature of activated carbon decided by the internal surface area. The activated carbons for commercial utilization are usually prepared from different sources of raw materials utilizing different processing methods. Major raw materials for the preparation of activated carbon are coal, petroleum, peat, wood and agricultural wastes. Among which, coal is the most commonly used precursor for activated carbon production due to the advantage of its availability and cost[11,12].

The activated carbon used as absorbent easily becomes saturated after sustained utilization. The waste adsorbents are generally either dumped as solid waste or sold in the secondary market. For example, in the silicon production industry, the tail gas comprises of large amounts of recyclable components such as hydrogen, hydrogen chloride, trichlorosilane and silicon tetrachloride. The hydrogen needs to be separated and recycled back to the reactor as its utilization in the reactor is very low due to low conversion of the chlorosilanes. Typical process of separation involves recovery of chlorosilane by bubbling absorption or refrigeration absorption as the primary step. After the chlorosilanes removal the HCl is separated from the hydrogen by activated carbon[13]. Although the process is well developed, there are still deficiencies such as the contamination of HCl with recycled hydrogen. Since the life cycle of activated carbon is very short, large amounts of spent activated carbon are generated in the production process. Hence it is imperative to regenerate the spent activated carbon for economic, environmental and energy benefits.

Over these years, a wide variety of reactivation techniques have been suggested and applied to deal with spent activated carbon[14,15], which could be thermal, wet oxidation[16,17], supercritical fluid extraction[18] and electrochemical regenerations[19]. Chemical regeneration method such as wet oxidation and supercritical fluid regenerations, could be carried out by decomposing the adsorbates using oxidizing chemical agents under subcritical or supercritical conditions[20], however, these techniques usually involve high pressure and temperature, which rendering them economical unfavorable. Thermal regeneration is the most widespread method which utilizes variety of activating agents such as steam, carbon dioxide or inert gases[21-23]. However, it should be noted that the conventional thermal regeneration technique demands larger processing time and are energy intensive, necessitating the need for alternative methods of thermal regeneration.

In recent years, microwave irradiation has been widely investigated due to its capability of molecular level heating, which leads to homogeneous and quick thermal reactions[24]. Comparing with traditional heating techniques, microwave heating provides additional advantages such as higher heating rates, possibility of selective heating, precise control of temperature, small equipment size and reduced waste[25]. The application of microwave heating technology in the industrial regeneration of spent activated carbon has indicated promising results[26,27].

The basic purpose of regeneration is to provide products with good adsorption capacity for reutilization, thus the influences of factors critical to the quality of product in the thermal regeneration process such as temperature, time and the flow rate of regeneration agent demands optimization. In most process development exercise, the optimization becomes complicated as there is more than one characteristic parameter that needs to be considered. In order to overcome this problem, response surface methodology (RSM) is utilized which is one of the relevant multivariate techniques which has the capability to perform multivariant experimental design, statistical modeling and process optimization[28-31]. The RSM technique has been widely utilized for process optimization for the preparation of activated carbon from different raw materials such as coconut shell and husk[32-34], oil palm fiber[35], Turkish lignite[36], sewage sludge[37], olive - waste cakes[38], Jatropha hull[39] and regeneration of activated carbon from spent catalyst made from vinyl acetate synthesis[40].

In the present study, microwave assisted heating regeneration of spent coal based activated carbon used in silicon production industry is attempted with steam as the regenerating agent. The effect of all three influencing parameters such as the regeneration temperature, regeneration time and steam flow rate were assessed to identify the optimum conditions with the objective function of maximizing iodine adsorption capacity and yield. The improvement in the textural characteristics of the regenerated activated carbon in comparison with the spent activated carbon is made based on the nitrogen BET adsorption isotherm and SEM analysis. The suitability of regenerated carbon for liquid phase application is assessed based on the adsorption of methylene blue dye molecule, by experimentally generating the adsorption isotherms at different temperatures. Further the kinetics of methylene blue adsorption is also estimated by varying the initial concentration of the methylene blue solution. The adsorption isotherms are modeled based on the popular adsorption isotherms, while the kinetic data are modeled using first - order and the second - order kinetics equations and an intraparticle diffusion model.

2 Materials and methods

2.1 Regeneration of spent activated carbon

Spent coal based activated carbons, with iodine number of about 800mg/g loaded with impurities such as HCl and SiO_2 is washed with deionized water for several times to remove any foreign materials and HCl. The completeness of washing is ensured by testing the wash liquor with $AgNO_3$ solution until no precipitation formed. The washed spent carbon is dried in an air oven at 105℃, until it is completely dry. The dried carbons are sieved to the particle size of 2mm to 3mm and stored in airtight container for further experimentation. The proximate analysis of spent activated carbon is shown in Table 1. The ash content is high (12.71%) which is primarily made of the contents such as SiO_2 and Fe_2O_3 contributing to 48.36% and 12.09% respectively.

Table 1 Proximate analysis of spent activated carbon

Sample	Moisture (wt.)/%	Volatile Matter (wt.)/%	Fixed carbon (wt.)/%	Ash (wt.)/%
Spent coal based activated carbon	2.89	1.62	82.78	12.71

The regeneration process is carried out in a self - made microwave tubular furnace, which utilizes a single - mode continuous controllable power for the experiments and is shown in Fig. 1. The microwave frequency is 2.45GHz, while the output power could be set to a maximum of 3000W. The activation temperature is controlled by the input microwave power during the activation process, which is measured by nichrome - nickel silicon armor type thermoelement, placed such that it touched the material. The thermo element has the dimension of 8mm diameter and a length of 450mm, with the temperature range of 0 - 1250℃ and a measurement precision of ± 0.5℃. The details of the microwave tubular furnace are also reported in our recent publication[39]. The experiments are initiated by keeping 15g of the spent activated carbon into the reactor followed by setting the reactor to the desired temperature along with the N_2 flow rate at 200cm^3/min. Steam is allowed into the reactor after the reactor attained the desired temperature, to initiate the regeneration process. The effectiveness of microwave heating can be evidenced from the short duration of heating time(8 – 10min) as compared to conventional heating(120 – 150min). The regeneration time reported in Table 2 corresponds to the time the material is left in the reactor upon reaching the desired temperature. The completion of activation process is marked by terminating the steam flow and switching to the nitrogen flow until the activated carbon is cooled to room temperature. The regeneration process is carried out by varying the regeneration temperature(900 – 1000℃), the steam flow rate(2 – 3g/min), the regeneration time(50 – 70min). The ranges of experimental conditions mentioned above are based on the preliminary experiments, prior to a process optimization exercise.

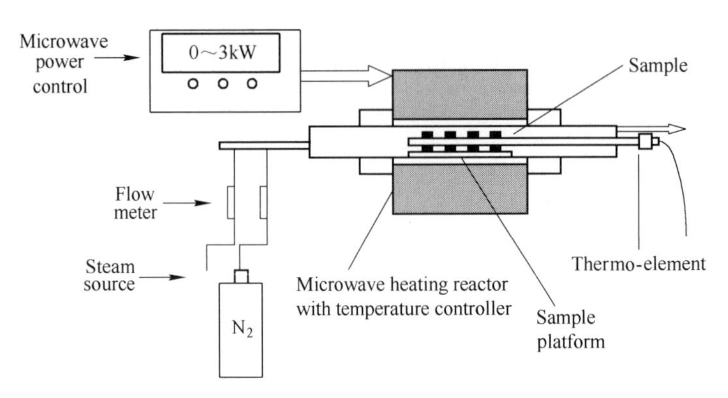

Fig. 1 Schematic of microwave heating equipment

The product is subject to characterization of iodine number, yield and the BET surface area. The yield is defined as grams of activated carbon per gram char utilized for activation. The iodine adsorption capacity is represented as iodine number which indicates milligrams of iodine adsorbed by a gram of activated carbon(mg/g), using the standard testing methods of PR China(GB/T 12496.8—1999)[41] for testing acting activated carbons. The BET surface area, average pore size distribution are estimated using the surface area analyzer, Autosorbe 1 – C made by Quantachrome Instruments, USA. The pore structure of the carbon char and the activated carbon is characterized by nitrogen adsorption at 77K with an accelerated surface area and porosimetry system(Autosorb – 1 – C, Quantachrome) adhering to the details reported in our earlier paper[39]. The microstructures are analyzed by the Scanning Electron Microscope(SEM, Philips XL30ESEM – TMP).

2.2 Design of experiments

A central composite design(CCD) of RSM is utilized to optimize the experimental parameters with a minimum number of experiments and to analyze the interactions between the parameters, the total number of the experiment is 20 including 6 center points. The details of the experiment conditions are provided in Table 2. The dependant variables selected are regeneration temperature(X_1), regeneration time(X_2) and steam flow rate(X_3). The experimental data are analyzed using statistical software Design Expert version 7.1.5(STAT – EASE Inc., Minneapolis, USA). The activated carbons are characterized for the iodine absorption capacity and yield; the results are compiled as well in Table 2. The activated carbon with highest iodine number is characterized for BET surface area and pore size distribution.

Table 2 Experimental design matrix and results

Run	X_1/℃	X_2/min	X_3/g · min^{-1}	Y_1/mg · g^{-1}	Y_2/%
1	900. 00	50. 00	2. 00	900	80. 24
2	1000. 00	50. 00	2. 00	950	77. 30
3	900. 00	70. 00	2. 00	960	74. 04
4	1000. 00	70. 00	2. 00	995	72. 37
5	900. 00	50. 00	3. 00	953	75. 41
6	1000. 00	50. 00	3. 00	1060	66. 19
7	900. 00	70. 00	3. 00	1050	65. 56
8	1000. 00	70. 00	3. 00	990	60. 22
9	865. 91	60. 00	2. 50	1011	83. 28
10	1034. 09	60. 00	2. 50	899	63. 01
11	950. 00	43. 18	2. 50	942	78. 73
12	950. 00	76. 82	2. 50	1025	70. 69
13	950. 00	60. 00	1. 66	945	78. 63
14	950. 00	60. 00	3. 34	1000	55. 9
15	950. 00	60. 00	2. 50	1110	68. 04
16	950. 00	60. 00	2. 50	1105	68. 36
17	950. 00	60. 00	2. 50	1107	68. 20
18	950. 00	60. 00	2. 50	1102	68. 73
19	950. 00	60. 00	2. 50	1104	68. 56
20	950. 00	60. 00	2. 50	1108	68. 11

2. 3 Adsorption of methylene blue

Methylene blue(MB) , purchased from Sigma – Aldrich is used as adsorbate to determine the adsorption capacity a regenerated activated carbon. Batch adsorption experiments are performed in a set of erlenmeyer flasks(250mL) , each contains 100mL of different initial concentrations of methylene blue(200 – 400mg/L) , together with 0. 1g of regenerated activated carbon of particle size around 300μm. A gas bath thermostatic oscillator is utilized to maintain the desired temperature for desired adsorption time. The concentrations of methylene blue in the supernatant solutions are measured by a spectrophotometer(UV – 2550 Shimadzu, Japan) at 668nm, before and after the adsorption process.

The procedure for estimation of kinetic of adsorption of methylene blue is basically identical to those of equilibrium tests, except the fact that the liquid samples were taken at intervals of time, with the zero time corresponding to the time the activated carbon is charged into the dye solution. The calculation equations for adsorbed amount of MB at equilibrium and time t are obvious and not provided in the present paper.

3 Results and discussion

3.1 Iodine adsorption

The most important characteristic of an activated carbon is its adsorption capacity, which is characterized by the surface area and pore size which are highly influenced by the process of preparation. Among all the three parameters chosen in the present study, steam flow rate is found to have the most significant influence on iodine adsorption capacity, while the activation temperature has the least influence on iodine adsorption capacity. Fig. 2 shows the three - dimensional response surfaces of the combined effect of activation temperature and steam flow rate on the iodine number, at an activation time of 130min. As can be seen, the iodine number increases with an increase in the activation temperature and steam flow rate, with the maximum iodine number of about 1050mg/g. The iodine number reaches an asymptote at high steam flow rate corresponding to steam flow rate in excess of 2. 25g/L. Although the effect of regeneration temperature didn't seem to be as influencing as the steam flow rate, it does exhibit an asymptote at high temperatures. The change in activation temperature and steam flow rate leads to a change in the activation energy for carbon - steam reaction, which would affect the rate of reaction for the variety of conversion, the change in conversion is indicative of modifications in the carbon porous structure.

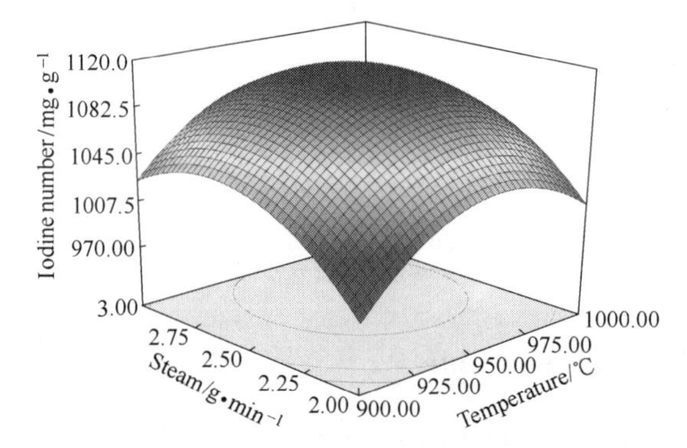

Fig. 2 Three - dimensional response surface plot of Iodine uptake: Effect of regeneration temperature and steam flow rate on the iodine number (regeneration time: 65min)

Fig. 3 shows the three - dimensional response surfaces of the combined effect of regeneration time and steam flow rate on the iodine number, at a regeneration temperature of 980℃. As can be seen, the iodine number of regenerated carbon increases with the increase regeneration temperature and time. Similar to the trend with steam flow rate and the regeneration temperature, the regeneration time as well shows a possible peak, evidenced from the presence of asymptote at high regeneration time.

An attempt has been made to compare the iodine number of the regenerated carbon in the present study with the typical values reported in literature. The iodine number of the regenerated carbon is compared with the typical values reported in literature. Chen, et al. [42] have reported a maximum iodine number of 1000mg/g for lignite activated with steam for 4h. In addition they reported steam

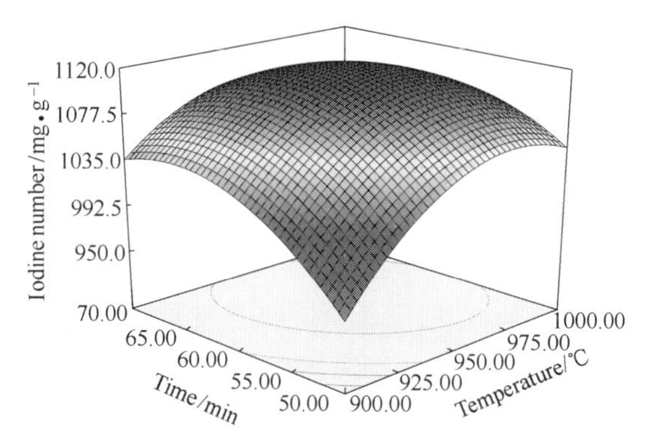

Fig. 3 Three – dimensional response surface plot of Iodine uptake: Effect of regeneration temperature and regeneration time on the iodine number(steam flow rate: 2.6g/min)

flow rate to be the significant parameter affecting the iodine number of activated carbon. Linares – Solano, et al. [43] have reported a surface area of $1000m^2/g$ for activated carbon prepared from bituminous coal by steam activation, with the activated carbon exhibiting narrow pores (below 0.7nm) and wider porosity (super micro and meso porosity). Marten – Gulln, et al. [44] have reported temperature and steam flow rate to be most significant parameters for preparation of activated carbon from bituminous coal. Li and Peng[45] reported an iodine number of 1085mg/g for coconut shell precursor, activated using steam and microwave heating, while Xia and Peng[46] reported a maximum iodine number of 1061mg/g using tobacco stems as precursor with microwave assisted steam activation. Ania, et al. [21,26,27] compared the conventional regeneration with the microwave regeneration and reported better qualities of the regenerated carbon using microwave irradiation.

3.2 Activated carbon yield

The yield of activated carbon is also an important parameter as it quantifies the amount of final product. Steam flow rate shows the most significant effect on activated carbon yield as compared with the rest of the two parameters. Table 2 shows the yield of activated carbon for each of the experiments. Fig. 4 shows the three dimensional plot of effect of activation temperature and steam flow rate while Fig. 5 shows the effect of activation time and temperature on the yield of activated carbon. The yield of activated carbon decreases with the increase of all the three parameters, with the highest yield corresponding to the lowest point of the all three parameters. It should be noted that the highest yield corresponds to activated carbon with lowest iodine number. An increase in activation temperature, activation time or the steam flow rate increases the extent of carbon – steam reaction and hence the yield of activated carbon decreases with increase in anyone of the parameters. Corresponding to the iodine number above 1000mg/g, the yield of the activated carbon is found to be around 60%. The yields have been reported in literature to be widely varying with Chen, et al. [42] have reported a highest yield of 32.5% from lignite while Linares – Solano, et al. [43] have reported a yield of 50% from bituminous coal. Marten – Gulln, et al. [44] have reported a yield of 40% for activated carbon prepared from bituminous coal, while Xia and Peng[46] have reported a yield of 30.83% by microwave heating with relatively low steam flow rate. Pis, et al. [47] have reported a yield of 52% for four kinds of coals, while Gryglewicz, et al. [48] have reported a yield of 50% for activated carbon prepared from oil agglomerated bituminous coal by steam activation. Zabaniotou, et al. [49] reported that the amount of carbon exhausted was found to depend on the extent of carbon – steam reaction, which was reported to depend on the regeneration temperature, regeneration time and the steam flow rate.

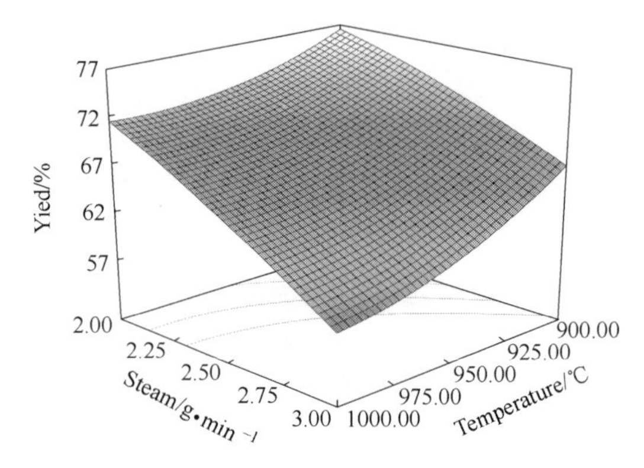

Fig. 4 Three – dimensional response surface plot of yield: Effect of regeneration temperature and steam flow rate on the iodine number (regeneration time: 65min)

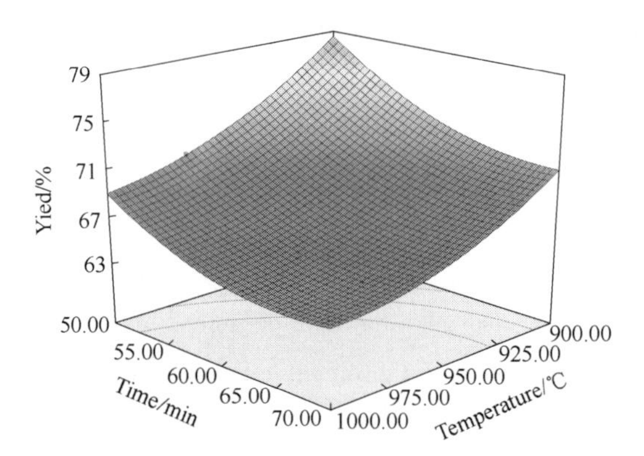

Fig. 5 Three – dimensional response surface plot of yield: Effect of regeneration temperature and regeneration time on the iodine number (steam flow rate: 2.6g/min)

3.3 Development of regression model

The design together with the response values from the experiments are shown in Table 2. Runs 15 – 20 at the center point are repeated to determine the experimental error. The iodine adsorption and carbon yield are utilized in the quadratic model suggested by the software. The polynomial regression equation is developed using CCD. The final empirical models in terms of coded factors for iodine adsorption (Y_1) and carbon yield (Y_2) are shown in Eqs. (1) and (2) respectively,

$$Y_1 = -21823.02 + 38.89X_1 + 90.82X_2 + 1078.29X_3 - 0.04X_1X_2 - 0.02X_1^2 - 0.39X_2^2 - 171.82X_3^2 \quad (1)$$

$$Y_2 = +765.20 - 1.26X_1 - 3.52X_2 + 53.03X_3 + 6.41E^{-004}X_1^2 + 0.02X_2^2 \quad (2)$$

The suitability of model equation is evaluated using the correlation coefficients(R^2), which are 0.9444 and 0.9314 respectively. The proximity of R^2 value to unity, indicate the suitability of the model equation. Both the R^2 values of Iodine adsorption and carbon yield are relatively high indicating, good agreement between experimental data and the model prediction.

Checking the adequacy of model is an important part of the data analysis procedure, since it would result in poor or misleading results if the fit is inadequate[50]. Thus, the analysis of variance (ANOVA) is carried out to justify the adequacy of the model. The ANOVA for the quadratic model of Iodine adsorption is shown in Table 3, where the F - value of 18.87 and Prob > F less than 0.0001 prove that the model is significant. The value of model terms Prob > F less than 0.05 indicates that the model terms are significant. In this case X_1, regeneration temperature, X_2, regeneration time, X_3, steam flow rate and interaction parameters of X_1X_2, X_1^2, X_2^2, X_3^2 are significant model terms. The ANOVA of the quadratic model of activated carbon yield is shown in Table 4. An F - value of 15.09 and Prob > F value of 0.0001 prove that the model is significant. The value of the model terms Prob > F less than 0.05 indicates that all the three process parameters, X_1, X_2, X_3 and interaction parameters of X_1^2, X_2^2 are significant. The results show that the model is suitable to predict the carbon yield within the range of factors studied.

Table 3 Analysis of variance(ANOVA) for response surface quadratic model for Iodine uptake

Source	Sum of squares	Degree of freedom	Mean square	F value	Prob > F
Model	96336.24	9	10704.03	18.87	<0.0001
X_1	9571.23	1	9571.23	16.87	0.0021
X_2	6268.52	1	6268.52	11.05	0.0077
X_3	9568.91	1	9568.91	16.87	0.0021
X_1X_2	3240.13	1	3240.13	5.71	0.0380
X_1X_3	36.13	1	36.13	0.064	0.8059
X_2X_3	406.13	1	406.13	0.72	0.4173
X_1^2	31863.26	1	31863.26	56.17	<0.0001
X_2^2	21994.10	1	21994.10	38.77	<0.0001
X_3^2	26591.15	1	26591.15	46.88	<0.0001

Table 4 Analysis of variance(ANOVA) for response surface quadratic model for activated carbon yield

Source	Sum of squares	Degree of freedom	Mean square	F value	Prob > F
Model	857.82	9	95.31	5.09	0.0001
X_1	202.87	1	202.87	32.11	0.0002
X_2	119.83	1	119.83	18.97	0.0014
X_3	419.40	1	419.40	66.38	<0.0001
X_1X_2	3.12	1	3.12	0.49	0.4979
X_1X_3	10.13	1	10.13	1.60	0.2342
X_2X_3	3.12	1	3.12	0.49	0.4979

Continues Table 4

Source	Sum of squares	Degree of freedom	Mean square	F value	Prob > F
X_1^2	37.12	1	37.12	5.87	0.0358
X_2^2	55.27	1	55.27	8.75	0.0143
X_3^2	6.93	1	6.93	1.10	0.3197

Fig. 6 shows the comparison of the predicted iodine number versus the experimental iodine number of activated carbon, while Fig. 7 shows the predicted value versus the experimental values for yield of activated carbon. The experimental iodine numbers and activated carbon yield are the measured data of a particular experimental run while the predicted values are evaluated from the model. As can be seen, the predicted values match well with the experimental values, indicating the ability of the model to successfully capture the correlation between the process variables and the iodine adsorption capacity.

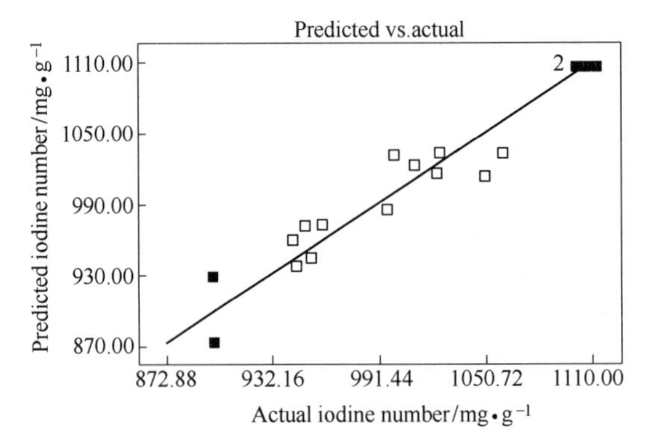

Fig. 6 Predicted vs. experimental adsorption uptake on iodine

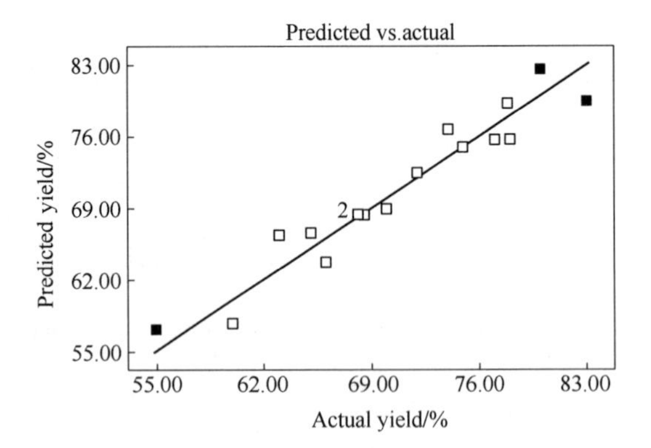

Fig. 7 Predicted vs. experimental activated carbon yield

3.4 Process optimization

It is of general commercial interest to regenerate carbon with highest yield and iodine number. It is established experimentally that the iodine uptake and carbon yield responds opposite to each other with the process parameter. In order identify the optimum conditions, the function of desirability is applied using Design Expert software version,7. 1. 5(STAT - EASE Inc. ,Minneapolis,USA). The experimental conditions with highest desirability are selected with the help of software. The optimum condition for preparation of activated carbon is found to be a regeneration temperature of 950℃, regeneration time of 60min and steam flow rate of 2. 5 g/min, with the iodine adsorption of 1101mg/g and carbon yield of 69%. The repeat experimental runs are conducted to ensure the accuracy of the optimized conditions, resulted in an average iodine number of 1103mg/g and average carbon yield of 68. 5%.

3. 5 Pore structure and surface area analysis

The nitrogen adsorption isotherm estimated using the Autosorb instrument is shown Fig. 8. It can be seen observed that the volume adsorbed increases sharply at low relative pressure, which indicates the process of filling the micro pores. The adsorption capacity continues to increase with increase of the relative pressure up to a p/p_0 value of 1. The trend of the adsorption isotherm pertains to type II isotherm under the IUPAC classification of isotherms, based on the progressive increase in the adsorption capacity beyond the relative pressure of 0. 1. The cumulative pore volume plot shown in Fig. 9 indicuates that the significant quantity of pores in the microporous region, with the average

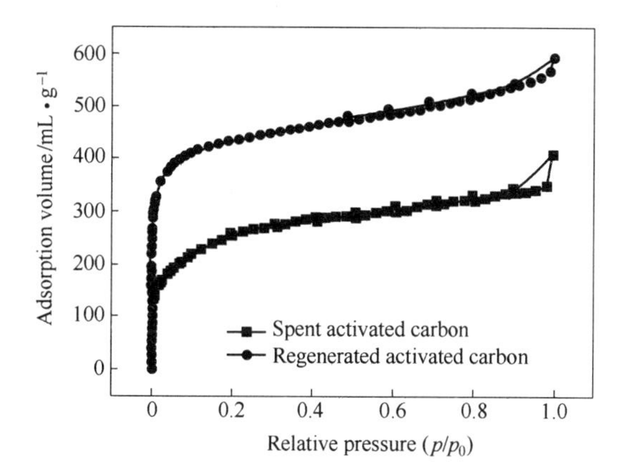

Fig. 8 Nitrogen adsorption isotherm of the spent and regenerated activated carbon

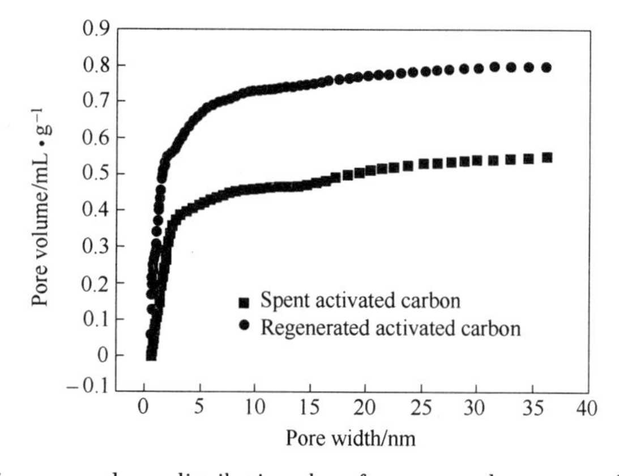

Fig. 9 Cumulative pore volume distribution chart for spent and regenerated activated carbon

pore diameter estimated to be 2.64nm. The surface area of the regenerated carbon is estimated to be $1302m^2/g$, while the total pore volume is 0.86mL/g.

The pore volume corresponding to the micropore and mesopore along with average pore diameter of the spent and regenerated activated carbon is shown in Table 5. A comparison of the quality of activated carbon with the spent activated carbon shows a significant increase in the pore volume, the micropore volume and the surface area attributed to the regeneration process. The increase in the textural characteristics is due to the regeneration process, which accounts for both creation of micropores and clearing of the pores blocked in the spent activated carbon.

Table 5 Details of pore structure of spent activated carbon vs. regenerated activated carbon

Subject	Regenerated activated carbon	Spent activated carbon
Pore volume/$mL \cdot g^{-1}$	0.86	0.53
Average pore diameter/nm	2.64	2.5
Micropore volume/%	69.27	51.93
Mesopore volume/%	30.73	48.06
Surface area/$m^2 \cdot g^{-1}$	1302	838

3.6 Microscopic structure analysis of the activated carbon

The microscopic structure of the spent activated carbon and the regenerated activated carbon prepared using steam regeneration are shown in Fig. 10. As can be seen in Fig. 10(a), the surface of the spent activated carbon is covered by impurities and the pore structure could not be seen clearly. After regeneration, as can be seen in Fig. 10(b), the impurities on the surface of the activated carbon are removed and the clear pore structure is visible.

(a) (b)

Fig. 10 SEM images of the spent and regenerated activated carbon

(a) Spent activated carbon; (b) Regenerated activated carbon

3.7 Adsorption of methylene blue

3.7.1 Adsorption isotherms

The adsorption isotherms were generated following the procedure as stated in the experimental section, in accordance with the literature [51] by varying the initial concentration of the methylene blue at three different temperatures of 298K, 303K and 308K. The increase in equilibrium adsorption capacity with increase in the initial concentration of the dye solution is well understood from the basic concepts of the equilibrium adsorption isotherms. The experimental data are plotted in accordance with the popular Langmuir and Freundlich adsorption isotherms as shown in Figs. 11 and 12. The estimated adsorption isotherm model parameters are presented in Table 6. The adsorption equilibrium data matches well with the Langmuir isotherm model than the Freundlich isotherm. The equilibrium adsorption increases with increase in the adsorption temperature, with the maximum equilibrium adsorption of 385mg/g. A maximum adsorption capacity denotes the potential of the regenerated carbon for commercial liquid phase adsorption, as product with methylene blue number

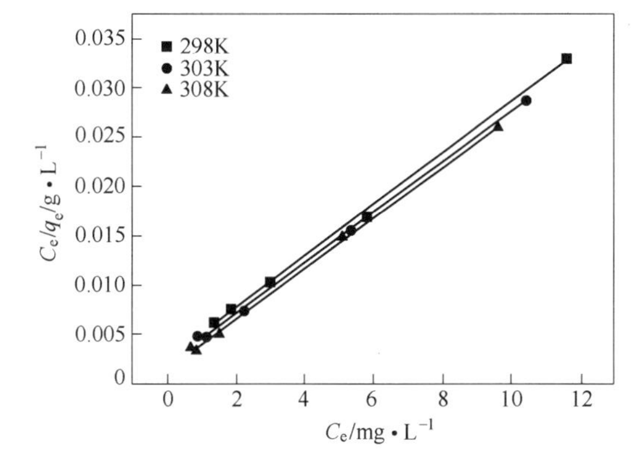

Fig. 11 Langmuir isotherms for methylene blue dye adsorption onto regenerated activated carbon at different temperatures

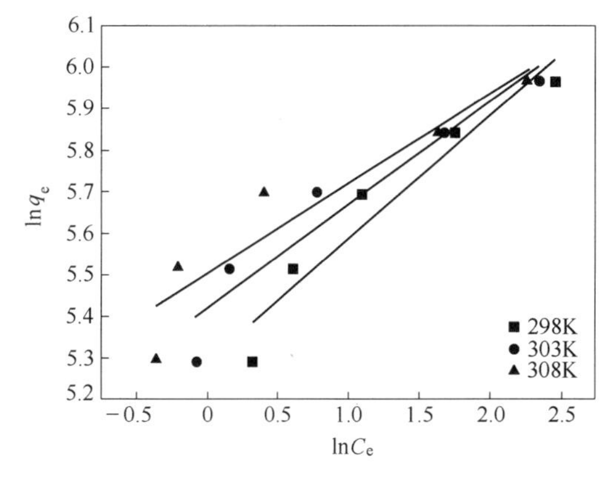

Fig. 12 Freundlich isotherms for methylene blue dye adsorption onto regenerated activated carbon at different temperatures

in excess of 200mg/L have good market realization. Only a few recent studies using chemical activation methods using biomass have reported methylene blue number in excess of 400 mg/g[51-54].

Table 6 Isotherm parameters for removal of methylene blue by regenerated activated carbon at different temperatures

Isotherms	Parameters	Temperature		
		298K	303K	308K
Langumir	Q_0/mg · g^{-1}	384.02	384.16	385.20
	b/L · mg^{-1}	1.23	1.37	1.86
	R^2	0.99	0.99	0.99
Freundlich	$1/n$	0.30	0.25	0.22
	K_F/mg · (g(L/mg)$^{1/n}$)$^{-1}$	198.15	225.90	245.45
	R^2	0.92	0.91	0.88

3.7.2 Adsorption kinetics

The adsorption kinetic experiments were performed by taking liquid samples at known intervals of time as stated in the experimental section. The kinetics of dye adsorption was estimated for different initial concentrations of the dye solution ranging from 200mg/L to 400 mg/L, and was found that the rate of adsorption increase with increase in the initial concentration of the dye solution. The dye adsorption kinetics is modeled using the popular first order[55], second order[56] and the intraparticle diffusion[57] models. As these models are widely used in literature to model the kinetics of adsorption, the details of the model equations are not presented here and can be referred from the referred literature. Figs. 13 – 15 is the plot of the adsorption kinetics plotted in accordance with the first order, second order and the intra particle diffusion model. The model parameters and the appropriateness of the model in reflecting the trend of the experimental data could be evidenced from the model parameters and R^2 values shown in Table 7 and Table 8. The high R^2 value close to 1 for the second order kinetic model suggests the rate of adsorption reflects a second order kinetics.

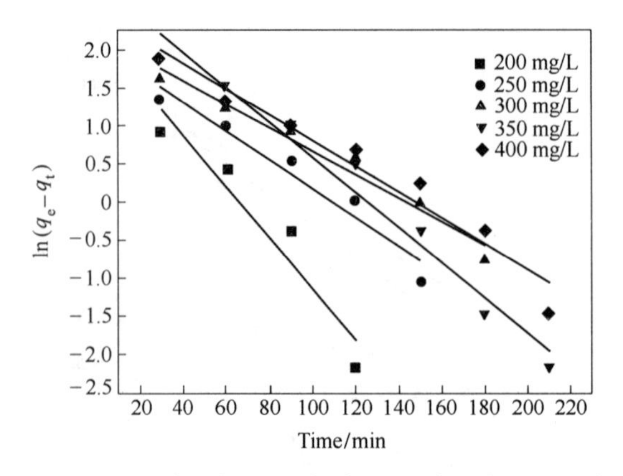

Fig. 13 Pseudo – first – order kinetics for adsorption of methylene blue dye onto regenerated activated carbon at 298K

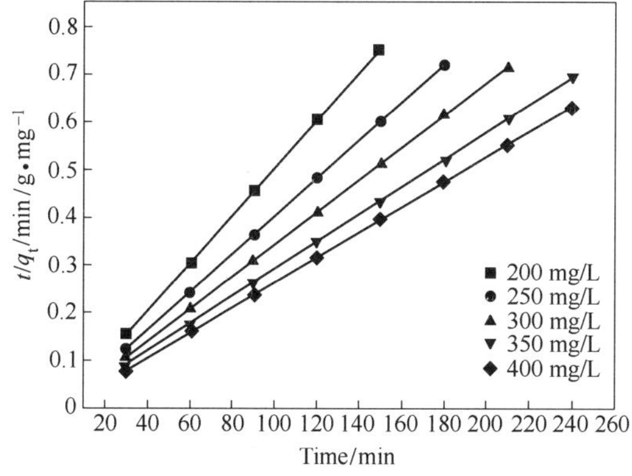

Fig. 14 Pseudo - second - order kinetics for the adsorption of methylene blue onto regenerated activated carbon at 298K

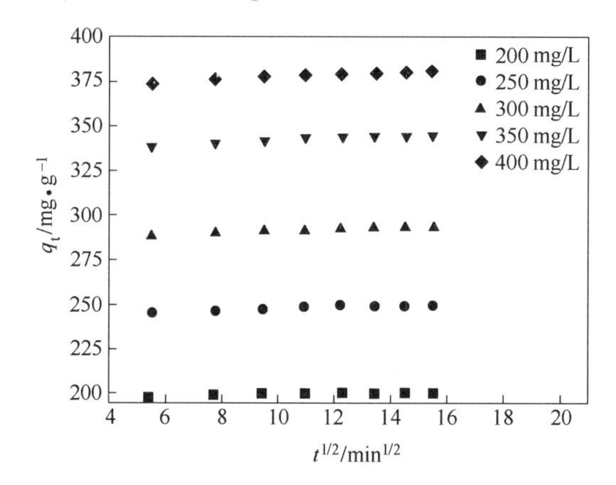

Fig. 15 Intraparticle diffusion model for the adsorption of methylene blue onto regenerated activated carbon at 298K

Table 7 Kinetic parameters for the removal of methylene blue of different initial concentrations by regenerated carbon at 298K

Initial concentration /mg · L^{-1}	$q_{e,exp}$ /mg · g^{-1}	First - order kinetic model			Second - order kinetic model		
		k_1/min	$q_{e,cal}$/mg · g^{-1}	R^2	k_2	$q_{e,cal}$/mg · g^{-1}	R^2
200	199. 305	0. 0335	8. 09	0. 9215	0. 0083	200	1
250	249. 189	0. 0191	9. 14	0. 9418	0. 0050	250	1
300	293. 494	0. 0154	9. 18	0. 9706	0. 0037	294. 11	0. 99
350	344. 904	0. 0231	18. 36	0. 9710	0. 0031	344. 82	0. 99
400	380. 388	0. 0170	12. 31	0. 9526	0. 0030	384. 61	0. 99

Table 8 Intraparticle diffusion model parameters for different initial MB concentrations at 298K

Initial concentration /mg · L^{-1}	$q_{e,exp}$ /mg · g^{-1}	Intraparticle diffusion model			
		$q_{e,cal}$/mg · g^{-1}	K_p/mg · (g · min$^{1/2}$)$^{-1}$	C	R^2
200	199. 305	199. 76	0. 25	195. 89	0. 83
250	249. 189	249. 41	0. 41	243. 46	0. 94
300	293. 494	293. 92	0. 52	285. 87	0. 97
350	344. 904	345. 82	0. 68	335. 29	0. 94
400	380. 388	380. 91	0. 63	371. 16	0. 94

4 Conclusions

The response surface methodology(RSM) is used to optimize the process conditions, for the microwave heating assisted regeneration of spent activated carbon. The influences of the three major parameters, regeneration temperature, regeneration time and CO_2 flow rate on the properties of the regenerated carbon are investigated using analysis of variance(ANOVA), to identify the significant parameters. The experimental data of the adsorption capacity and yield are found to agree satisfactorily with the model predictions. The regenerated carbon has been characterized and its adsorption isotherms and kinetics have been investigated, using methylene blue as model compound. The results are summarized as follows:

(1) The optimum conditions for preparation of activated carbon has been identified to be an regeneration temperature of 950℃, regeneration time of 60min and steam flow rate of 2.5g/min. The optimum conditions result in an activated carbon with an iodine number of 1103mg/g and a yield of 68.5% respectively. The BET surface area evaluated using nitrogen adsorption isotherm for the optimal sample correspond to $1302m^2/g$, with the pore volume of $0.86cm^3/g$. The activated carbon is hetero porous with the micropore volume contributing to 69.27%. The results of this study indicate that microwave assisted regeneration of activated carbon by steam activation is a feasible and an effective way for reusing the spent activated carbon. A comparison of the quality of the regenerated carbon with the fresh activated carbon as reported in literature, indicate the regenerated carbon to have higher yield and better adsorption capacity, rendering the adoption of the process for commercial production.

(2) The regenerated activated carbon presents a high affinity toward the solute, due to the faster uptake of methylene blue at low concentrations. The amount adsorbed onto the regenerated activated carbon at equilibrium increased from 199mg/g to 380mg/g as the methylene blue concentration increased from 200mg/L to 400mg/L. The equilibrium data were described better by the Langmuir isotherm model than the Freundlich isotherm. Methylene blue numbers in excess of 400mg/L are very rarely reported in literature and it corresponds to activation using chemical activation using biomass based precursors. The high methylene blue adsorption capacity highlights its potential application for adsorption high molecular weight compounds typically suitable for liquid phase adsorption, which also proves that the application of microwave assisted heating is feasible and efficient.

(3) The adsorption kinetics results demonstrate that the adsorption process of methylene blue on the regenerated activated carbon follows closely a pseudo - second - order kinetic model. The regenerated carbon exhibit excellent textural characteristics evidenced by the high surface area and an adsorbent with equal proportion of micro and meso pores. The high yield along with good textural characteristics has number of commercial applications substantiating the process optimization exercise for regeneration of spend coal from silicon industry.

Acknowledgements

The authors would like to express their gratitude to the National Natural Science Foundation of China (No. 51004059/E041601) and Natural Science Foundation of Yunnan Province(No. 14051157) for financial support.

References

[1]Henning K D,Schäfer S. Impregnated activated carbon for environmental protection[J]. Gas separation & purification,1993,7(4):235 - 240.

[2]Mazyck D W,Cannon F S. Overcoming calcium catalysis during the thermal reactivation of granular activated carbon:Part Ⅰ. Steam - curing plus ramped - temperature N_2 treatment[J]. Carbon, 2000, 38(13): 1785 - 1799.

[3]Walker G M,Weatherley L R. Textile wastewater treatment using granular activated carbon adsorption in fixed beds[J]. Separation Science and Technology,2000,35(9):1329 - 1341.

[4]Gurrath M,Kuretzky T,Boehm H P,et al. Palladium catalysts on activated carbon supports:influence of reduction temperature,origin of the support and pretreatments of the carbon surface[J]. Carbon,2000,38(8): 1241 - 1255.

[5]Mudoga H L,Yucel H,Kincal N S. Decolorization of sugar syrups using commercial and sugar beet pulp based activated carbon s[J]. Bioresource Technology,2008,99(9):3528 - 3533.

[6]Holtz R D,Oliveira S B,Fraga M A,et al. Synthesis and characterization of polymeric activated carbon - supported vanadium and magnesium catalysts for ethylbenzene dehydrogenation[J]. Applied Catalysis A:General, 2008,350(1):79 - 85.

[7]Li W,Peng J,Zhang L,et al. Preparation of activated carbon from coconut shell chars in pilot - scale microwave heating equipment at 60kW[J]. Waste Management,2009,29(2):756 - 760.

[8]Ioannidou O,Zabaniotou A. Agricultural residues as precursors for activated carbon production—a review[J]. Renewable and Sustainable Energy Reviews,2007,11(9):1966 - 2005.

[9]Mui E L K,Ko D C K,McKay G. Production of active carbons from waste tyres—a review[J]. Carbon,2004,42 (14):2789 - 2805.

[10]Yin C Y,Aroua M K,Daud W M A W. Review of modifications of activated carbon for enhancing contaminant uptakes from aqueous solutions[J]. Separation and Purification Technology,2007,52(3):403 - 415.

[11]Ahmadpour A,Do D D. The preparation of active carbons from coal by chemical and physical activation [J]. Carbon,1996,34(4):471 - 479.

[12]Bansal R C,Donnet J B,Stoekcli H F. Active Carbon[M]. New York:Decker,1988.

[13]Liu J J. Purification in recovering hydrogen during production of polycrystalline silicon[J]. China Nonferrous Metallurgy,2000,29:17 - 20.

[14]Bagreev A,Rahman H,Bandosz T J. Thermal regeneration of a spent activated carbon previously used as hydrogen sulfide adsorbent. [J]. Carbon,2001,39(9):1319 - 1326.

[15]Okoniewska E,Lach J,Kacprzak M,et al. The trial of regeneration of used impregnated activated carbons after manganese sorption[J]. Desalination,2008,223(1):256 - 263.

[16]Shende R V,Mahajani V V. Wet oxidative regeneration of activated carbon loaded with reactive dye [J]. Waste Management,2002,22(1):73 - 83.

[17]Lee S K,Chung M S,Oh W J,et al. Wet regeneration of impregnated activated carbon by solvent extraction

[J]. Jpn. Kokai Tokkyo Kohk, Jpn. Pat, 1999, 11147708: A2.

[18] Chihara K, Oomori K, Oono T, et al. Supercritical CO_2 regeneration of activated carbon loaded with organic adsorbates[J]. Water Science and Technology, 1997, 35(7): 261 - 268.

[19] Clifford A L, Dong D F, Mumby J A, et al. Chemical and electrochemical regeneration of activated carbon[P]. US Pat, 1997, 5702587.

[20] Richard S, Horng R S, Tseng I C. Regeneration of granular activated carbon saturated with acetone and isopropyl alcohol via a recirculation process under H_2O_2/UV oxidation[J]. Journal of Hazardous Materials, 2008, 154: 366 - 372.

[21] Ania C O, Parra J B, Menendez J A, et al. Effect of microwave and conventional regeneration on the microporous and mesoporous network and on the adsorptive capacity of activated carbons[J]. Microporous and Mesoporous Materials, 2005, 85(1): 7 - 15.

[22] Maroto - Valer M M, Dranca I, Clifford D, et al. Thermal regeneration of activated carbons saturated with ortho - and meta - chlorophenols[J]. Thermochimica Acta, 2006, 444(2): 148 - 156.

[23] Sabio E, Gonzalez E, Gonzalez J F, et al. Thermal regeneration of activated carbon saturated with p - nitrophenol[J]. Carbon, 2004, 42(11): 2285 - 2293.

[24] Yuen F K, Hameed B H. Recent developments in the preparation and regeneration of activated carbons by microwaves[J]. Advances in Colloid and Interface Science, 2009, 149(1): 19 - 27.

[25] Jones D A, Lelyveld T P, Mavrofidis S D, et al. Microwave heating applications in environmental engineering—a review[J]. Resources, Conservation and Recycling, 2002, 34(2): 75 - 90.

[26] Ania C O, Menéndez J A, Parra J B, et al. Microwave - induced regeneration of activated carbons polluted with phenol. A comparison with conventional thermal regeneration[J]. Carbon, 2004, 42(7): 1383 - 1387.

[27] Ania C O, Parra J B, Menendez J A, et al. Microwave - assisted regeneration of activated carbons loaded with pharmaceuticals[J]. Water Research, 2007, 41(15): 3299 - 3306.

[28] Bezerra M A, Santelli R E, Oliveira E P, et al. Response surface methodology(RSM) as a tool for optimization in analytical chemistry[J]. Talanta, 2008, 76(5): 965 - 977.

[29] Fu J, Zhao Y, Wu Q. Optimising photoelectrocatalytic oxidation of fulvic acid using response surface methodology[J]. Journal of Hazardous Materials, 2007, 144(1): 499 - 505.

[30] Gönen F, Aksu Z. Use of response surface methodology(RSM) in the evaluation of growth and copper(II) bioaccumulation properties of Candida utilis in molasses medium[J]. Journal of Hazardous Materials, 2008, 154(1): 731 - 738.

[31] Secula M S, Suditu G D, Poulios I, et al. Response surface optimization of the photocatalytic decolorization of a simulated dyestuff effluent[J]. Chemical Engineering Journal, 2008, 141(1): 18 - 26.

[32] Gratuito M K B, Panyathanmaporn T, Chumnanklang R A, et al. Production of activated carbon from coconut shell: optimization using response surface methodology [J]. Bioresource Technology, 2008, 99(11): 4887 - 4895.

[33] Tan I A W, Ahmad A L, Hameed B H. Optimization of preparation conditions for activated carbons from coconut husk using response surface methodology[J]. Chemical Engineering Journal, 2008, 137(3): 462 - 470.

[34] Tan I A W, Ahmad A L, Hameed B H. Preparation of activated carbon from coconut husk: optimization study on removal of 2,4,6 - trichlorophenol using response surface methodology[J]. Journal of Hazardous Materials, 2008, 153(1): 709 - 717.

[35] Hameed B H, Tan I A W, Ahmad A L. Optimization of basic dye removal by oil palm fibre - based activated carbon using response surface methodology[J]. Journal of Hazardous Materials, 2008, 158(2): 324 - 332.

[36] Karacan F, Ozden U, Karacan S. Optimization of manufacturing conditions for activated carbon from Turkish

lignite by chemical activation using response surface methodology[J]. Applied Thermal Engineering,2007,27(7):1212 - 1218.

[37] Rio S, Faur - Brasquet C, Coq L L, et al. Experimental design methodology for the preparation of carbonaceous sorbents from sewage sludge by chemical activation—application to air and water treatments[J]. Chemosphere,2005,58(4):423 - 437.

[38] Baçaoui A, Yaacoubi A, Dahbi A, et al. Optimization of conditions for the preparation of activated carbons from olive - waste cakes[J]. Carbon,2001,39(3):425 - 432.

[39] Duan X H, Srinivasakannan C, Peng J H, et al. Preparation of activated carbon from Jatropha hull with microwave heating: optimization using response surface methodology[J]. Fuel Processing Technology,2011,92(3):394 - 400.

[40] Zhang Z, Peng J, Qu W, et al. Regeneration of high - performance activated carbon from spent catalyst: optimization using response surface methodology[J]. Journal of the Taiwan Institute of Chemical Engineers,2009,40(5):541 - 548.

[41] Liu X, Quan X, Bo L, et al. Temperature measurement of GAC and decomposition of PCP loaded on GAC and GAC - supported copper catalyst in microwave irradiation[J]. Applied Catalysis A: General,2004,264(1):53 - 58.

[42] Chen W, Liu Z H, Fan Y Q. Study on preparation and properties of activated carbon from lignite[J]. Coal Conversion,2004,27:62 - 64.

[43] Linares - Solano A, Martin - Gullon I, Salinas - Martinez de Lecea C, et al. Activated carbons from bituminous coal: effect of mineral matter content[J]. Fuel,2000,79(6):635 - 643.

[44] Martin - Gullon I, Asensio M, Font R, et al. Steam - activated carbons from a bituminous coal in a continuous multistage fluidized bed pilot plant[J]. Carbon,1996,34(12):1515 - 1520.

[45] Li W, Peng J, Zhang L, et al. Preparation of activated carbon from coconut shell chars in pilot - scale microwave heating equipment at 60kW[J]. Waste Management,2009,29(2):756 - 760.

[46] Xia H, Peng J, Zhang L, et al. Study on the preparation of granular activated carbon from tobacco stems by microwave radiation and water vapor[J]. Chemical Engineering(China),2007,1:012.

[47] Pis J J, Mahamud M, Pajares J A, et al. Preparation of active carbons from coal: Part Ⅲ: activation of char[J]. Fuel Processing Technology,1998,57(3):149 - 161.

[48] Gryglewicz G, Grabas K, Lorenc - Grabowska E. Preparation and characterization of spherical activated carbons from oil agglomerated bituminous coals for removing organic impurities from water[J]. Carbon,2002,40(13):2403 - 2411.

[49] Zabaniotou A, Stavropoulos G, Skoulou V. Activated carbon from olive kernels in a two - stage process: Industrial improvement[J]. Bioresource Technology,2008,99(2):320 - 326.

[50] Körbahti B K, Rauf M A. Determination of optimum operating conditions of carmine decoloration by UV/H_2O_2 using response surface methodology[J]. Journal of Hazardous Materials,2009,161(1):281 - 286.

[51] Raposo F, De La Rubia M A, Borja R. Methylene blue number as useful indicator to evaluate the adsorptive capacity of granular activated carbon in batch mode: influence of adsorbate/adsorbent mass ratio and particle size[J]. Journal of Hazardous Materials,2009,165(1):291 - 299.

[52] Hameed B H, Din A T M, Ahmad A L. Adsorption of methylene blue onto bamboo - based activated carbon: kinetics and equilibrium studies[J]. Journal of Hazardous Materials,2007,141(3):819 - 825.

[53] Tan I A W, Ahmad A L, Hameed B H. Adsorption of basic dye on high - surface - area activated carbon prepared from coconut husk: equilibrium, kinetic and thermodynamic studies[J]. Journal of Hazardous Materials,2008,154(1):337 - 346.

[54] Altenor S, Carene B, Emmanuel E, et al. Adsorption studies of methylene blue and phenol onto vetiver roots activated carbon prepared by chemical activation [J]. Journal of Hazardous Materials, 2009, 165 (1): 1029 - 1039.

[55] Thinakaran N, Panneerselvam P, Baskaralingam P, et al. Equilibrium and kinetic studies on the removal of Acid Red 114 from aqueous solutions using activated carbons prepared from seed shells[J]. Journal of Hazardous Materials, 2008, 158(1): 142 - 150.

[56] Largergren S. Zur theorie der sogenannten adsorption geloster stoffe. Kungliga Svenska Vetenskapsakademiens [J]. Handlingar, 1898, 24: 1 - 39.

[57] Ho Y S, McKay G. Sorption of dye from aqueous solution by peat[J]. Chemical Engineering Journal, 1998, 70 (2): 115 - 124.

Pilot – scale Extraction of Zinc from the Spent Catalyst of Vinyl Acetate Synthesis by Microwave Irradiation

Wei Li, Jinghui Peng, Libo Zhang, Zebiao Zhang, Lei Li, Shimin Zhang, Shenghui Guo

Abstract: A pilot – scale experiment of extracting zinc from spent catalyst of vinyl acetate synthesis by pretreatment using microwave irradiation was investigated. A variety of factors that affect zinc leaching were evaluated systematically. The experiment results indicated that the pretreatment with microwave irradiation could improve the zinc leaching rate greatly. Taking into consideration the experimental results and the industry production practice, the optimum conditions were obtained as follows: the temperature of pretreatment with microwave irradiation was 950℃; the leaching agent was a mixture of ammonium bicarbonate, ammonia and water; the dose of leaching agent was 2.5 times more than the theoretical dose; the ratio of liquid to solid was 2∶1; leaching time at room temperature was 3h; the residue was washed 4 times. The average zinc extraction was 93.45% under the optimal conditions. The surface morphology and compositions of spent catalyst before and after microwave irradiation, and before and after being leached using the leaching agents were observed and characterized by SEM and XRD techniques. The results showed that the spent catalyst untreated by microwave irradiation was mainly composed of zinc acetate and activated carbon, but that of being treated by microwave irradiation was mainly composed of carbon and zinc oxide, and furthermore the pores of activated carbon were opened. The zinc oxide was adsorbed onto the surface and pores of spent catalyst, so the leaching agents could react with zinc oxide effectively. The spent catalyst after being leached was mainly composed of carbon; indicating the extraction of zinc from spent catalyst was very effective.

Keywords: spent catalyst; microwave irradiation; leaching; vinyl acetate

1 Introduction

Fresh catalyst used in vinyl acetate production contains up to 35wt. % zinc acetate on the surface of activated carbon. After its use, the catalyst becomes a waste that contains approximately 26wt. % zinc occurring in acetate, oxide or a metallic form as well as organic compounds, which are by – products of vinyl acetate synthesis. The catalyst is composed of granular activated carbon of high mechanical resistance and porosity. Spent catalysts are, on one hand, dangerous wastes, but, on the other hand, they are valuable secondary materials.

Efficient regeneration systems are required to permit a wider application of carbon adsorption processes and to ensure their economic feasibility[1]. This has motivated companies to develop methods for regenerating and reusing saturated or deactivated activated carbon and valuable metals. Over the years, a wide variety of regeneration techniques have been suggested and applied. These are based either on desorption, induced by increasing the temperature or by displacement with solvents, or on decomposition induced by thermal, chemical, catalytic or microbiological processes[2,3]. By far the most extensively used technique is thermal regeneration under steam or an inert atmosphere. Unfortunately, the regeneration process is time consuming and after successive heating and cooling cycles, the carbon becomes damaged. Eventually, it turns to dust and is properly disposed of. However, in the case of regeneration of spent carbon sorbents and catalysts, only chemical methods are to be considered. Chemical methods of regenerating spent carbon sorbents and catalysts consist mainly of removing metals and their compounds bound on the surface of the carbon via extraction with organic and inorganic solvents as well as supercritical solvents[4-13].

Microwaves are now being used in various technological and scientific fields in order to heat dielectric materials[14-17]. The main advantage of using microwave heating is that the treatment time can be considerably reduced, which in many cases represents a reduction in the energy consumption as well. In addition, the consumption of gases used in the treatment can be reduced. Microwave - induced chemical reactions can be used to solve the above problems associated with conventional surface heating because microwave heating is both internal and volumetric heating. Therefore, the tremendous thermal gradient from the interior of the char particle to its cool surface allows the microwave - induced reaction to proceed more quickly and effectively at a lower bulk temperature, resulting in energy savings and shortening the processing time[18]. In fact, microwave - induced chemical reactions have been used in a large number of applications, such as pyrolysis of high volatile bituminous coal with nitrogen and coal gasification in water vapor[19-21]. It has been found that char is a good receptor of microwave energy[19]. By applying a certain amount of microwave energy, the char may reach the minimum reaction conditions. Yet, in the particular case of carbon materials, there are relatively few publications that describe the use of microwaves for producing and regenerating activated carbon[22-24]. The results are very promising due to the rapid heating of the activated carbon by microwave energy. In addition, microwave technology allows the carbon to be recycled and reused a large number of times. This technique does not damage the carbon; rather, it increases the surface area allowing more contaminates to adhere, thereby increasing the value. Microwave energy was found to preserve the porous structure of the initial samples to a great extent, as compared to samples treated in the electric furnace[25]. However, relatively little information was available in the literature on the extraction of zinc from spent catalyst (zinc acetate/activated carbon) of vinyl acetate synthesis pretreated by microwave irradiation.

The objective of this paper was to investigate the feasibility of zinc extraction from spent catalyst of vinyl acetate synthesis pretreated by microwave irradiation, using NH_4HCO_3 and $NH_3 \cdot H_2O$ as leaching agents. X - ray diffraction and SEM analyses were conducted in order to characterize the employed spent catalysts treated and untreated by microwave irradiation. The effects of microwave pretreatment temperature, leaching time, ratio of liquid to solid, washing times, and dose of leaching agents on the leaching rate of zinc were evaluated systematically.

2 Experimental

2.1 Materials and apparatus

Spent catalyst of vinyl acetate synthesis was obtained from a chemical plant in Yunnan province,

China. The composition of the spent catalyst was presented in Table 1. NH_4HCO_3, $NH_3 \cdot H_2O$ and H_2O were used as leaching agents.

Table 1 Components of the spent catalyst (%)

Element	C	Zn	Fe	Si	Mg	Ca
wt. /%	81.71	8.76	0.5	0.15	0.007	0.015

Scanning electron microscopy (SEM, EMPA – 1600) was used to investigate the microstructure of the catalyst. XRD patterns were obtained with a Rigaku diffractometer (D/max 2500) using CuK_α radiation; the thin powder samples were placed onto an oriented monocrystalline quartz plate and scanned from 10 to 80° (2θ) at the speed of 5/min.

2.2 Features of pilot – scale microwave heating equipment

A microwave system typically consists of a generator to produce the microwaves, a waveguide to transport the microwaves and an applicator (usually a cavity) to manipulate microwaves for a specific purpose and a control system (tuning, temperature, power, etc.). A schematic diagram of the pilot – scale microwave heating equipment is shown in Fig. 1. The equipment used in present study mainly contained several parts as follows:

(1) Microwave can be generated by either thermionic devices such as magnetron, klystron, back wave tube (BWT), and gyrotron; or solid – state devices such as microwave transistors and diodes. Among these generators the magnetron is most widely used for the industrial applications due to their availability and low cost. The main components of the magnetron include filament, inside cylindrical cathode, outside tubular anode, magnet, exit passage, antenna, vane, and strip ring. Power conversion and control are realized by power regulator. The main components in the power regulator are power switches. The first component is the power unit where microwaves are generated at the required frequency band. Magnetron tubes are used to generate microwave power. The energy is coupled into the applicator (cavity) through a slot, an array of resonant slots, where the material is subjected to intense microwave fields, and to which any additional ancillary process equipment such as pumps for operation under moderate vacuum conditions, steam or hot air injection, must be connected. The power supply of the microwave heating equipment was sixty magnetrons at 2.45GHz frequency and 1kW power, which was cooled by water circulation. A control system maintains the inert atmosphere in the reactor cavity of the microwave unit, whose size was accurately calculated by microwave theories addressing materials that preferably absorb microwave energy.

(2) Raw materials were charged into a special ceramic tube inclining in the reactor cavity by a screw feeder and those products were discharged by turning of the tube.

(3) The rate of charge varying from 0 to 70kg/h was controlled by rotational speed of the screw and the rate of discharge ranging from 0 to 0.9kg/min. was controlled by the tube's turning speed.

(4) Steam was introduced into a stainless steel pipe at the bottom of the ceramic tube and distributed through many proportionally spacing holes, which could uniformly activate raw materi-

als. The flow rate of steam was controlled by adjusting the pressure of the steam between 0 and 0. 3MPa.

(5) The temperature measurement system, consisted of a temperature indicator and thermocouple fixed in the middle of the stainless steel pipe. Thermocouple provides feedback information to the control panel that controls the power to the magnetron, controlling the temperature of the sample during the regeneration process in order to prevent the sample from overheating.

(6) The input power of microwaves, rate of charging and discharging, flow rate of steam and the atmosphere were controlled by a control system, which is a on – line automatic control system.

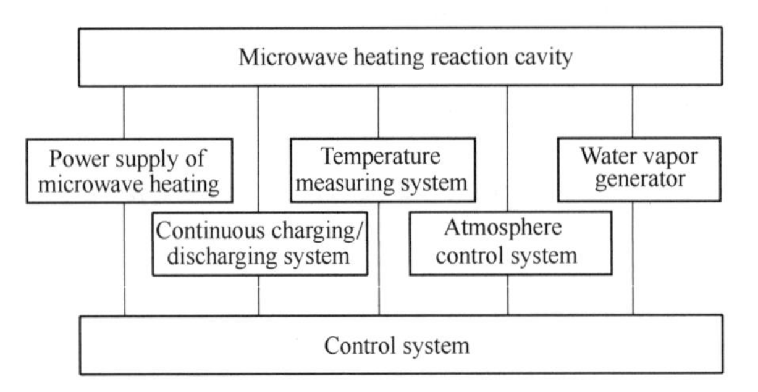

Fig. 1 Schematic diagram of the pilot – scale microwave heating equipment

For the present study, the experimental equipment for the microwave pretreatment of spent catalysts consisted basically of a microwave magnetron of a maximum output power 6000W (using only 1/10 of all magnetrons) at 2450MHz and a single mode cavity where the sample was exposed to microwave heating. The process parameters used in the present study were as follow: the rotation speed was 50r/min; the rate of discharge was 0. 4kg/min; the steam pressure was 0. 005Pa.

2. 3 Methods

Weighed a certain amount of spent catalyst (1kg, 5kg, 8kg, 10kg) pretreated using the self – made microwave heating equipment and un – pretreated by microwave irradiation, added leaching reagents (every 1kg spent catalyst adding 387. 5g NH_4HCO_3 and 900mL $NH_3 \cdot H_2O$) according to the defined ratio of liquid to solid (2∶1), and then stirring to leach in the agitation tank (1 – 5h) at room temperature. Agitation (360 – 720r/min) was provided by a magnetic stirrer that enabled adequate dispersion of the samples without evaporation loss of the solution, and then filtering using vacuum filter, washing (0 – 5 times using leaching reagents), and calculating the leaching rate of zinc.

The zinc extraction was determined by analyzing zinc concentration in solution and in residue using EDTA complexometric titration method.

3 Results and discussions

3. 1 Temperature rise curve of different mass spent catalysts pretreated by microwave irradiation

According to the interaction with microwave, materials can be categorized into three principal groups: transparent which are low loss materials, where microwaves pass through without any losses; conductors where microwaves are reflected and cannot penetrate; and absorbing which are high loss materials, where microwaves are absorbed depending on the value of the dielectric loss factor. We investigated the relationship between temperature and microwave heating time in order to know if the spent catalyst can be heated by microwave heating.

Different masses of spent catalysts (5kg, 8kg, 10kg) were treated using the self-made microwave heating equipment (Fig. 1) and the temperature rise curves were shown in Fig. 2. We could find that the temperature of the samples increased with increasing microwave irradiation time. As for 10kg sample, the temperature reached 1000℃ after being treated 4h, then controlling the microwave power and held at this temperature 30 - 45min, we found that the temperature nearly kept at 1000℃, indicating that the self-made microwave heating equipment could be controlled well and used for further study, and the spent catalysts used in the present study were microwave absorbing materials.

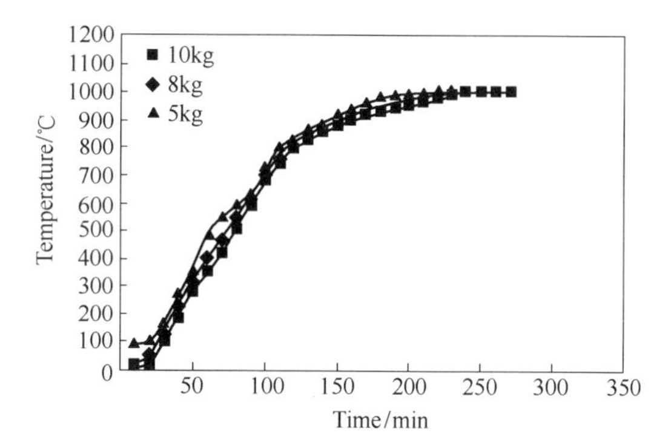

Fig. 2 Temperature rise curves of different mass spent catalysts pretreated by microwave irradiation

3.2 Effect of different pretreatment modes on the leaching rate

The effect of different pretreatment modes on the leaching rate is presented in Table 2. It was shown that the leaching rate of spent catalyst untreated by microwave irradiation was much lower than that of spent catalyst treated by microwave irradiation, indicating that microwave irradiation had an obvious effect on extraction of zinc. Fresh catalyst would deactivate after being used in a period of time, because the surface of the catalyst was polluted due to the formation of a large amount of coking, organic compounds which blocked the micropores of carbon[26]. So, the contact area between the leaching agents and spent catalyst was small during the leaching process, giving rise to a lower leaching rate. For the spent catalyst after being treated by microwave irradiation, the volatile organic compounds adsorbed onto the carbon porosity were eliminated by microwave irradiation, and the pores were opened, so the contact area between the leaching agent and spent catalyst and the leaching rate were increased subsequently. With microwave irradiation time of 3h, a maximum zinc extraction of 95.65% was obtained.

Table 2 The effect of different pretreatment modes on the leaching rate

Time/h	Zinc extraction/%	Zinc extraction/%
	Untreated by microwave irradiation	Treated by microwave irradiation
1	30.24	91.04
2	32.08	91.43
3	37.04	95.65
4	35.94	89.44
5	33.56	88.84

3.3 Characterization of spent catalyst untreated by microwave irradiation

Microtexture of activated carbon is often evaluated by X - ray diffraction (XRD) measurement, as well as TEM/SEM observation. In general, TEM/SEM brings information mainly on the microscopic crystallographic structure, while XRD measurement introduces the averaged knowledge about the structure of the sample. Characterization of activated carbon is effectively carried out by combination of these techniques[27].

In the present study, the spent catalyst untreated by microwave irradiation was characterized by XRD and SEM techniques, as shown in Figs. 3 and 4, respectively. It was observed from Fig. 3 that the spent catalyst was mainly composed of zinc acetate and activated carbon. It could be seen from Fig. 4 that the black particles were activated carbon and the white particles were zinc acetate adsorbed on the surface or pores of activated carbon. There was a certain amount of porosity in the spent catalyst, but adsorbates blocked the pores of the catalyst, causing a low leaching rate of zinc.

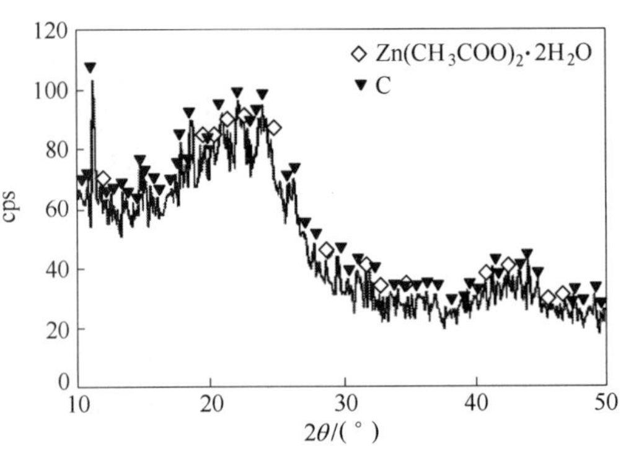

Fig. 3 XRD patterns of spent catalyst untreated by microwave irradiation

Fig. 4 SEM image of spent catalyst untreated by microwave irradiation

3.4 Characterization of spent catalyst pretreated by microwave irradiation

The spent catalyst was pretreated by microwave irradiation to 950℃ and also characterized by XRD and SEM techniques, as shown in Figs. 5 and 6, respectively. It was found that the spent catalyst after being pretreated by microwave irradiation was mainly composed of zinc oxide and activated carbon. Compared to Fig. 3, the peak of zinc acetate disappeared in Fig. 5, while the diffraction peak of zinc oxide, which was obtained by the pyrolysis of zinc acetate heated by microwave irradiation appeared. From the SEM image in Fig. 6, we could observe that the particles in the spent catalyst disappeared after being pretreated by microwave irradiation, adsorbates inside the pores were removed, and the pores of activated carbon were opened. The zinc oxide was adsorbed onto the surface and pores of spent catalyst, so the leaching agents (NH_4HCO_3 and $NH_3 \cdot H_2O$) could react with zinc oxide effectively, enhancing the zinc extraction.

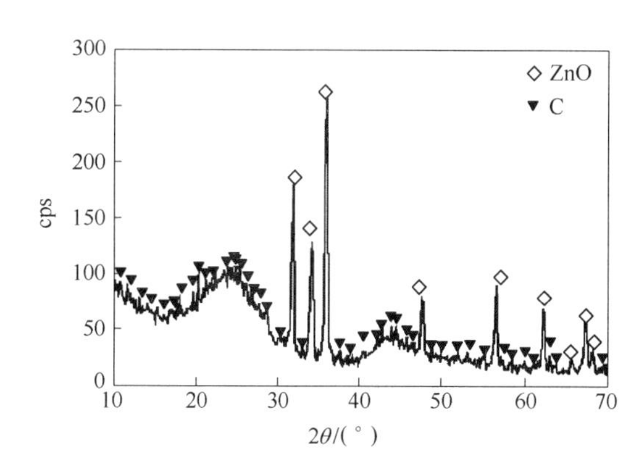

Fig. 5 XRD patterns of spent catalyst after microwave irradiation pretreatment

Fig. 6 SEM image of spent catalyst after microwave irradiation pretreatment

3.5 Effect of the temperature of microwave pretreatment on zinc extraction

Nine portions of spent catalyst were weighed, and pretreated using different microwave temperatures. The charge included 387.5g NH_4HCO_3 and 900mL $NH_3 \cdot H_2O$ according to a 2:1 of ratio of liquid to solid; leaching time was 3h; the residue was washed 4 times. The effect of microwave pretreatment temperature on the extent of zinc extraction is presented in Table 3. It is indicated that the extraction increased with increasing microwave pretreatment temperature. XRD characterization (Fig. 3) showed that the spent catalyst was mainly composed of activated carbon and zinc acetate. When the temperature reached 400℃, zinc oxide, carbon dioxide and water were formed due to the decomposition of zinc acetate. As the temperature rose, the organic compounds transformed into volatile species, subsequently the pores of activated carbon were opened, so that the leaching agent could react with zinc oxide inside the activated carbon, increasing the leaching rate. When the temperature reached about 1000℃, zinc oxide inside activated carbon was reduced to zinc by carbon, and evaporated, causing a lower zinc extraction. So, the microwave pretreatment temperature was chosen to be 950℃ in order to increase the extent of zinc leaching and to decrease zinc loss. With microwave irradiation temperature of 950℃, a maximum zinc extraction of 90% was obtained.

Table 3 Effect of the temperature of microwave irradiation pretreatment on the extent of leaching

Temperature/℃	450	550	650	750	850	950	1050	1150	1250
Zinc extraction/%	54. 5	68. 5	71. 2	76. 6	84	90	89. 89	88	85

3.6 Effect of leaching time on zinc extraction

The best charge was 1000g spent catalyst that had been pretreated by microwave irradiation, 387. 5g NH_4HCO_3 and 900mL $NH_3 \cdot H_2O$ according to a 2∶1 of ratio of liquid to solid. Leaching time were 1h, 2h, 3h, 4h or 5h and the residues were washed 4 times. Table 4 shows the effect of leaching time on the extent of leaching. We could see that zinc extraction increased slightly with increasing leaching time from 1h to 2h, the maximum zinc extraction was obtained when the leaching time was 3h, and then the leaching rate decreased with increasing leaching time. The leaching agent may react with zinc oxide inside activated carbon according to the following reaction equation.

$$ZnO + NH_4HCO_3 + (i-1)NH_3 \rightleftharpoons Zn(NH_3)_iCO_3 + H_2O \quad (1)$$

This is a reversible reaction, the complex is $Zn(NH_3)_iCO_3$ ($i = 1, 2, 3, 4$), and as ammonia in solution increases the value of i increases and the complex becomes more stable. As leaching time increased during the stirring process in the tests, the leaching agents reacted with zinc oxide adsorbed onto activated carbon to form the complex $Zn(NH_3)_iCO_3$ ($i = 1, 2, 3, 4$), transferring into liquid phase, so that the separation of zinc from activated carbon could be realized effectively. But, as leaching time increased the ammonia would evaporate and the concentration of ammonia in solution decreased, the value of i decreased, the complex $Zn(NH_3)_iCO_3$ ($i = 1, 2, 3, 4$) became unstable, the zinc oxide was adsorbed onto activated carbon again, giving rise to a lower extent of leaching.

Table 4 Effect of leaching time on the extent of leaching

Leaching time/h	1	2	3	4	5
Zinc extraction/%	91. 04	91. 43	95. 65	89. 44	88. 84

3.7 Effect of the ratio of liquid to solid on the leaching rate

In the baseline test, 1000g spent catalyst pretreated by microwave irradiation was combined with 387. 5g NH_4HCO_3 and 900mL $NH_3 \cdot H_2O$ according to a 2∶1 of ratio of solid to liquid; leaching time was 3h and the residue was washed 4 times. Table 5 shows the effects of the ratio of liquid to solid on the extent of zinc leaching. During the experimental process, the spent catalyst containing zinc oxide and activated carbon still had a certain adsorption capacity. When the ratio of liquid to solid was less than 1. 5, a portion of solution would be adsorbed onto the activated carbon, leading to viscous slurry, which was difficult to stir, this resulted in decrease in the contact area between leaching agents and the spent catalyst, slowing down the chemical reactions. When the ratio of liquid to solid was increased to a certain value, the leaching rate increased slightly, so the increase of ratio of liquid to solid had no significance to the reaction due to the completion of leaching process. So, the optimum ratio of liquid to solid was chosen to be 2∶1 for present study.

Table 5 Effect of the ratio of liquid to solid on the extent of leaching

Ratio of liquid to solid	1.5	2	3	4	5	6
Zinc extraction/%	85.46	89.04	89.54	89.44	89.69	89.94

3.8 Effect of the extent of washing on the extent of leaching

In each leaching test 1000g spent catalyst pretreated by microwave irradiation was combined with 387.5g NH_4HCO_3 and 900mL $NH_3 \cdot H_2O$ according to a 2∶1 of ratio of solid to liquid. The leaching time was 3h. The effect of the numbers of wash stages on the extent of zinc leaching is shown in Table 6. Zinc extraction was limited by the extent of washing if the residue was washed fewer than four times. In the present study, the optimum number of washes was chosen to be 4.

Table 6 Effect of washing on zinc extraction

Number of washes	0	1	2	3	4	5
Zinc extraction/%	78.68	85.78	88.56	91.34	95.65	95.9

3.9 Effects of the dose of leaching agents on the leaching rate

1000g spent catalyst being pretreated by microwave irradiation was combined with a certain amount of NH_4HCO_3, $NH_3 \cdot H_2O$ and water according to a 2∶1 of ratio of solid to liquid. The leaching time was 3h and the residue was washed 4 times. The effects of the amount of NH_4HCO_3 and $NH_3 \cdot H_2O$ on the leaching rate are illustrated in Tables 7 and 8, respectively. It could be seen that the extent of leaching was lower when the leaching agent NH_4HCO_3 was less than 387.5g and $NH_3 \cdot H_2O$ was less than 900mL.

Table 7 Effect of the amount of NH_4HCO_3 on the extent of leaching

Amount of ammonium bicarbonate/g	240	310	387.5	465	544	600
Zinc extraction/%	87.54	90.04	92.86	92.66	92.96	92.96

Table 8 Effect of the amount of $NH_3 \cdot H_2O$ on the leaching rate

Amount of ammonia/mL	600	720	900	1080	1260
Zinc extraction/%	88.54	88.65	91.04	90.51	91.14

3.10 Characterization of spent catalyst after being leached

Experimental conditions were as follows: the ratio of liquid to solid 2∶1; stirring at room temperature 3h; the filter residue washed with water 4 times and then dried. The spent catalyst with microwave irradiation pretreatment after being leached was characterized by XRD and SEM techniques, which are shown in Figs. 7 and 8, respectively. We could see obviously from Fig. 7 that the spent catalyst pretreated by microwave irradiation after being leached was mainly composed of activated carbon; compared to Fig. 5, the diffraction peak of zinc oxide disappeared in Fig. 7, and only carbon peaks appeared. This was because of the formation of complex$[Zn(NH_3)_4]CO_3$ through the reaction between ZnO adsorbed onto the surface or pores of the activated carbon and leaching agents ($NH_3 \cdot H_2O$ and NH_4HCO_3), giving rise to ZnO desorbed from activated carbon into liquid phase to realize the separation between zinc oxide and carbon effectively. It was shown in the SEM image (Fig. 8) that the particles adsorbed onto the activated carbon after extraction of zinc and adsorbates in the macropores of activated carbon were removed, but the micropores could not be seen, indicating the micropores were not opened completely.

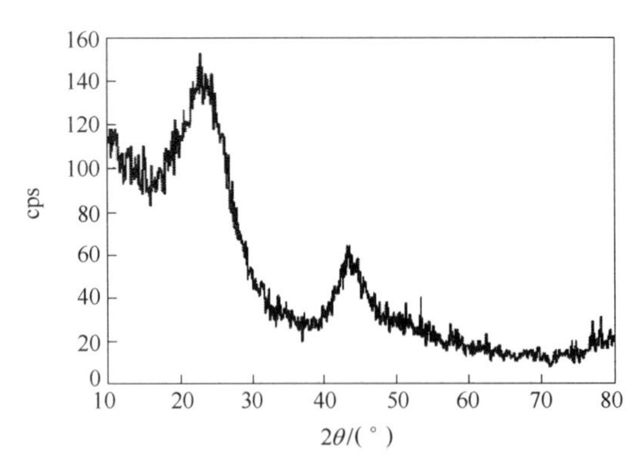

Fig. 7 XRD pattern of spent catalyst by microwave irradiation after being leached

Fig. 8 SEM image of spent catalyst pretreated by microwave irradiation after being leached

3.11 The results of the optimal conditions

Through the above experiments the optimal conditions were obtained as follows: microwave pretreatment temperature: 950℃; leaching agent: NH_4HCO_3 and $NH_3 \cdot H_2O$; dose of leaching agents: 2.5 times the theoretical amount; ratio of liquid to solid: 2∶1; leaching time at room temperature: 3h; washing stages: 4. The average result of the extent of zinc extraction under optimal conditions was 93.45%, as listed in Table 9.

Table 9 Experimental results under optimal conditions

Test	Zinc content in spent catalyst/%	Zinc content in leach residue/%	Zinc extraction/%
1	10.04	0.44	95.62
2	8.96	0.77	91.41
3	10.32	0.69	93.31
Average	9.77	0.63	93.45

4 Conclusions

From the experimental results, the following conclusions could be drawn:

(1) Spent catalyst of vinyl acetate synthesis pretreated by microwave irradiation exhibits enhancement of the leaching of zinc.

(2) Spent catalyst vinyl acetate synthesis untreated by microwave irradiation was mainly composed of zinc acetate and activated carbon. There was a certain amount of porosity in spent catalyst, but adsorbates blocked the pores of the catalyst, giving rise to a low leaching rate of zinc.

(3) Spent catalyst of vinyl acetate synthesis after being pretreated by microwave irradiation contained zinc oxide, which was obtained by the pyrolysis of zinc acetate and thus spent catalyst was mainly composed of zinc oxide and carbon. Adsorbates inside the pores were removed, and the pores of activated carbon were opened. The zinc oxide was adsorbed onto the surface and pores of carbon, so that the contact area with leaching agent was increased.

(4) Zinc oxide adsorbed onto the carbon could be leached out easily using $NH_3 \cdot H_2O$ and NH_4HCO_3.

(5) Spent catalyst was mainly composed of carbon after being pretreated by microwave irradiation and leached using $NH_3 \cdot H_2O$ and NH_4HCO_3, and the particles adsorbed onto the activated carbon after extraction of zinc and adsorbates in the macropores of activated carbon were removed, realizing the separation between the zinc oxide and carbon effectively.

Acknowledgements

Financial support for this work from specialized research fund for International Cooperation Project of Science and Technology in China(2006GH01) is gratefully acknowledged.

References

[1] Mattson J S, Mark H B. Activated carbon: surface chemistry and adsorption from solution[M]. New York: M. Dekker, 1971.

[2] Dranca I, Lupascu T, Vogelsang K, et al. Utilization of thermal analysis to establish the optimal conditions for regeneration of activated carbons[J]. Journal of Thermal Analysis and Calorimetry, 2001, 64(3): 945 - 953.

[3] Sheintuch M, Matatov - Meytal Y I. Comparison of catalytic processes with other regeneration methods of activated carbon[J]. Catalysis Today, 1999, 53(1): 73 - 80.

[4] Cline S R, Reed B E. Lead removal from soils via bench - scale soil washing techniques[J]. Journal of Environmental Engineering, 1995, 121(10): 700 - 705.

[5] Higashidate S, Yamauchi Y, Saito M. Enrichment of eicosapentaenoic acid and docosahexaenoic acid esters from esterified fish oil by programmed extraction - elution with supercritical carbon dioxide[J]. Journal of Chromatography A, 1990, 515: 295 - 303.

[6] Hu H, Guo S. Kinetic study of supercritical extraction of oil shales[J]. Fuel Processing Technology, 1992, 31(2): 79 - 90.

[7] Mishra V S, Mahajani V V, Joshi J B. Wet air oxidation[J]. Industrial & Engineering Chemistry Research, 1995, 34: 2 - 48.

[8] Mvndale V D, Joglekar H S, Kalam A, et al. Regeneration of spent activated carbon by wet air oxidation[J]. The Canadian Journal of Chemical Engineering, 1991, 69(5): 1149 - 1159.

[9] Reed B E, Carriere P C, Moore R. Flushing of a Pb(Ⅱ) contaminated soil using HCl, EDTA, and $CaCl_2$[J]. Journal of Environmental Engineering, 1996, 122(1): 48 - 50.

[10] Smith R M. Sample preparation perspectives: supercritical fluid extraction of natural products. LC - GC International, 1996, 9: 8 - 15.

[11] Tan C S, Liou D C Desorption of ethyl acetate from activated carbon by supercritical carbon dioxide [J]. Industrial & Engineering Chemistry Research, 1988, 27(6): 988 - 991.

[12] Wang S, Elshani S, Wai C M. Selective extraction of mercury with ionizable crown ethers in supercritical carbon dioxide[J]. Analytical Chemistry, 1995, 67(5): 919 - 923.

[13] Wilhelm A, Hedden K. A non - isothermal experimental technique to study coal extraction with solvents in liquid and supercritical state[J]. Fuel, 1986, 65(9): 1209 - 1215.

[14] Appleton T J, Colder R I, Kingman S W, et al. Microwave technology for energy - efficient processing of waste [J]. Applied Energy, 2005, 81(1): 85 - 113.

[15] Jones D A, Lelyveld T P, Mavrofidis S D, et al. Microwave heating applications in environmental engineering—a review[J]. Resources, Conservation and Recycling, 2002, 34(2): 75 - 90.

[16] Thostenson E T, Chou T W. Microwave processing: fundamentals and applications[J]. Composites Part A: Applied Science and Manufacturing, 1999, 30(9): 1055 - 1071.

[17] Venkatesh M S, Raghavan G S V. An overview of microwave processing and dielectric properties of agri - food materials[J]. Biosystems Engineering, 2004, 88(1): 1 - 18.

[18] Guo J, Lua A C. Preparation of activated carbons from oil - palm - stone chars by microwave - induced carbon dioxide activation[J]. Carbon, 2000, 38(14): 1985 - 1993.

[19] Djebabra D, Dessaux O, Goudmand P. Coal gasification by microwave plasma in water vapour[J]. Fuel, 1991, 70(12): 1473 - 1475.

[20] Fu Y C, Blaustein B D, Sharkey Jr A G. Reaction of coal with nitrogen in a microwave discharge[J]. Fuel, 1972, 51(4): 308 - 311.

[21] Marsh H, Crawford D, O'Grady T M, et al. Carbons of high surface area. A study by adsorption and high resolution electron microscopy[J]. Carbon, 1982, 20(5): 419 - 426.

[22] Cha C Y. Microwave induced reactions of SO_2 and NO_x decomposition in the char - bed [J]. Research on Chemical Intermediates, 1994, 20(1): 13 - 28.

[23] Fang C S, Lai P M C. Microwave regeneration of spent powder activated carbon [J]. Chemical Engineering Communications, 1996, 147(1): 17 - 27.

[24] Kong Y, Cha C Y. Microwave - induced regeneration of NO_x - saturated char[J]. Energy & Fuels, 1996, 10(6): 1245 - 1249.

[25] Ania C O, Menéndez J A, Parra J B, et al. Microwave - induced regeneration of activated carbons polluted with phenol. A comparison with conventional thermal regeneration[J]. Carbon, 2004, 42(7): 1383 - 1387.

[26] Dabek L. Sorption of zinc ions from aqueous solutions on regenerated activated carbons[J]. Journal of Hazardous Materials, 2003, 101(2): 191 - 201.

[27] Yoshizawa N, Maruyama K, Yamada Y, et al. XRD evaluation of CO_2 activation process of coal - and coconut shell - based carbons[J]. Fuel, 2000, 79(12): 1461 - 1466.

Chapter Ⅲ

New Technology of Ultrasonic Metallurgy

A Comparison of the Conventional and Ultrasound – Augmented Leaching of Zinc Residue Using Sulfuric Acid

Xin Wang, Shaohua Ju, Dajin Yang, Jinhui Peng, C. Srinivasakannan

Abstract: This paper attempts to capture the difference between the ultrasound – assisted and conventional sulfuric acid leaching of zinc from zinc residue, having a Zn content of 12.31%, along with other metallic compounds such as Fe, Pb, SiO_2. The leaching temperature, sulfuric acid concentration, particle size, liquid/solid ratio and the ultrasound power have been chosen as parameters for investigation. The shrinking core model is utilized for analyzing the rate controlling step in the leaching process. Only a maximum of 67% of zinc could be leached using conventional process, while 80% could be leached with the ultrasound augmentation. For both the processes the rate controlling step was identified to be the diffusion through the product layer. The reaction order with respect to the sulfuric acid concentration was found to be 1.33 and 0.94, while the activation energy being are 13.07kJ/mol and 6.57kJ/mol, for the conventional and ultrasound assisted leaching process. The raw as well as the leached residue were characterized using XRD and SEM/EDX analysis.

Keywords: kinetic; zinc; leaching; sulphuric acid; ultrasound – augmented; conventional leaching

1 Introduction

Currently, vast majority of metallic zinc is manufactured by the Roast – Leach – Electrowinning (RLE) process[1], which produces a large amount of leach residues as insoluble substance forms a layer on the zinc surface during the acid leaching process, resulting in wastage of large amount of valuable metals[2]. The zinc residue is categorized as hazardous waste, further creating disposal problems to the operating plants[3]. High demand for this valuable material has promoted the ancillary industries to utilize the secondary sources such as zinc ash, zinc dross and leach residues as potential valuable sources[4].

In the process of leaching, many insoluble materials are concentrated in the residue, with part of zinc present in the form of zinc ferrite, which contribute for high zinc losses in hydrometallurgical processes[5]. Altundogan and Safarzadeh have reported recovery of valuable metals from the zinc plant residue[6,7]. The kinetics of recovering zinc from residues using sulfuric acid dissolution has been reported in literature [8-11]. The effect of acid concentration and leaching temperature along with the reaction order and the activation energy of the leaching of zinc residue have also been reported in literature[12-15].

Over the past decades, many techniques have been developed for improving the recovery of valuable metals. They include application of microwave[16], ultrasound[17], pressure[18] and mechanical activation[19]. The use of ultrasound is reported to increase the leaching yield as well as shorten the duration of leaching due to well known cavitation effects, rendering it to be a popular method in the metal processing industry. Balasubrahmanyam Avvaru, et al. [20] have reported a significant increase in the leaching efficiency with the use of ultrasonic as compared to conventional mixing methods in the process of extraction of uranium (a radioactivity element) from Narwapahar uranium ore with nitric acid and sulphuric acid being the leaching agent. Hursit, et al. [21] have reported enhanced faster recovery of zinc from smithsonite ore with ultrasound assisted leaching. Swamy, et al. [22] have designed a dual frequency ultrasonic leaching equipment which was reported to increase the recovery by 46% as compared to a conventional agitation process. Salim Öncel, et al. [23] have reported a 96.6% yield of silver from solid residue using microwave assisted leaching from thiourea. A 20% increase in the % yield of TiO_2 was reported from red mud as compared to conventional leaching methods under identical conditions[24]. An overall compilation of the effectiveness of ultrasound assisted leaching can be inferred from the comparison with the conventional leaching process[24]. Bese[25] has reported % yields for ultrasound – assisted leaching to be 89.28% for copper, 51.32% for zinc, 69.87% for cobalt, and 1.11% for iron vs. 80.41% for copper, 48.28% for zinc, 64.52% for cobalt, and 12.16% for iron, in absence of ultrasound.

The XRD and SEM were utilized to identify the difference in morphology and phase constitution of the initial residue in comparison with the ultrasonic – assisted leached residue and conventional leached residue. All the influencing parameters such as the ultrasound power (80 – 240W), leaching temperature (55 – 85℃), particle size (53 – 104μm), sulphuric acid concentration (40 – 170g/L) and liquid/solid (3 – 6) have been investigated on both the forms of leaching. The kinetics of leaching process is modeled using the shrinking core model (SCM), in addition to the estimation of the order of the reaction and the activation energy.

2 Materials and methods

2.1 Materials

The zinc residues used for the experiments were supplied by a metallurgical plant of Yunnan province. Prior to use in the leaching experiments the samples were washed, dried, sieved to different size ranges. X – ray diffraction analysis (XRD) of the zinc residue shows the mineralogical content of the sample (Fig. 1).

XRD analysis shows the $ZnFe_2O_4$ and $PbSO_4$ and SiO_2 as the major components in the sample. The chemical composition and particle size distribution of the zinc residues were also analyzed using an atomic absorption spectrometer and the results are shown in Table 1[26].

Table 1 Chemical analysis of different sieve fractions of zinc residue

Serial number	Size/μm	Content/%	Element/%			
			Zn	Fe	Pb	SiO_2
1	104 – 89	9	12.68	21.22	12.26	6.99
2	89 – 74	11	12.80	20.36	12.38	6.95
3	74 – 61	16	12.48	21.75	12.50	7.03

Continues Table 1

Serial number	Size/μm	Content/%	Element/%			
			Zn	Fe	Pb	SiO_2
4	61 - 53	18	12.36	21.64	12.19	7.00
5	<53	30	12.58	21.15	12.07	7.10

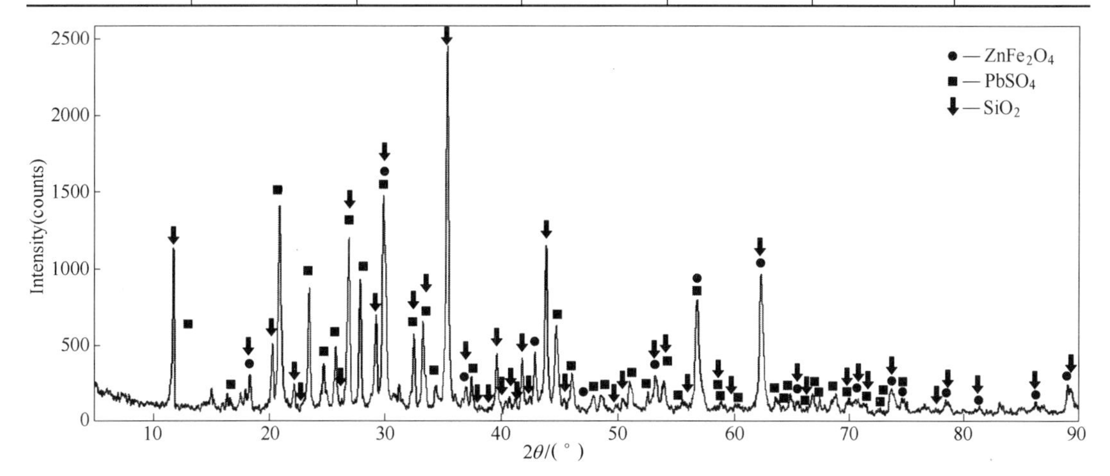

Fig. 1 XRD patterns of zinc plants residues

2.2 Experimental procedure

Both the conventional and ultrasound - assisted leaching test have been carried out in a 1000mL capped - beaker in thermostatically controlled water bath, equipped with sampling device and thermometer (temperature controlled within ± 0.1℃). In the case of ultrasound - assisted leaching process the mechanical stirrer was replaced with a ultrasonic transducers. The ultrasonic transducer generates ultrasonic waves with a frequency of 20kHz at varying powers of 80W, 160W and 240W. (The apparatus was manufactured by Guangzhou Hengda Ultrasonic Electric Technological Ltd., China.). The schematic diagram of the ultrasound and conventional leaching experimental setup is shown in Fig. 2.

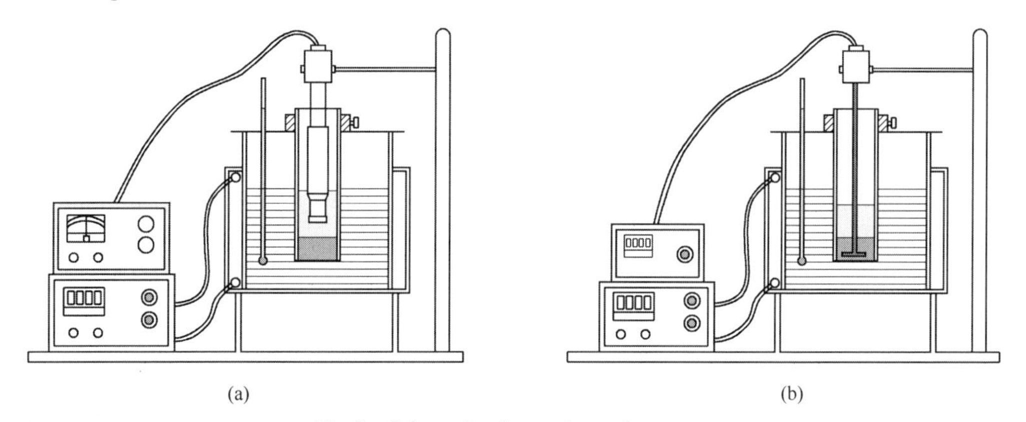

Fig. 2 Schematic of experimental set - up

(a) Ultrasound - assisted leaching set - up; (b) Conventional leaching set - up

Deionized water is utilized as a diluent with sulphuric acid to prepare various concentrations of acid solution. The dependent variables were varied in the range: acid solution/zinc residues weight ratio of 3 to 6, the acid concentration between 40g/L to 170g/L, the ultrasound power from 80W to 240W, and the leaching temperature in the range of 55℃ to 85℃. At selected time intervals, a small known amount(5mL) of slurry is withdrawn and quickly filtered. The zinc content of the filtered sample was estimated using atomic absorption spectrometer. The Zn recovery was estimated, using the following equation:

$$Zn_{recovery} = \frac{Zn_{T_0} - Zn_T}{Zn_{T_0}}$$

The zinc concentration was analyzed by atomic absorption spectrometer. The analyses of the solid residues were carried out by SEM – EDS (XL30ESEM – TMP scanning electron microscope, Philips, Holland) and XRD (Brukerd – advance diffractometer, Germany) with a step size of 0.02° and ranging from 10° to 90° at 30mA and 40kV.

3 Results and discussion

Reaction between zinc ferrite and sulfuric acid can be written as follows[9,27]:

$$2ZnFe_2O_4 + 8H_2SO_4 \longrightarrow 2Fe_2(SO_4)_3 + 2ZnSO_4 + 8H_2O$$

Based on the above reaction, effect of variables ultrasound power, leaching temperature, acid concentration, particle size and liquid/solid ratio were assessed on the leaching rate.

3.1 Kinetics analysis

Solid – fluid heterogeneous reactions are common in chemical and hydrometallurgical processes. In order to determine the kinetic parameters and rate controlling step of Zn leaching process, the popular shrinking core model was utilized[28].

It is well known that in any solid – fluid reaction systems, the reaction rate may either be controlled by the diffusion through the product layer or by the chemical reaction at the surface of the material[29].

If the reaction rate is controlled by diffusion through the product layer, the SCM can be simplified through the integrated form as represented by Eq. (1), as follows:

$$1 - \frac{3}{2}X - (1 - X)^{2/3} = k_d t \tag{1}$$

If the reaction rate is controlled by the surface reaction, the SCM can be simplified to the following form as represented by Eq. (2):

$$1 - (1 - X)^{1/3} = k_r t \tag{2}$$

where X is the fraction reacted; k_r is the kinetic parameter for reaction control; k_d the kinetic parameter for product diffusion control; t reaction time (min). The kinetic parameters are estimated based on the linear relationship between the left hand side and right hand side of the equation. The temperature dependence of the reaction was estimated based on the Arrhenius Eq. (3):

$$k_d = A\exp\left(\frac{-E_a}{RT}\right) \tag{3}$$

where, A is frequency factor; E_a the activation energy of the reaction; R the universal gas constant; T is absolute temperature.

3.1.1 Effect of ultrasound power

The leaching efficiencies of zinc with respect to ultrasound powers at an acid concentration of 140g/L, particle size of 53 – 61μm, Liquid/solid of 4∶1 and leaching temperature of 65℃ is shown in Fig. 3.

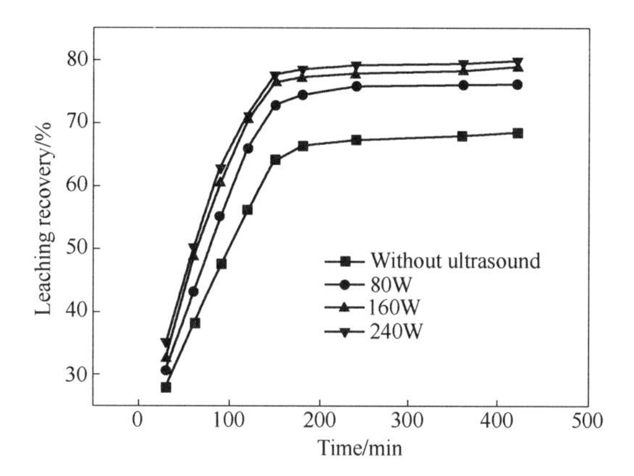

Fig. 3 Effect of ultrasound power on leaching recovery

As can be seen from the Fig. 3, leaching recovery of zinc increased with the increase in the ultrasonic power, however only up to a leaching duration of 180min, beyond which no significant increase was observed. The rate of leaching was found to increase with increase in the ultrasound power, with maximum % leaching at the highest of the ultrasound power. However, there is no significant increase was observed with the increase in power from 160W to 240W. The ultrasound augmented leaching zinc could achieve a recovery of 80%, while the recover only 70% clearly indicating the effectiveness of the conventional leaching. This could be attributed to the fact that the ultrasound reduces the mass transfer resistance for the movement of the ions from the surface of zinc particle into liquid, which is known to be controlled by diffusion resistance[30]. In Addition the ultrasound incites liquid cavitations' that blows out the surface of zinc particle covered by insoluble substance like zinc ferrite($ZnFe_2O_4$) and lead sulfate($PbSO_4$), quartz(SiO_2), generating a highly reactive surface[24].

3.1.2 Effect of leaching temperature

The effect of leaching temperature is assessed in the range of 55 – 85℃, comparing conventional leaching with ultrasound augmented leaching as shown in Fig. 4. The process parameters are held at a ultrasound power of 160W, acid concentration of 170g/L, particle size of 74 – 89μm and liquid/solid ratio of 4∶1 for the ultrasound assisted leaching (Fig. 4(a)), while the parameters are held at acid concentration of 140g/L, particle size of 61 – 53μm, liquid/solid of 4∶1 for conventional leaching (Fig. 4(b)).

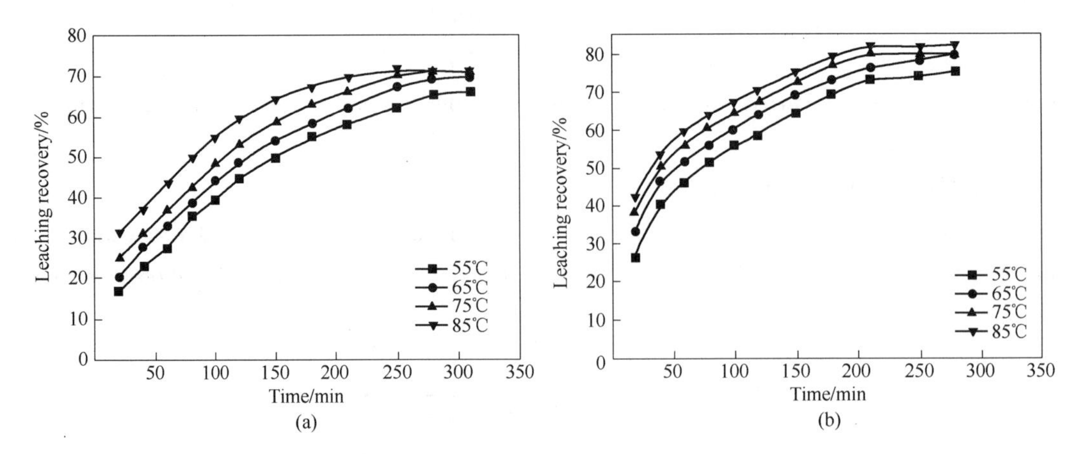

Fig. 4 Effect of leaching temperature on leaching recovery

(a) Conventional leaching; (b) Ultrasound - assisted leaching

An increase in temperature clearly improves the % recovery of Zn, for both the modes of leaching. However, recovery using ultrasound augmented leaching is higher than the conventional leaching under the identical process conditions an increase in temperature is known to increase the molecular movement or reduce the diffusional resistance, which could result in an increased leaching rate and hence an increased recovery. Further an increase in leaching temperature would reduce the activation energy for the reaction, which could additionally enhance the rate of reaction.

3.1.3 Effect of leaching acid concentration

The effect of sulfuric acid concentration is assessed at different concentrations ranging from 30g/L to 170g/L comparing the conventional and ultrasound - assisted leaching. The ultrasound power is held at 160W, leaching temperature at 65℃, particle size of 74 - 89μm and liquid/solid 4∶1 (Fig. 5(b)), while for the conventional leaching, leaching temperature at 75℃, acid concentration at 170g/L, particle size at 53 - 61μm and liquid/solid ratio of 4∶1 (Fig. 5(a)).

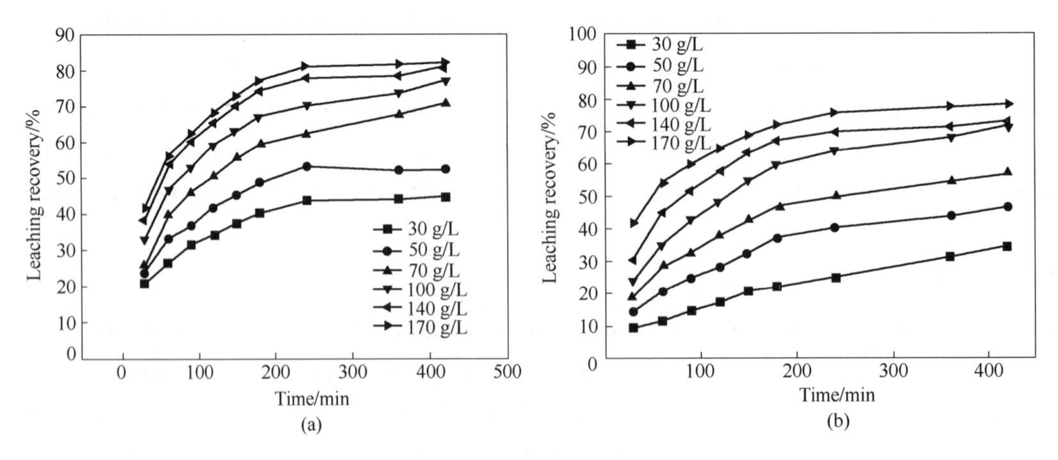

Fig. 5 Effect of acid concentration on leaching recovery

(a) Ultrasound - assisted leaching; (b) Conventional leaching

As seen in Fig. 5, an increase in acid concentration increases the dissolution rate of zinc, with the percentage recovery higher for ultrasound - augmented leaching as compared to conventional leaching. During the process of leaching the sulfuric acid is continuously consumed as it gets converted into product, resulting in a possible sulfuric acid devoid situation at low acid concentrations and high leaching durations, limiting the maximum recovery. An increase in concentration is known to drive the reaction of zinc residue with sulfuric acid, facilitating an increase in leaching. In the case of ultrasound assisted leaching the zinc molecules are known to absorb the ultrasonic energy and vibrate at the balance site more vehemently[21]. Additionally ultrasound can generate tiny bubbles in the solution around residue particles congregate, amalgamate, grow quickly and collapse transitorily resulting in high temperature and pressure within the tiny bubbles, would facilitate breaking the limit of the reaction.

3.1.4 Effect of particle size

The effect of five different particle sizes(+ 104μm to 84μm, + 89μm to 74μm, - 74μm to 61μm, + 61μm to 53μm and - 53μm) at an ultrasound power of 160W, at a leaching temperature of 65℃, acid concentration 140g/L, particle size of 74 - 89μm and liquid/solid 4:1, is shown in Fig. 6(a). Fig. 6(b) shows the effect of particle size using conventional leaching at a leaching temperature of 75℃, acid concentration of 170g/L, particle size of 53 - 61μm, and liquid/solid ratio of 4:1.

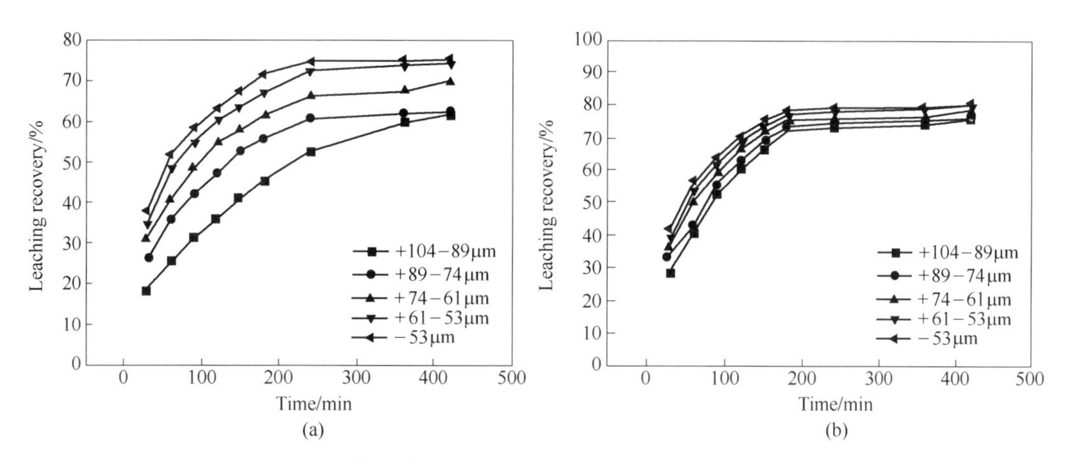

Fig. 6 Effect of leaching particle size on leaching recovery
(a) Conventional leaching; (b) Ultrasound - assisted leaching

As can be seen from Fig. 6, both the plots show an increase in the recovery with reduction in particle size, however the effect was very significant in the conventional leaching as compare to the ultrasound assisted leaching. The insignificant effect of ultrasound could be due to the ability of the ultrasound to renew the reaction surface through its ability to blow out the insoluble solid film layer.

3.1.5 Effect of liquid/solid ratio

The effect of liquid/solid ratio is examined covering the range from 3:1 to 6:1, both with conventional an ultrasound assisted leaching process. In the case of ultrasound assisted leaching, an ultrasound power of 160W is utilized with the other parameters held at a leaching temperature of 65℃, acid concentration of 140g/L, particle size of 74 - 89μm, while with the conventional leaching, a leaching temperature of 75℃, acid concentration of 170g/L, particle size of 53 - 61μm.

Fig. 7 shows that an increase in the liquid solid ratio facilitates an improvement in the recovery, for both the conventional as well as the ultrasound assisted process. However a significant improvement was observed until an liquid/solid ratio of 6:1 for the conventional process, while the effect was significant only up to an liquid/solid ratio of 5:1, with the ultrasound assisted process. Overall the recovery was found to be higher with the ultrasound assisted process as compared to conventional process. An increase in the liquid/solid ratio is expected to improve the mass transfer conditions due to reduced solution viscosity, facilitating a improved recovery.

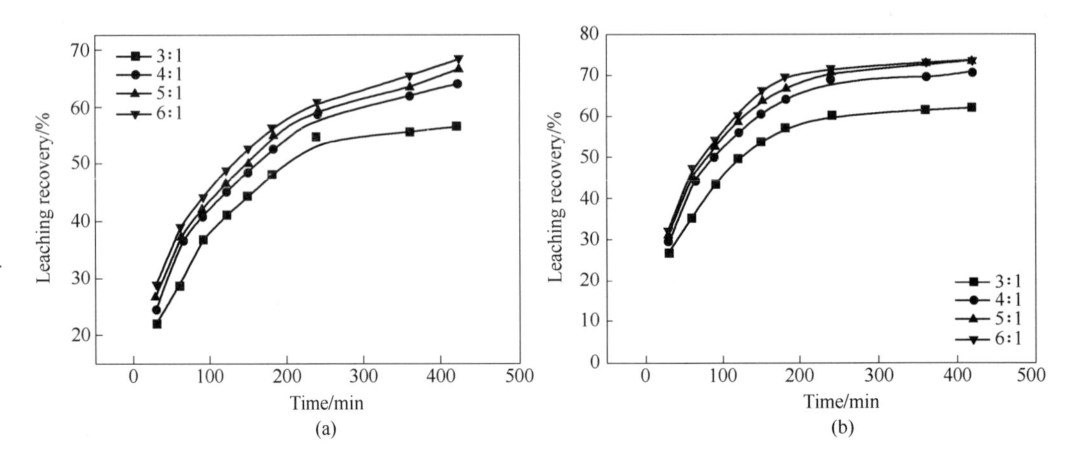

Fig. 7 Effect of liquid/solid ratio on leaching recovery

(a) Conventional leaching; (b) Ultrasound - assisted leaching

3.2 Kinetics of leaching

Solid - fluid heterogeneous reactions are common in chemical and hydrometallurgical processes. To determine the kinetic parameters and rate controlling step of Zn leaching process, the popular shrinking core model was utilized. The Eqs. (1) and (2) are applied for the results obtained from the Fig. 4. Fig. 8 presents the $1-(1-X)^{1/3}$ versus time plot according to chemical reaction control. Fig. 9 shows the $1-2/3X-(1-X)^{2/3}$ versus time plot according to diffusion control process. The results show that the straight lines with the regression coefficient R^2 in Fig. 9 are closer to 1 than in Fig. 8. So Eq. (2) is found to fit the data best, and the zinc leaching from this plant residue is controlled by diffusion through product layer with both of the ultrasound augmented and conventional acid leaching augmented and conventional acid leaching.

The apparent rate constants (k_d) is calculated as slopes of the straight lines. Using the k_d obtained by application of Eq. (3), the Arrhenius plot is used showing the variation of $\ln k_d$ vs. $1/T$, where T is temperature in the range of 55℃ to 85℃, and yields the straight line shows in Fig. 10 (a) and (b). For conventional acid leaching, the activation energy of zinc leaching in sulfuric acid is calculated to be 13.07 kJ/mol.

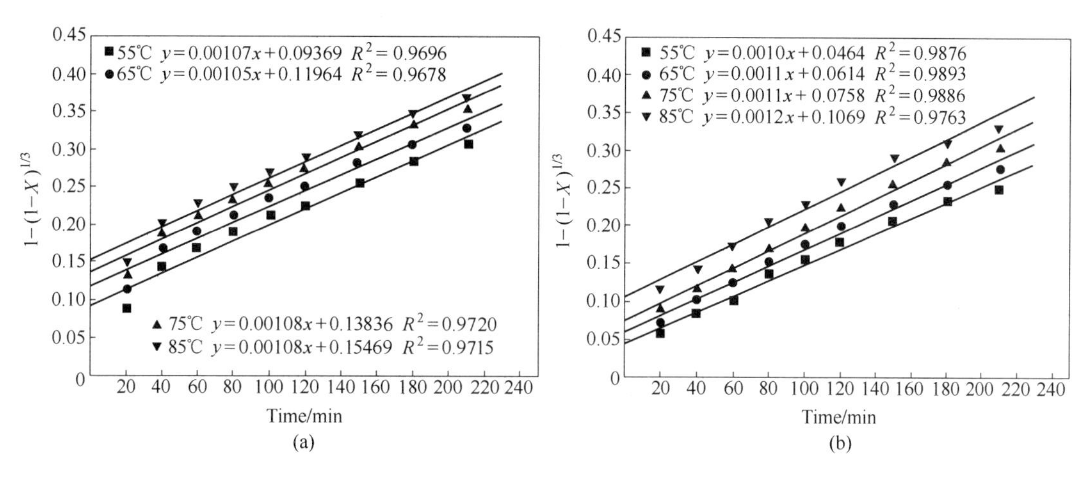

Fig. 8 The variation of $1-(1-X)^{1/3}$ with time at various temperatures

(a) Conventional leaching; (b) Ultrasound - assisted leaching

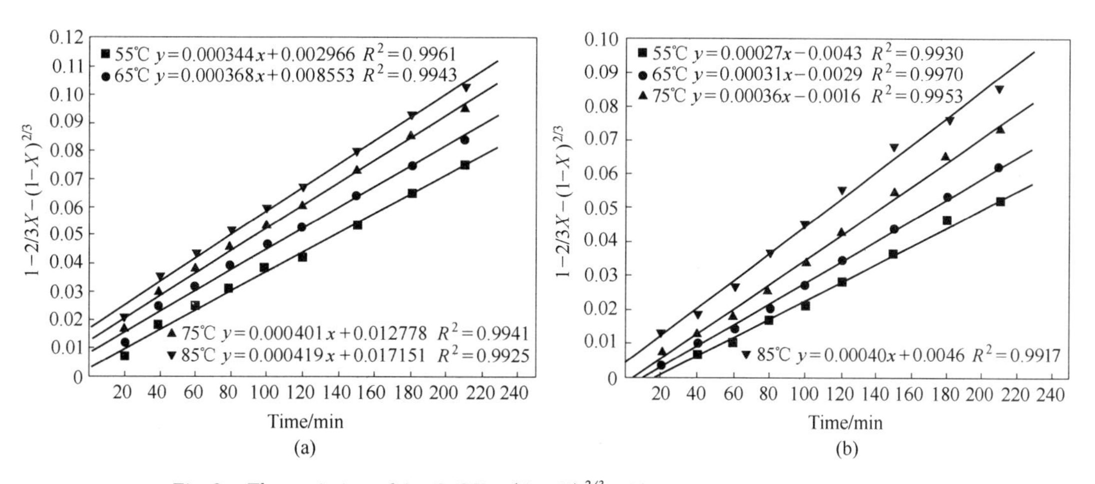

Fig. 9 The variation of $1-2/3X-(1-X)^{2/3}$ with time at various temperatures

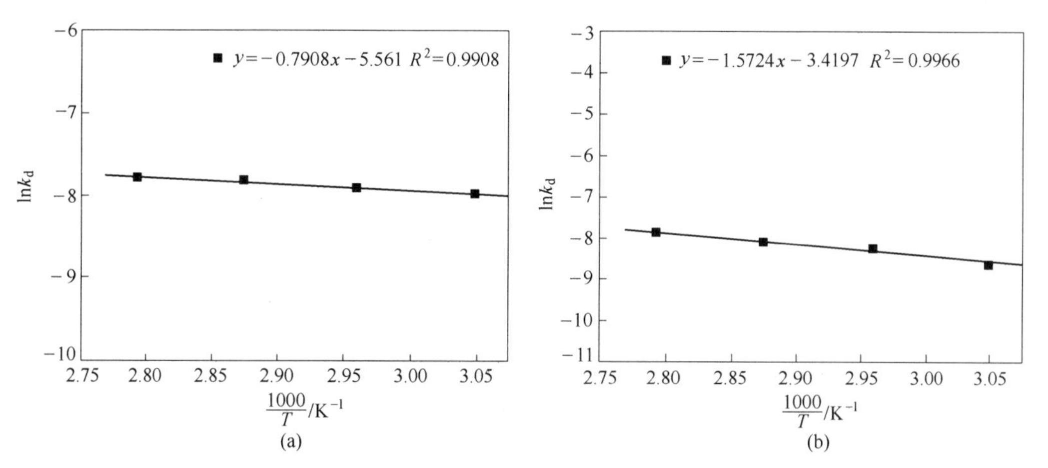

Fig. 10 Arrhenius plot of reaction rate against reciprocal of temperature

The rate constants are slightly different and activation energy of ultrasound – assisted leaching is calculated from the Arrhenius plot as 6. 57kJ/mol. The activation energy of zinc with ultrasound – assisted is lower than the conventional leaching, it means that ultrasound – assisted leaching is easy to accomplish in same conditions. However, the values of the rate constants for conventional leach method are bigger than those of the ultrasound augmented leach. The reason is maybe that the ultrasound – augmented leaching is a fast reaction, it can increase the leaching recovery within a relatively short period of time. So, we can see in the Fig. 4, the first sampling time is 30min, the ultrasound – augmented leaching recovery is obviously bigger than conventional leaching recovery. But as time goes on, compared with conventional leaching, the ultrasound – augmented leaching reaction is near the end of the reaction, and has a high recovery. This result clearly confirmed that the zinc leaching process is controlled by diffusion of ions through a product layer. The similar results are observed by many authors[31-33].

3.3 Order of reaction

To determine the order of reaction with respect to sulfuric acid concentration, the experiment data on the effect of sulfuric acid concentrations is obtained from Fig. 5, are applied to the selected kinetic model. The values of rate constants at different sulfuric acid concentrations are firstly calculated using Eq. (3) for both ultrasound – assisted and conventional leaching, and the results are shown in Table 2.

Table 2 Parameters of the order of reaction for both ultrasound – assisted and conventional leaching

Concentration	Ultrasound – assisted leaching		Conventional leaching	
	k_d	R^2	k_d	R^2
30g/L	0.0035	0.9967	0.0013	0.9816
50g/L	0.0056	0.9937	0.0032	0.1955
70g/L	0.0087	0.9923	0.0053	0.9968
100g/L	0.0117	0.9926	0.0101	0.9965
140g/L	0.0152	0.9928	0.0116	0.9936
170g/L	0.0174	0.9958	0.0123	0.9919

Also, a plot of $\ln k_d$ vs. $\ln[H_2SO_4]$ is obtained as shown in Fig. 11. The order of the reaction with respect to sulfuric acid is proportional to a 0. 94 power $[H_2SO_4]^{0.94}$ with a correlation coeffcient of 0. 9921 in the ultrasound – assisted leaching condition, and a 1. 33 power $[H_2SO_4]^{0.94}$ with a correlation coeffcient of 0. 9557 in the conventional leaching condition. The result shows that the acid concentration of conventional leaching experiment plays an important role than ultrasound – assisted leaching experiment.

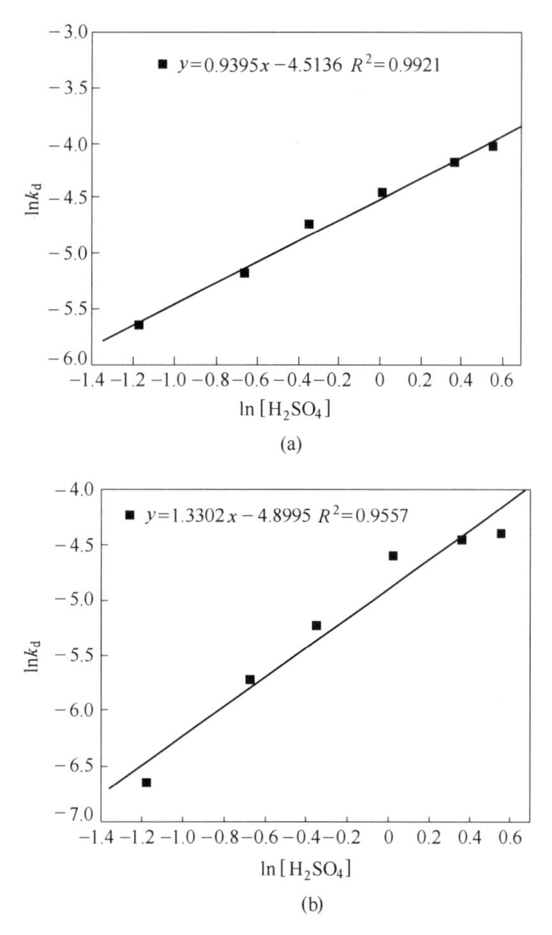

Fig. 11 Plot for the determination of reaction order with respect to $\ln[H_2SO_4]$

3.4 Morphology of the leached residue

The SEM – EDS of the solid particles present's a rough surface, various shapes and different size generated by the zinc plant's previous leach process that led to the small particles found on the surface of the larger ones, as observed in Fig. 12(a), (b). The SEM is sustained by EDS analysis, which for marked points in Fig. 12(c) is shown in Fig. 12(d), (e).

After the leaching progress, the micrographs of the different morphology with ultrasound – assisted leaching and conventional leaching are shown in the Fig. 13(a), (b). As observed in the Fig. 13(a), (b). leached residue of conventional leaching show a progressive increase in the roughness of the particles. Also an increase in the amount of insoluble is covering the particle surfaces, and the particles present their surface completely covered by a film layer. However, after the leaching progress of ultrasound – assisted, the sample is distinguished by it smooth and porous surface, because of ultrasonic radiation can blow out the insoluble substance of the solid surface.

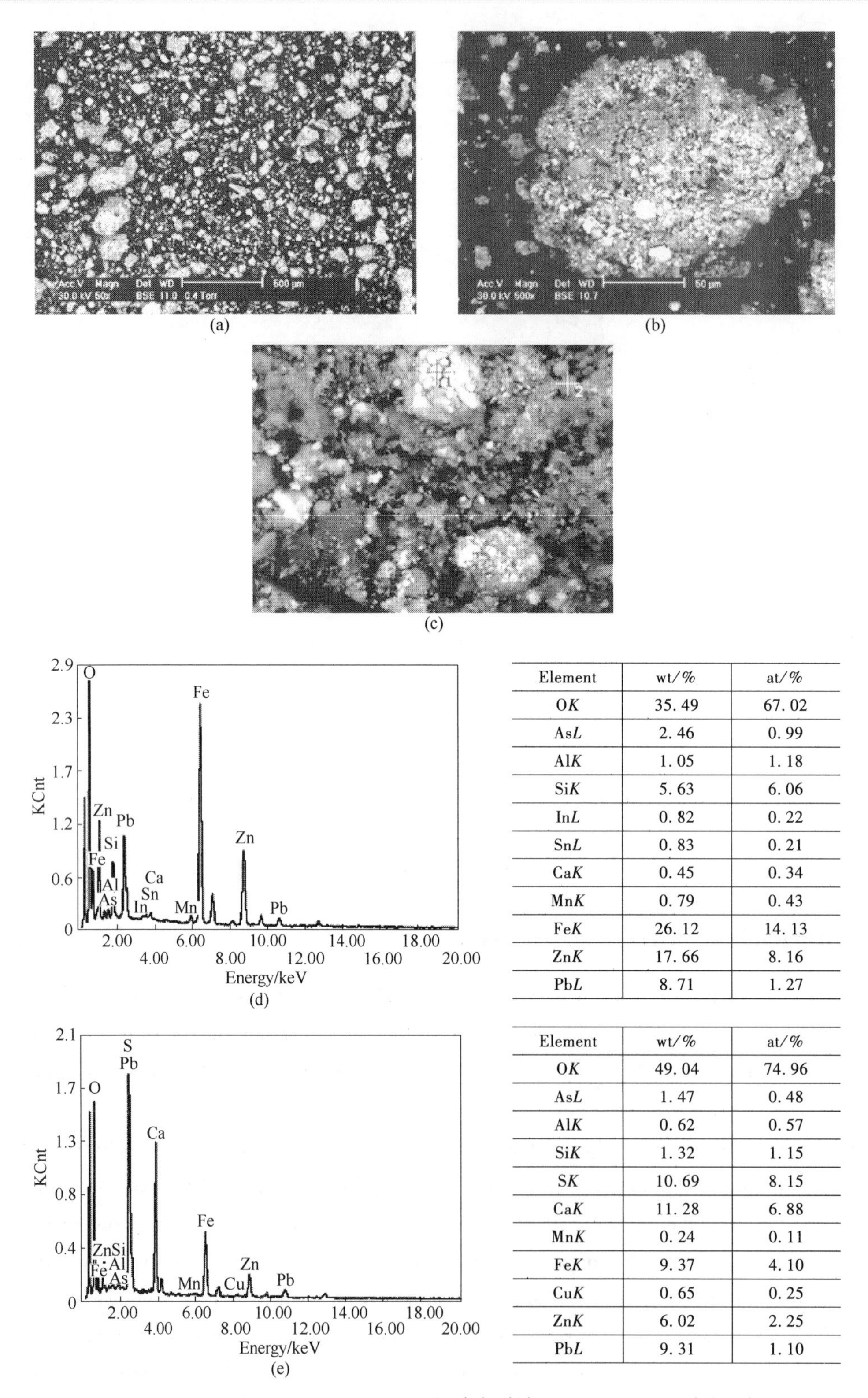

Element	wt/%	at/%
O*K*	35.49	67.02
As*L*	2.46	0.99
Al*K*	1.05	1.18
Si*K*	5.63	6.06
In*L*	0.82	0.22
Sn*L*	0.83	0.21
Ca*K*	0.45	0.34
Mn*K*	0.79	0.43
Fe*K*	26.12	14.13
Zn*K*	17.66	8.16
Pb*L*	8.71	1.27

Element	wt/%	at/%
O*K*	49.04	74.96
As*L*	1.47	0.48
Al*K*	0.62	0.57
Si*K*	1.32	1.15
S*K*	10.69	8.15
Ca*K*	11.28	6.88
Mn*K*	0.24	0.11
Fe*K*	9.37	4.10
Cu*K*	0.65	0.25
Zn*K*	6.02	2.25
Pb*L*	9.31	1.10

Fig. 12 SEM micrograph of zinc plant residue(a),(b) and EDS patterns(c)-(e) of zinc plant residue((d) for point 1,(e) for point 2)

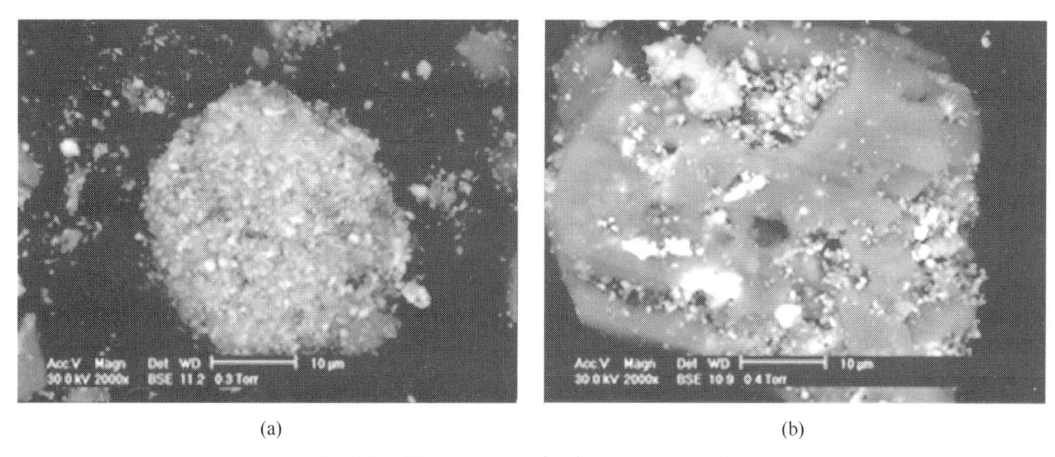

(a) (b)

Fig. 13 SEM micrograph of zinc plant residue

(a) Conventional leaching; (b) Ultrasound - assisted leaching

4 Conclusions

Based on the results obtained in this research, the following conclusions could be summarized:

The obtained results show that leaching of about 67% of zinc is achieved at 75℃ for 250min with 140g/L, sulphuric acid concentration using 53 - 61μm particle size and liquid/solid ratio 4 with conventional leaching; and leaching of above 80% of zinc is achieved at the optimum leaching conditions as ultrasound power 160W, leaching temperature of 65℃ for 180 min with 140g/L sulphuric acid concentration using particle size 74 - 89μm. It can be seen that the reaction rate is very sensitive to temperature in the range of 55 - 85℃. However, a little increase in dissolution efficiency is observed in the last stage of the leaching. According to acid concentration, it is shown that the recovery of zinc increased obviously up to 140g/L, and increasing liquid/solid ratio and decreasing particle size can enhance the recovery of zinc.

Both of the ultrasound augmented and conventional acid leaching experimental parameters are best fitted by diffusion control kinetic model. The activation energies of the diffusion controlled are 13.07kJ/mol and 6.57kJ/mol, and the reaction orders with sulphuric acid concentrations are approximately 1.33 and 0.94, respectively. According to the test of kinetic, the result shows that leaching recovery of zinc with ultrasound augmented is higher than the conventional leaching. The main reason, presumably lies in the ultrasonic radiation incites liquid cavitations that can blow out the cover of the solid surface increasing the reactive surface, and the radiation obviously increase the zinc leaching recovery by increasing the ultrasound power. In addition, the activation energy of zinc with ultrasound augmented is lower than the conventional leaching; it means that the ultrasound - augmented leaching is easy to accomplish in same conditions. This result clearly confirmed that the zinc leaching process is controlled by diffusion of ions through a product layer.

References

[1] DeSouza A D, Pina P S, Leão V A. Bioleaching and chemical leaching as an integrated process in the zinc in-

dustry[J]. Minerals Engineering,2007,20(6):591 – 599.

[2]Altundogan H S,Erdem M,Orhan R. Heavy metal pollution potential of zinc leach residues discarded in Cinkur plant[J]. Turkish Journal of Engineering and Environmental Sciences,1998,22(3):167 – 178.

[3]Isinkaye M O. Distribution of heavy metals and natural radionuclides in selected mechanized agricultural farmlands within Ekiti State,Nigeria[J]. Arabian Journal for Science and Engineering,2012,37(5):1483 – 1490.

[4]Jha M K,Kumar V,Singh R J. Review of hydrometallurgical recovery of zinc from industrial wastes[J]. Resources,Conservation and Recycling,2001,33(1):1 – 22.

[5]Turan M D,Altundogan H S,Tümen F. Recovery of zinc and lead from zinc plant residue[J]. Hydrometallurgy,2004,75(1):169 – 176.

[6]Safarzadeh M S,Moradkhani D,Ilkhchi M O,et al. Determination of the optimum conditions for the leaching of Cd – Ni residues from electrolytic zinc plant using statistical design of experiments[J]. Separation and Purification Technology,2008,58(3):367 – 376.

[7]Altundogan H S,Tümen F. Removal of phosphates from aqueous solutions by using bauxite. I:effect of pH on the adsorption of various phosphates[J]. Journal of Chemical Technology and Biotechnology,2002,77(1):77 – 85.

[8]Núñez C,Viñals J. Kinetics of leaching of zinc ferrite in aqueous hydrochloric acid solutions[J]. Metallurgical Transactions B,1984,15(2):221 – 228.

[9]Elgersma F,Kamst G F,Witkamp G J,et al. Acidic dissolution of zinc ferrite[J]. Hydrometallurgy,1992,29(1):173 – 189.

[10]Elgersma F,Witkamp G J,Van Rosmalen G M. Kinetics and mechanism of reductive dissolution of zinc ferrite in H_2O and D_2O[J]. Hydrometallurgy,1993,33(1):165 – 176.

[11]Langová Š,Leško J,Matýsek D. Selective leaching of zinc from zinc ferrite with hydrochloric acid[J]. Hydrometallurgy,2009,95(3):179 – 182.

[12]Bobeck G E,Su H. The kinetics of dissolution of sphalerite in ferric chloride solution[J]. Metallurgical Transactions B,1985,16(3):413 – 424.

[13]Perez I P,Dutrizac J E. The effect of the iron content of sphalerite on its rate of dissolution in ferric sulphate and ferric chloride media[J]. Hydrometallurgy,1991,26(2):211 – 232.

[14]Xie F,Li H,Ma Y,et al. The ultrasonically assisted metals recovery treatment of printed circuit board waste sludge by leaching separation[J]. Journal of Hazardous Materials,2009,170(1):430 – 435.

[15]Crundwell F K. The influence of the electronic structure of solids on the anodic dissolution and leaching of semiconducting sulphide minerals[J]. Hydrometallurgy,1988,21(2):155 – 190.

[16]Xia H Y,Peng J H,Niu H,et al. Non – isothermal microwave leaching kinetics and absorption characteristics of primary titanium – rich materials[J]. Transactions of Nonferrous Metals Society of China,2010,20(4):721 – 726.

[17]Li C,Xie F,Ma Y,et al. Multiple heavy metals extraction and recovery from hazardous electroplating sludge waste via ultrasonically enhanced two – stage acid leaching[J]. Journal of Hazardous Materials,2010,178(1):823 – 833.

[18]Dorfling C,Akdogan G,Bradshaw S M,et al. Determination of the relative leaching kinetics of Cu,Rh,Ru and Ir during the sulphuric acid pressure leaching of leach residue derived from Ni – Cu converter matte enriched in platinum group metals[J]. Minerals Engineering,2011,24(6):583 – 589.

[19]Zhao Z,Zhang Y,Chen X,et al. Effect of mechanical activation on the leaching kinetics of pyrrhotite[J]. Hydrometallurgy,2009,99(1):105 – 108.

[20]Avvaru B,Roy S B,Ladola Y,et al. Sono – chemical leaching of uranium[J]. Chemical Engineering and Pro-

cessing: Process Intensification, 2008, 47(12): 2107 - 2113.

[21] Wattoo M H S, Tirmizi S A, Quddos A, et al. Aerosol - assisted chemical vapor deposition of thin films of cadmium sulfide and zinc sulfide prepared from bis(dibutyldithiocarbamato) metal complexes[J]. Arabian Journal for Science and Engineering, 2011, 36(4): 565 - 571.

[22] Swamy K M, Narayana K L. Intensification of leaching process by dual - frequency ultrasound[J]. Ultrasonics Sonochemistry, 2001, 8(4): 341 - 346.

[23] Öncel M S, Ince M, Bayramoglu M. Leaching of silver from solid waste using ultrasound assisted thiourea method[J]. Ultrasonics Sonochemistry, 2005, 12(3): 237 - 242.

[24] Şayan E, Bayramoglu M. Statistical modeling and optimization of ultrasound - assisted sulfuric acid leaching of TiO_2 from red mud[J]. Hydrometallurgy, 2004, 71(3): 397 - 401.

[25] Bese A V. Effect of ultrasound on the dissolution of copper from copper converter slag by acid leaching[J]. Ultrasonics Sonochemistry, 2007, 14(6): 790 - 796.

[26] Derradji E F, Benmeziane F, Maoui A, et al. Evaluation of salinity, organic and metal pollution in groundwater of the Mafragh watershed, NE Algeria[J]. Arabian Journal for Science and Engineering, 2011, 36(4): 573 - 580.

[27] Filippou D, Demopoulos G P. Steady - state modeling of zinc - ferrite hot - acid leaching[J]. Metallurgical and Materials Transactions B, 1997, 28(4): 701 - 711.

[28] He S, Wang J, Yan J. Pressure leaching of synthetic zinc silicate in sulfuric acid medium[J]. Hydrometallurgy, 2011, 108(3): 171 - 176.

[29] Zhang Y, Li X, Pan L, et al. Studies on the kinetics of zinc and indium extraction from indium - bearing zinc ferrite[J]. Hydrometallurgy, 2010, 100(3): 172 - 176.

[30] Yadawa P K. Computational study of ultrasonic parameters of hexagonal close - packed transition metals Fe, Co, and Ni[J]. Arabian Journal for Science and Engineering, 2012, 37(1): 255 - 262.

[31] Souza A D, Pina P S, Leão V A, et al. The leaching kinetics of a zinc sulphide concentrate in acid ferric sulphate[J]. Hydrometallurgy, 2007, 89(1): 72 - 81.

[32] Safarzadeh M S, Moradkhani D, Ojaghi - Ilkhchi M. Kinetics of sulfuric acid leaching of cadmium from Cd - Ni zinc plant residues[J]. Journal of Hazardous Materials, 2009, 163(2): 880 - 890.

[33] Abdel - Aal E A. Kinetics of sulfuric acid leaching of low - grade zinc silicate ore[J]. Hydrometallurgy, 2000, 55(3): 247 - 254.

Leaching Kinetics of Zinc Residues Augmented with Ultrasound

Xin Wang, C. Srinivasakannan, Xinhui Duan, Jinhui Peng, Dajin Yang, Shaohua Ju

Abstract: The leaching kinetics of zinc residue, having total Zn content of 12.31%, along with other metallic components such as Fe and Pb, leached using sulfuric acid augmented with ultrasound is presented in the present paper. The effects of variables such as the leaching temperature, sulfuric acid concentration, particle size, liquid/solid ratio and the ultrasound power have been assessed. The results show the maximum recovery of zinc is 80% at an ultrasound power of 160W, leaching temperature of 65℃, sulfuric acid concentration of 1.4mol/L, particle size range of 74 – 89μm and liquid/solid ratio of 4. The kinetics of leaching is modeled using shrinking core model and the rate controlling step is identified to be the diffusion through the product layer. The raw and the leached residue are characterized using XRD and SEM/EDX analysis. The activation energy is estimated to be 6.57kJ/mol, while the order of reaction with respect to sulfuric acid concentration is 0.94 and particle size is 0.12 respectively.

Keywords: zinc; zinc plant residues; leaching kinetic; ultrasound – assisted; shrinking core model

1 Introduction

Zinc is one of the most popular and versatile metal that finds wide application, including plating, coating and alloying with other metals. Commercially zinc can be rated as the third most common nonferrous metal after aluminum and copper[1-3]. Currently, the vast majority of metallic zinc is produced by the Roast – Leach – Electro – winning (RLE) process[4]. This process generates large amount of leach residue as insoluble substance, which still contains significant amount of zinc[5]. The zinc residue is categorized as hazardous waste, further creating disposal problems to the operating plants[6].

The high demand for zinc has attracted the interest of industry to utilize the secondary sources such as zinc ash, zinc dross and leach residues as potential valuable sources [7]. In the process of leaching, many insoluble materials are concentrated in the residue, while the zinc is present in the form of zinc ferrite[8]. Altundogan, et al. and Safarzadeh, et al. [9] reported recovery of the valuable metals from the zinc plant residue. Additionally, the kinetics of zinc recovery from the residue using sulfuric acid dissolution has also been reported in the literature[10-13]. It has been reported that the high temperature and high acid concentration can increase the leaching rate, and the activation energy for these reactions are reported[14-16]. However, ultrasound assisted zinc leaching has been reported rarely in open literature.

Although RLE process is industrially established, the accumulation of insoluble substances on the ore surface restrains the liberation of zinc ions. Over the past decade, different unconventional ways of zinc recovery have been attempted, which include application of microwave[17], ultrasound[18], pressure[19], mechanical[20] methods. The use of ultrasound is reported to increase the leaching yield as well as shorten the duration of leaching due to the well - known cavitation effect, rendering it to be a popular method in the metal processing industry. Avvaru, et al. [21] have reported a significant increase in the leaching efficiency with ultrasound as compared to conventional mixing methods in the process of extraction of uranium (an radioactivity element) from Narwapahar uranium ore with nitric acid and sulphuric acid being the leaching agents. Similarly, Hursit, et al. [22] have reported enhanced zinc leaching kinetics from smithsonite ore with ultrasound. Swamy and Narayana[23] have designed a dual frequency ultrasonic leaching equipment which was reported to increase the recovery by 46% as compared to a conventional agitation process. Öncel, et al. [24] have reported a 96.6% yield of silver from solid residue using microwave assisted leaching with thiourea. A 20% increase in the % yield of TiO_2 was reported from red mud as compared to conventional leaching methods under identical conditions by Sayan[25]. The following compilation well evidences the effectiveness of the ultrasound assisted leaching as compared to conventional process. Bese[26] have reported % yields for ultrasound - assisted leaching to be 89.28% for copper, 51.32% for zinc, 69.87% for cobalt, and 1.11% for iron vs. 80.41% for copper, 48.28% for zinc, 64.52% for cobalt, and 12.16% for iron, in the absence of ultrasound. An overall increased effectiveness of ultrasound assisted leaching can be inferred in comparison with the conventional leaching process.

In the present study, the effects of parameters such as ultrasound power, leaching temperature, particle size, sulphuric acid concentration and liquid/solid ratio were assessed on the leaching kinetics of zinc. The kinetic data was modeled using the popular Shrinking Core Model (SCM), which accounts for chemical reaction and product layer diffusion control. In addition the order of the reaction was estimated, using a semi - empirical equation, for the diffusion - controlled process. The leached samples characteristics were described using the advanced analytical instruments such as X - ray diffraction (XRD) and scanning electron microscope (SEM - EDS)[27].

2 Materials and methods

2.1 Materials

The zinc residues used for the experiments were supplied by a metallurgical plant of Yunnan province, China. Prior to the experiments, the samples were washed, dried and sieved to different size ranges. X - ray diffraction analysis (XRD) of the zinc residue shows the mineralogical content of the sample (Fig. 1).

XRD analysis shows the $ZnFe_2O_4$ and $PbSO_4$ and SiO_2 as the major components in the sample. The chemical composition of the different size ranges of zinc residues were also analyzed using an atomic absorption spectrometer and the results are shown in Table 1[28].

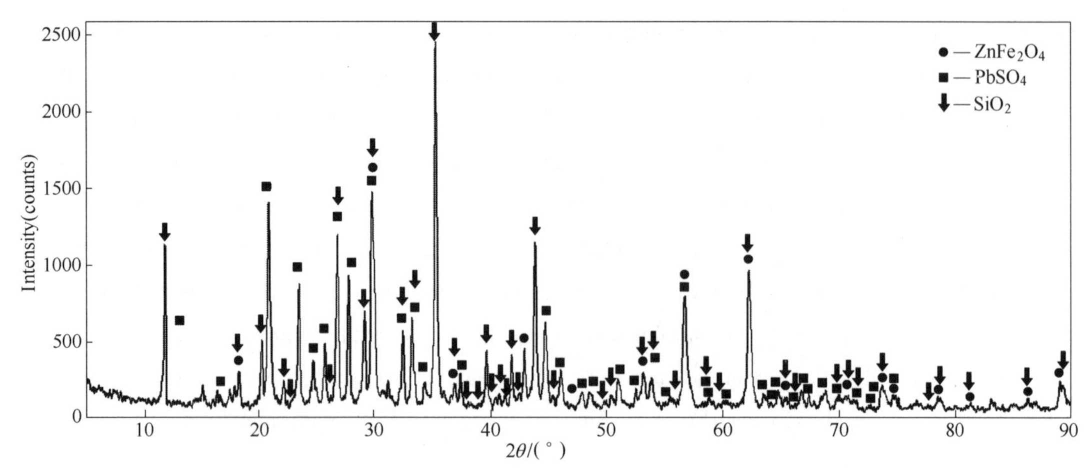

Fig. 1 XRD patterns of zinc plants residues

Table 1 Chemical analysis of different sieve fractions of zinc residues

Serial number	Size/μm	Content/%	Element/%			
			Zn	Fe	Pb	SiO_2
1	104 - 89	9	12.68	21.22	12.26	6.99
2	89 - 74	11	12.80	20.36	12.38	6.95
3	74 - 61	16	12.48	21.75	12.50	7.03
4	61 - 53	18	12.36	21.64	12.19	7.00
5	< 53	30	12.58	21.15	12.07	7.10

2.2 Methods

The ultrasound - assisted leaching experiments were carried out in a 1000mL, lidded beaker in a thermostatically controlled water bath, within a precision of ±0.1℃, equipped with sampling device. The ultrasonic transducer is positioned inside the beaker and connected to an ultrasonic generator which generates ultrasonic waves with a frequency of 20kHz at different power outputs of 80W, 160W and 240W (Guangzhou Hengda Ultrasonic Electric Technological Ltd., China). A schematic of the experimental setup is shown in Fig. 2.

Deionized water is utilized as a diluent with sulphuric acid to prepare various concentrations of acid solution. The dependent variables were varied in the range: acid solution/zinc residues weight ratio (3 to 6), the acid concentration (0.3mol/L to 1.7mol/L), the ultrasound power (80W to 240 W), the particle size range (<53 to 89 - 104) and the leaching temperature (55℃ to 85℃). The rate of leaching was estimated by withdrawing a small sample (< 5mL). The zinc content of the filtered

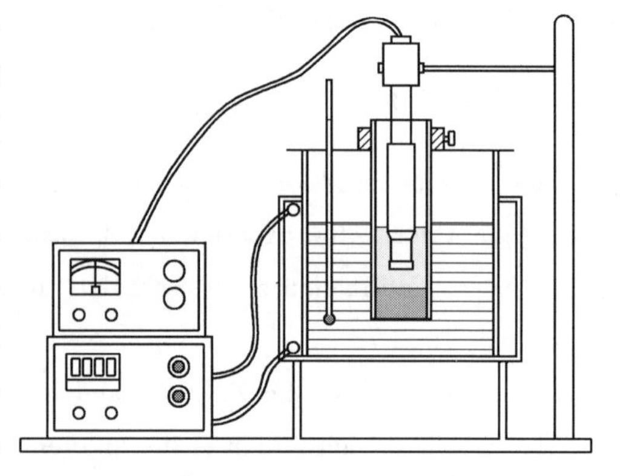

Fig. 2 Schematic of experimental set - up

sample was estimated using atomic absorption spectrometer, and the variations in the experiments were estimated to be less than ±2%. The Zn recovery was estimated, using the equation:

$$Zn_{recovery} = \frac{Zn_{T_0} - Zn_T}{Zn_{T_0}}$$

The analysis of solid residue was carried out using SEM - EDS (XL30ESEM - TMP scanning electron microscope, Philips, Holland) and XRD (Brukerd - advance diffractmeter, Germany) with a step size of 0.02°, ranging from 10° to 90° at 30mA and 40kV.

3 Results and discussion

Reaction between zinc ferrite and sulfuric acid can be written as follows[11,29-32]:

$$2ZnFe_2O_4 + 8H_2SO_4 \longrightarrow 2Fe_2(SO_4)_3 + 2ZnSO_4 + 8H_2O$$

Based on the above reaction, the effects of variables ultrasound power, leaching temperature, acid concentration, particle size and liquid/solid ratio on the leaching rate were assessed.

3.1 Kinetics analysis

Solid - fluid heterogeneous reactions are common in chemical and hydrometallurgical processes. In order to determine the kinetic parameters and rate controlling step of Zn leaching process, the popular shrinking core model was utilized[33].

In the solid - fluid reaction systems, the reaction rate may be usually controlled by the following step: diffusion through the product layer or the chemical reaction at the surface of the material.

If the reaction is controlled by diffusion through the product layer, it will have an integrated rate equation, as follows[32]:

$$1 - \frac{3}{2}X - (1 - X)^{2/3} = k_d t \tag{1}$$

If the reaction is controlled by chemical reaction on the surface, it will have an integrated rate equation as follows[32]:

$$1 - (1 - X)^{1/3} = k_r t \tag{2}$$

where X is the fraction reacted; k_r is the kinetic parameter for reaction control; k_d is the kinetic parameter for product diffusion control; t is the reaction time (min). The plot of left hand side of Eq. (1) and (2) against the time (t) is expected to be linear in order to identify the appropriate controlling mechanism. The slope of the line is the rate constant k_d or k_r, while the temperature dependence of the reaction rate constant can be calculated using the Arrhenius equation:

$$k_d = A\exp\left(\frac{-E_a}{RT}\right) \tag{3}$$

where A is frequency factor; E_a is the activation energy of the reaction; R is the universal gas constant and T is the absolute temperature.

3.2 Effects of parameters

3.2.1 Effect of ultrasound power

The leaching efficiency of zinc at different ultrasound powers of 80W, 160W, 240W is shown in

Fig. 3, while the other parameters are held constant at acid concentration of 1.0mol/L, particle size of 74 – 89μm, liquid/solid of 4∶1 and leaching temperature of 65℃. The results indicate an increase in zinc recovery with increase in ultrasonic power, up to a reaction time of 180min, with the maximum recovery of around 78%. It can also be noted that the effect is not significant beyond an ultrasonic power of 160W. An increase in the ultrasonic power can contribute to the reduction in external mass transfer resistance through the product layer facilitating a higher rate of reaction. The ion movement from the surface of zinc residues into liquid phase is limited by solid – liquid phase diffusion transfer, due to the resistance offered by the product layer and the insoluble impurities. The ultrasonic power is known to incite liquid cavitation that can blow off the solid surface, promoting turbulence creating a highly reactive surface[16].

3.2.2 Effect of leaching temperature

The effects of leaching temperature was assessed in the range from 55℃ to 85℃, while the other parameters are held constant at an acid concentration of 1.4mol/L, particle size of 74 – 89μm, liquid/solid of 4∶1 and ultrasound power of 160W. Fig. 4 shows the significant effect of leaching temperature on the zinc recovery. It can be noted from Fig. 4 that the rate of recovery is insignificant beyond duration of 200min, irrespective of the leaching temperature. The maximum recovery corresponds to a temperature of 85℃, which concurs with the earlier literature reports[30,34 – 38].

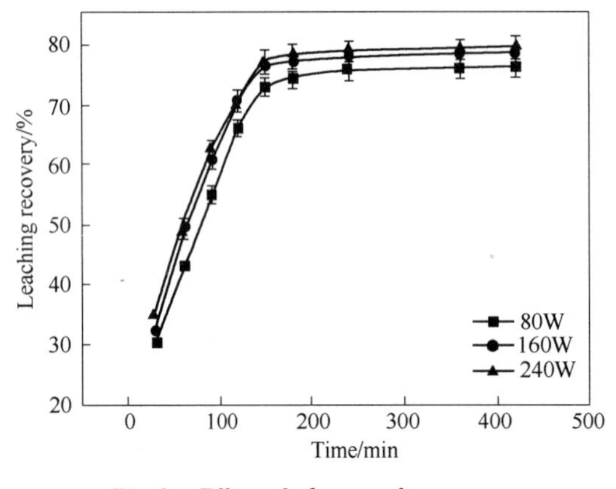

Fig. 3 Effect of ultrasound power on leaching recovery(±2 Error)

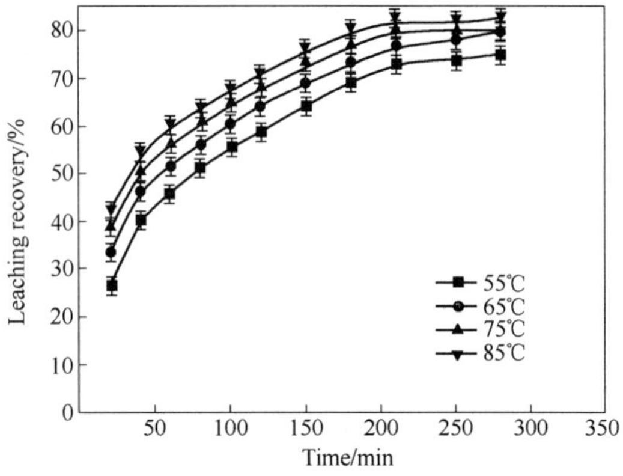

Fig. 4 Effect of leaching temperature on leaching recovery(±2 Error)

The two forms of simplified shrinking core models that account for diffusion through product layer and surface chemical reaction were tested with the experimental data shown in Figs. 4 – 6. present the fit of model equation with the experimental data, in order to identify the rate controlling step. The diffusion control shrinking core model is found to fit the experimental data much better than the reaction controlled version, evidenced through the high R^2 value. Hence it can be concluded that at all leaching temperatures the process is controlled by the rate of diffusion through the product layer.

Fig. 7 presents the plot of $\ln k_d$ with respect to $1/T$, covering the temperature range of 55℃ to 85℃ in order to evaluate the activation energy for the leaching process. The goodness of the fit of the experimental data with the Arrehenius equation can be evidenced from Fig. 7. The activation energy was calculated to be 6.57kJ/mol, which further evidence the ultrasound - assisted leaching process being controlled by diffusion of ions through a product layer. Similar observations have been reported in open literature on conventional leaching process[10-14].

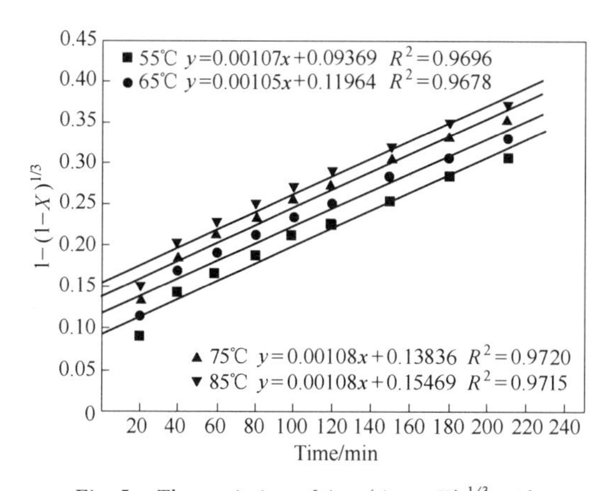

Fig. 5 The variation of $1-(1-X)^{1/3}$ with time at various temperatures

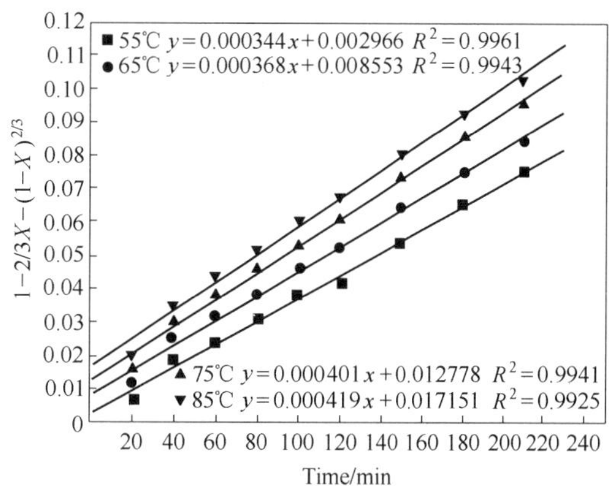

Fig. 6 The variation of $1-2/3X-(1-X)^{2/3}$ with time at various temperatures

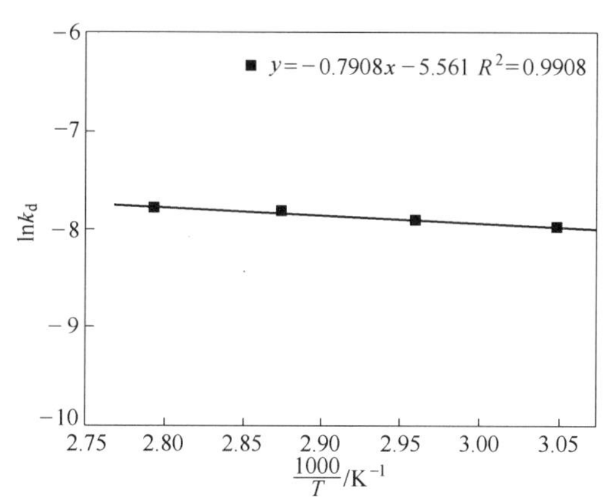

Fig. 7 Arrhenius plot of reaction rate against reciprocal of temperature

3.2.3 Effect of leaching acid concentration

Fig. 8 presents the effect of increase in sulfuric acid concentration, with the other leaching parameters being held at temperature of 65℃, particle size of 74 - 89μm, liquid/solid of 4:1 and ultrasound power of 160W. An increase in the acid concentration is found to increase the rate of leaching, with the maximum recovery corresponding to the highest of acid concentration. The maximum recovery in excess of 80% was achieved at the highest acid concentration of 1.7mol/L. However, it should be noted that at acid concentration higher than 1.4mol/L, the increase in % recovery is not significant. The acid concentrations of 0.3mol/L to 0.5mol/L virtually correspond to an acid deficiency situation. The decrease in rate of leaching with time could be attributed to the reduction in acid concentration, due to the consumption in the reaction. A low recovery at low acid concentrations could be due to the concentration of acid less than the stoichiometric requirement for reaction. In order to estimate the order of the reaction with respect to the acid concentration, the plot of ln k_d vs. ln[H_2SO_4] was utilized as shown in Fig. 9. An order of the reaction with respect of sulfuric acid concentration of 0. 94 was estimated with the correlation coefficient(R^2) being 0. 9921.

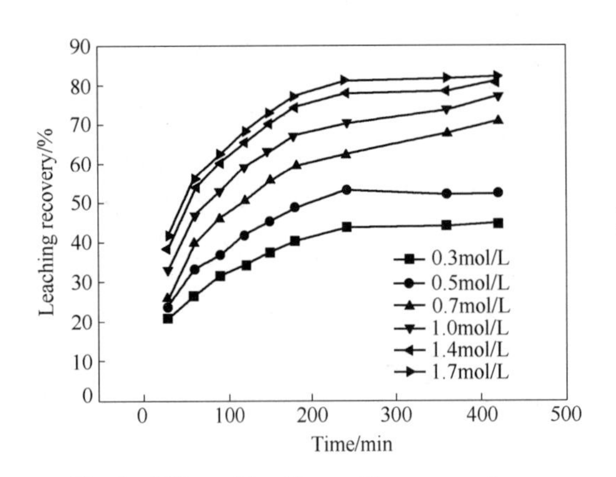

Fig. 8 Effect of leaching acid concentration on leaching recovery(±2 Error)

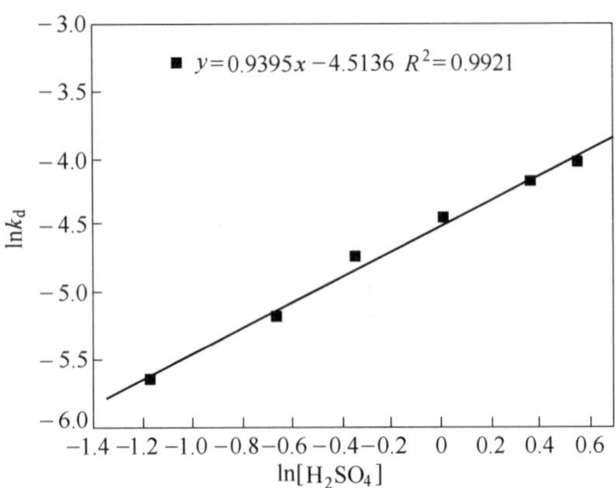

Fig. 9 Plot for the determination of reaction order with respect to ln[H_2SO_4]

3. 2. 4 Effect of particle size

Fig. 10 shows the effects of particle size, with the other leaching parameters held at constant at a leaching temperature of 65℃, acid concentration of 1. 4mol/L, liquid/solid ratio of 4 and microwave power of 160W. In general it is expected that a reduction in the particle size will increase the rate of reaction owing to the increase in area of contact between the reactants. Fig. 10 clearly indicates an increase in the rate as well as overall recovery with decrease in the particle size, with the maximum recovery of about 80% for the particle size range of 53μm to 61μm. The plot of ln k_d vs. lnr_0 shown in Fig. 11, facilitates the estimation of the order of reaction with respect to particle size and it was estimated to be 0. 12, with the R^2 being 0. 9824.

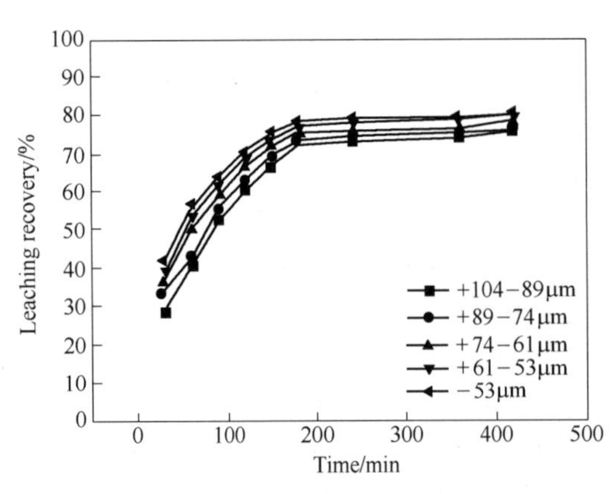

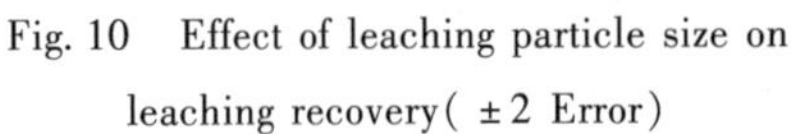
Fig. 10 Effect of leaching particle size on leaching recovery(±2 Error)

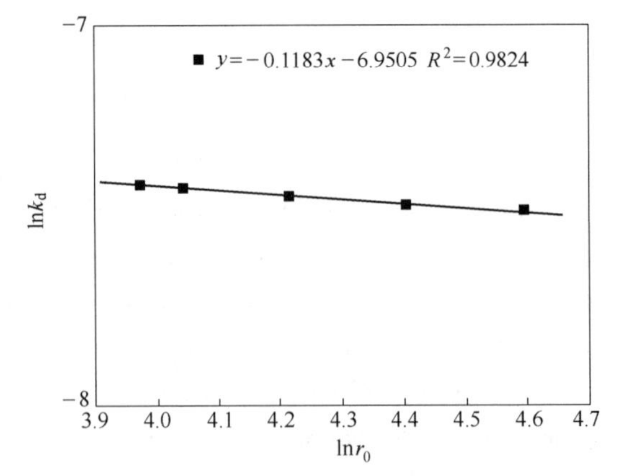

Fig. 11 Plot of lnk_d against lnr_0

3. 2. 5 Effect of liquid/solid ratio

The effects of liquid/solid ratio was assessed in the range from 3 to 6, with the other leaching parameters held at a leaching temperature of 65℃, acid concentration of 1.4mol/L, a particle size of 89 – 74μm and a microwave power of 160W. As can be seen from the Fig. 12, an increase in the liquid/solid ratio enhances zinc recovery effectively until a liquid/solid ratio of 4. An increase in the liquid content is expected to reduce the viscosity of the mixture there by facilitating better mixing, contributing to the reduction in the diffusional mass transfer resistance. Additionally a liquid solid ratio less than four could be a acid deficient condition, less than the stoichiometric requirement for the reaction.

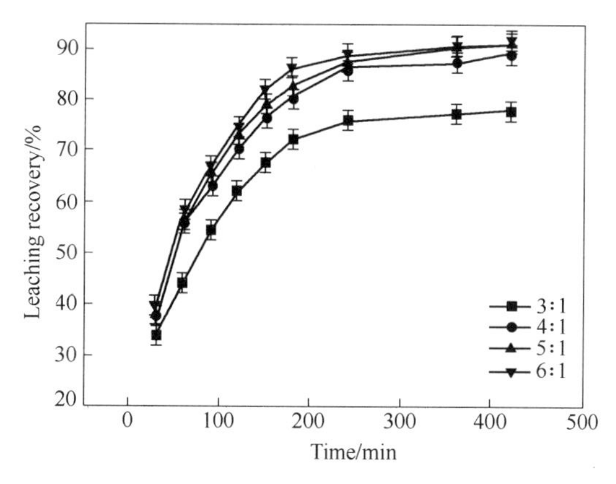

Fig. 12 Effect of liquid/solid ratio on leaching recovery(±2 Error)

Although the reaction rate with respect to liquid/solid ratio can be estimated with the the plot of $\ln k_d$ vs. ln(L/S) it is not meaningful as one reactant limits the extent of reaction, typically reflecting the effect of acid concentration.

Based on reaction kinetics of the Zn leaching, a kinetic equation for the diffusion controlled process using the shrinking core model was developed and is presented in the form of following equation

$$1 - \frac{3}{2}X - (1 - X)^{2/3} = k_d t = k_0 [H_2SO_4]^{0.94} \times r_0^{-0.12} \exp\left(\frac{-6.57}{RT}\right) t$$

The effects of ultrasound power, temperature, sulfuric acid concentration and particle size on the kinetics of ultrasound – assisted zinc leaching is assessed and the kinetics is modeled using the shrinking core model. The results show 80% of zinc recovery at the optimum leaching conditions at ultrasound power 160W, leaching temperature of 65℃ for 180min time with 1.4mol/L sulfuric acid concentration using 74 – 89μm particle size.

3.3 Morphology of the residues

The SEM – EDS of zinc residue present a rough surface, different shapes and sizes observed in Fig. 13(a), (b). The SEM is supplemented by EDS analysis, for marked points in Fig. 13(c) is shown in Fig. 13(d), (e). The EDS mapping corroborated the presence of Zn and the large amount of elemental, Pb and Fe in the initial residue. Insoluble substances such as $ZnFe_2O_4$ and $PbSO_4$, SiO_2 covered on the surface of zinc particles, render the process diffusion controlled through insoluble substance layer.

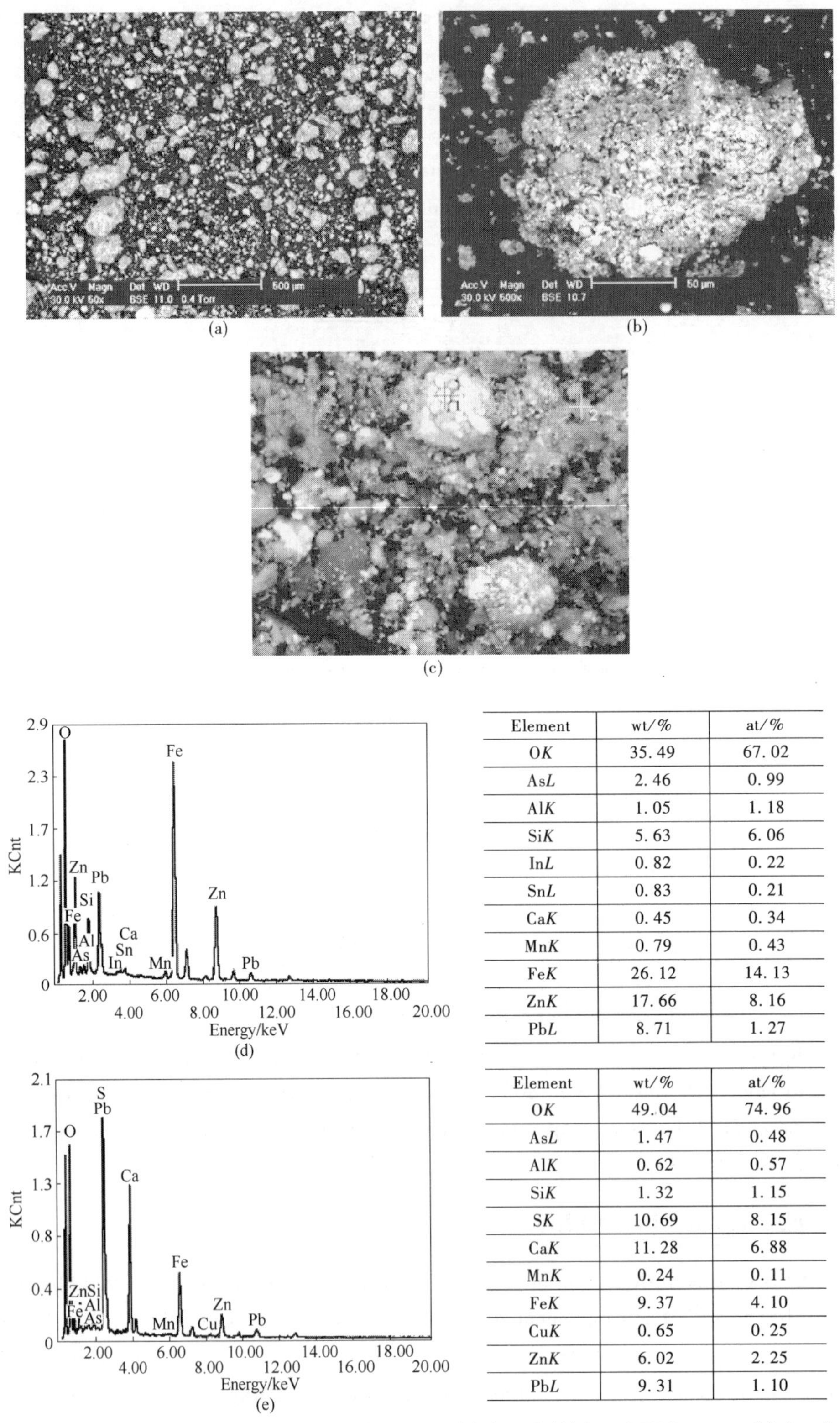

Element	wt/%	at/%
O*K*	35.49	67.02
As*L*	2.46	0.99
Al*K*	1.05	1.18
Si*K*	5.63	6.06
In*L*	0.82	0.22
Sn*L*	0.83	0.21
Ca*K*	0.45	0.34
Mn*K*	0.79	0.43
Fe*K*	26.12	14.13
Zn*K*	17.66	8.16
Pb*L*	8.71	1.27

Element	wt/%	at/%
O*K*	49.04	74.96
As*L*	1.47	0.48
Al*K*	0.62	0.57
Si*K*	1.32	1.15
S*K*	10.69	8.15
Ca*K*	11.28	6.88
Mn*K*	0.24	0.11
Fe*K*	9.37	4.10
Cu*K*	0.65	0.25
Zn*K*	6.02	2.25
Pb*L*	9.31	1.10

Fig. 13 SEM micrograph of initial zinc plant residues((a) and (b)) and EDS patterns((c) – (e)) of initial zinc plant residues(the(d) for point 1,(e) for point 2)

As shown in Fig. 14(a),(b),after the leaching,the sample is distinguished by its smooth surface,due to cleanup activity of the surface of solid by the ultrasonic radiation. The Fig. 14(b) is supplemented by EDS analysis,which is shown in Fig. 14(c).

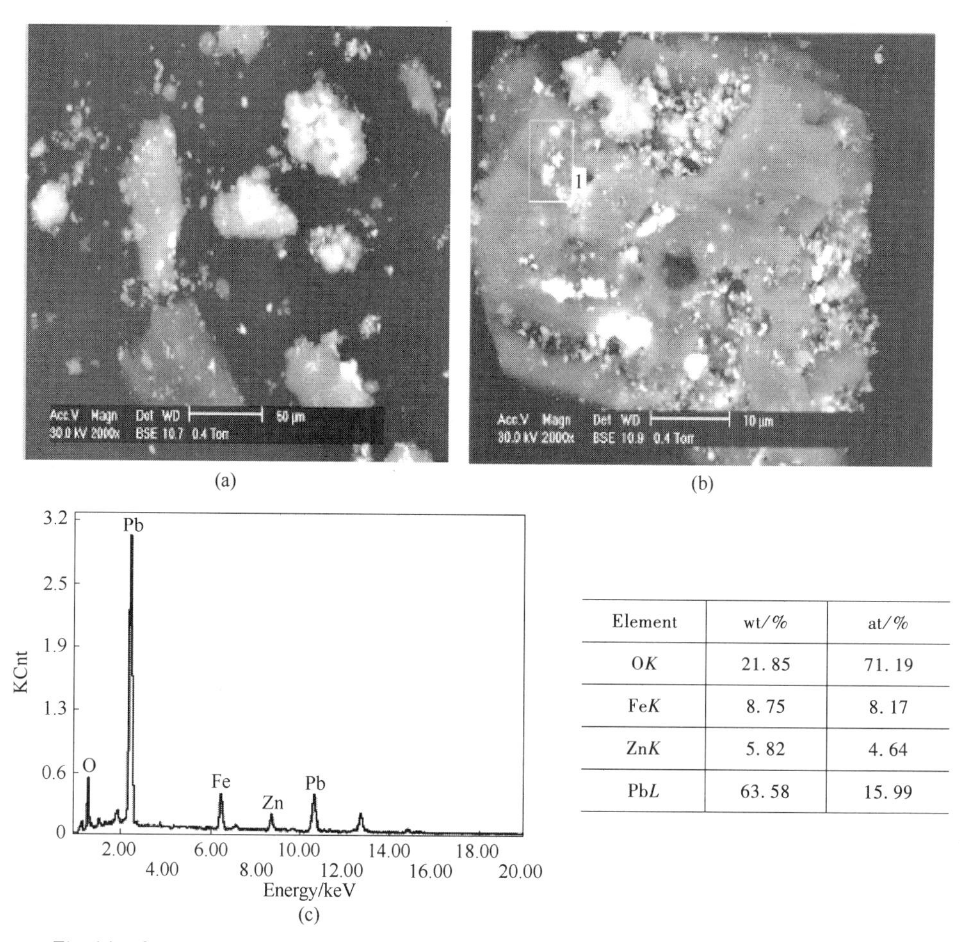

Element	wt/%	at/%
O*K*	21.85	71.19
Fe*K*	8.75	8.17
Zn*K*	5.82	4.64
Pb*L*	63.58	15.99

Fig. 14 SEM micrograph of leached zinc plant residues((a) and (b)) and EDS patterns (c) of leached zinc plant residues((c) for region 1)

The process of leaching has significantly lowered the Zn content in the residue. Additionally large proportion of Fe was also removed in the process of leaching.

The effects of ultrasound power, temperature, sulfuric acid concentration, particle size and liquid/solid ratio on the kinetics of ultrasound - assisted zinc leaching were assessed. The results indicate 80% removal of zinc under the optimum conditions of ultrasound power 160W, leaching temperature of 65℃ ,with 1.4mol/L sulfuric acid concentration using 74 - 89μm particle size and liquid/solid ratio of 4. The raw as well as leached residue was characterized using XRD and SEM/EDX analysis.

References

[1]Babu M N,Sahu K K,Pandey B D. Zinc recovery from sphalerite concentrate by direct oxidative leaching with ammonium,sodium and potassium persulphates[J]. Hydrometallurgy,2002,64(2):119 - 129.

[2]Porter F. Zinc Handbook:Properties,Processing and Use in Design[M]. New York:Marcel Dekker Inc. ,1991:5.

[3]Kim T H,Senanayake G,Kang J G,et al. Reductive acid leaching of spent zinc - carbon batteries and oxidative precipitation of Mn - Zn ferrite nanoparticles[J]. Hydrometallurgy,2009,96(1):154 - 158.

[4]DeSouza A D,Pina P S,Leão V A. Bioleaching and chemical leaching as an integrated process in the zinc industry[J]. Minerals Engineering,2007,20(6):591 - 599.

[5]Altundogan H S,Erdem M,Orhan R. Heavy metal pollution potential of zinc leach residues discarded in Cinkur plant[J]. Turkish Journal of Engineering and Environmental Sciences,1998,22(3):167 - 178.

[6]Wang Q. Industrial Solid Waste Disposal and Recycling (fourth ed.)[M]. Peking:China Environmental Science Press,2006.

[7]Jha M K,Kumar V,Singh R J. Review of hydrometallurgical recovery of zinc from industrial wastes[J]. Resources,Conservation and Recycling,2001,33(1):1 - 22.

[8]Turan M D,Altundogan H S,Tümen F. Recovery of zinc and lead from zinc plant residue[J]. Hydrometallurgy,2004,75(1):169 - 176.

[9]Safarzadeh M S,Moradkhani D,Ilkhchi M O,et al. Determination of the optimum conditions for the leaching of Cd - Ni residues from electrolytic zinc plant using statistical design of experiments[J]. Separation and Purification Technology,2008,58(3):367 - 376.

[10]Núñez C,Viñals J. Kinetics of leaching of zinc ferrite in aqueous hydrochloric acid solutions[J]. Metallurgical Transactions B,1984,15(2):221 - 228.

[11]Elgersma F,Kamst G F,Witkamp G J,et al. Acidic dissolution of zinc ferrite[J]. Hydrometallurgy,1992,29(1):173 - 189.

[12]Elgersma F,Witkamp G J,Van Rosmalen G M. Kinetics and mechanism of reductive dissolution of zinc ferrite in H_2O and D_2O[J]. Hydrometallurgy,1993,33(1):165 - 176.

[13]Langová Š,Leško J,Matýsek D. Selective leaching of zinc from zinc ferrite with hydrochloric acid[J]. Hydrometallurgy,2009,95(3):179 - 182.

[14]Bobeck G E,Su H. The kinetics of dissolution of sphalerite in ferric chloride solution[J]. Metallurgical Transactions B,1985,16(3):413 - 424.

[15]Perez I P,Dutrizac J E. The effect of the iron content of sphalerite on its rate of dissolution in ferric sulphate and ferric chloride media[J]. Hydrometallurgy,1991,26(2):211 - 232.

[16]Xie F,Li H,Ma Y,et al. The ultrasonically assisted metals recovery treatment of printed circuit board waste sludge by leaching separation[J]. Journal of Hazardous Materials,2009,170(1):430 - 435.

[17]Xia H Y,Peng J H,Niu H,et al. Non - isothermal microwave leaching kinetics and absorption characteristics of primary titanium - rich materials[J]. Transactions of Nonferrous Metals Society of China,2010,20(4):721 - 726.

[18]Li C,Xie F,Ma Y,et al. Multiple heavy metals extraction and recovery from hazardous electroplating sludge waste via ultrasonically enhanced two - stage acid leaching[J]. Journal of Hazardous Materials,2010,178(1):823 - 833.

[19]Dorfling C,Akdogan G,Bradshaw S M,et al. Determination of the relative leaching kinetics of Cu,Rh,Ru and Ir during the sulphuric acid pressure leaching of leach residue derived from Ni - Cu converter matte enriched in platinum group metals[J]. Minerals Engineering,2011,24(6):583 - 589.

[20]Zhao Z,Zhang Y,Chen X,et al. Effect of mechanical activation on the leaching kinetics of pyrrhotite[J]. Hydrometallurgy,2009,99(1):105 - 108.

[21]Avvaru B,Roy S B,Ladola Y,et al. Sono - chemical leaching of uranium[J]. Chemical Engineering and Pro-

cessing:Process Intensification,2008,47(12):2107 - 2113.

[22] Hursit M, Lacin O, Sarac H. Dissolution kinetics of smithsonite ore as an alternative zinc source with an organic leach reagent[J]. Journal of the Taiwan Institute of Chemical Engineers,2009,40(1):6 - 12.

[23] Swamy K M, Narayana K L. Intensification of leaching process by dual - frequency ultrasound[J]. Ultrasonics Sonochemistry,2001,8(4):341 - 346.

[24] Öncel M S, Ince M, Bayramoglu M. Leaching of silver from solid waste using ultrasound assisted thiourea method[J]. Ultrasonics Sonochemistry,2005,12(3):237 - 242.

[25] Sayan E, Bayramoglu M. Statistical modeling and optimization of ultrasound - assisted sulfuric acid leaching of TiO_2 from red mud[J]. Hydrometallurgy,2004,71(3):397 - 401.

[26] Bese A V. Effect of ultrasound on the dissolution of copper from copper converter slag by acid leaching[J]. Ultrasonics Sonochemistry,2007,14(6):790 - 796.

[27] Leiva C, Ahumada I, Sepúlveda B, et al. Polychlorinated biphenyl behavior in soils amended with biosolids [J]. Chemosphere,2010,79(3):273 - 277.

[28] Bailey N T, Wood S J. A comparison of two rapid methods for the analysis of copper smelting slags by atomic absorption spectrometry[J]. Analytica Chimica Acta,1974,69(1):19 - 25.

[29] Filippou D, Demopoulos G P. Steady - state modeling of zinc - ferrite hot - acid leaching[J]. Metallurgical and Materials Transactions B,1997,28(4):701 - 711.

[30] Zhang Y, Li X, Pan L, et al. Studies on the kinetics of zinc and indium extraction from indium - bearing zinc ferrite[J]. Hydrometallurgy,2010,100(3):172 - 176.

[31] Ramachandra Sarma V N, Deo K, Biswas A K. Dissolution of zinc ferrite samples in acids[J]. Hydrometallurgy,1976,2(2):171 - 184.

[32] Ryczaj K, Riesenkampf W. Kinetics of the dissolution of zinc - magnesium ferrites in sulphuric acid solutions related to zinc leach processes[J]. Hydrometallurgy,1983,11(3):363 - 370.

[33] He S, Wang J, Yan J. Pressure leaching of synthetic zinc silicate in sulfuric acid medium[J]. Hydrometallurgy,2011,108(3):171 - 176.

[34] Kim J, Kaurich T A, Sylvester P, et al. Enhanced selective leaching of chromium from radioactive sludges[J]. Separation Science and Technology,2006,41(1):179 - 196.

[35] Dutrizac J E. The leaching of sulphide minerals in chloride media[J]. Hydrometallurgy,1992,29(1):1 - 45.

[36] Dutrizac J E. The kinetics of sphalerite dissolution in ferric sulphate - sulphuric acid media[J]. Lead & Zinc, 2005,5:833 - 851.

[37] Aydogan S, Aras A, Canbazoglu M. Dissolution kinetics of sphalerite in acidic ferric chloride leaching[J]. Chemical Engineering Journal,2005,114(1):67 - 72.

[38] Dutrizac J E, MacDonald R J C. The dissolution of sphalerite in ferric chloride solutions[J]. Metallurgical Transactions B,1978,9(4):543 - 551.

Chapter Ⅳ

New Technology of Microfluidic

Microfluidic Solvent Extraction and Separation of Cobalt and Nickel

Lihua Zhang, Jinhui Peng, Shaohua Ju, Libo Zhang, Linqing Dai, Nengsheng Liu

Abstract: The extraction and separation of cobalt from sulphate solution containing Ni^{2+} and Co^{2+} by the process of microfluidic extraction was investigated on a counter – current flow interdigital micromixer with channels of 40μm width, which has two opposite inlets and an upwards outlet. Meanwhile, the comparative batch extraction experiments were conducted in separatory funnels. The effects of pH and flow rates or contact time on the microfluidic and batch experiments were studied using an aqueous solution containing 73.09g/L of nickel and 2.44g/L of cobalt and 20vol% PC88A diluted with 260# solvent naphtha. In addition, cobalt extraction isotherms (Mc – Cabe Thiele) were constructed to determine the number of stages. The results of percentage extraction and seperation factor of microfluidic extraction was better than that of batch extraction. The features of the microreactors, i. e. large specific surface area and short diffusion distance were effective for the efficient extraction and separation of cobalt from nickel.

Keywords: microfluidic; solvent extraction; separation; cobalt; nickel

1 Introduction

Cobalt and nickel are amongst the most important nonferrous metals. The extraction and separation of cobalt and nickel from sulphate, chloride, and ammoniacal solutions have been of interest to hydrometallurgists for a long time[1-3]. The very similar physical and chemical properties of these two metals have made their separation a challenging task. Many methods such as solvent extraction, liquid membrane, ion exchange, and precipitation have been conducted for the separation or preconcentration of cobalt and nickel[4-7]. Among these, solvent with organophosphorus acids has attracted much attention.

The organophosphorus reagents (D2EHPA, PC88A, and Cyanex272) were widely used to extract cobalt and nickel in sulphate solutions. Out of the three types of extractants, D2EHPA is the least selective and cyanex272, though most selective, is the most expensive. Therefore, the phosphonic acid based reagent PC88A was used in many Chinese companies for its relatively cheap price. However, there are some deficiencies by using the traditional technology, such as: (1) low extraction efficiency of a single stage and large extractant consumption; (2) complex operation for multistage of extraction and stripping; (3) prone to emulsification; and (4) significant footprint size and fire hidden trouble. Microfluidic extraction may be capable to overcome these disadvantages in extraction and separation of cobalt and nickel.

Microfluidic technology has attracted much attention in the fields of analytical chemistry, chemical synthesis, chemical engineering and biotechnology[8-11]. In microfluid systems, reagents and starting materials typically driven through microchannels on the order of 10 - 1000μm, and mass/heat transfer is enhanced by promoting contact between very thin fluid reactant layers and realized rapid micromixing. Due to the small dimensions, chemical processes in microchannels can be different from those in macroscale processes. Secondary phenomena become significant in microreactors when the characteristic length decreases, i. e. mass diffusion, surface condition and heat conduction. Compared to conventional batch reactors, the unique operating characteristics of microreactors are as follows: (1) high reaction efficiency due to high surface area to volume ratio, rapid mass and heat transfer; (2) high degree of chemical selectivity due to precise control of the reaction temperature and time; (3) mini footprint size, save operation and friendly environment; and (4) fast and direct amplification via "numbering - up" parallel processing without scale - up effect[12-14].

As the above unique advantages, microfluidic technology for the extraction process have emerged in the past few years. Shekhar, et al. [15] mentioned a simple but effective micro - mixer - settler made of a rotated helical coil of 100μm i. d. micro - tubing for nuclear solvent extraction. Nearly 100% efficiency was observed in extraction as well as stripping procedures. Syouhei, et al. [16] investigated a micro solvent extraction system for the separation of lanthanides. The microchannel is fabricated on a PMMA plate with 100μm width and 100μm depth. The effective separation of lanthanides (Pr/Nd and Pr/Sm) can be achieved on the micro extraction chip. The phase separation of the aqueous and organic phases after extraction can be carried out by changing the cross section of the micro flow channel. Fukiko, et al. [17] performed the extraction of three rare earth metals (yttrium, europium and lanthanum) with PC88A as the extractant dissolved in toluene on a microreactor fabricated on a silicon wafer (20mm × 40mm) by photolithography and wet etching methods. The aqueous and organic phases successfully kept an aqueous - organic interface in a microchannel, 92μm in depth and 300 - 434μm in width, at volumetric flow rates from $5.6 \times 10^{-10} m^3/s$ to $2.8 \times 10^{-9} m^3/s$. The metals are satisfactorily extracted by the microreactor in the residence time of 0.7s. Osamu, et al. [18] used slug flow channel microreactor for liquid - liquid extraction of cesium from cesium nitrate solution. The results indicated that the Cs^+ extraction rate was significantly increased with the slug flow microreactor, compared to conventional batch extraction, and extraction equilibrium was achieved within 40s. Craig, et al. [19] invistigaed microfluidic extraction of copper from particle - laden solutions. Though in the presence of silica nanoparticles, the formation of particle - stabilised emulsions was prevented. Thus, microfluidic solvent extraction shows great promise for handling particle - laden solutions of industrial relevance.

Above literatures indicated that the microfluidic technology was applicable to be used in the solvent extraction process and was useful as a simple test plant for the construction of an efficient separation process. In this article, extraction and separation of cobalt and nickel in sulphate solution both in microfluidic and batch extraction were presented. The effects of equilibrium pH and the flow rates/contact time were investigated. In addition, the Mc - Cabe Thiele diagrames were constructed to determine the number of stages required at a chosen volume phase ratio.

2 Experimental

2.1 Materials

The initial aqueous was a synthetic solution with 73.09g/L of nickel and 2.44g/L of cobalt, using AR grade $NiSO_4 \cdot 6H_2O$ and $CoSO_4 \cdot 7H_2O$ dissolved in pure water, concentrations similar to those of a real sulphate purification. The aqueous pH was adjusted by the addition of concentrated sodium hydroxide(5mol/L) and sulfuric acid solutions(98%) to the appropriate value 5.0 ± 0.1 measured by PHS - 3D pH meter.

Industrial grade extractant, PC88A (C16H35O3P, 2 - ethylhexyl phosphonic acid mono - 2 - ethylhexyl ester) and diluent, 260# solvent naphtha(C_{11} - C_{17} alkanes mixtures primarily, and containing 4wt% aromatic hydrocarbons) were simultaneously supplied by Aoda Chemical Co., Ltd. China. Both were used without further purification. The extractant PC88A was pre - neutralized by 10mol/L NaOH aqueous solution. Organic solution was prepared by dissolving 20% v/v Na - PC88A in 80% v/v 260# solvent naphtha to form a single phase.

2.2 Apparatus and procedures

The microfluidic experiments characterizing the extraction effciency were performed in a counter - current flow interdigital micromixer (stainless steel) with 40mm width channels fabricated by IMM, Germany, as shown in Fig. 1(a). It consists of a mixing element, two opposite inlets and one upwards outlet. The feed flows including aqueous and organic solution were introduced to a mixing element inside the micromixer via the two inlets, from opposite directions. As shown in Fig. 2, constant flow pumps were used to feed the two fluids into the microchannels, which were provided by Yanshang Instrument Factory, China, and the flow velocity was in the range of 0.1 - 40 mL/min. The two fluids flow through the interdigital channels and then flow upwards into a slit, which is perpendicular to the interdigital structure, as shown in Fig. 1(b)[20], where, the mixing and extraction reaction took place, the most of Co in the aqueous was extracted to the organic phase by Na - PC88A, and most of Ni was still remained in the aqueous phase. The mixed fluid was separated and clarified to form two clear phases in the settler, the upper dark blue Co loaded organic phase and the bottom green Ni raffnate. The microreactor was immersed in a water bath to keep the constant temperature at 25℃ ±0.2℃.

The comparative batch extraction tests were carried out by contacting equal volumes of the organic and aqueous phases, i.e. phase ratio O∶A = 1∶1, in 125mL pear shape separatory funnels. Then the system was placed horizontally in the roundtrip thermostatic water bath oscillator (WHY - 2) to mix and react at a certain time, and the oscillation intensity was 200r/min. Subsequently, the mixture was separated after 30min to form bottom aqueous phase containing Ni and upper organic phase loading Co.

For all microfluidic and batch experiments, the raffinate was taken to determine the concentra-

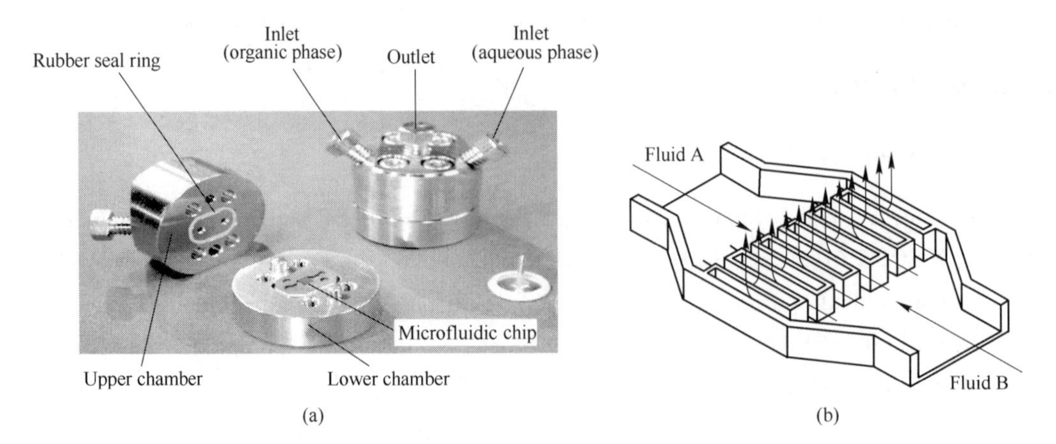

Fig. 1 Schematic diagram of microreactor

(a) Microreactor; (b) Mixing and reacting principle

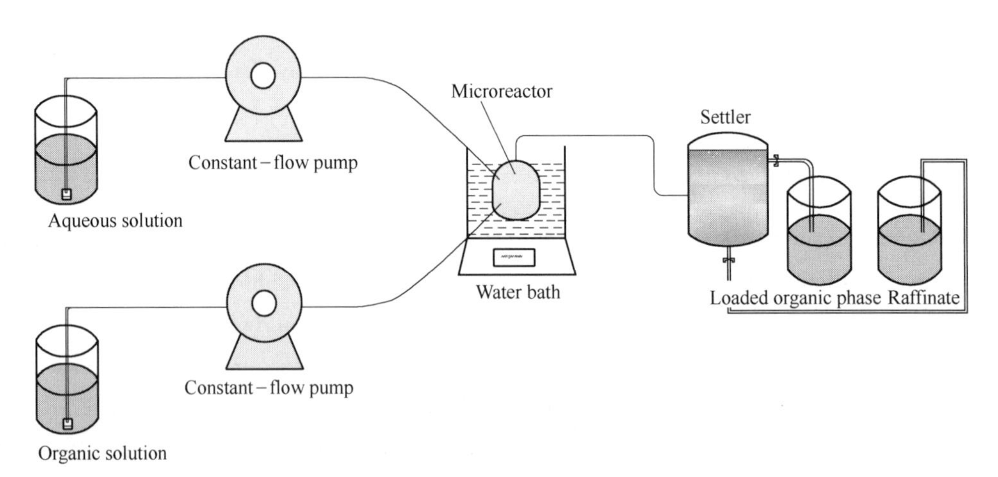

Fig. 2 Schematic of the microfluidic experimental setup

tions of the metals Ni and Co by inductively coupled plasma - atomic emission spectroscopy (AA240FS ICP - AES spectrophotometer). Those in the top loaded organic phase were calculated in accordance with mass balance.

3 Results and discussion

The extractant PC88A was converted to the sodium salt by the addition of sodium hydroxide. The neutralization reaction can be written as:

$$Na_{aq}^{+} + 1/2(HR)_{2org} \longrightarrow NaR_{org} + H_{aq}^{+} \quad (1)$$

According to Sarangi, et al.[21], the neutral form of the extractant exists as monomer, whereas the acidic form as a dimmer, both forms take part in the extraction. In acidic sulfate media, the extraction of divalent metals by mixtures of acidic extractants can be presented by the following equilibrium[22]:

$$M_{aq}^{2+} + R_{org}^{-} + 2(HR)_{2,org} \rightleftharpoons MR_2 \cdot 3HR_{org} + H_{aq}^{+} \quad (2)$$

where the distribution ratio, D, was calculated as the concentraion of metal present in the organic phase to that in the aqueous phase at equilibrium, as shown in the flowing equation:

$$D = \frac{[M]_{org}}{[M]_{aq}} \quad (3)$$

From the D values, the percentage extraction E and separation factor of cobalt and nickel $\beta_{Co/Ni}$ were calculated using the flowing equations:

$$E = \frac{[M]_{org} \cdot V_{org}}{[M]_{org} \cdot V_{org} + [M]_{aq} \cdot V_{aq}} \times 100\% \quad (4)$$

$$\beta_{Co/Ni} = \frac{D_{Co}}{D_{Ni}} = \frac{[Co]_{org} \cdot [Ni]_{aq}}{[Co]_{aq} \cdot [Ni]_{org}} \quad (5)$$

where M represented the corresponding metal cobalt and nickel respectively; V represented the volume. The subscript aq and org denoted the aqueous and organic phases respectively.

3.1 Effect of equilibrium pH

The extraction of Co with a cation exchange type of extractant, PC88A, was pH dependent and involved the release of protons from the extractant during metal transfer from the aqueous phase to the organic phase. As a result, the pH of the aqueous phase decreased. pH values would be raised by the addition of alkali solution or by using a partially saponified PC88A for effective metal extraction. We followed the later methodology of partial saponification of PC88A for Co separation studies.

As in the microfluidic and batch experiments, the extraction of cobalt and nickel from aqueous sulphate medium with pH = 5.0 was studied using the sodium salt of PC88A (20% v/v) which was neutralised with concentrated NaOH to 25% - 85%, correspongding to the equilibrium pH changing in the range 3.80 - 5.30, respectively. The flow rates of organic and aqueous phase were the same as 25mL/min in microfluidic extraction. The time of mixing and reaction in batch extraction was 10min which was enough to ensure complete cobalt extraction reaction. By analyzing the raffinate, data shown in Fig. 3 were obtained.

The percentage extraction of cobalt and nickel increased with increasing neutralisation of the solvent obviously due to a change in the equilibrium pH value in both microfluidic and batch extraction (Fig. 3 (a)). The optimum equilibrium pH of the aqueous phase appears to be in the range 4.9 - 5.3 with about more than 72% Co extraction in the batch extraction and more than 85% Co extraction in the microfluidic extraction. The separation factor, β (Fig. 3 (b)) was low due to high co - extraction of nickel. Highest separation factor of 33 in microfluidic extraction at an equilibrium pH of ~5.3 were obtained, whereas in case of batch extraction it was 13 at an equilibrium pH of ~5.0.

The cobalt extraction ratio in microfluidic extraction was much greater than that in batch extraction under the same conditions. The reason was that components were mixed through intensive oscillation in batch extraction, while in microstructures this process was mainly realized through diffusion. Multilamination phenomenon happened in the interdigital micromixer of IMM. The micromixer

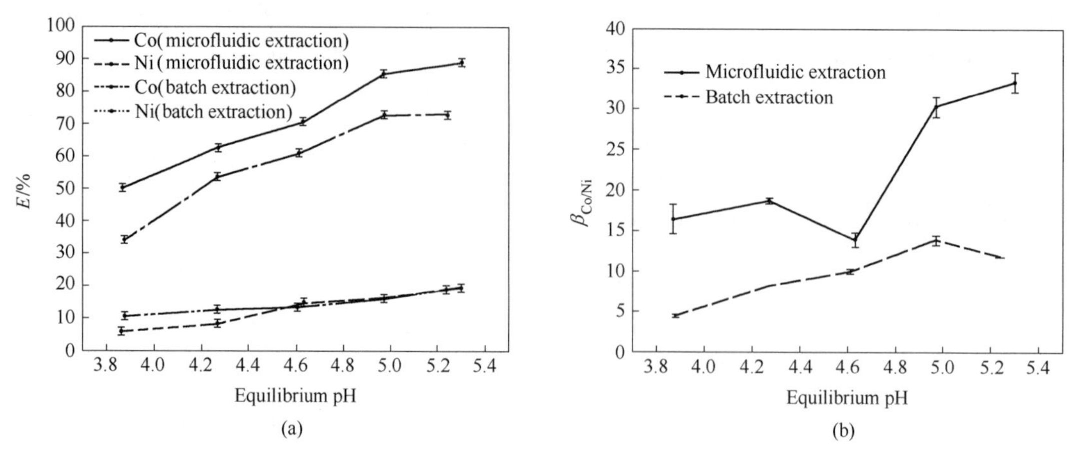

Fig. 3 Effect of equilibrium pH on the extraction of Co and Ni

produced thin liquid lamellae and guided them to contact and flow through chambers. Splitting the inlet streams into substreams and recombining them increases the contact surface between the two fluids causing diffusion to occur faster[23]. Despite laminar flows in microstructures, molecules had short paths to exceed the surface of organic – aqueous interface so that a nearly complete mixture was achieved within a few seconds, and often as little as a few milliseconds[24]. The features of the micromixer, e. g. , large specific surface area and short diffusion distance were effective for the efficient extraction and separation of Co from Ni.

During the course of microfluidic extraction, cobalt was preferentially extracted and the co – extraction of nickel was not negligible resulting in 5. 78% – 19. 30% extraction with increase equilibrium pH from 3. 8 – 5. 3, that was not very different from batch extraction results.

3. 2 Effect of flow rate

For study the influence of flow rate on the extraction of cobalt and nickel, the aqueous solution and the organic phase were contacted at a 1:1 phase ratio in microfluidic extraction, and the other reaction conditions were shown in section 2. 1. The volumetric flow rate of the aqueous phase, V_{aq}, was equal to that of the organic phase, V_{org}, the flow rate of 5mL/min, 10mL/min, 15mL/min, 20mL/min, 25mL/min and 30mL/min was adapted to investigate the influence on the extraction and separation of Co and Ni.

The relationship between the extraction degree of cobalt and nickel, E, and the flow rate was shown in Fig. 4(a). As the flow rate increased, the extraction ratio of the cobalt increased as well, but the extraction ratio of the nickel almost kept constant. This suggested that exchange reaction between Co in the organic phase and Ni in the aqueous phase reached the extraction equilibrium state. When the flow rate decreased less than 15mL/min, the percentage extraction of cobalt was almost linear growth with the flow rate increased. However, when the flow rate was greater than 15mL/min, the percentage extraction of cobalt was slowdown in growth trend. The percentage extraction had a maximum at intermediate to high flow rates and decreased at very high flow rates

(>25mL/min). Fig. 4 (b) showed the results of the seperation factor of cobalt and nickel. The $\beta_{Co/Ni}$ was also increased with the increase of flow rate <25mL/min. But with the further increase of flow rate, $\beta_{Co/Ni}$ was reduced in contrast. The highly effective seperation can be achieved at high flow rate of 25mL/min and the extraction rate of cobalt was 85.49%. If the flow rate was too high, it was difficult to seperate of the aqueous and organic phases in the settler.

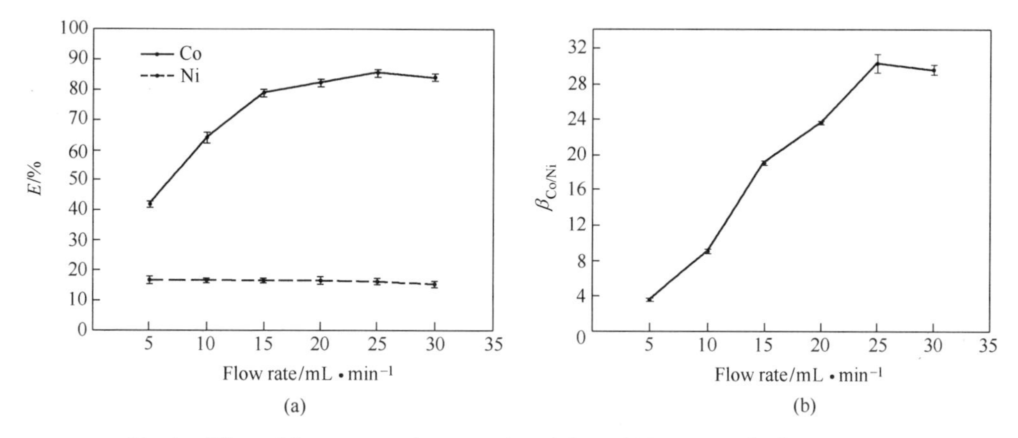

Fig. 4 Effect of flow rate on the extraction of Co and Ni in microfluidic extraction

Since the internal volume of the microreactor is on the order of 8μL (volume of the chamber), residence times were less than 0.019s at the flow rate of 25mL/min. Despite these extremely short residence time, the thermodynamic equilibrium could also be reached. The results of batch extraction was shown in Fig. 5. In these batch experiments, the extraction equilibrium was reached when the contact time was more than 3min, this time was much larger than that in the microreactor.

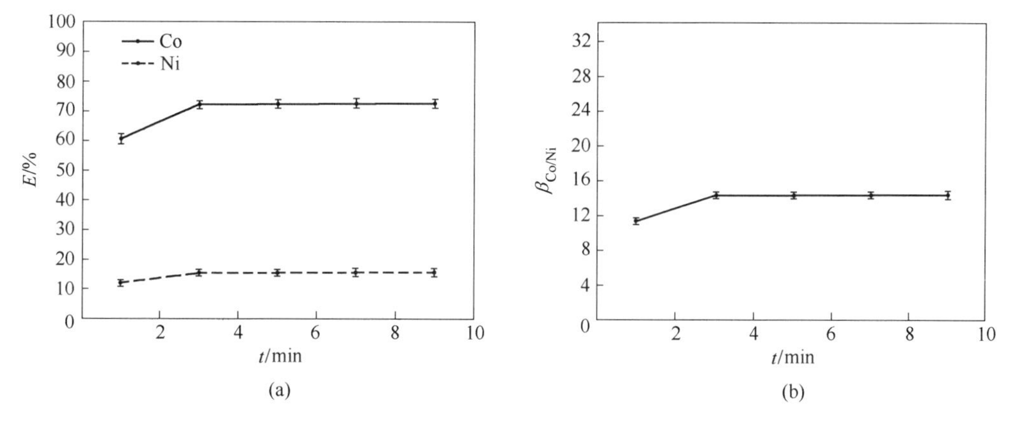

Fig. 5 Effect of contact time on the extraction of Co and Ni in batch extraction

In microfluidic extraction, since the residence time "t" is inversely related to the volume flow, one would expect maximum extraction efficiency at low flow rates. However, the results were in the opposite direction. According to Benz, et al. [25], the diffusion (diffusion coefficient, D) from a sphere of radius, r, into a well-stirred volume was a complex function of time, t, described by two parameters:

$$\frac{C}{C_0} = f\left[\frac{Dt}{r^2}, \frac{k \cdot V_E}{V_R}\right] \tag{6}$$

With k being the partition coefficient and V_E/V_R the ratio of extract to raffinate volume, which was set to one. In this function, only the first parameter, Dt/r^2, was time dependent.

The droplet diameter, $2r$, as a measure of diffusional length, almost linear decreased with increasing flow rate, and the mass transfer surface area greatly increased. Although the residence time was shorter, the mass transfer efficiency was linear increased. Furthermore, it was expected that at very high flow rates the drop size approached a minimum value, and hence, the diffusional length was not further decreased. As a result, the extraction efficiency reached a maximum value at high flow rates as shown in Fig. 4(a). In this case, the extraction efficiency had a maximum at intermediate to high flow rates and decreased again at very high flow rates.

3.3 Cobalt extraction isotherm

To determine the number of stages required at a chosen volume phase ratio, the Mc – Cabe Thiele diagrames were constructed with fresh aqueous solution to simulate the first stages of counter – current extraction (fresh aqueous solution to meet loaded organic solution). Both in microfluidic and batch extraction tests, the same organic with 70% saponified 20% v/v PC88A was contacted with fresh aqueous solution with A/O ratio 1∶1 at pH = 5.0. In batch extraction, the shak – out time of these two phases was 5min at every tests, and the microfluid flow rates of these two phases were both 25mL/min. From the extraction isotherm, shown in Fig. 6(a), it was observed that quantitative extraction of Co(>99.9%) was achieved in seven stages in batch extraction, and that was achieved in four stages in microfluidic extraction, shown in Fig. 6(b). The results showed that it reduced extraction stages of cobalt by using microfluidic technology.

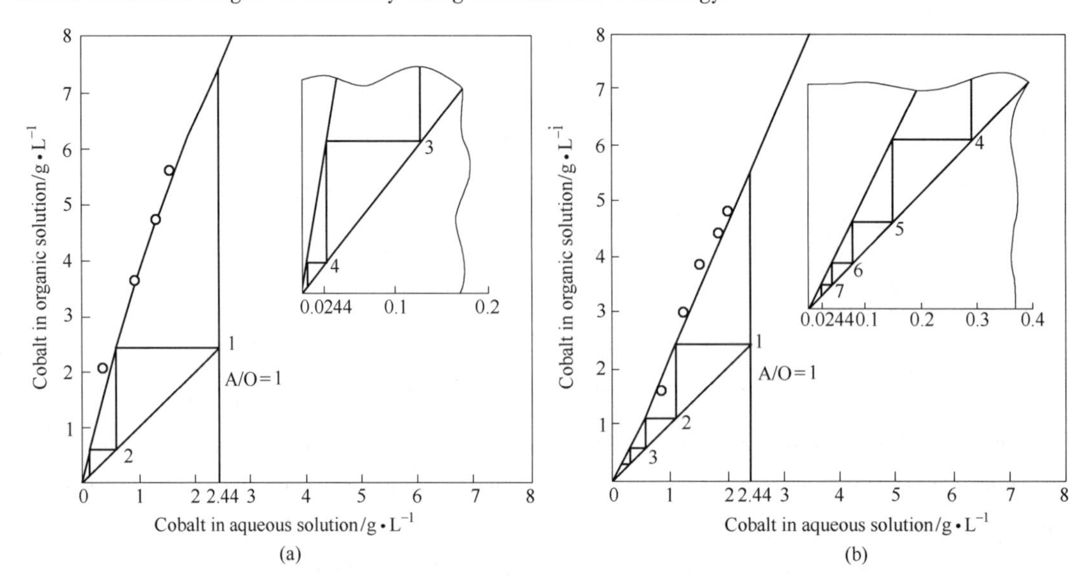

Fig. 6 Mc – Cabe Thiele plot for cobalt extraction

(a) Microfluidic extraction; (b) Batch extraction

4 Conclusions

Microfluidic extraction had shown great promise for extraction and separation of cobalt from nickel in sulphate solutions. The main conclusions were summarized as follows:

(1) In all the same equilibrium pH from 3.8 to 5.4, the cobalt percentage extraction and separation factor in microfluidic extraction were much greater than that in batch extraction. The co - extraction of nickel was not negligible both in these two experiments.

(2) The Co extraction was significantly affected by the flow rates of micromixer, compared to conventional batch extraction, and extraction equilibrium was achieved less than 0.02s.

(3) It reduced extraction stages of cobalt by using microfluidic technology. Quantitative extraction of Co(>99.9%) was achieved in seven stages in batch extraction, but that was achieved in four stages in microfluidic extraction.

Acknowledgements

The authors gratefully acknowledge the Joint Funds of the National Natural Science Foundation of China for financial support of this research.

References

[1] Higashi - Osaka A F, Nara I M, Tegukayamani K N. US Pat., 4196076, 1980.

[2] Luo L, Wei J H, Wei G Y, et al. Extraction studies of cobalt(Ⅱ) and nickel(Ⅱ) from chloride solution using PC88A[J]. Transactions of Nonferrous Metals Society of China, 2006, 16(3): 687 - 692.

[3] Bhaskara Sarma P V R, Nathsarma K C. Extraction of nickel from ammoniacal solutions using Lix87QN[J]. Hydrometallurgy, 1996, 42(1): 83 - 91.

[4] Devi N B, Nathsarma K C, Chakravortty V. Separation and recovery of cobalt(Ⅱ) and nickel(Ⅱ) from sulphate solutions using sodium salts of D2EHPA, PC88A and Cyanex272[J]. Hydrometallurgy, 1998, 49(1): 47 - 61.

[5] Kumbasar R A, Sahin I. Separation and concentration of cobalt from ammoniacal solutions containing cobalt and nickel by emulsion liquid membranes using 5,7 - dibromo - 8 - hydroxyquinoline (DBHQ) [J]. Journal of Membrane Science, 2008, 325(2): 712 - 718.

[6] Mendes F D, Martins A H. Selective sorption of nickel and cobalt from sulphate solutions using chelating resins [J]. International Journal of Mineral Processing, 2004, 74(1): 359 - 371.

[7] William H F, Thomas S R. Analytical Chemistry, 1954, 26: 1648 - 1649.

[8] Ouyang X, Besser R S. Development of a microreactor - based parallel catalyst analysis system for synthesis gas conversion[J]. Catalysis today, 2003, 84(1): 33 - 41.

[9] Mills P L, Quiram D J, Ryley J F. Microreactor technology and process miniaturization for catalytic reactions—a perspective on recent developments and emerging technologies[J]. Chemical Engineering Science, 2007, 62(24): 6992 - 7010.

[10] Stone B M, DEmello A. FOCUS life, the universe and microfluidics[J]. Lab on a Chip, 2002, 2(4): 58N - 65N.

[11] Matsuura S, Ishii R, Itoh T, et al. Immobilization of enzyme - encapsulated nanoporous material in a microreactor and reaction analysis[J]. Chemical Engineering Journal, 2011, 167(2): 744 - 749.

[12]Ehrfeld W,Golbig K,Hessel V,et al. Characterization of mixing in micromixers by a test reaction:single mixing units and mixer arrays[J]. Industrial & Engineering Chemistry Research,1999,38(3):1075 - 1082.

[13]Pennemann H,Hessel V,Löwe H. Chemical microprocess technology—from laboratory - scale to production [J]. Chemical Engineering Science,2004,59(22):4789 - 4794.

[14]Pennemann H,Hardt S,Hessel V,et al. Micromixer based liquid/liquid dispersion[J]. Chemical Engineering & Technology,2005,28(4):501 - 508.

[15]Kumar S,Kumar B,Sampath M,et al. Development of a micro - mixer - settler for nuclear solvent extraction [J]. Journal of Radioanalytical and Nuclear Chemistry,2012,291(3):797 - 800.

[16]Syouher N,Yasuyuki T,Kazuharu Y. Ars Separatoria Acta,2006,4,18 - 26.

[17]Fukiko K,Jun - ichi V,Masahiro G. Solvent Extraction Research and Development,Japan,2003,10,93 - 102.

[18]Tamagawa O,Muto A. Development of cesium ion extraction process using a slug flow microreactor[J]. Chemical Engineering Journal,2011,167(2):700 - 704.

[19]Priest C,Zhou J,Sedev R,et al. Microfluidic extraction of copper from particle - laden solutions[J]. International Journal of Mineral Processing,2011,98(3):168 - 173.

[20]Löb P,Pennemann H,Hessel V,et al. Impact of fluid path geometry and operating parameters on l/l - dispersion in interdigital micromixers[J]. Chemical Engineering Science,2006,61(9):2959 - 2967.

[21]Sarangi K,Reddy B R,Das R P. Extraction studies of cobalt(Ⅱ) and nickel(Ⅱ) from chloride solutions using Na - Cyanex 272:separation of Co(Ⅱ)/Ni(Ⅱ) by the sodium salts of D2EHPA,PC88A and Cyanex272 and their mixtures[J]. Hydrometallurgy,1999,52(3):253 - 265.

[22]Devi N B,Nathsarma K C,Chakravortty V. Sodium salts of D2EHPA,PC - 88A and Cyanex - 272 and their mixtures as extractants for cobalt(Ⅱ)[J]. Hydrometallurgy,1994,34(3):331 - 342.

[23]Jähnisch K,Hessel V,Löwe H,et al. Chemistry in microstructured reactors[J]. Angewandte Chemie International Edition,2004,43(4):406 - 446.

[24]Löb P,Drese K S,Hessel V,et al. Steering of liquid mixing speed in interdigital micro mixers - from very fast to deliberately slow mixing[J]. Chemical Engineering & Technology,2004,27(3):340 - 345.

[25]Benz K,Jäckel K P,Regenauer K J,et al. Utilization of micromixers for extraction processes[J]. Chemical Engineering & Technology,2001,24(1):11 - 17.

Synthesis of Copper Nanoparticles by a T - shaped Microfluidic Device

Lei Xu, Jinhui Peng, C. Srinivasakannan, Libo Zhang, Di Zhang, Chenhui Liu, Shixing Wang, Amy Q. Shen

Abstract: The copper nanoparticles were prepared by reduction of metal salt solutions with sodium borohydride in a T - shaped microfluidic device at room temperature. The influence of flow rates on copper particle diameter, morphology, size distribution, and elemental compositions has been investigated. Experimental results demonstrated that copper nanoparticles were uniform in size distribution, without being oxidized. With the increase in fluid flow rate the copper nanoparticles mean diameter increased, and the surface plasmon resonance absorptions of copper nanoparticle exhibited slight blue - shifting.

Keywords: microfluidic; copper; nanoparticles; solution reduction

1 Introduction

In recent years, microreactors have gained much attention in nanoparticle synthesis due to the advantage of precise control of reaction and mixing conditions. The benefits from microreactors are high - surface - area - to - volume ratio, tunable inner - wall properties, flow - orientation and flexible - structure designs of the micromixers[1-6]. In addition, the key advantages of micro - reaction technology are diffusive exchange of heat and mass within small length scales. These features have enabled the use of microreactors to improve the reaction efficiency and control the synthesis of nanomaterials[4-12].

Metal nanoparticles have been widely used in catalysis, optoelectronics, photovoltaic technology, information storage, environmental technology, and biosensors, to name a few. The need for low - cost, scalable, and dispersible nanomaterials drives new research in the field of nanocrystal synthesis[13,14]. The preparation of Cu nanoparticles has become an intensive area of scientific research as Cu nanoparticles exhibit excellent physical and chemical properties such as high electrical conductivity and chemical activity. Cheaper Cu have been considered possible replacements for Ag and Au particles in some potential applications such as high electrical conductivity and chemical activity[12-16]. Existing chemical procedures to synthesize Cu nanoparticles include, radiation, thermal reduction, microemulsion, sonochemical reduction, vacuum vapor deposition, metal vapor synthesis, laser ablation, and aqueous reduction methods[13-20]. Among these methods, aqueous reduction method is most widely employed because of its simple operation procedure, high yield and quality, limited equipment requirements and ease of control[15]. Liu, et al. reported preparation of Cu nanoparticles with $NaBH_4$ by aqueous reduction method[15]. Song, et al. described controlled growth of Cu nanoparticles by a tubular microfluidic reactor[1].

The processing conditions to synthesize nanoparticles are of particular importance for the control of particle size, particle shape, aggregation and composition of Cu nanoparticles. The rate of nucleation and the rate of particle growth depends in different ways on the local concentrations of educts[8]. In the present work, Cu nanoparticles were prepared by aqueous reduction method, using $NaBH_4$ as reducing agent in T - shaped microfluidic chip.

2 Experimental method

2.1 Materials

Copper(Ⅱ) sulfate pentahydrate ($CuSO_4 \cdot 5H_2O$, purity ≥98%), Sodium borohydride ($NaBH_4$, purity ≥99%), Polyvinylpyrrolidone (PVP, average molecular weight 360000), Ammonium hydroxide solution ($NH_3 \cdot H_2O$, 28.0% - 30.0%) and Sodium hydroxide (NaOH, 99%) were purchased from Sigma - Aldrich (St. Louis, MO, USA). Sylgard 184 silicone elastomer kits including poly (Polydimethylsiloxane, PDMS) and a curing agent were purchased from Dow Corning Co. (Midland, MI).

2.2 Fabrication of chip - based microfluidic reactor

The PDMS microfluidic devices were fabricated by a soft lithography technique as described in literature[10-21]. The PDMS prepolymer and the curing agent are mixed in a 10∶1 ratio (v/v), and poured onto a silicon wafer patterned with SU - 8 photoresist. After degassing under vacuum in a desiccator for an hour, the PDMS material was baked for 2h at 65℃ in an oven. The PDMS replicas and the glass slide were then bonded after oxygen plasma treatment and placed in an oven (65℃) for 2 days before experiments.

2.3 Synthesis of copper nanoparticles

The flowchart of the experimental process is shown in Fig. 1. Prior to carrying out the experiments, 0.2mol/L $CuSO_4$ solution and 0.4mol/L $NaBH_4$ solution were prepared, and argon gas was bubbled through both solutions for 30min. A certain proportion of ammonium hydroxide solution was added to the $CuSO_4$ solutions complexant, which turned the color of solutions from blue to dark blue. Then the pH of $CuSO_4$ solutions was adjusted to 12 by the addition of NaOH solution. The PVP at a concentration of 3.2g/L was added to the $CuSO_4$ complex solution as a dispersant.

The microfluidics chip used to prepare Cu nanoparticles is shown in Fig. 2. T - shaped microchannel was used to introduce two aqueous solutions flowing into the microchannel. The microfluidic synthesis was performed at room temperature (25℃ ±2℃). This microfluidic device was placed on an inverted microscope (Leica DM IRB) to track the synthesis of Cu nanoparticles. The complex solution $CuSO_4$ and the aqueous mixture containing $NaBH_4$ were pumped to the microchannel using digitally controlled syringe pumps (PHD 2000 Infusion, Harvard Apparatus, USA). The experiments were conducted at different flow rates covering the range 10mL/h to 30mL/h.

The general chemical reaction to produce Cu nanoparticlesis based on the following equation[15].

$$4Cu^{2+} + BH_4^- + 8OH^- \longrightarrow 4Cu + B(OH)_4^- + 4H_2O \quad (1)$$

When $NH_3 \cdot H_2O$ is adopted as complexant, the reduction reaction can be represented by the following equations,

$$Cu^{2+} + 4NH_3 \cdot H_2O \longrightarrow Cu(NH_3)_4^{2+} + 4H_2O \quad (2)$$

$$4Cu(NH_3)_4^{2+} + BH_4^- + 8OH^- \longrightarrow 4Cu + B(OH)_4^- + 16NH_3 + 4H_2O \quad (3)$$

In theory, the stoichiometric ratio of Cu ions to $NaBH_4$ is 4: 1. In addition, Liu, et al. , have reported a reduction in the average size of the Cu nanoparticles with increase in $NaBH_4$ concentration[15]. In general the reducing agents are added far in excess of stoichiometric requirement to promote the reaction sufficiently.

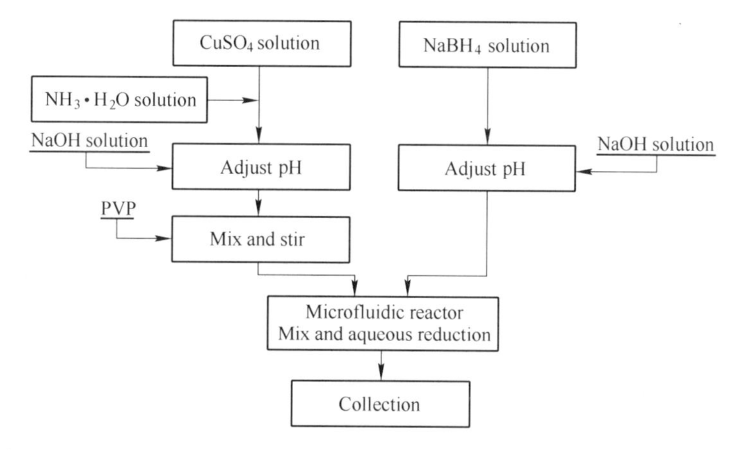

Fig. 1 Flowchart of Cu nanoparticles preparation processes by aqueous reduction with sodium borohydride

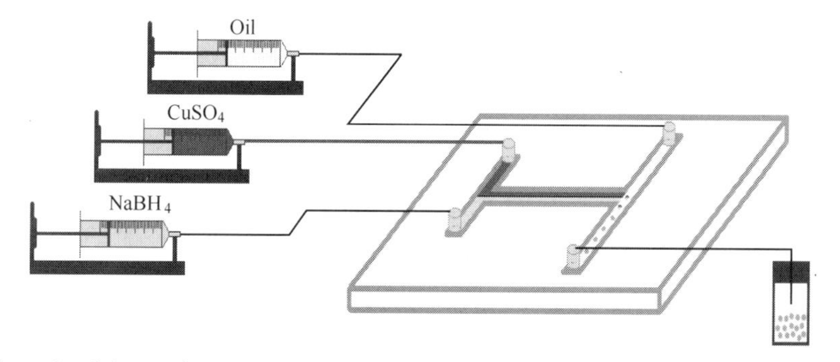

Fig. 2 Schematic of the synthesis processes of the copper nanoparticles by T - shaped microfluidic chip at room temperature (the length, width and height of the channel are 10mm, 200μm and 30μm, respectively)

2. 4 Material characterizations

The Cu nanoparticles size and morphology are characterized by using transmission electron microscopy (TEM, 2100F, 200kV, JEOL). For the TEM investigations the material was deposited by drying its suspension on a copper - grid - supported, perforated, transparent carbon foil. Elemental compositions of the copper nanoparticles were obtained by using energy - dispersive X - ray spectrometer(EDS), operated at 200kV. The UV - vis spectrometry was used to verify the chemical composition of nanoparticles and the average size was estimated using a spectrometer(UV - 2450, SHIMADZU).

3 Results and discussion

The influence of flow rates in the microfluidic flow on final copper particle diameter, morphology, size distribution, and elemental compositions was examined. Fig. 3 shows the size, morphology and elemental compositions of the copper nanoparticles at flow rate 10mL/h. From Fig. 3(a) and (b), copper nanoparticles are spherical, uniform particle size, with good dispersion. The particle size distribution shows the mean diameter of 8.95nm(Fig. 3(c)). Fig. 3(d) presents the energy dispersive spectrometer which indicates the absence of oxidation in the process of copper nanoparticles synthesis, proving the effectiveness of the PVP dispersant as an antioxidant.

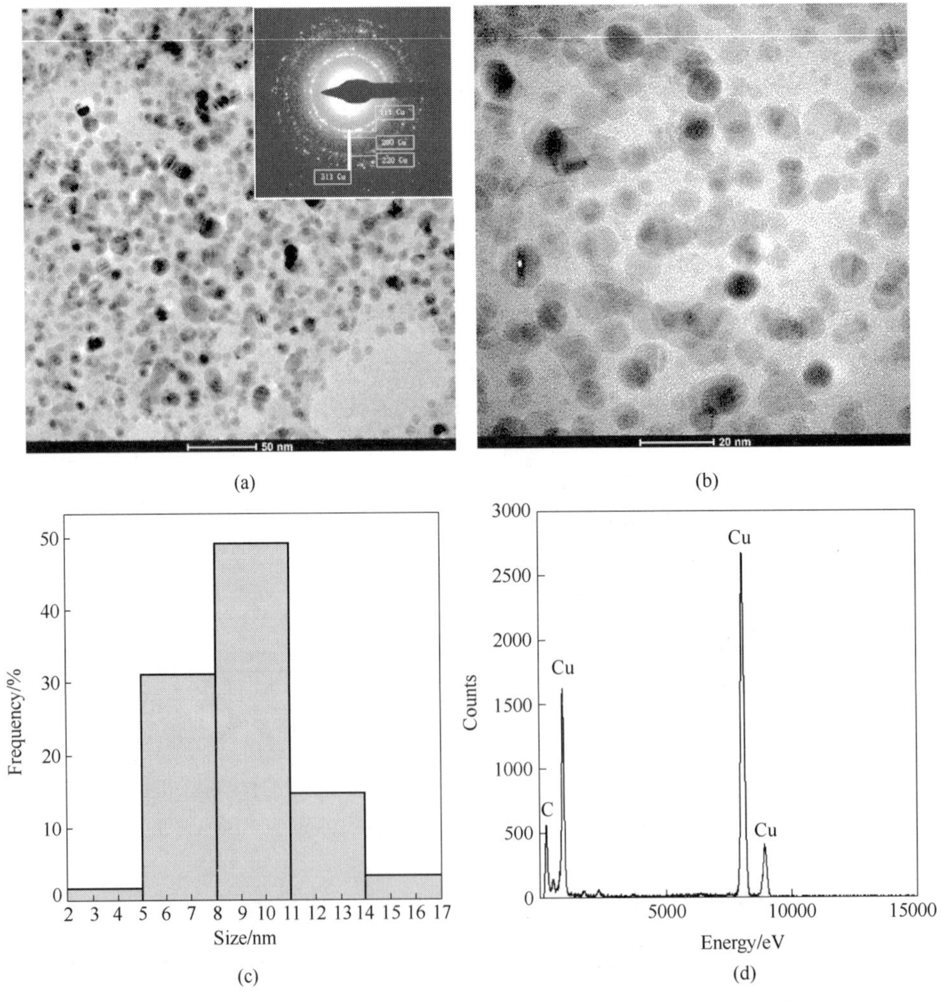

Fig. 3 The TEM image ((a) and (b)) and the particle size distribution(c) and the EDS spectrum (d) of the copper nanoparticles prepared by aqueous reduction with sodium borohydride in the T - shaped microfluidic chip at the flow rate of 10mL/h

Fig. 4 shows the size and morphology of the copper nanoparticles at a flow rate of 20mL/h and 30mL/h. The mean particle diameter of copper nanoparticles was estimated to be 9. 18nm and 14. 15nm, respectively. An increase in the flow rate clearly yields an increase in the mean particle diameter. Wei, et al., have reported synthesis and thermal conductivity of microfluidic copper nanofluids, cupric - sulfate($CuSO_4$) reduced by hydrazine - hydrate(N_2H_4). Their results showed an increase in mean particle diameter with an increase in the flow rate, except for the case of N_2H_4 molar concentration 0. 02mol/L[12], which are in agreement with the results of the present work. Liu, et al., have reported preparation of very fine Cu nanoparticles of 37nm, at the following conditions: concentrations of Cu^{2+} and $NaBH_4$ being 0. 2mol/L and 0. 4mol/L respectively, the dripping rate of 50mL/min, gelatin concentration of 1%, pH at 12 and solution temperature at 313K[15]. In the present work, the mean particle diameter as low as 9nm could be achieved utilizing a T - shaped microfluidic chip.

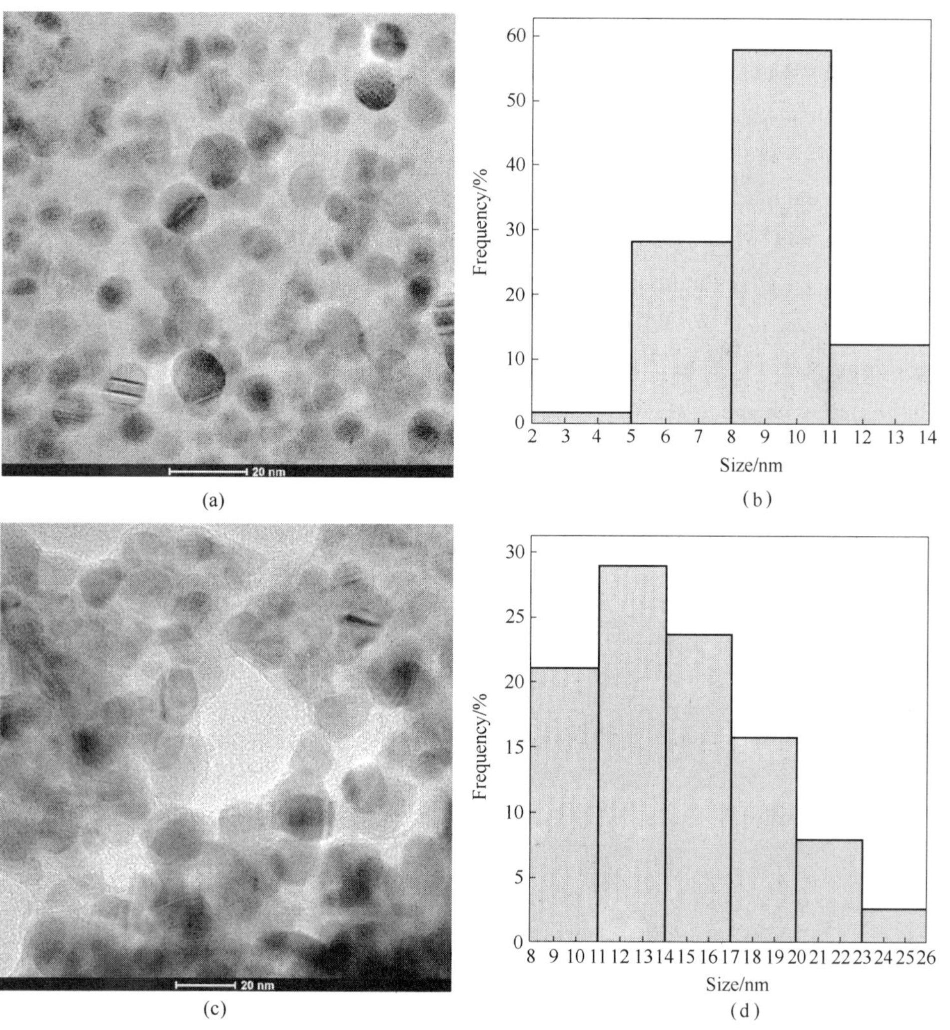

Fig. 4 The TEM image((a) and (c)) and the particle size distribution((b) and (d)) of the copper nanoparticles prepared by aqueous reduction with sodium borohydride in the T - shaped microfluidic chip at the flow rate of 20mL/h and 30mL/h, respectively

Comparing Fig. 3 and Fig. 4, the effect of flow rate on the appearance of copper nanoparticles prepared by microfluidic chips can be demonstrated. The particle size distribution as well as the mean particle size at a flow rate of 20mL/h compare closely with the flow rate of 10mL/h. However, at the flow rate of 30mL/h, the particle size distribution is skewed to the right with the mean particle diameter much higher than the lower flow rates. The proportion of particles in the lower size range is significantly higher than the particles at the higher size range. The non uniformity in the size is evident from the particle size distribution.

Song, et al., have reported the growth of nearly monodispersed Cu nanoparticles of 135.6nm ± 11.4nm in a tubular microfluidic reactor. Such a large particle size was enabled through continued controlled growth of the initially formed Cu nanoparticle seeds in a tubular microfluidic reactor[1]. In the present work, Cu nanoparticles were prepared using aqueous reduction which include four distinct stages: the formation of a supersaturated solid solution, nucleation, growth and aggregation, as proposed by LaMer and Dinegar[1,22]. The solute consisting of ions are formed by chemical reactions. When the solute concentration attains a critical concentration, the formation of spontaneous nuclei occurs, which reduces solute concentration, halts further nucleation and freezes the number of nuclei that are formed. The nucleation stage is followed by the growth of the particles. The growth of particles from the limited nuclei proceeds until all of the solute has been consumed. Therefore, in order to obtain a narrow particle size distribution, the nucleation time should be minimized.

In the T - shaped microfluidic chip, a basic characteristic of the micro - scale fluid flow is laminar (Fig. 5). The reaction is mass transfer controlled with the rate of diffusion of ions to the interphase governing the rate of reaction. Therefore, when the reaction proceeded to 60s, the layer of reaction to be darker than the reaction proceeded to 30s. The diffusion of ions to the interphase is rather slow in laminar flow and hence at faster flow rates the diffusion of ions to the interphase is limited, which results in formation of fewer Cu nanoparticles nuclei. The formed nuclei continue to grow resulting in relatively larger and non uniform particle size. In contrast, at low flow rates the diffusion of Cu ions to the interphase is relatively larger resulting in higher nuclei formation. The

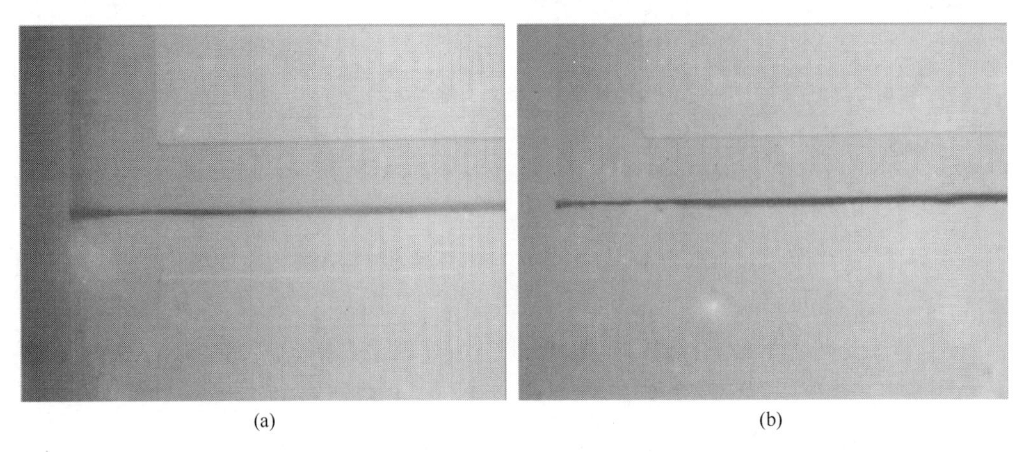

(a) (b)

Fig. 5 Optical microscope image of copper nanoparticles formed at different times in the T - shaped microfluidic chip

(a) The reaction proceeded to 30s; (b) The reaction proceeded to 60s continued growth of solute over the larger number of nuclei results in smaller and more uniform size of the Cu nanoparticles.

Variations of the size, shape and elemental compositions of Cu nanoparticles prepared by microfluidic chip at different flow rate definitely affect their localized surface plasmon resonance absorption characterized by the UV – vis spectroscopy. As shown in Fig. 6, with the increase in fluid flow rate, the copper nanoparticles diameter increases and their surface plasmon resonance absorptions exhibit slight blue – shifting, as shown by the peak at 575nm for the 8. 95nm Cu nanoparticles, the peak at 567nm for the 9. 18nm Cu nanoparticles and the peak at 564nm for the 14. 15nm Cu nanoparticles respectively. Many authors have associated the blue – shift of the surface plasmon resonance of the metallic nanoparticles with their decreasing size[13,23,24], which may not be the case always. Yang, et al. showed contrary data for spherical Cu nanoparticles[25], which agrees with the results of the present work, supported by the TEM images. This could be due to the combined effects of variations of their size, size distribution, shape, aggregation, surface roughness and crystallinity, and the surface – coated surfactant[19,26-28].

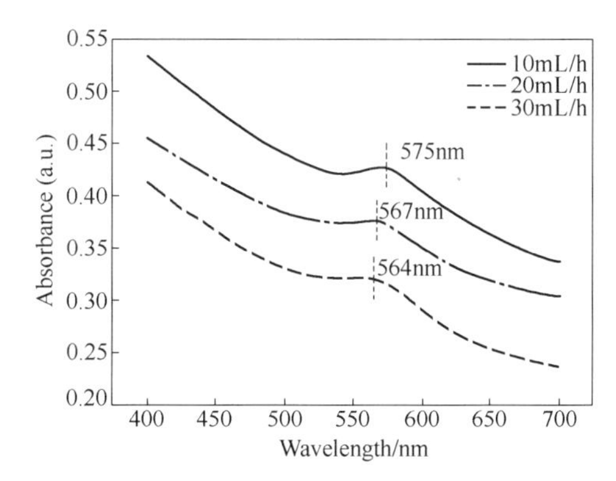

Fig. 6 UV – vis absorption of Cu nanoparticles prepared by the flow rate of 10mL/h, 20mL/h and 30mL/h

4 Conclusions

In summary, copper nanoparticles can be prepared by reduction of metal salt solutions with sodium borohydride in the T – shaped microfluidic chip at room temperature. PVP as a dispersant has better dispersion and antioxidant effects. The prepared copper nanoparticles were uniform in size distribution without being oxidized. The mean particle diameter increased with increase in the fluid flow rate, with the maximum mean of 14. 15nm at a fluid flow rate of 30mL/h, however with a non uniform particle size distribution (distribution skewed to the right). At low flow rates the particle size distribution was found to be normal, however with a smaller mean particle diameter. An increase in the particle size could be achieved with increase in the flow rate. Additionally, with increase in particle size the surface plasmon resonance absorptions exhibited slight blue – shifting.

Acknowledgements

This work was financially supported by the National Natural Science Foundation of China (grant

No. 51204081 and U1302271), and by China Scholarship Council (No. 2011853521), and by Yunnan Provincial Science and Technology Innovation Talents scheme——Technological Leading Talent of China(No. 2013HA002), and by Applied Basic Research Project of Yunnan Province of China(grant No. 2013FZ008), and by Yunnan Provincial Department of Education Research Fund of China(grant No. 2013Z118).

References

[1] Song Y, Li R, Sun Q, et al. Controlled growth of Cu nanoparticles by a tubular microfluidic reactor[J]. Chemical Engineering Journal, 2011, 168(1): 477 - 484.

[2] Sounart T L, Safier P A, Voigt J A, et al. Spatially - resolved analysis of nanoparticle nucleation and growth in a microfluidic reactor[J]. Lab on a Chip, 2007, 7(7): 908 - 915.

[3] Demello A J. Control and detection of chemical reactions in microfluidic systems[J]. Nature, 2006, 442 (7101): 394 - 402.

[4] Zhao B, Moore J S, Beebe D J. Surface - directed liquid flow inside microchannels[J]. Science, 2001, 291 (5506): 1023 - 1026.

[5] He S, Kohira T, Uehara M, et al. Effects of interior wall on continuous fabrication of silver nanoparticles in microcapillary reactor[J]. Chemistry Letters, 2005, 34(6): 748 - 749.

[6] Park J I, Saffari A, Kumar S, et al. Microfluidic synthesis of polymer and inorganic particulate materials[J]. Annual Review of Materials Research, 2010, 40: 415 - 443.

[7] Zhao C X, He L, Qiao S Z, et al. Nanoparticle synthesis in microreactors[J]. Chemical Engineering Science, 2011, 66(7): 1463 - 1479.

[8] Wagner J, Tshikhudo T R, Koehler J M. Microfluidic generation of metal nanoparticles by borohydride reduction [J]. Chemical Engineering Journal, 2008, 135: S104 - S109.

[9] Parisi J, Su L, Lei Y. In situ synthesis of silver nanoparticle decorated vertical nanowalls in a microfluidic device for ultrasensitive in - channel SERS sensing[J]. Lab Chip, 2013, 13(8): 1501 - 1508.

[10] Ali M A, Solanki P R, Patel M K, et al. A highly efficient microfluidic nano biochip based on nanostructured nickel oxide[J]. Nanoscale, 2013, 5(7): 2883 - 2891.

[11] Song Y, Doomes E, Prindle J, Tittsworth R, Hormes J, Kumar C S R. The Journal of Physical Chemistry B, 2005, 109, 9330 - 9338.

[12] Wei X, Wang L. Synthesis and thermal conductivity of microfluidic copper nanofluids[J]. Particuology, 2010, 8 (3): 262 - 271.

[13] Tokarek K, Hueso J L, Kustrowski P, et al. Green synthesis of chitosan - stabilized copper nanoparticles[J]. European Journal of Inorganic Chemistry, 2013(28): 4940 - 4947.

[14] Lignier P, Bellabarba R, Tooze R P. Scalable strategies for the synthesis of well - defined copper metal and oxide nanocrystals[J]. Chemical Society Reviews, 2012, 41(5): 1708 - 1720.

[15] Liu Q, Zhou D, Yamamoto Y, et al. Preparation of Cu nanoparticles with $NaBH_4$ by aqueous reduction method [J]. Transactions of Nonferrous Metals Society of China, 2012, 22(1): 117 - 123.

[16] Raspolli Galletti A M, Antonetti C, Marracci M, et al. Novel microwave - synthesis of Cu nanoparticles in the absence of any stabilizing agent and their antibacterial and antistatic applications[J]. Applied Surface Science, 2013, 280: 610 - 618.

[17] Joshi S S, Patil S F, Iyer V, et al. Radiation induced synthesis and characterization of copper nanoparticles [J]. Nanostructured Materials, 1998, 10(7): 1135 - 1144.

[18] Dhas N A, Raj C P, Gedanken A. Synthesis, characterization, and properties of metallic copper nanoparticles [J]. Chemistry of Materials, 1998, 10(5): 1446 - 1452.

[19] Yu W, Xie H, Chen L, et al. Synthesis and characterization of monodispersed copper colloids in polar solvents [J]. Nanoscale Research Letters, 2009, 4(5): 465 - 470.

[20] Park B K, Jeong S, Kim D, et al. Synthesis and size control of monodisperse copper nanoparticles by polyol method[J]. Journal of Colloid and Interface Science, 2007, 311(2): 417 - 424.

[21] Wang Q, Zhang D, Yang X, Xu H, Shen A Q, Yang Y. Green Chemistry, 2013, 15, 2222 - 2229.

[22] LaMer V K, Dinegar R H. Theory, production and mechanism of formation of monodispersed hydrosols[J]. Journal of the American Chemical Society, 1950, 72(11): 4847 - 4854.

[23] Dang T M D, Le T T T, Fribourg - Blanc E, et al. Synthesis and optical properties of copper nanoparticles prepared by a chemical reduction method[J]. Advances in Natural Sciences: Nanoscience and Nanotechnology, 2011, 2(1): 015009.

[24] Schwartzberg, Adam M, et al. Synthesis, characterization, and tunable optical properties of hollow gold nanospheres[J]. The Journal of Physical Chemistry B, 2006, 40(110): 19935 - 19944.

[25] Hung L I, Tsung C K, Huang W, et al. Room - temperature formation of hollow Cu_2O nanoparticles[J]. Advanced Materials, 2010, 22(17): 1910 - 1914.

[26] Moores A, Goettmann F. The plasmon band in noble metal nanoparticles: an introduction to theory and applications[J]. New Journal of Chemistry, 2006, 30(8): 1121 - 1132.

[27] Braundmeier Jr A J, Arakawa E T. Effect of surface roughness on surface plasmon resonance absorption[J]. Journal of Physics and Chemistry of Solids, 1974, 35(4): 517 - 520.

[28] Song Y, Elsayed - Ali H E. Aqueous phase Ag nanoparticles with controlled shapes fabricated by a modified nanosphere lithography and their optical properties [J]. Applied Surface Science, 2010, 256 (20): 5961 - 5967.

Solvent Extraction of In^{3+} with Microreactor from Leachant Containing Fe^{2+} and Zn^{2+}

Yaqian Wei, Jinhui Peng, Lei Xu, Shaohua Ju, Shenghui Guo,
Libo Zhang, Lihua Zhang, Linqing Dai

Abstract: The microreactor has been developed for a wide range of applications because of many advantages, such as high mass transfer efficiency, low energy consumption, and it provides a closed and safe system. The application of Microfluidic technique in the traditional hydrometallurgy extraction process is expected to overcome the difficulties such as co-extracting of impurities, large consumption of extractant and hidden fire risks. In this study, the extraction and separation efficiency of In^{3+} from a complex sulfate solution containing impurities, such as Fe^{2+} and Zn^{2+}, were studied. The microfluidic extraction was carried out in a Pyrex™ microchip, and the organic phase was prepared with the extractant di(2-ethylhexyl) phosphoric acid (D2EHPA) diluted in 260# kerosene. The results shown that with only 0.55s contacting of the organic and aqueous phase, the extraction ratio of In^{3+} can reach to 90.80% while only 0.16% of Fe^{2+} and 0.22% of Zn^{2+} were co-extracted, and the average mass transfer speed of In^{3+} was calculated as high as $0.34g/(m^2 \cdot s)$. Compared with the traditional mixing settler process, microfluidic extraction has the advantages of higher extraction rate of In^{3+}, lower trend of co-extraction of the impurities and emulsification.

Keywords: indium; mass transfer; microreactor; solvent extraction

1 Introduction

Various microreactors with micron channel structures, the internal diameters in the range of 10 - 500μm, have been widely utilized in the field of chemical engineering and analysis fields for intensifying due to their advantages over macro scale reactors, such as higher ratio of contacting interface to volume of the flows, shorter diffusion distance, and the feasibilities of precisely controlling the temperature of the reaction. It is the rapid development of advanced micro scale manufacturing technology that pushed efficiently the application of microreactors in the above fields[1].

Thus, microfluidics process usually embodies a much smaller chemical device, shorter reaction time, higher efficiency[2], lower consumption of materials and energy[3] and safer operation condition. Another advantages of micro fluidics process is that it can easily scale up by simply "numbering up". Many different microfluidic procedures were applied to organic synthesis[4-6] and material preparation[7-9]. Micro fluidics devices now can generate tones of product per year[10].

Solvent extraction with mixing settler is a traditional key process for separation and purification of some metal ions in metallurgical solution system[11,12]. Although mixing settler has some merits like simple structure, low cost, effective to some extent, there are still some problems, such as low efficiency[13], too much extraction stages, emulsifying easily[14-17], great place occupation and fire risks[18].

For example, in a zinc plant in Yunnan province, China, the extraction procedure of In^{3+} ion by D2EHPA contains altogether 14 steps. They are 2 - step of extraction, 3 - step of eluting impurities (mainly Fe^{2+}, Zn^{2+}), 4 - step of stripping with hydrochloric acid, 4 - step of stripping iron and 1 - step of stripping chloride ion. Actually, the impurities co - extracted in the organic phase increase the complexity of the extraction process.

Very few researchers have explored the potential of using microreactor in the field of solvent extraction of metal ions. In Australia, the possibility of extraction Cu^{2+} from a particle - laden solution had been investigated, with a positive result[19]. In Japan, it was studied that Cs^{+} extraction from a complex solution in stable slug flows by inserting a piece of glass bead into a microchannel, resulting in rapid separation that compared favorably with conventional batch experiments[20].

In this paper, the extraction and separation efficiency of In^{3+} from a complex leaching solution containing impurities, such as Fe^{2+}, Zn^{2+} and SO_4^{2-}, was studied in a laminar flow microreactor. The process of diffusion and mass transfer in microreactor were also analyzed, and the extraction effects were compared with the traditional extraction.

2 Experimental

2.1 Material and device

The experimental material, an aqueous solution in sulfuric system with a pH value of 0.44, was obtained from a hydrometallurgy zinc plant in Yunnan province, China. Its chemical composition was analyzed and listed in Table 1. The organic phase for solvent extraction contains 30% of di(2 - ethylhexyl) phosphoric acid(D2EHPA) and 70% of 260# kerosene. The main extraction equation is as follow[21]:

$$(In^{3+})_{aq} + (3HR_2PO_4)_{org} = [In(R_2PO_4)_3]_{org} + (3H^+)_{aq} \tag{1}$$

Table 1 Chemical composition and the pH value of the complex solution

Elements unit	$In^{3+}/g \cdot L^{-1}$	$Fe^{2+}/g \cdot L^{-1}$	$Zn^{2+}/g \cdot L^{-1}$	pH value
Data	3.17	3.4	52.8	0.44

The microchip used in this study is shown in Figs. 1 and 2. It was made by Pyrex™ glass(Institute of Microchemical Technology, Japan), produced with combinated methods of photolithography, wet - etching and thermal bonding. The extraction reaction takes place in a 120mm long microchannel(160μm × 40μm), which was showed in Fig. 1(b), with a guide structure for obtaining a stable flow interface of the two immiscible liquid - liquid phases.

2.2 Procedures

During the experiments of microfluidic extraction, the aqueous and organic phases were respectively

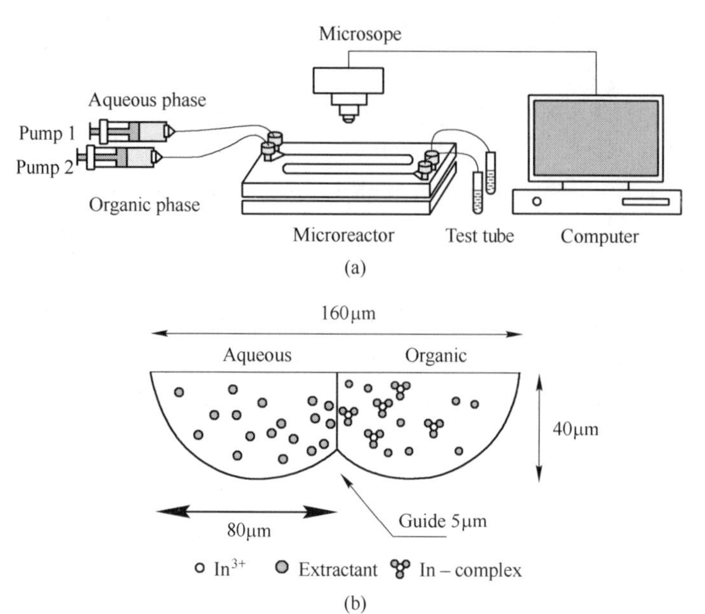

Fig. 1 Schematic of microreactor system used in this study

(a) Microreactor System; (b) Cross – section of the extraction channel

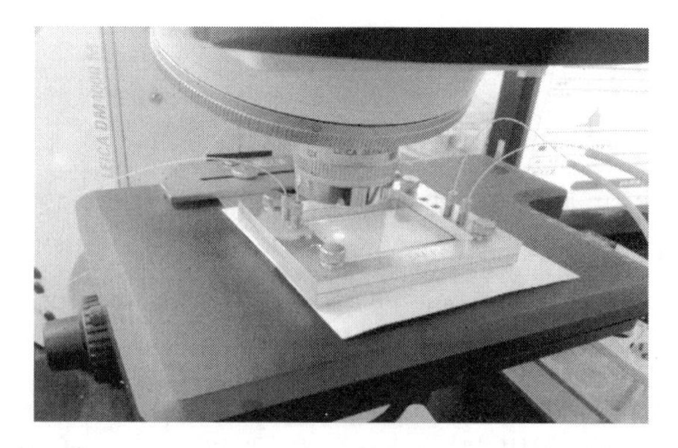

Fig. 2 Microreactor channel conditions monitored by an optical microscopy

pumped into the Pyrex glass microchip by two constant flow pumps (HLB – 4015, Yansan). The flow rate of the aqueous phase ranged from 1mL/h to 6mL/h at a fixed organic/aqueous flow rate ratio of 0.8. Flow stability in the microchip was monitored with an optical microscopy (Leica DM 4000M). The cross – section of the extraction channels includes a guide structure which helps to maintain the position of the liquid – liquid interface. The extraction channel terminates at a second Y – junction at which the two phases are separated and flow out of the microchip for sample collection and analysis.

The experiments of conventional extraction were carried out in a 250mL separating funnel. The phase ratio was set as O: A = 2: 1, 1: 1, 1: 2 and 1: 4, respectively. Each time, 40mL solution was added to the organic phase (30% D2EHPA in 260# kerosene) in a 250mL separating funnel. Then, the funnel was stirred on a shaking machine for 5min at a speed of 200r/min. After separation, the aqueous phase was diluted and analyzed by inductively coupled plasma atomic emission spectrometry (ICP - AES).

2.3 Analysis method

The concentration of In^{3+}, Fe^{2+} and Zn^{2+} in aqueous phase before and after the extraction was determinated by Inductively Coupled Plasma Atomic Emission Spectrometry (Leeman ICP - AES PS1000) The acidity of the aqueous solution determinated by an acidity meter. Eq. (2) is an expression of extraction rate, where the C_{ao} and C_{al} signify the element contains before and after extraction, respectively.

$$E = \frac{C_{ao} - C_{al}}{C_{ao}} \times 100 \tag{2}$$

3 Results and discussion

3.1 Stability of liquid - liquid interface

The liquid - liquid profile is well formed due to the large interfacial tension between the two immiscible fluid phases. The pictures of aqueous and organic stream were taken by an optical microscopy in Fig. 3. A stable liquid - liquid interface was observed during this experiment under the condition of optimal flow rate ratio (organic/aqueous), $R = 0.80$, and flow was laminar in all cases. Compared with the emulsions easily appeared in conventional extractions, this type of microfluidics extraction radically abandoned formation of emulsions of oil - in - water[22].

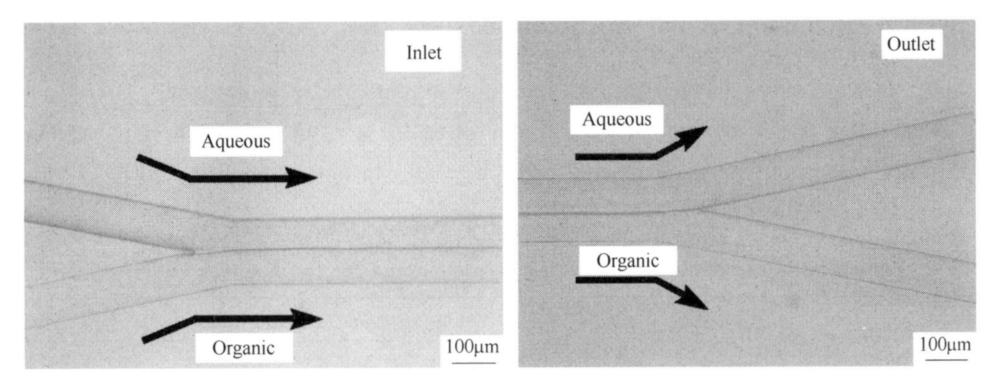

Fig. 3 Merging and diverging of aqueous and organic streams

3.2 Calculation of contact time in the microchannel

The changes of geometry structure of a microfluidic system have great influence on the flow pattern and the shape of the interface between L - L phases[23]. According to the micro channel geometry in Fig. 4, the cross - section area of the aqueous channel, S_a, was estimated by Eq. (3). And the volume of aqueous channel, V_a, was calculated by Eq. (4).

$$S_a = \int_0^{80} [\sqrt{59.4^2 - (x - 23.9)^2} - 19.4]\,dx \tag{3}$$

$$V_a = S_a \times L \tag{4}$$

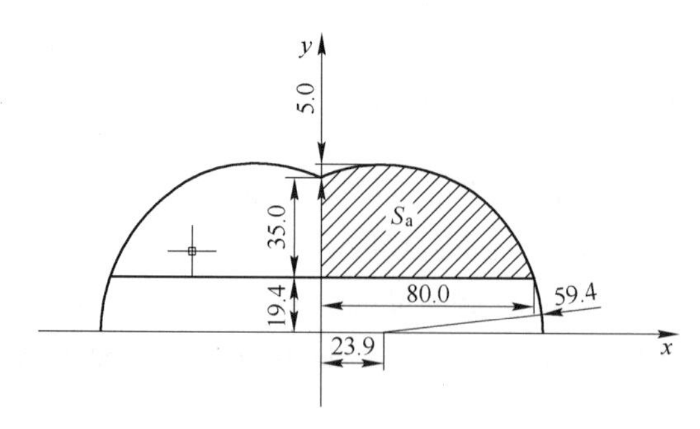

Fig. 4 The actual size of cross – section(μm)

In Eq. (4), L is the total length of the channel, which is equal to 12cm. It is calculated that V_a is equal to $3.08 \times 10^{-10} m^3$. When the flow rate of the aqueous phase was 2mL/h, which means volumetric flow rate of the aqueous phase $v = 5.56 \times 10^{-10} m^3/s$, and the phase ratio $R = 0.8$. "t" means the residence time of the aqueous phase in the microchannel, which can be expressed as Eq. (5):

$$t = V_a / v \tag{5}$$

The value of t is calculated as 0.55s, which indicated that the extractability could be measured for a short contact time of both the phase in microreactor. While in traditional solvent extraction plants with a mixing settler system, the extraction time is usually about 4min (1min of mixing and 3min of phase separating).

3.3 Micro extraction of In^{3+} in a complex solution

A complex solution with a composition in Table 1 was used in the experiment of micro extraction. When the flow rate of the aqueous phase was set as 2mL/h and that of the organic phase was set as 1.6mL/h, the raffinate was collected and analyzed with method of ICP – AES. The analysis result is shown in Table 2.

Table 2 The chemical composition of the original solution and the raffinate after micro – extraction

Composition	In^{3+}	Fe^{2+}	Zn^{2+}
Original solution/g · L^{-1}	3.17	3.46	52.82
Raffinate/g · L^{-1}	0.29	3.41	51.65
Extraction rate/%	90.8	0.16	0.22

It is shown in Table 2 that the extraction rate of indium ion is much higher than that Fe^{2+} and Zn^{2+}.

In an industry plant, after 2 – step extraction of In^{3+} ion by D2EHPA, Fe^{2+} and Zn^{2+} concentration in organic phase is much higher. Thus a 3 – step of eluting process was needed to remove the above impurities.

The microreactor provides a good separation effect of indium and impurities. The main reason lies in that the mass transfer speed of indium is much higher than Fe^{2+} and Zn^{2+}[24].

During the microchannel extraction, the flow pattern is much different with the mixing settler system. The mass transfer of the former is by diffusion through a stable interface, which may lead to better selection of metal ions.

For checking the co - extraction effects of Fe^{2+} and Zn^{2+} during traditional mixing settler system, the serials experiment of extraction with separating funnel were conducted. The relationship between extraction rate and phase ratio of this experiments is shown in Fig. 5.

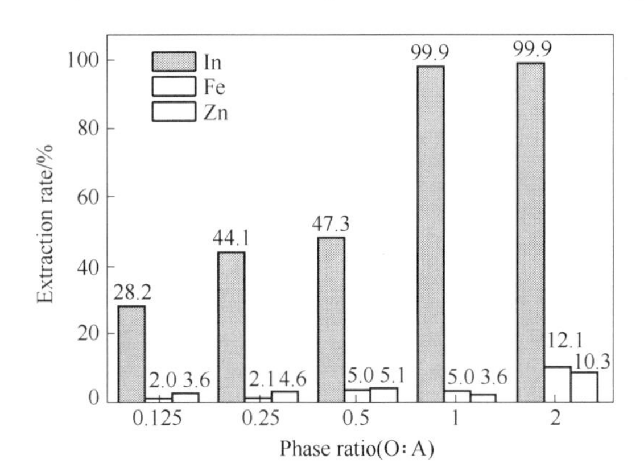

Fig. 5 Extraction rate in different phase ratio during separating funnel experiments

From Fig. 5, we can see that co - extraction ratio of Fe^{2+} and Zn^{2+} is about one magnitude higher than micro - extraction.

3.4 Surface - to - volume ratio of the microchannel and its mass transfer character

The aqueous - organic interfacial area $A(m^2)$ can be calculated with follow equation:

$$A = (d - h_g) \times L \tag{6}$$

where d is height of microchanel; h_g is height of the guide structure. Thus, A is calculated to be $4.2 \times 10^{-6} m^2$. In the contact time of 0.55s, $3.08 \times 10^{-10} m^3$ of solution involved in the reaction and 0.98×10^{-6}g of indium transferred into the organic phase calculated by the data shown in Table 2. And the mass transfer speed can be as high as $0.34 g/(m^2 \cdot s)$.

The surface - to - volume ratio is about $6.8 \times 10^3 m^2/m^3$ for this channel, which is much higher than that of the traditional mixing settler process. The surface - to - volume ratio of conventional scale laboratory or industrial reactors are usually as low as $100 - 1000 m^2/m^3$ [25].

4 Conclusions

The extraction of indium from a complex solution with D2EHPA was conducted in a glass laminar flow micro reactor. According to the result, the conclusions are draw as following:

(1) In traditional mixing settler process, the operation time of a single stage is about 4min, while in microfluidic process, the time can be as short as 0.55s.

(2) In microfluidic process, the extraction rate of indium is very high and the average mass transfer speed can be as high as $0.34 g/(m^2 \cdot s)$.

(3) In mixing settler process, the co - extractions of Fe^{2+} and Zn^{2+} into the organic phase can be as high as 5.0% and 3.6% respectively, while in the microfluidic system, their co - extractions are only 0.16% and 0.22%, which shown a great benefit of reduction of the eluting stages after the extraction.

Acknowledgements

Foundation item: Projects (No. U1302271 and No. 51264015) supported by the National Natural Science Foundation of China.

References

[1] Ehrfeld W, Hessel V, Haverkamp V. Wiley - VCH Verlag GmbH & Co. KGaA, 2000.

[2] Huh Y S, Jeon S J, Lee E Z, et al. Microfluidic extraction using two phase laminar flow for chemical and biological applications[J]. Korean Journal of Chemical Engineering, 2011, 28(3): 633 - 642.

[3] Hessel V, Löwe H, Schönfeld F. Micromixers—a review on passive and active mixing principles [J]. Chemical Engineering Science, 2005, 60(8): 2479 - 2501.

[4] Yamamoto T, Shimizu Y, Ueda T, et al. Application of micro - reactor chip technique for millisecond quenching of deuterium incorporation into 70S ribosomal protein complex[J]. International Journal of Mass Spectrometry, 2011, 302(1): 132 - 138.

[5] Fletcher P D I, Haswell S J, Pombo - Villar E, et al. Micro reactors: principles and applications in organic synthesis[J]. Tetrahedron, 2002, 58(24): 4735 - 4757.

[6] Yamada M, Kobayashi J, Yamato M, et al. Millisecond treatment of cells using microfluidic devices via two - step carrier - medium exchange[J]. Lab on a Chip, 2008, 8(5): 772 - 778.

[7] Peela N R, Kunzru D. Oxidative steam reforming of ethanol over Rh based catalysts in a micro - channel reactor [J]. International Journal of Hydrogen Energy, 2011, 36(5): 3384 - 3396.

[8] Kralj J G, Sahoo H R, Jensen K F. Integrated continuous microfluidic liquid - liquid extraction[J]. Lab on a Chip, 2007, 7(2): 256 - 263.

[9] Yang K, Chu G, Shao L, et al. Micromixing efficiency of viscous media in micro - channel reactor[J]. Chinese Journal of Chemical Engineering, 2009, 17(4): 546 - 551.

[10] Pennemann H, Hessel V, Löwe H. Chemical microprocess technology—from laboratory - scale to production [J]. Chemical Engineering Science, 2004, 59(22): 4789 - 4794.

[11] Kordosky G A. Copper recovery using leach/solvent extraction/electrowinning technology: forty years of innovation, 2.2 million tonnes of copper annually[J]. Journal of the South African Institute of Mining and Metallurgy, 2002, 102(8): 445 - 450.

[12] Sarangi K, Reddy B R, Das R P. Extraction studies of cobalt(Ⅱ) and nickel(Ⅱ) from chloride solutions using Na - Cyanex 272: separation of Co(Ⅱ)/Ni(Ⅱ) by the sodium salts of D2EHPA, PC88A and Cyanex 272 and their mixtures[J]. Hydrometallurgy, 1999, 52(3): 253 - 265.

[13] Sole K C, Cole P M, Feather A M, et al. Solvent extraction and ion exchange applications in Africa's resurging uranium industry: a review[J]. Solvent Extraction and Ion Exchange, 2011, 29(5 - 6): 868 - 899.

[14] Qi Z, Renman R, Jiankang W E N, et al. Influences of solid particles on the formation of the third phase crud during solvent extraction[J]. Rare Metals, 2007, 26(1): 89 - 96.

[15] Olette M. Interfacial phenomena and mass transfer in extraction metallurgy[J]. Steel Res., 1988, 59(6): 246 - 256.

[16] Ritcey G M. Crud in solvent extraction processing—a review of causes and treatment[J]. Hydrometallurgy, 1980, 5(2): 97 - 107.

[17] Fortes M C B, Benedetto J S. Separation of indium and iron by solvent extraction[J]. Minerals Engineering, 1998, 11(5): 447 - 451.

[18] Pang C F. Henan Science, 2010, 28, 290 - 292(in Chinese).

[19] Priest C, Zhou J, Sedev R, et al. Microfluidic extraction of copper from particle - laden solutions [J]. International Journal of Mineral Processing, 2011, 98(3): 168 - 173.

[20] Tamagawa O, Muto A. Development of cesium ion extraction process using a slug flow microreactor [J]. Chemical Engineering Journal, 2011, 167(2): 700 - 704.

[21] Lee M S, Ahn J G, Lee E C. Solvent extraction separation of indium and gallium from sulphate solutions using D2EHPA [J]. Hydrometallurgy, 2002, 63(3): 269 - 276.

[22] Su Y, Zhao Y, Chen G, et al. Liquid - liquid two - phase flow and mass transfer characteristics in packed microchannels [J]. Chemical Engineering Science, 2010, 65(13): 3947 - 3956.

[23] Aota A, Mawatari K, Takahashi S, et al. Phase separation of gas - liquid and liquid - liquid microflows in microchips[J]. Microchimica Acta, 2009, 164(3 - 4): 249 - 255.

[24] Zhang C Q, Zhou J Z, Zhou X Z, et al. Study of non - equilibrium extraction separating indium and iron [J]. Nonferrous Met. (China), 1995, 47(1): 78 - 82.

[25] Zhao Y C, Zheng H C, Shen J N, Chen G W, Yuan Q. Sciencepaper Online, 2008, 3(3): 157 - 169(in Chinese).

About the Laboratory

The Key Laboratory of Unconventional Metallurgy, Ministry of Education, which was founded on the basis of the Institute of Microwave Application, Kunming University of Science and Technology, which is mainly focusing on the great urgent questions of the comprehensive utilization of mineral resources in the metallurgical industry, energy – saving and emission – reduction and increase in value of deep processing of metallurgical products, and acquiring original innovations and proprietary intellectual property rights in the fields of unconventional metallurgy, culturing urgent innovation talents for Yunnan province and nation, strengthening scientific and technical innovation of metallurgical industry, boosting induced effect of metallurgical industry to the national economy and social development. The main research areas include microwave metallurgy, application of microfluidic technology on metallurgy, new technique of high gravity metallurgy, applications of microwave and plasma in material fields, new technology of resource comprehensive utilization and R & D of higher temperature microwave reactors.

At present, the key laboratory is undertaking more than 90 projects mainly including Key Project of National Nature Science Foundations, 973 Program, International Cooperation Project of Ministry of Science and Technology, Project of Yunnan Provincial College Cooperation, and Enterprise Commitment Projects, etc. So far, we have published more than 600 papers and 5 books/monographes, applied for 219 patents with more than 100 granted. The laboratory was awarded with 1 the Second Prize of National Technology Invention, and more than 10 the First and Second Prizes of Nature Science of Yunnan Province.

The laboratory has developed and made independently new microwave metallurgical higher temperature reactors, and built up microwave higher temperature production line with characteristics of large – scale, continuous and automation at home and abroad firstly, and transferred about 82 items of high quality reactors, production line and related technology to the Instituto Nacional del Carbón (INCAR), the Spanish Scientific Council (CSIC), which creating outstanding economic benefit and prominent effect of energy – saving and emission – reduction, enlarging social influence.

The laboratory has in succession built up international collaboration relationships with the Institute of Polymer Technology and Materials Engineering, now the Department of Materials of Loughborough University, Britain; Karlsruhe Research Center, Germany; the Instituto Nacional del Carbón (INCAR) from the Spanish Scientific Council (CSIC); Materials Processing and Recycling Laboratory, Tokyo University, Japan; and Department of Chemical and Materials Engineering, University of Albert, Canada etc. We have carried out substantial exchange and cooperation with above – mentioned universities and institutes in new application fields of microwave energy technology.

http://um.kmust.edu.cn/client/index.asp